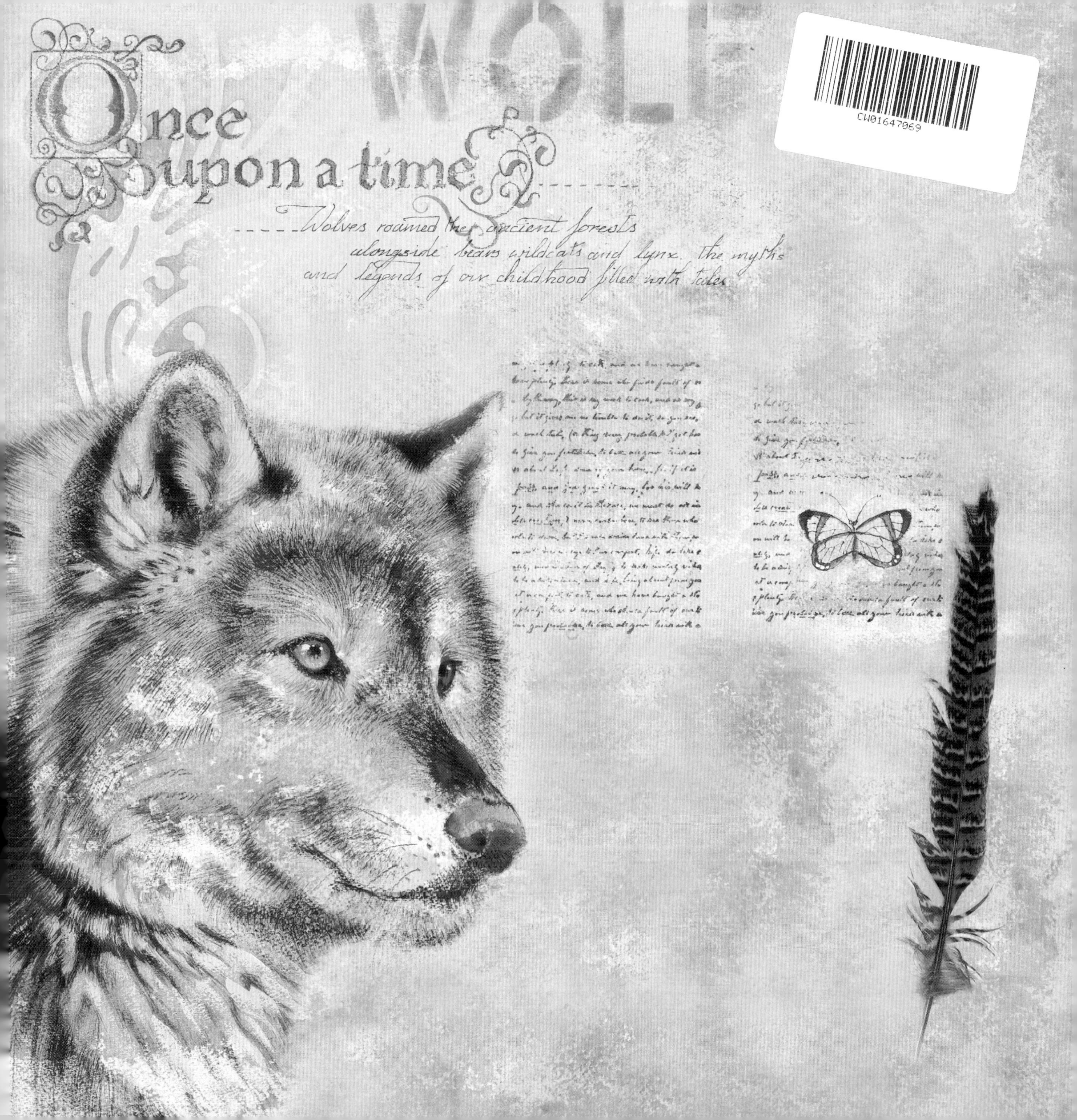
WOLF
Once upon a time
Wolves roamed the ancient forests alongside bears wildcats and lynx. the myths and legends of our childhood filled with tales

Way of the Wolf

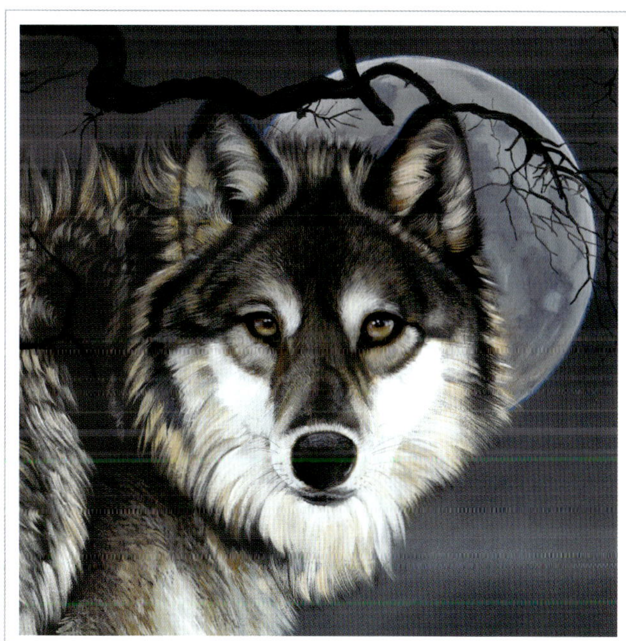

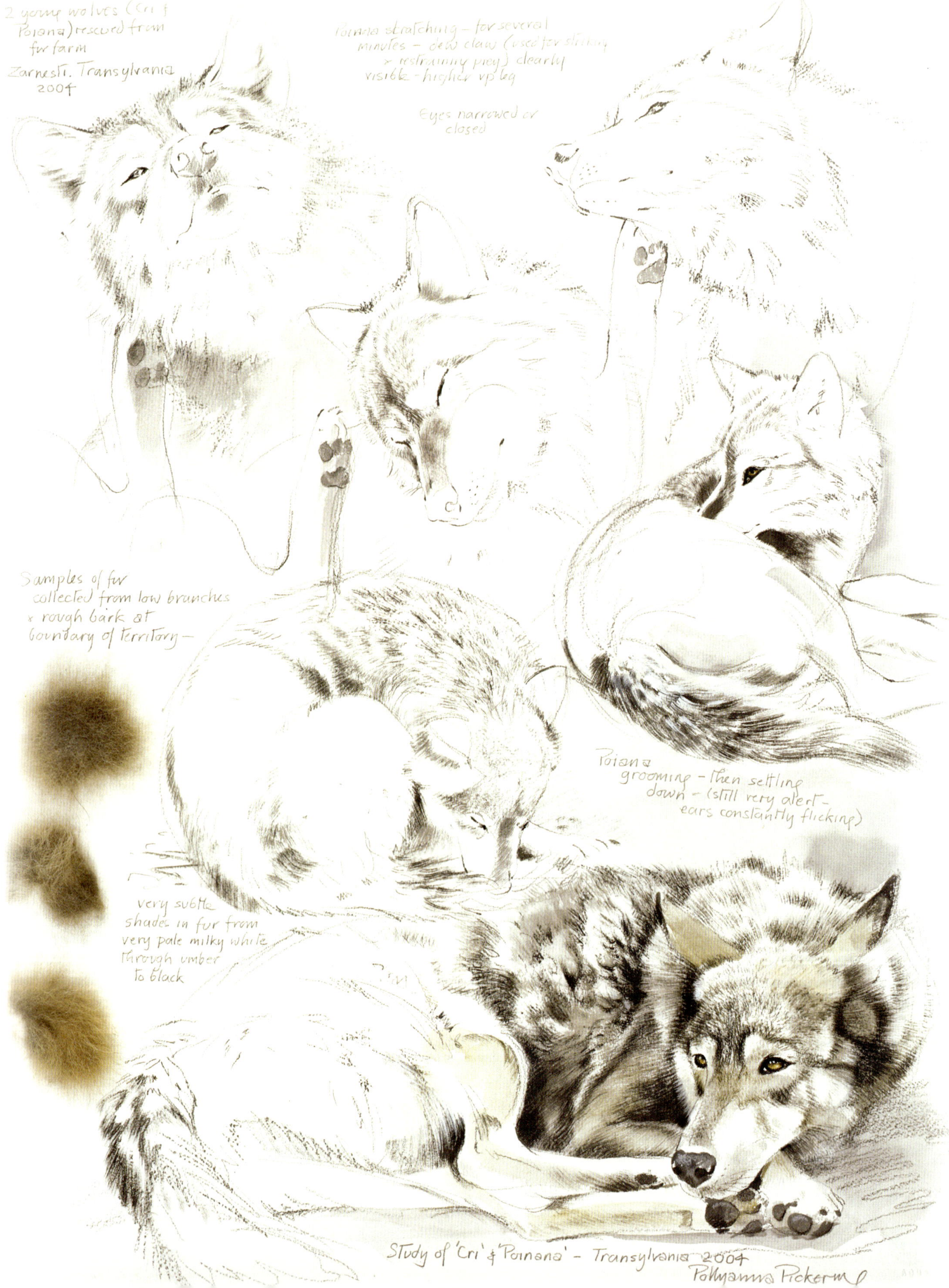

2 young wolves (Cri & Poiana) rescued from fur farm
Zarnesti. Transylvania 2004
Poiana stratching – for several minutes – dew claw (used for striking & restraining prey) clearly visible – higher up leg
Eyes narrowed or closed
Samples of fur collected from low branches & rough bark at boundary of territory –
Poiana grooming – then settling down – (still very alert – ears constantly flicking)
very subtle shades in fur from very pale milky white through umber to black
Study of 'Cri' & 'Poinana' – Transylvania 2004
Pollyanna Pickering

Way of the Wolf

The paintings of

Pollyanna Pickering

Text and photography by

Anna-Louise Pickering

Preface by Graham Norton

Foreword by Count Tibor Kalnoky

Foreword by Professor Claudio Sillero

Otter House (Licensing) Limited

For Sue Stainton

Your friendship and creativity mean so much to us

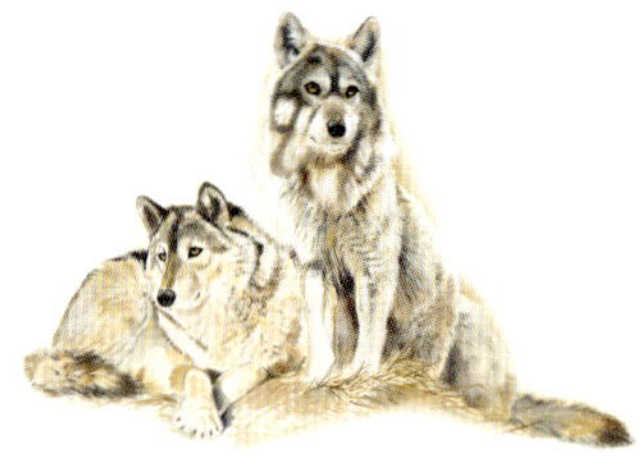

First published in Great Britain in 2013 by
Otter House,
a division of Otter House Limited
Exe Box, Matford, Exeter EX2 8FD

Designed by Sue Stainton and Victoria Selby

British Library Cataloguing in Publication Data
A catalogue record for this book is available from the British Library.

ISBN 978 0 9529369 4 7

Printed and bound in China by Wai Man Book Binding (China) Ltd.

Preface

In my line of work I have encountered many exotic creatures; Sophia Loren has slapped me, Tony Curtis has waltzed me around a dance floor and Mark Wahlberg has pinched my nipple, but nothing compares to the profound honour of meeting the wolves of Ethiopia.

Almost seven years ago the BBC was making a series of programmes about endangered species. They sent through the choice of animals and locations and while some of them sounded much more exciting, even glamorous, there was something about Ethiopian wolves that stopped me in my tracks. Firstly I have an enormous love of dogs, but had never heard of these creatures, and secondly I knew this was a journey I would never make by myself. Of course there was the third element which was where I imagined myself swathed like Lawrence of Arabia watching giant dogs prowl the sandy desert dunes. Well, I made the right choice but for all the wrong reasons. The wolves are far more like large foxes and the desert is in fact the very fertile Bale Mountains in the remote south-east corner of the country.

I have two abiding memories. The first is the beauty of the place. Unless you have experienced something similar you will think I'm mad, but one morning as I stood on a ridge watching the sun rise over the distant peaks, the combination of the light, the remoteness, and that sense of being one of the few people on the planet to have witnessed this view, all combined so that I found tears rolling down my face. The second thing is of course the wolves themselves. Because there are so very few of them and they are in such great peril one imagines the creatures will just be terrified and stressed, but when you encounter them they are as pristine and noble as they have been for thousands of years. It is precisely their ignorance of their dire circumstances that is so heartbreaking.

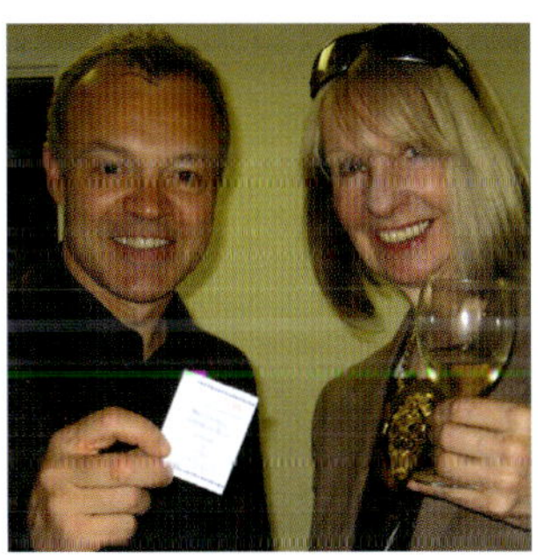

Graham Norton selects the winning ticket in the 2010 Pollyanna Pickering Foundation Prize Draw

I loved this book for bringing me back and reigniting all my happy memories of my time in Ethiopia, but please do read the wise words of my friend Claudio Sillero who explains far more eloquently and precisely than I ever could why it so very important that these creatures are saved. I know you have already bought a book but, if you can, please support the important work that Claudio and the Born Free Foundation are doing.

Long live the wolves.

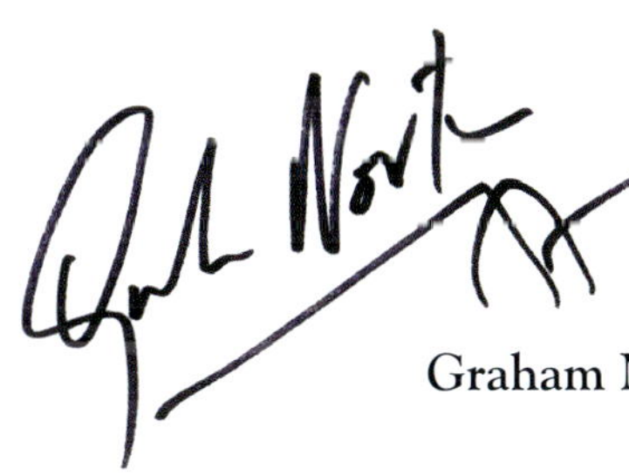

Graham Norton

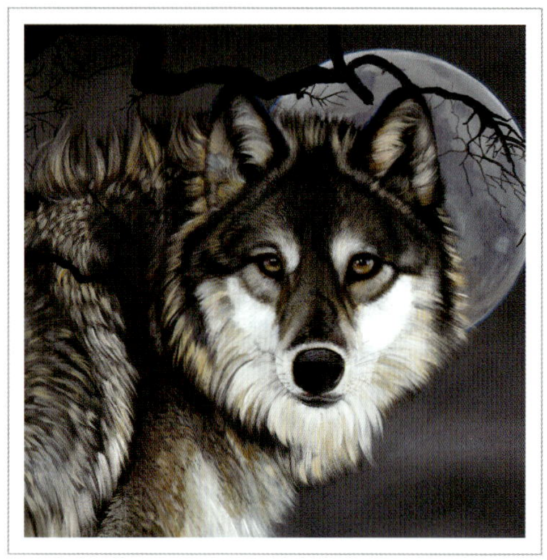

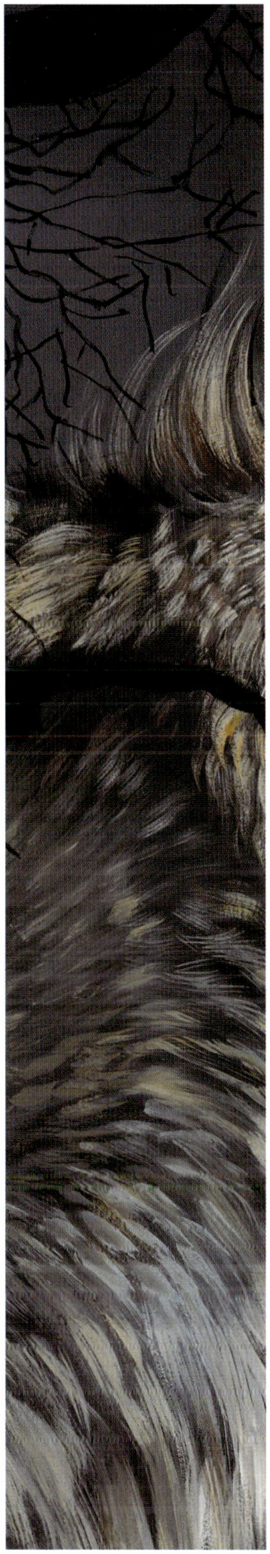

CONTENTS

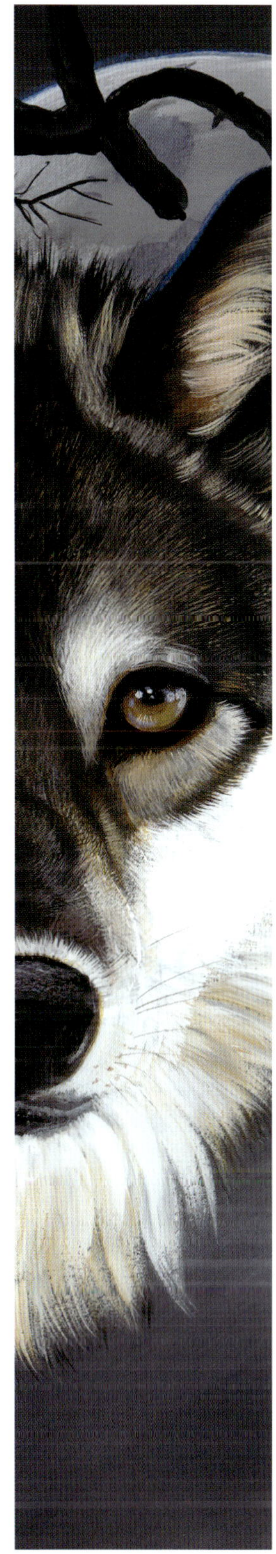

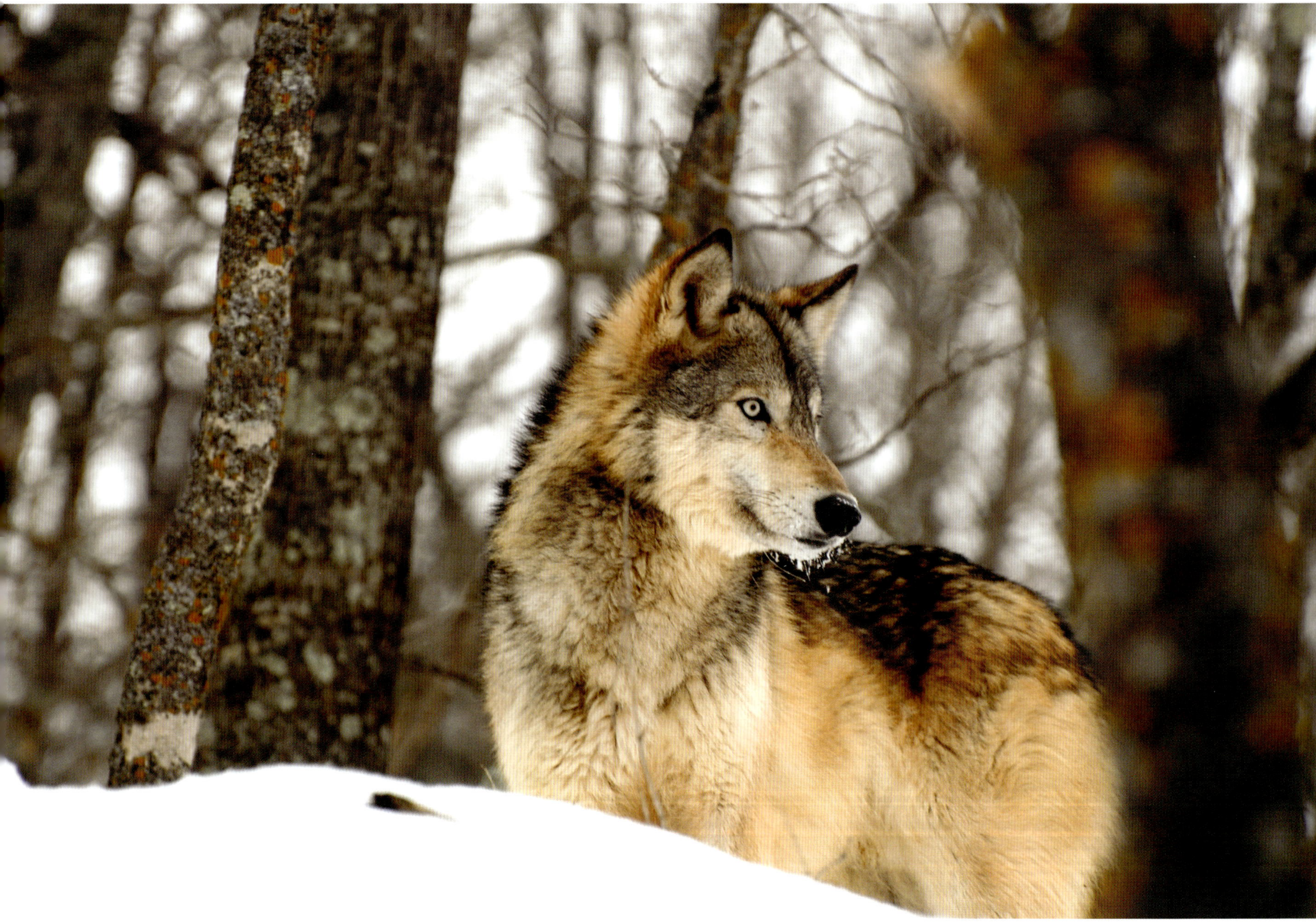

'On the ragged edge of the world I'll roam
And the home of the wolf will be my home'

Robert Service

Wolves are always with us.

Although wolves became extinct in Britain in the seventeenth century, they still surround us in our imagery, our literature and our fairy tales: "'Grandmother what big teeth you have.' 'All the better to eat you with!' At that the wicked wolf threw himself on Red Riding Hood and gobbled her up…'" For most of us, this is the very first time we will hear the word wolf – in a gruesome bedtime story warning us not to stray from the familiar path. The earliest stories we are told as children cast wolves as villains – savagely fierce, sly and cunning they are the shadowy monster lurking in the forest.

But later, alongside these fearsome tales, we warm to stories of wolves as nurturing rescuers who will rear human children as part of their own family pack – from Romulus and Remus, founders of ancient Rome, to the 'wolf children' of modern times. Many of us even live with a 'wolf' by our fireside – domestic dogs, the tame descendants of the wolf packs that once roamed our ancient and now vanished forests. And just occasionally, looking into their eyes, we might catch a brief glimpse of the wolf within.

Misrepresented in fiction, misunderstood and feared, the majority of accepted popular knowledge about wolves is in fact incorrect – based on centuries-old folklore. Wolves were hunted to extinction in Britain and globally they face a struggle for survival in many of the areas they once roamed in large numbers. Yet we remain fascinated by these enigmatic animals with their sharp intelligent eyes, thick luxuriant coats and by the myths and legends that surround them.

Strangely, though apparently so familiar, few of us have ever seen a wolf aside, perhaps, from a brief glimpse in the unnatural confines of a zoo. In the remote areas where wolves remain in the wild they are secretive, shy and elusive, sighted only briefly, moving in the shadows. Perhaps they have always existed more vividly in our imagination.

Inspired by the endless mystery and wonder of the natural world, Pollyanna Pickering has travelled the globe to study and paint endangered species in their natural habitats. Her resulting paintings inspire and inform – without uttering a word they speak to the observer of wild animals glimpsed in distant lands, bringing the beauty and wonder of the living world into people's homes.

Pollyanna has undertaken an ongoing series of journeys to gather material for a collection of paintings of the

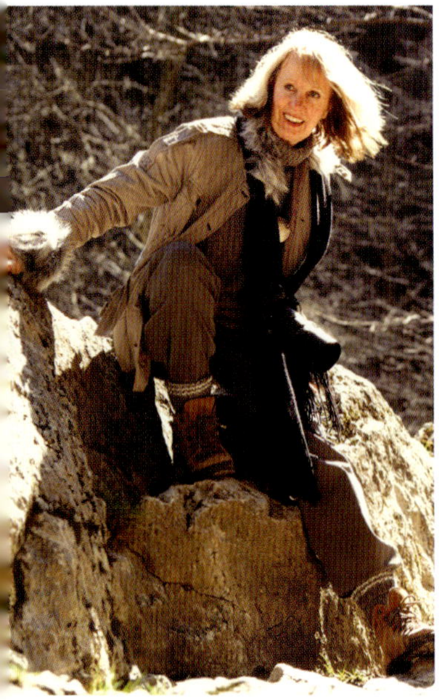

Pollyanna Pickering

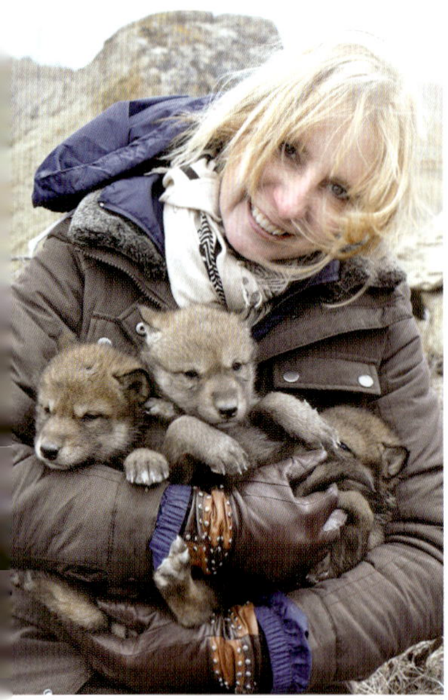

Anna-Louise Pickering

wolves of the world. She began by travelling through North America and Canada, trekking in Montana and the Yellowstone National Park where wolf packs were reintroduced in 1995. In Minnesota, home to the largest number of wild wolves in the contiguous United States, weeks of 4.00 am starts were rewarded with a glimpse of a lone wolf in the northern forests close to the international border with Canada.

The Mexican wolf is believed to be extinct in the wild, and certainly journeys on the Mexico/Texas borderlands proved fruitless – although in California and Arizona Pollyanna was rewarded with many sightings of their canine cousin, the coyote. On one memorable horse ride through the deserts of Arizona at dawn, an inquisitive coyote trotted along with Pollyanna's horse, always maintaining a safe distance but shadowing her for around twenty minutes as the sun rose. In the High Arctic, wolves padded unseen around Pollyanna's tent as she slept on the ice floes, their footprints clearly visible in the surrounding snow the following morning. A huge white wolf watched the camp from rocky cliffs, drawn by the lure of one of the female huskies in Pollyanna's dog team – an opportunity for Pollyanna to fill pages of her sketchpad with his image.

In 2004, Pollyanna turned her attention to the European wolf. The last remaining wild wolf in Britain was killed in Scotland more than three hundred years ago – reputedly shot by Sir Ewen Cameron of Lochiel in 1686. However, wolves can still be found in Russia, Scandinavia and much of mainland Europe. They survive in the Ukraine, Bosnia and Bulgaria, and the north-east of Poland. Small numbers remain in Bavaria and the Alpine regions of Germany, Italy and France, where recent sightings appear to confirm that wolves are also gradually returning to the Massif Central, a highland region far from the Alps. However, the highest density of European wolves can be found in Romania – in the Carpathian Mountains of Transylvania.

So this was to be Pollyanna's destination. As Pollyanna's daughter and business partner I arrange her expeditions to search for animals in the wild and accompany her to provide photographic documentation. Around the time we began to plan our journey to Romania the desperate plight of another group of wolves came to our attention. In the Bale Mountains, high in the uplands of Ethiopia, the last remaining Simien wolves were being threatened by an outbreak of rabies. These wolves are the only ones to be found on the entire continent of Africa, and are both the rarest carnivore in Africa and the rarest wolf in the world – a dubious double distinction. In 2004 the Ethiopian Wolf Conservation Programme carried out an emergency intervention to contain a rabies outbreak in the Web Valley subpopulation – one of three areas with high densities of wolves in this mountainous region. A total of seventy-two wolves were vaccinated and the spread of the disease was halted. Through the exhibition of paintings completed following her journey into Transylvania, Pollyanna was able to raise funds to purchase stocks of rabies vaccine for this project. Five years later she travelled to Ethiopia in search of these unique red wolves. Here she met the remarkable group of conservationists who have devoted their lives to save an animal few people have heard of, and even fewer will ever see.

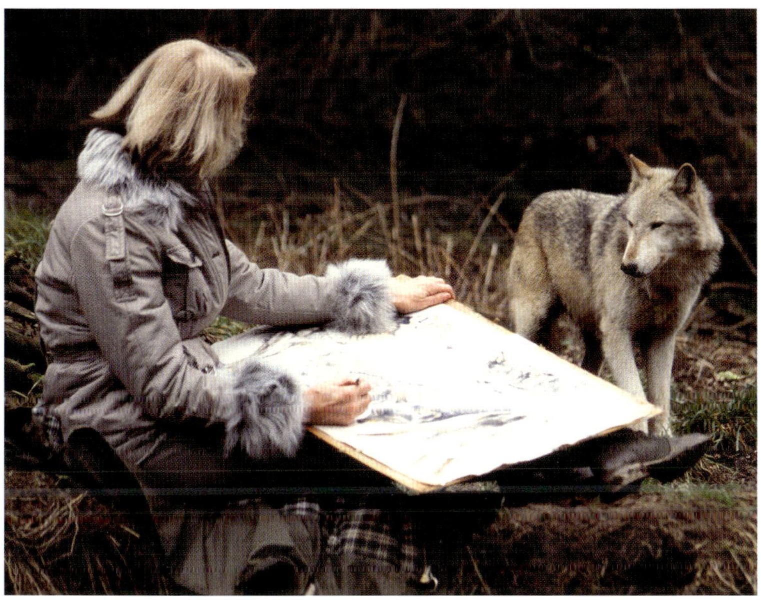

Pollyanna meets her fiercest critic!

This book contains extracts from the journals I kept throughout both of these journeys, alongside my photographs and Pollyanna's sketches, which together tell the background story of Pollyanna's extraordinary portraits of these enigmatic animals.

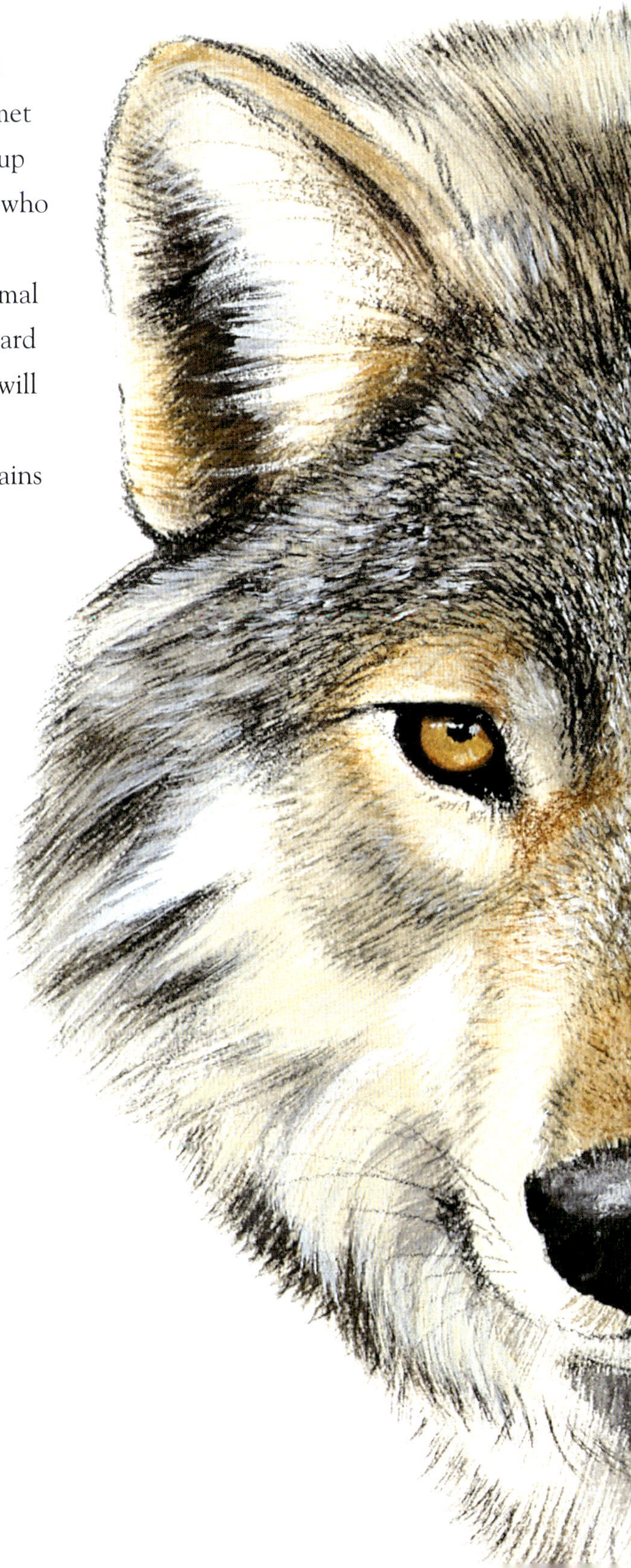

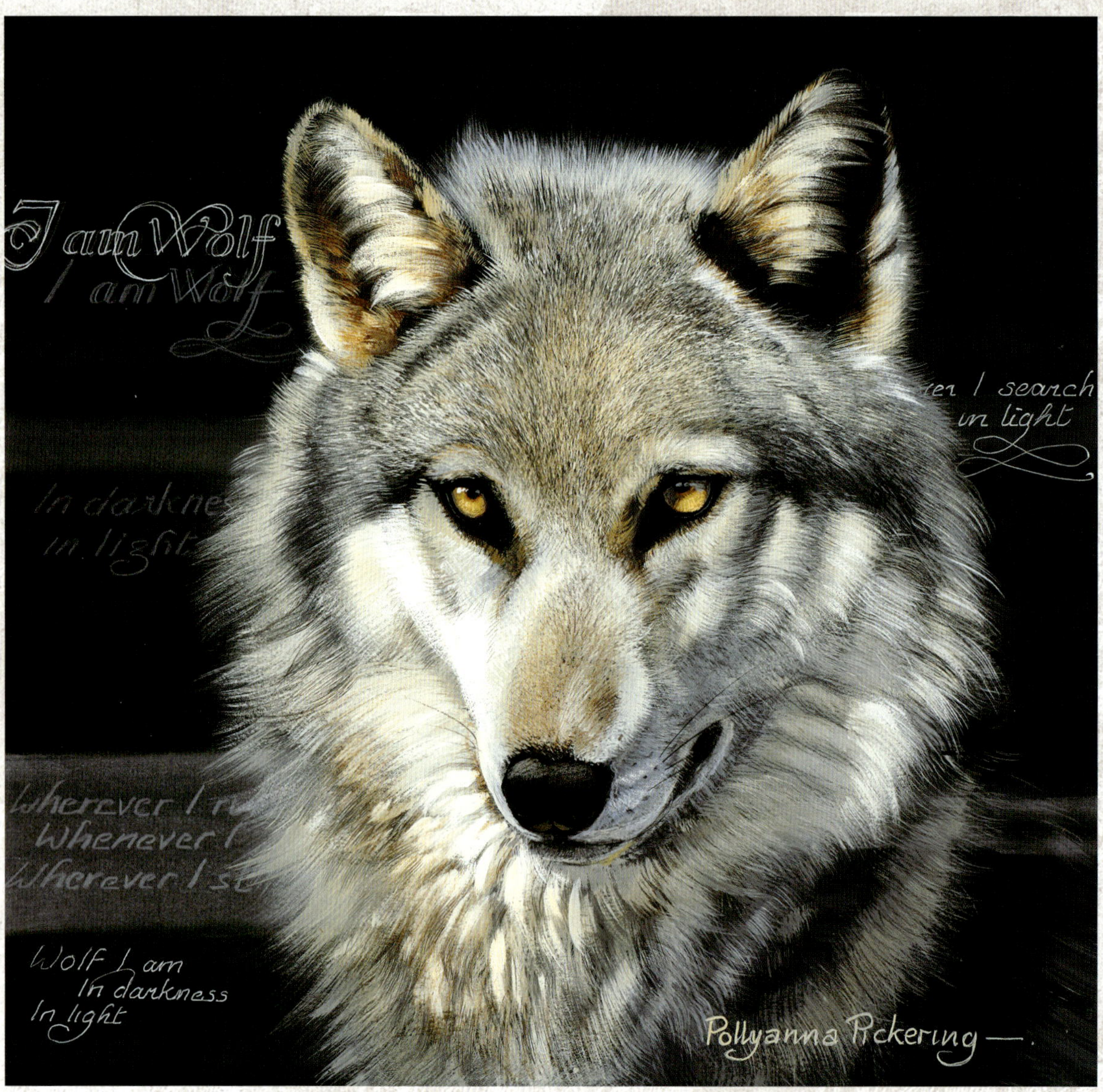
I am Wolf
I am Wolf
in light
In darkne
in light
Wolf I am
In darkness
In light
Pollyanna Pickering

THE FIRST JOURNEY

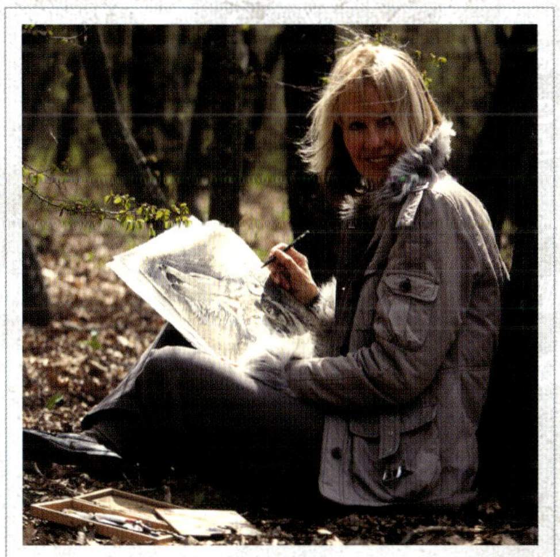

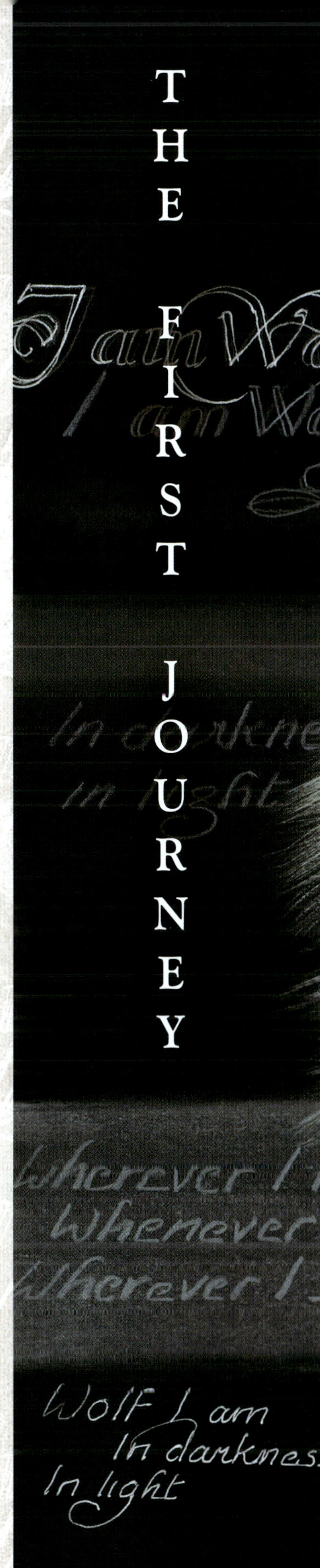

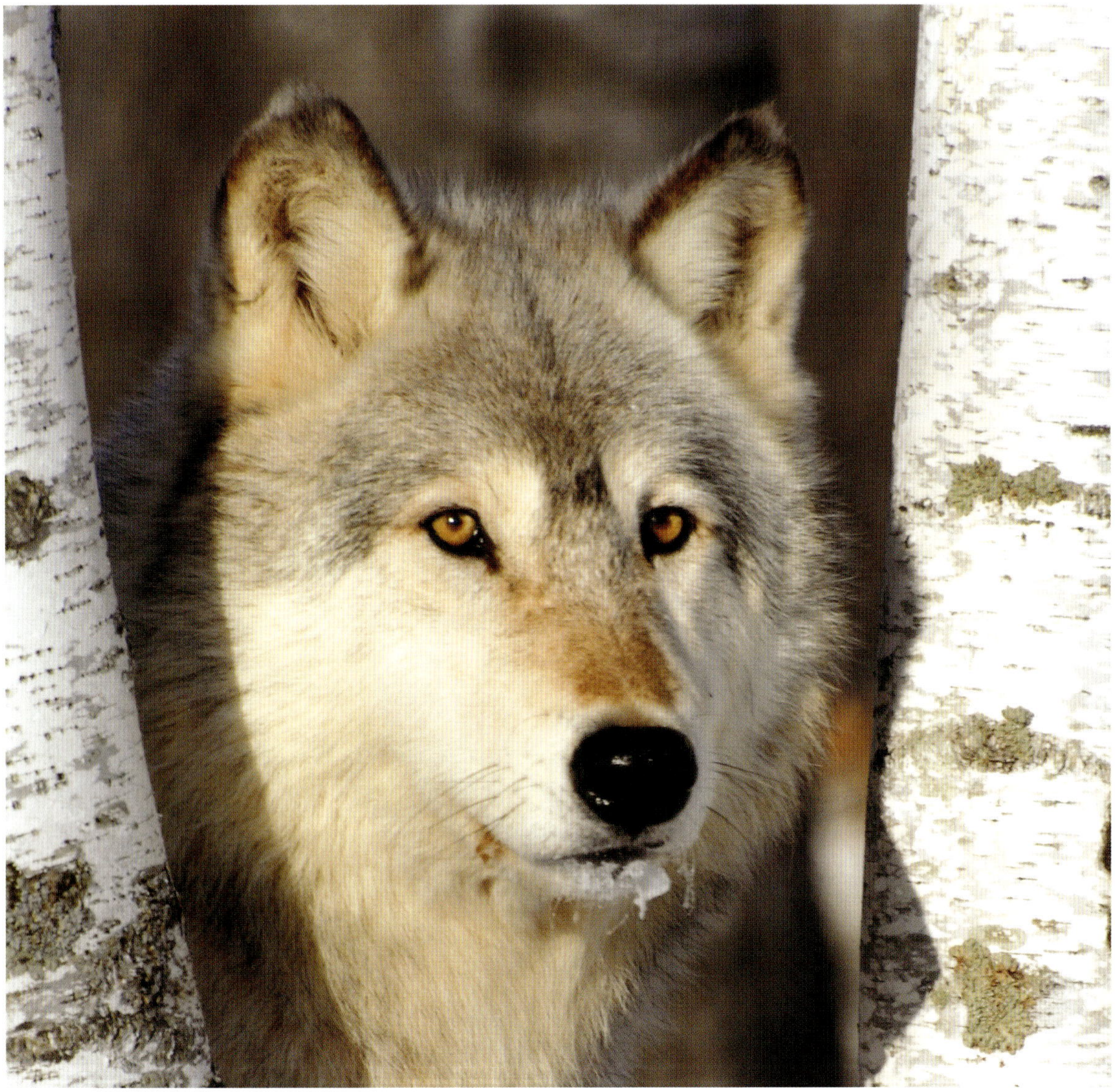

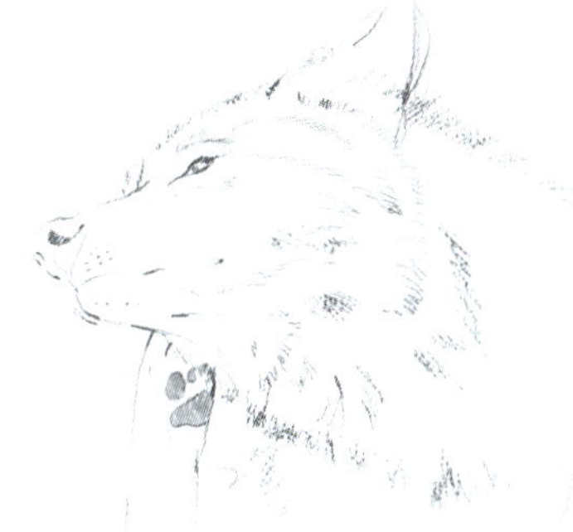

Foreword

When Pollyanna and Anna-Louise asked me to write a foreword to this book I did have one concern. Will I have enough time to read it all at the first place? I picked it up for a try and couldn't put it down anymore, reading straight through its captivating pages in one single go. I must admit I was most impressed by the depiction of my own theatrical appearance for dinner in the cellar that night – good to know it had such an effect although it was certainly not meant that way!

All in all, this book is not only an excellent introduction into everything about wolves. It also gives a realistic account of today's Transylvania and the larger Romania.

Wolves and Transylvania have one thing in common: they both suffer distorted images caused by blood-dripping fictions. I am keen on promoting the truth about us, which can be so much more fascinating than all those fantasies. This is why I am immensely grateful to the Pickering Ladies for their contribution – through this book – to convey an insight into the true life of wolves and into the reality of this unique region called Transylvania.

Reading this story and contemplating Pollyanna's realistic and inspirational pictures, one wishes to have accompanied these two Ladies on their adventures across the country, gradually making their way into the forests in our hills, where so many wolves and bears are still at home. Well, this is, for the most part, still possible. Not much has changed since – including the bad plumbing in our archaic villages: travelling to Transylvania, one can still become part of Wolf-Country.

Even without getting to see them face to face, one can feel their presence here. Especially in wintertime, from late October to early April, you can track their footprints in our snowy hills. If you are lucky, you can catch a glimpse of them at dusk or hear their distant howls, or at any rate listen to the villager's accounts on how their sheep were stolen by a pack despite the dogs defending them.

Follow Pollyanna and Anna-Louise Pickering in their tale of a very rare experience one might not think possible anymore in today's Europe.

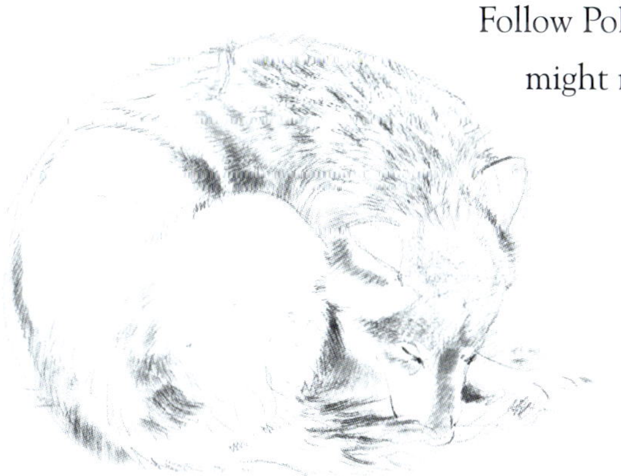

Count Tibor Kalnoky

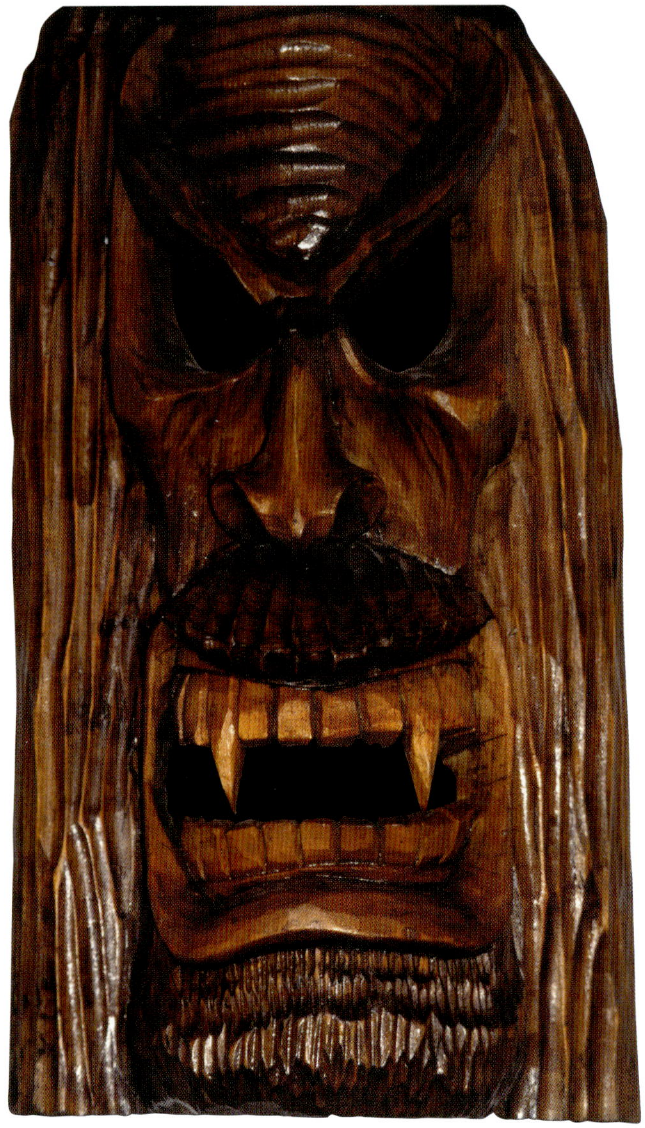

Prelude to the Journey

'The mouth, so far as I could see it...
was fixed and rather cruel-looking, with peculiarly sharp white teeth.'

JONATHAN HARKER'S JOURNAL, *DRACULA*

Transylvania is inextricably linked in our collective consciousness with Dracula. We may have been heading to this region to search for wolves, but thanks to Bram Stoker's famed novel, when we started telling our friends about our planned travels the universal response was to enquire if we had packed a crucifix/necklace of garlic cloves/phial of holy water/ freshly sharpened stakes.*

Although we were without this essential vampire slaying kit, we had added one extra item to our usual bags of sketching equipment, cameras and lenses – an emergency dental repair kit.

Journal – departure date minus two

Sunday morning, two days before we are due to fly to Romania. Rush round like a banshee all morning packing – panic when I cannot locate my favourite 500 mm lens, huge relief when I find it. Break for a sandwich. Pollyanna has mild toothache. We have mild concern. By evening Pollyanna has bad toothache and is starting to look like a hamster storing its entire winter provisions in one cheek. This can't be good.

Two days, two visits to the dentist each progressively more painful as he cuts away chunks of tissue from her jaw to remove the mystery infection. I attempt to console her with the thought that no matter how painful the treatment might be, the timing is very good –

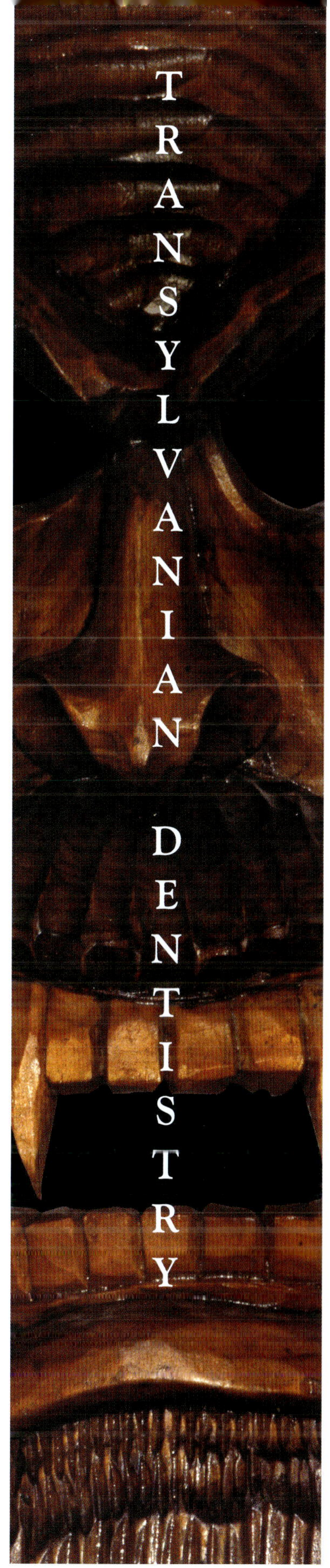

**Aside, that is, from one friend who asked if we had packed our fishnet stockings. His mind had immediately gone to the 'sweet transvestite from transsexual Transylvania' in the* Rocky Horror Show. *There's always one.*

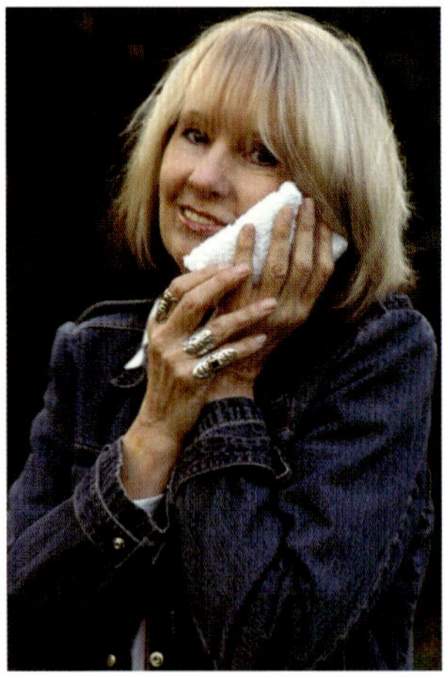

By evening Pollyanna has bad toothache and is starting to look like a hamster

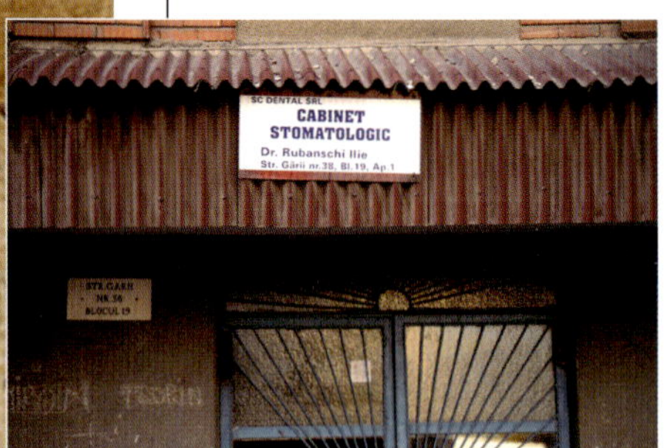

We take the earliest opportunity to check out the available dentists

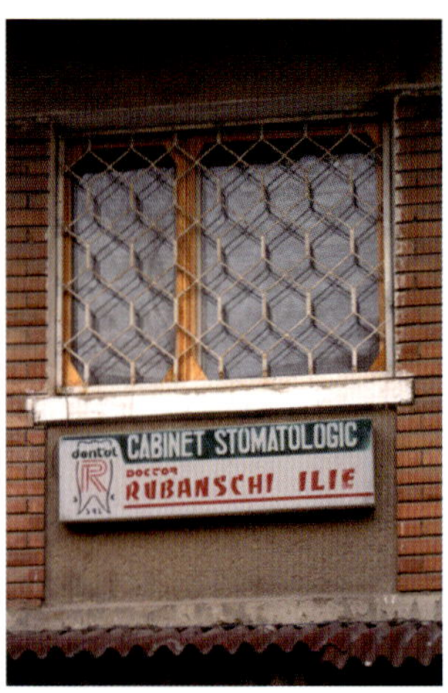

three days later and she could have found herself sitting in a waiting-room flipping nervously through a copy of *Vampyre Monthly*. We have discovered that putting the adjective 'Transylvanian' in front of almost any word makes it sound somewhat sinister and unnerving – but putting it in front of the word 'dentistry' has added spine-tingling connotations...

On our arrival in Romania we are still concerned that the infection may flare up again, so take the earliest opportunity to check out the available dentists. Reassuringly there are plenty of dental surgeries in the larger towns, often advertising themselves with signs depicting cheerily smiling cartoon teeth. However, our enquiries reveal two slightly worrying facts – the usual response to any dental complaint in Romania is to immediately extract the tooth causing problems, and the majority of dentists do not routinely stock or use anaesthetics.

If you are such a girl's blouse that you require this option you have to purchase your own privately and take it with you. (When Pollyanna returns to our dentist for a check-up on her return to the UK he is so alarmed to learn this that he promptly gives us both a syringe of dental anaesthetic, which now routinely travels with us as part of our increasingly comprehensive first-aid kit.)

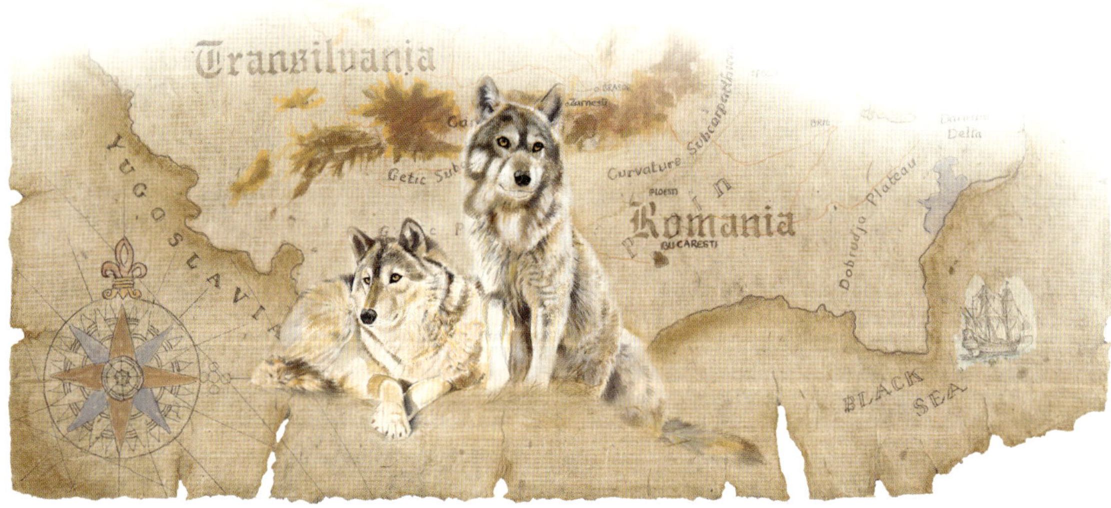

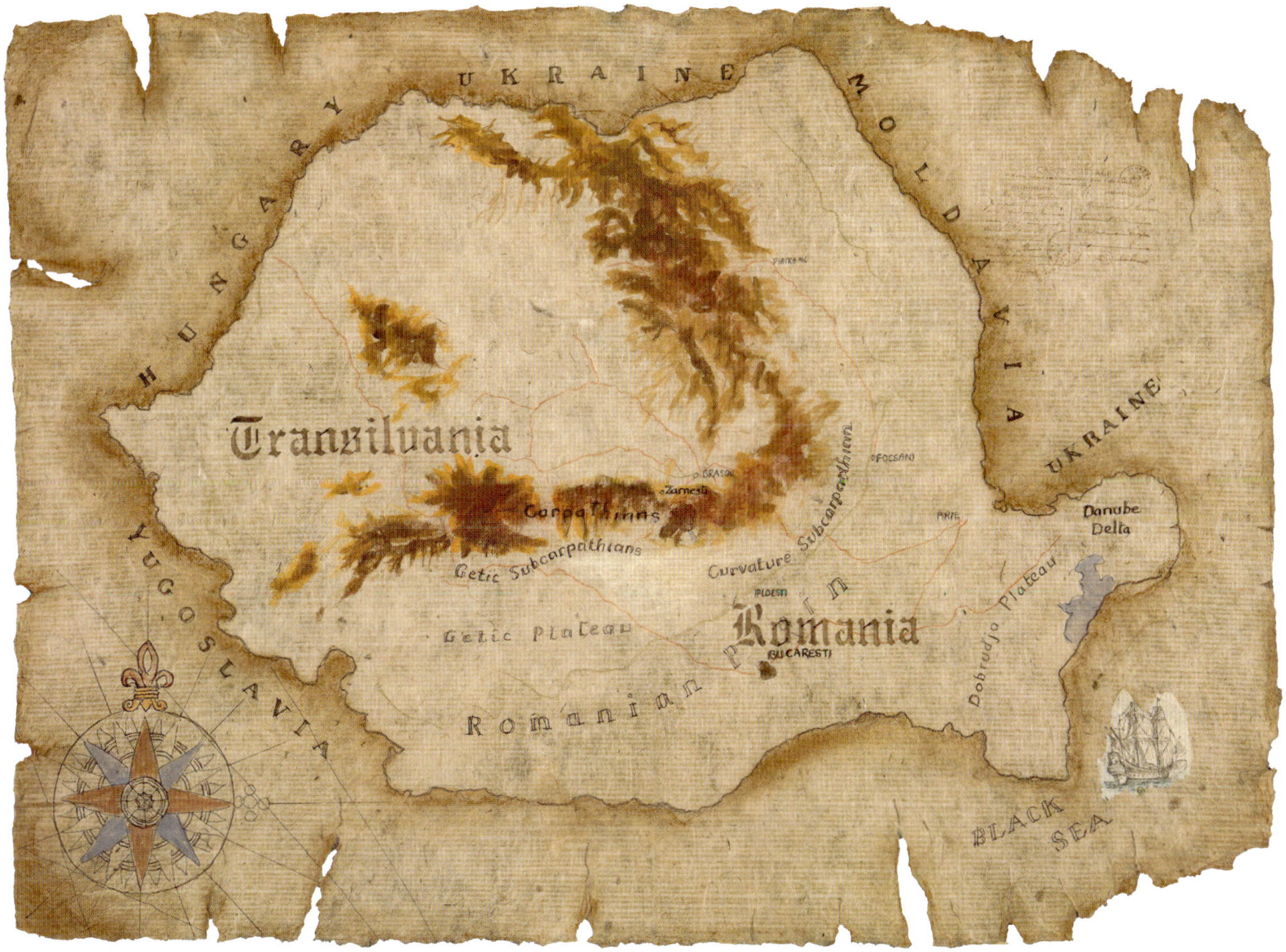

UKRAINE
HUNGARY
MOLDAVIA
UKRAINE
Transilvania
Carpathians
Getic Subcarpathians
Curvature Subcarpathians
Getic Plateau
Romania
BUCARESTI
Romanian plain
Dobrudja Plateau
Danube Delta
YUGOSLAVIA
BLACK SEA

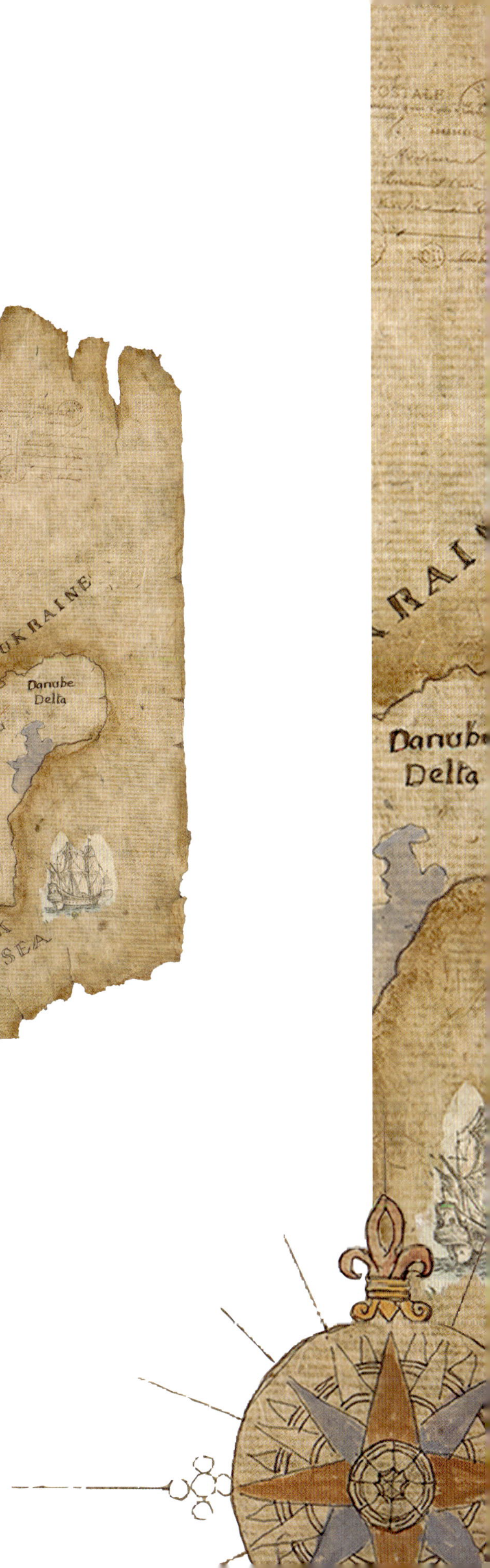

Danub
Delta

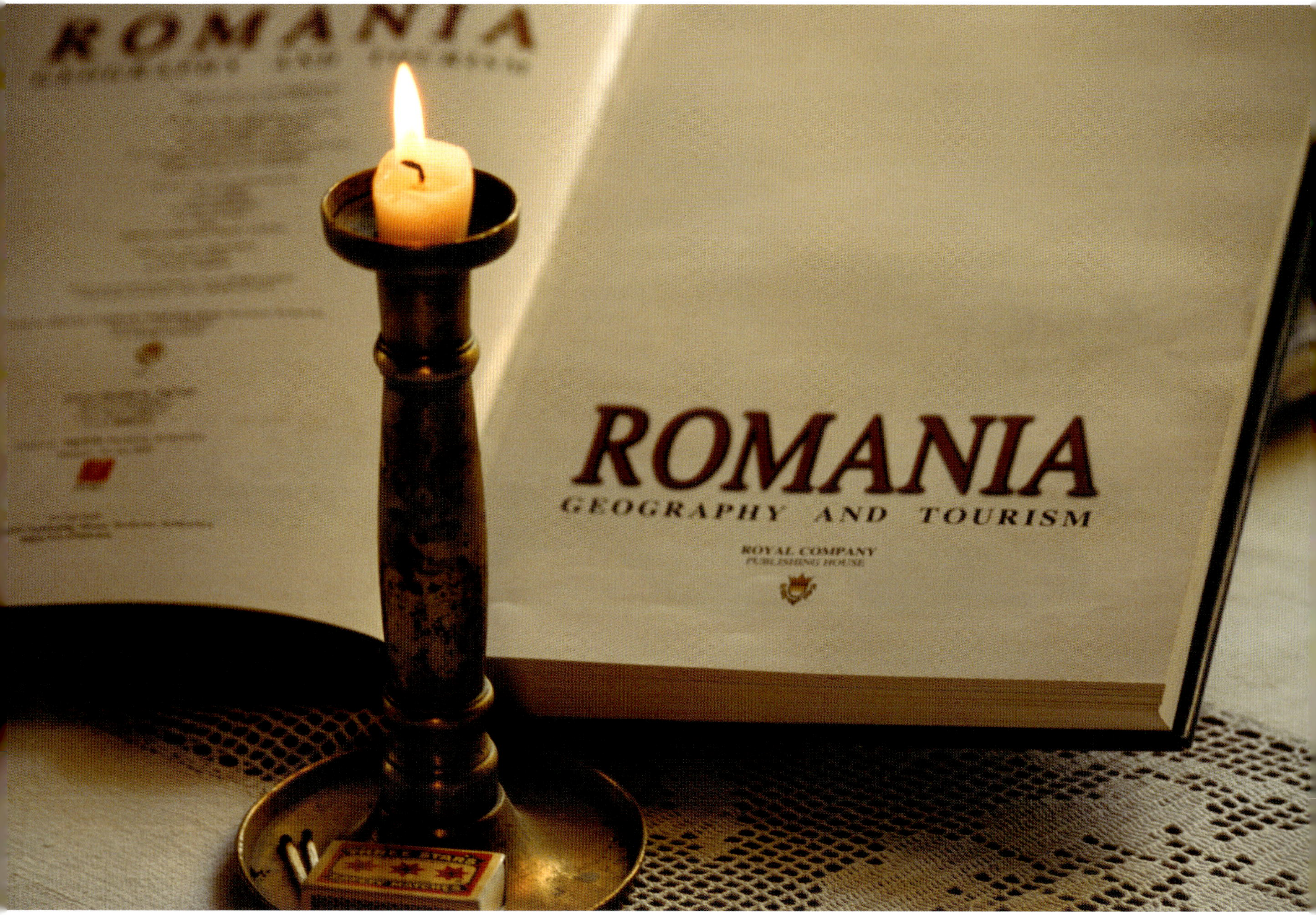
ROMANIA
ROMANIA
GEOGRAPHY AND TOURISM
ROYAL COMPANY
PUBLISHING HOUSE

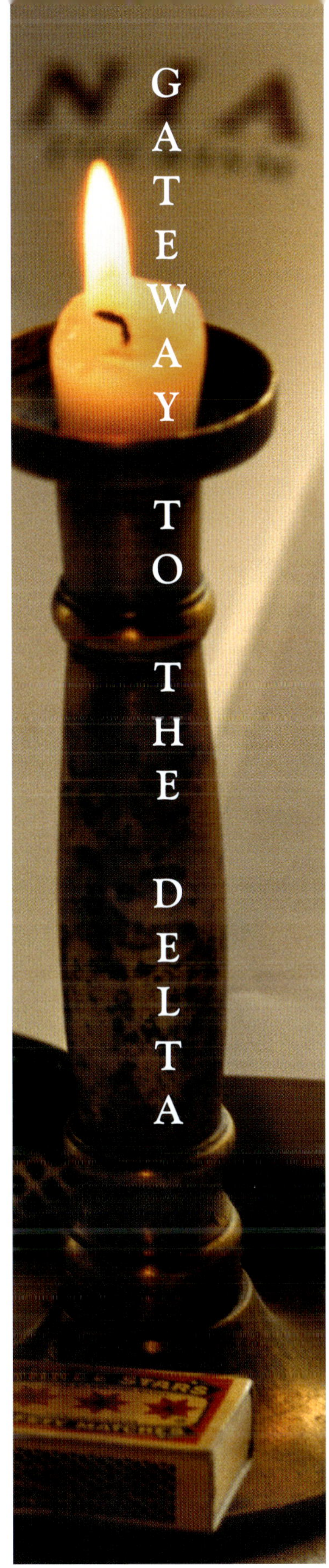

Chapter One

'I find that the district he named is in the extreme east of the country, just on the borders of three states, Transylvania, Moldavia and Bukovina, in the midst of the Carpathian Mountains; one of the wildest and least known portions of Europe...'

JONATHAN HARKER'S JOURNAL, *DRACULA*

Transylvania lies entirely within Romania. A region of fertile farmlands and thick forest, it is bordered to the south-east and west by the horseshoe of the Carpathian Mountains, which form a natural border. When his novel *Dracula* was published in 1897, author Bram Stoker described his chosen setting for the story as *'one of the wildest and least known portions of Europe'*. Even now, more than one hundred years later, Romania remains a relatively forgotten region of Central Europe. During nearly half a century of communist rule contact and communication with the rest of the world was strictly limited. The overthrow of the dictator Nicolae Ceauşescu in 1989 ushered in a period of rapid change, culminating in Romania's acceptance into the European Union in 2007, three years after our visit. In the midst of these political upheavals, and the trials inherent in the day to day life of the population, the country has had little opportunity to promote itself as a tourist destination.

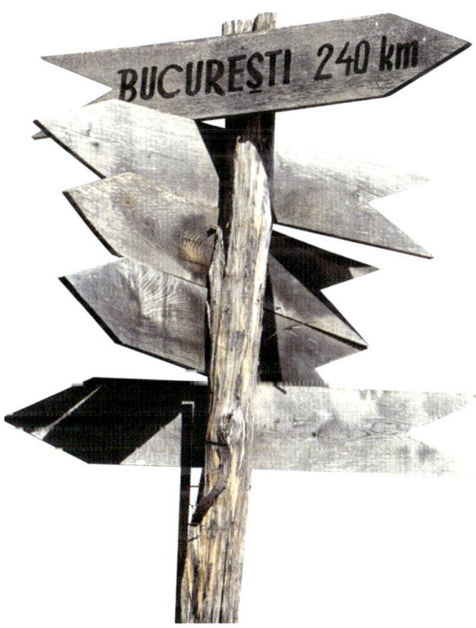

Unfortunately, the majority of news-reports emerging from the country in recent years have also left a largely negative impression: the heartrending images of orphaned and abandoned children starving in freezing cold orphanages; the rapid spread of Aids; the ongoing tensions with the gypsy population; tales of corruption at all levels of government; packs of stray dogs roaming streets of monotonous concrete apartment blocks. Such depressing images do little to tempt the casual visitor. Added to this fifty years of communism have left a legacy of bureaucracy, inefficiency and poor service which continue to ensure that travel through the country is not always an easy experience.

In recent years, however, new ecotourism ventures have provided the opportunity to visit one of the most unspoilt regions in Europe, offering a taste of a 'lost world'. In the remote land beyond the Carpathians an almost medieval way of life still survives. The protective semicircle of mountains has also created a natural 'ark' of relatively undisturbed forest habitat. Traditional farming methods, combined with a lack of intensive agriculture make this one of the best places to see wildlife which has long been extinct in more intensively cultivated regions. Transylvania is home to a greater concentration of large carnivores – bears, wolves and lynx – than any other country in Europe. And this is to be our ultimate destination.

For many years it seems that practically the only tourists who did venture into the country were Dracula obsessives lured by the possibility of 'paranormal horrific experiences on foreign soil' – as promised by the brochure of American company Dractours – who also offer 'rooms filled with ghosts' and the opportunity to attend a vampire wedding ceremony amongst their itineraries. Indeed when I first began researching this journey, the easiest way to reach Transylvania appeared to be hitching a lift on a coach full of enthusiastically gothic Americans sporting fake fangs and long capes. Luckily I was able to make other arrangements...

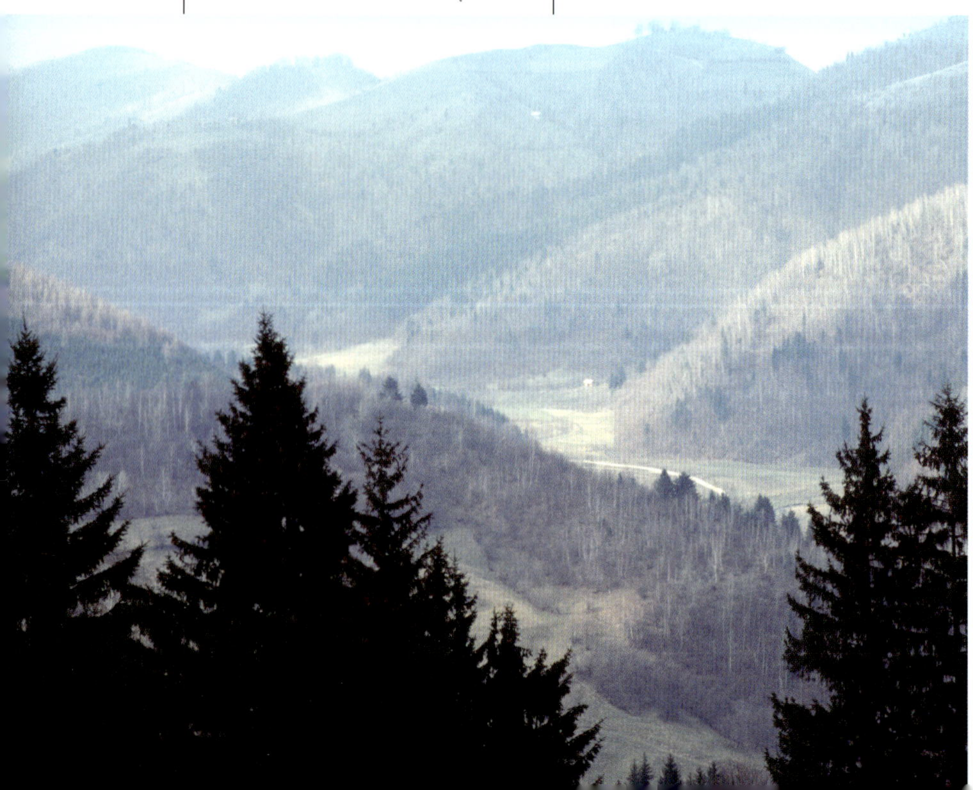

The Carpathian Mountains

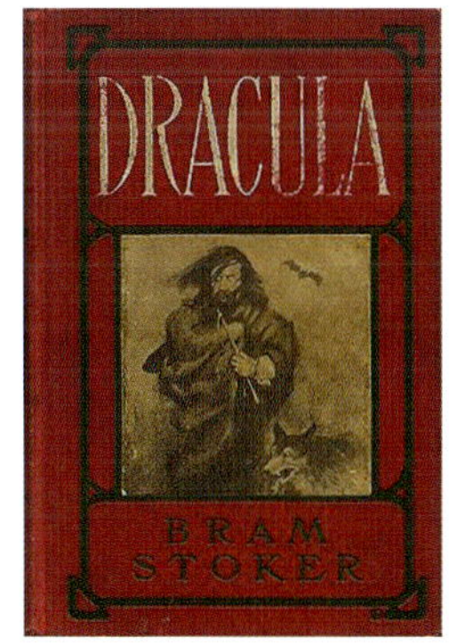

Journal: Day One

One of the most spine-tingling stories of all time, inspiration for countless blood curdling horror films, actually opens with the hero, Jonathan Harker, moaning about delays on the train: '*May 3. Bistritz.-- Left Munich at 8:35 P.M., on 1st May, arriving at Vienna early next morning; should have arrived at 6:46, but train was an hour late.*' Possibly not the most gripping tension-filled opening line ever written, but it does demonstrate that the English character hasn't changed in more than a century – aside from discussing the weather, moaning about the railways remains one of our favourite conversational topics. Pollyanna and I land in Budapest on time, after a smoothly uneventful flight, thus limiting our potential for making small talk. I fall asleep shortly after the in-flight meal is served and wake to a spectacular view of the snow-covered Carpathian Mountains on the right-hand side of the plane.

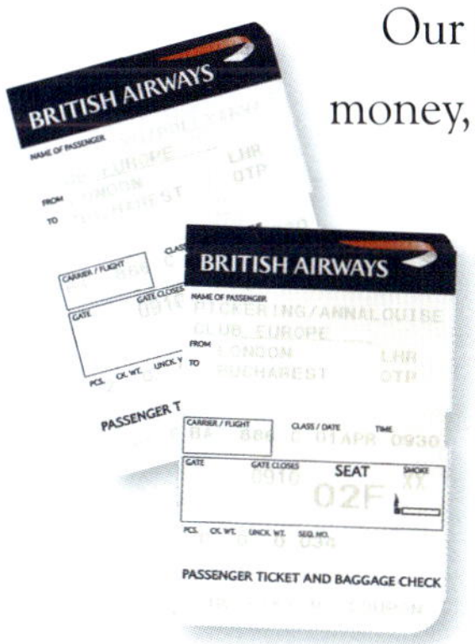

Our journey through the airport is equally easy – we pause only to change some money, handing over one hundred euros and receiving nearly one and a half million lei. Through into Arrivals clutching a letter telling me that the driver I have booked to collect us will be waiting 'towards the right-hand side'. There is no sign of anyone waiting for us, so I leave Pollyanna with the cases and go and explore. Eventually find a man looking worried and clutching a sign saying 'Pickery' on the far left of the hall. He introduces himself as Teodor and ushers us out to our waiting car, a Dacia, Romania's national car. The roads are full of these and, as we quickly discover on leaving the airport, equally crowded with horses and carts. We are soon driving through the suburbs of the capital city – rich and poor housing appears squeezed side by side, and there is clearly lots of new construction work in progress – and out into very flat countryside dotted with churches with matt silver domes. Teodor points out interesting things as we pass – 'church', 'horse', 'police car', 'accident', although these may also be the only things he knows the English words for (still vastly outstripping my Romanian).

Instead of heading directly north towards Transylvania, we are in fact driving east.

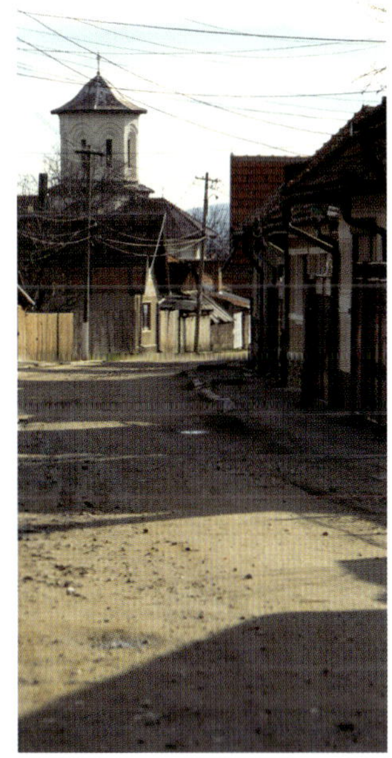

In the remote land beyond the mountains an almost medieval way of life still remains

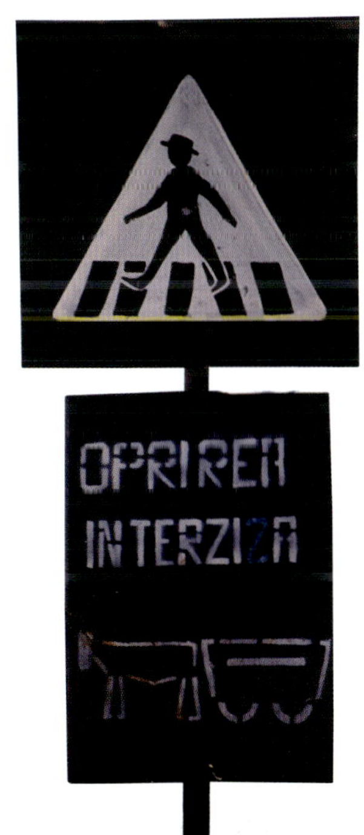

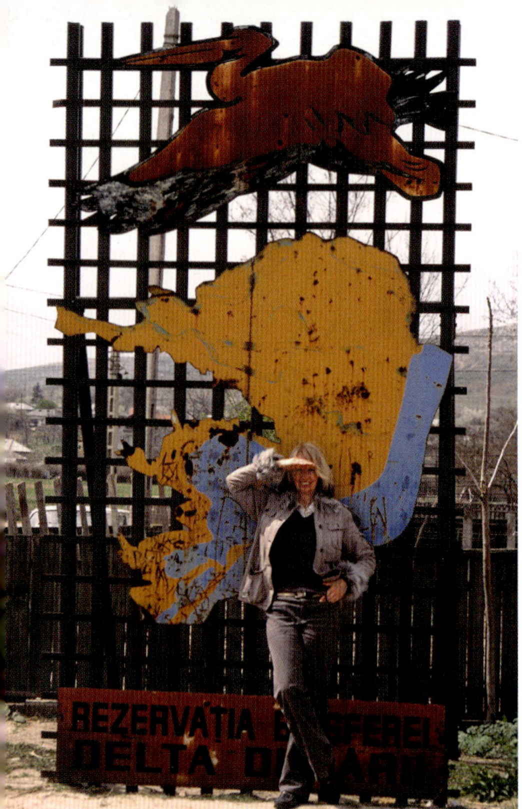

'Gateway to the Delta'

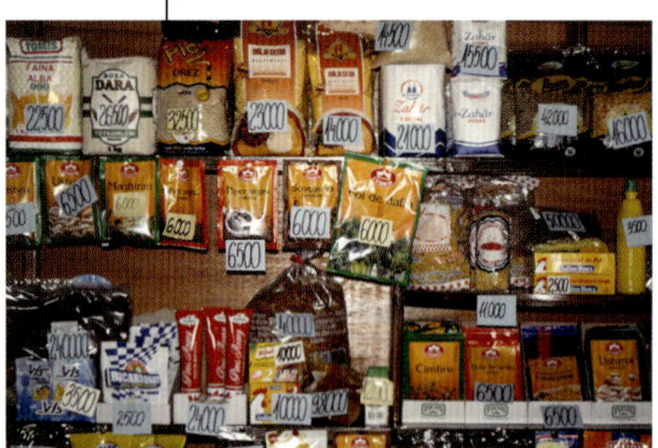

We still need to get to grips with the lei – we have spent 34,000 in one transaction, but as yet have no idea if this is 50p, £5.00 or £50.00

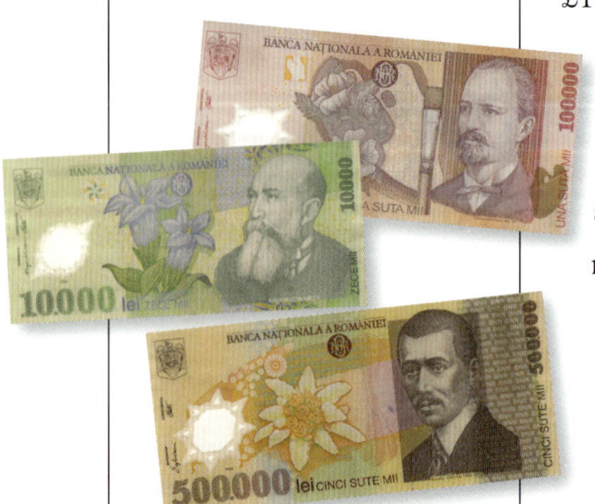

While we are so relatively close, Pollyanna cannot resist taking the opportunity to visit the Danube Delta, a paradise for bird watchers.

We break our long drive at the edge of the town of Tulcea, 'gateway to the Delta', to buy some water. We have been advised to treat the tap water with caution – although the locals drink it straight from the tap, foreigners are highly advised to boil it before drinking. The *Lonely Planet* guide warns, '*If you don't know for certain the water is safe, assume the worst*'. Even now only thirty-eight per cent of rural areas and sixty-two per cent of those living in towns have easy access to a supply of drinking water. Only seven per cent of rural houses have running water, most people still use wells and streams for their supply. Our shopping trip proves entertaining as we try to translate the labels and end up purchasing some bottles of 'de-carbonated' – which as the label suggests tastes exactly like sparkling water which has gone flat. We still need to get to grips with the lei – we have spent 34,000 in one transaction, but as yet have no idea if this is 50p, £5.00 or £50.00. We make a second stop for Teodor to buy cans of diesel for his tractor. He thrusts these on top of our luggage, so our bags now have a delicate aroma of oil, and are somewhat more flammable than before. As he completes his purchase we wander off into a Delta souvenir shack, where Pollyanna purchases a pair of stout hand-knitted hiking socks for 80,000 lei (10p? £100.00?). Teodor inspects them on our return to the car – he rubs the coarse wool between his fingertips and then sniffs the socks (Is this the usual sock buying protocol here? Should we have done this in the shop?) but seems to approve. He guesses the price at 100,000 lei, so we can be confident that we have not been ripped off even though we still have no idea how much we have spent. Spend the next few minutes creating a quick reference conversion table of lei to sterling, thus making our shopping easier as we travel.

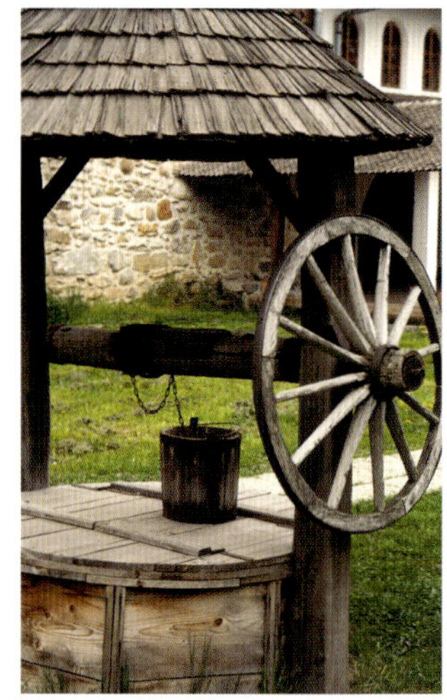

Most people still use wells and streams for their water supply

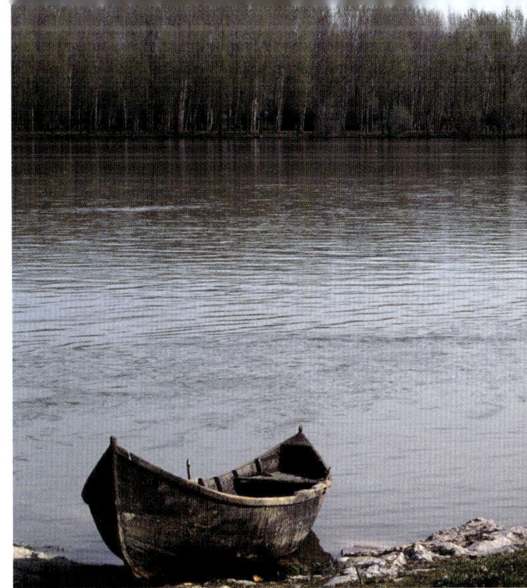

Finally we arrive at our destination – the Hotel Teo, built on the Sfantu Gheorge branch of the River Danube, just outside the village of Mahmudia. This village has grown up on the site of an ancient Roman walled city, Salsovia – which means 'Sun City' – although we arrive in light drizzle, at least the name provides some cause for optimism about the weather. In the lobby of the hotel we say our farewells to Teodor and thank him for getting us safely to our destination through the chaos that is the Romanian traffic.

Here we meet our companion for the first part of our journey. Colin Shaw is the co-founder of Roving Romania, one of the companies which has sprung up to offer travel into the newly open post-communist country. During my research for the trip I had made contact with Colin, a founder member of the Association of Ecotourism in Romania; he worked closely with the founders of the Carpathian Large Carnivore Project (CLCP). We have been communicating with this group about their study of wolves and other carnivores, and they have been very helpful in offering support and advice for our journey. For the next couple of weeks Colin will act as our driver and interpreter – having lived in Romania since 1994 he speaks and writes fluent Romanian and knows the country well. Colin is an excellent photographer and we had seen his photos in our *Lonely Planet* and *Bradt* guidebooks to Romania and Transylvania.

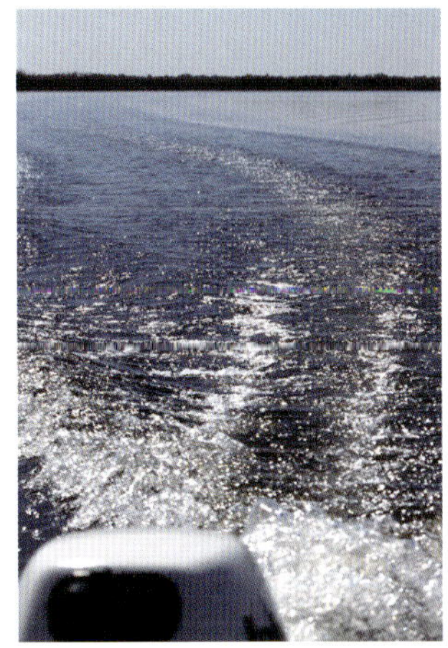

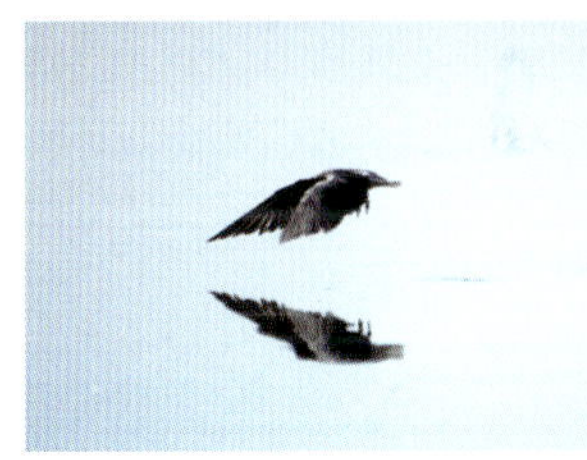

After a long day's drive we are very pleased to find that our rooms in the hotel are clean and warm, and we are just changing out of our grimy travel clothes before dinner, when Colin knocks on the door to announce that he has managed to arrange our first boat trip for that evening – a brief thirty-minute excursion to introduce us to the Delta habitat.

We change again into warm boots and put on most of our clothing – there is a distinct chill in the evening air. The hotel has its own small boats, which are available for their guests to explore the network of rivers which form the Delta, so we climb into a small motorboat. The Danube Delta covers 4152 square kilometres of waterlogged reed beds crossed by a labyrinth of rivulets and streams. We set out down a wide channel, the waves slapping

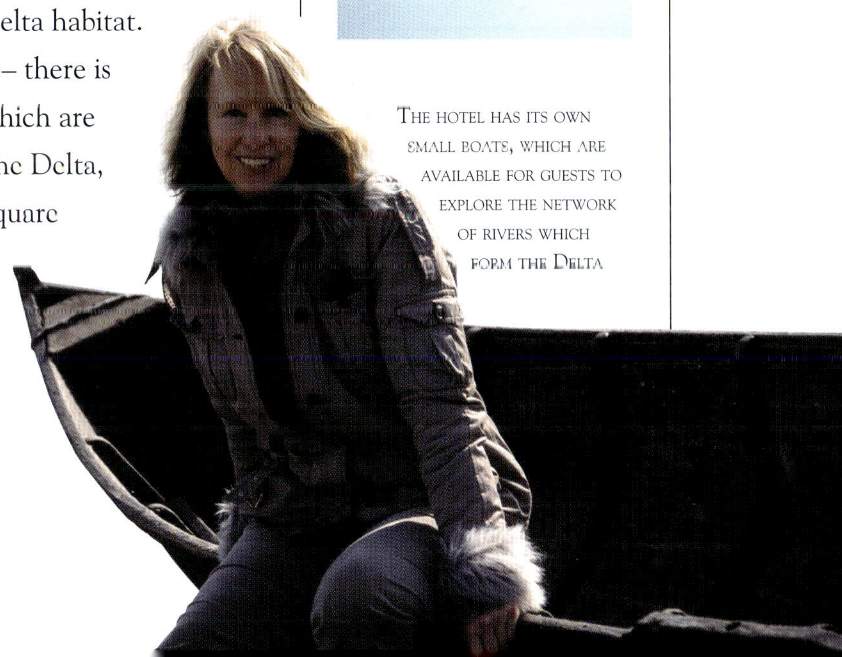

The hotel has its own small boats, which are available for guests to explore the network of rivers which form the Delta

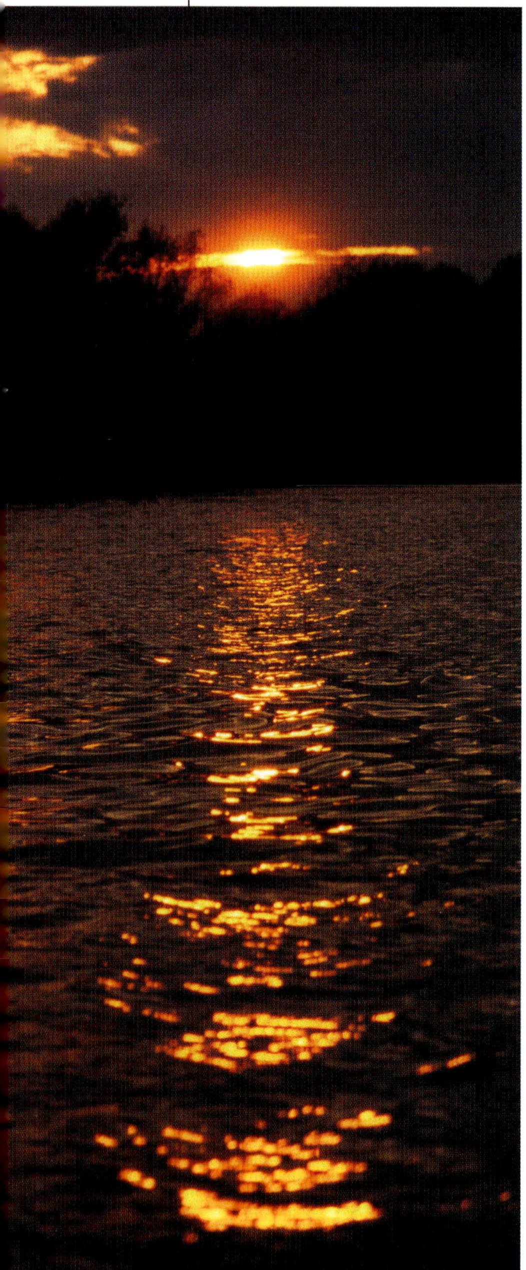

The sun is now starting to set. I capture a couple of shots in the fading light

against the side of the small boat and spraying us liberally with the famed 'blue' waters – which in this gloomy weather currently look muddy green. Our boatman Costica soon steers us away from the main channel into a smaller narrow waterway – willow branches dip into the waters on either side and oak trees entangled in lianas create a narrow corridor through which we pass. We have, of course, been lured to the Delta by the promise of the abundant birdlife – and almost immediately we spot a purple heron, motionless on a low branch. Grey herons stand like sentinels every few feet along the banks, a glossy ibis stands on spindly black legs in the shallows, and we quickly see whiskered terns, black-winged stilts and long-tailed ducks. Squacco herons seem entirely oblivious to the presence of our boat. However, we have arrived late in the day and although the drizzle has stopped and the skies cleared, the sun is now starting to set. I only capture a couple of shots in the fading light and within a short time it is even too dark for Pollyanna to sketch.

In the gloom the Danube looks like a set from a Tim Burton film, with twisted and gnarled branches looming towards us out of the half light. Very quickly it is also too dark for Costica to see the logs which float on the surface of the water – consequently we frequently get stuck and wait while he switches off the engine and uses an oar to push us free from the latest obstruction. Eventually we only have the light of the stars to navigate by and as low-hanging branches whip across my face, it becomes clear that Costica can't see anything at all. It is getting later and later, colder and darker, but we are rewarded by a spectacular view of the night sky – with no light pollution the stars shine brightly overhead. We can clearly make out the Milky Way, a pale shining cloud of billions of stars separated

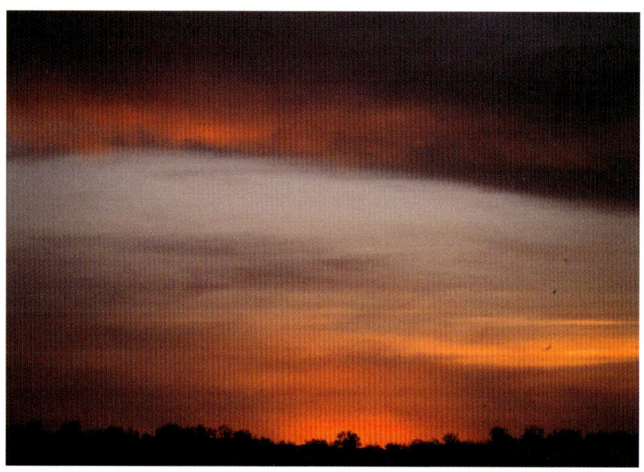

from us by an unimaginable distance of time and space. It is hard to believe that only this morning we were waiting in the departure lounge at Heathrow for our flight to be called – and now we are splashing through the Danube Delta under the glittering night sky.

However, Pollyanna is in no mood to appreciate the night sky. Tired from the lengthy day of travelling, increasingly cold and hungry she is huddled deep inside her coat, wincing slightly every time another stray branch lashes at her shoulders. We are both relieved when we spot the industrial zone close to our hotel – we can see the headlights of huge cranes burning through the gloom, slow moving prehistoric creatures working late into the night. Finally the lights of the hotel come into view across the dark water, and some three and a half hours after we set out, our thirty-minute boat trip comes to an end. Some two minutes after disembarking, we are sitting in the warm bar. We are swiftly presented with glasses of Tuica, the plum brandy which is enjoyed throughout the country. (Romania is the world's second largest producer of plums, after the USA, and as much as seventy-five per cent of the annual harvest is turned into Tuica – as we will discover on our travels.) If you have never come across this drink, it is a sort of industrial strength alcohol, best described as a cross between paint stripper and gasoline.

A small plate arrives containing one olive, a small piece of cheese, and a chunk of pork fat which resembles a piece of soap which has been left in the bath for too long. Pollyanna eats the olive and I nibble on the sheep's cheese – but we pass on the pork fat and instead head into the restaurant.

We are the only guests, and a sulky waitress slaps menus down in front of us. She doesn't seem best pleased that we have disturbed her evening. We fancy a nice warming bowl of soup. 'Soups Off'. So instead we have a nice warming toasted cheese sandwich with an accompaniment of microwaved tinned vegetables.

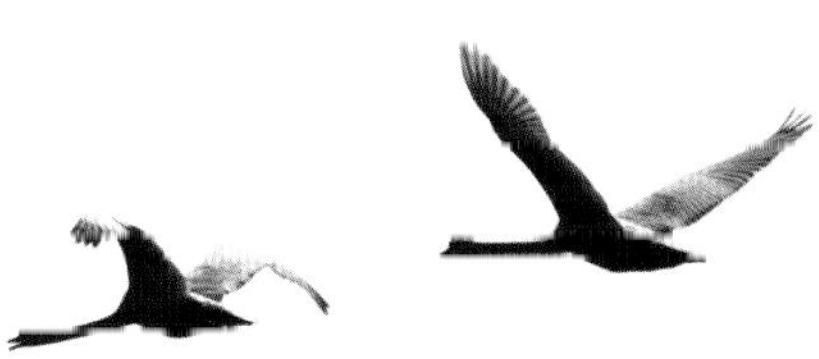

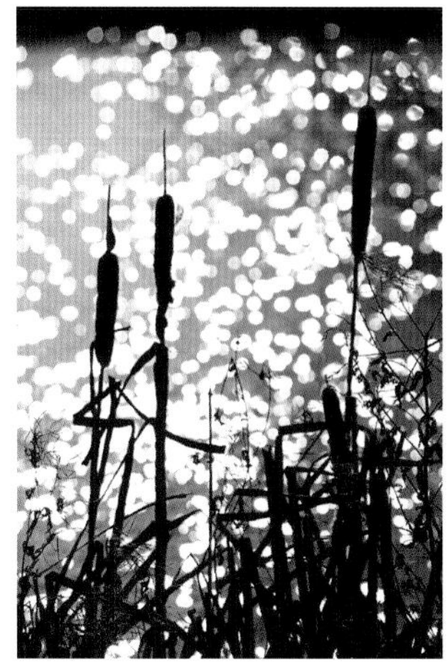

'In most cities the sky looks as though it has been emptied of stars, leaving behind a vacant haze that mirrors our fear of the dark and resembles the urban glow of dystopian science fiction. We've grown so used to this pervasive orange haze that the original glory of an unlit night—dark enough for the planet Venus to throw shadows on Earth—is wholly beyond our experience, beyond memory almost.'

VERLYN KLINKENBORG, *NATIONAL GEOGRAPHIC*

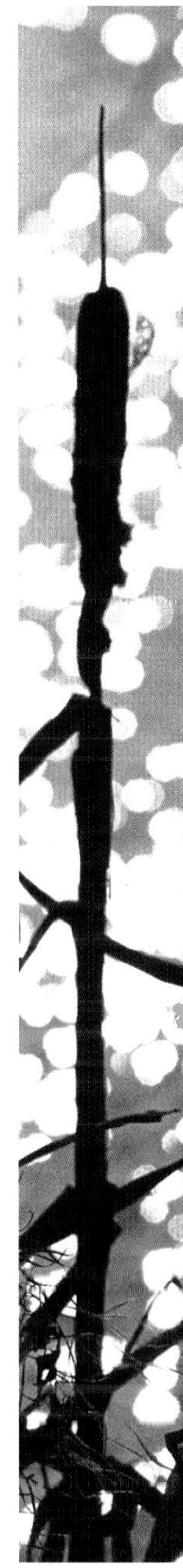

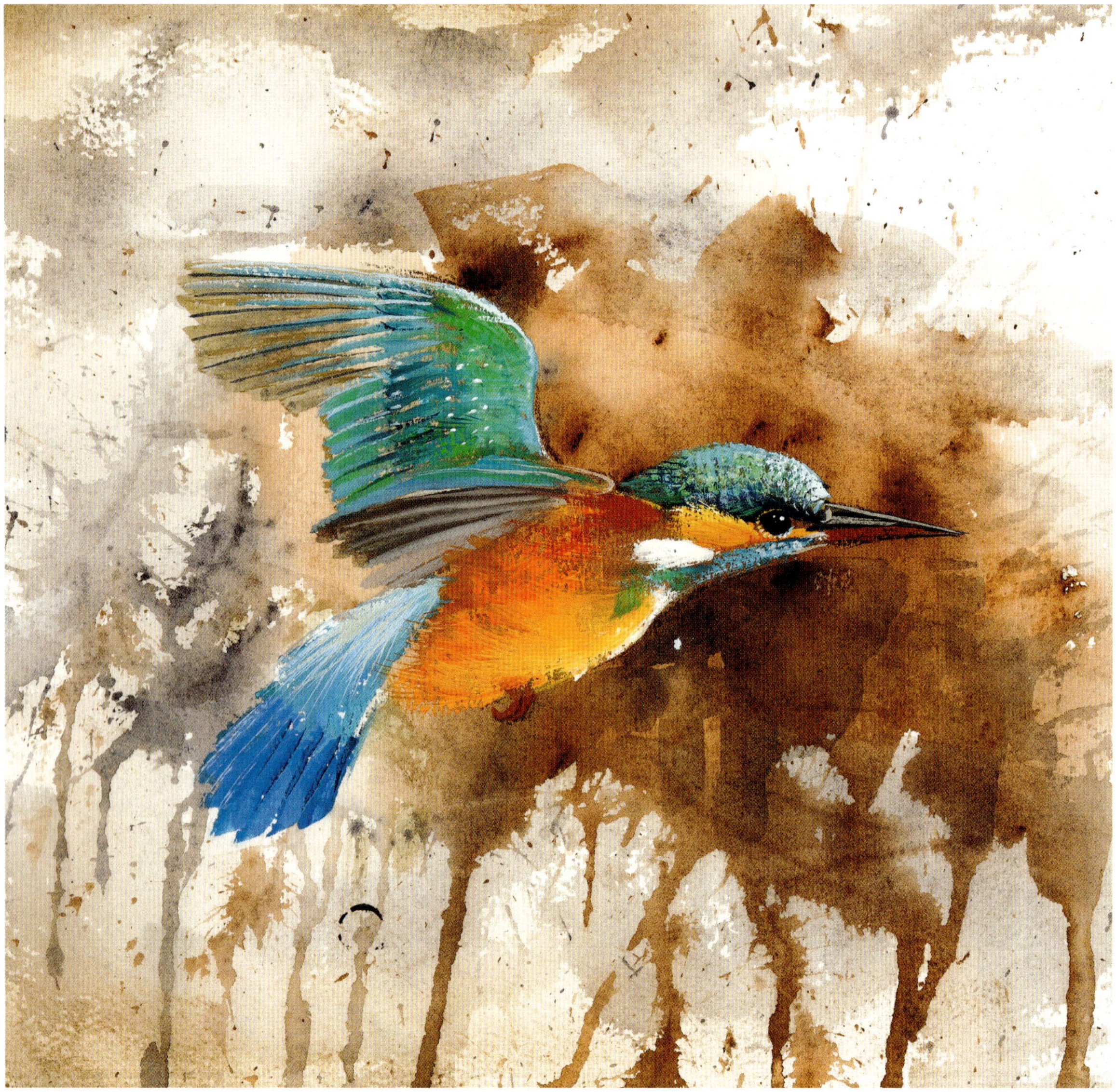

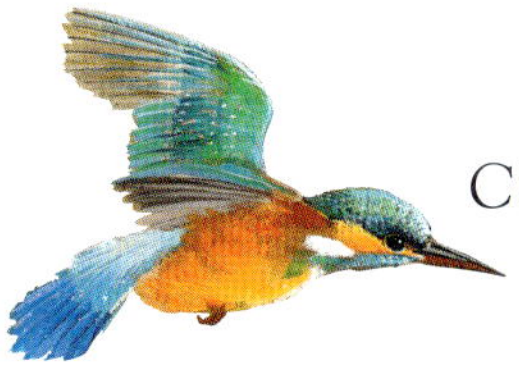

Chapter Two

'I can hear the waves lapping against the ship, and the water rushing by. Canvas and cordage strain and masts and yards creak. The wind is high... I can hear it in the shrouds, and the bow throws back the foam.'

JONATHAN HARKER'S JOURNAL, *DRACULA*

German poet Friedrich Hölderlin described the Danube as 'A refreshing, melodious river, sometimes foaming with high spirits, at other times dreaming serenely'. Rising from its source in Germany's Black Forest the waters from this mighty river flow 1,788 miles into the Black Sea. On the way it runs through, or forms part of, the border of nine countries, and courses through four capital cities: Vienna, Bratislava, Budapest and Belgrade. Eighty kilometres before reaching the coast the river splits into two then three branches to form the second largest and best preserved of Europe's deltas. Located halfway between the North Pole and the Equator, the Danube Delta covers 4,152 square kilometres of rivers, sand dunes, reed-beds, canals, marshes, tree-fringed lakes and islands: 3,446 square kilometres – more than eighty per cent – of this naturalist's paradise lie within Romania.

In 1991, the Romanian part of the Danube Delta became a UNESCO World Heritage Site and in 1998 the entire Delta was designated part of the world network of Biosphere Reserves. In fact the intricate pattern of islands, channels, streamlets and lakes combine to create a wildlife refuge which is home to the highest concentration of bird colonies in all of Europe. Remarkably the reserve has the third largest biodiversity in the world. Altogether, 3,450 bird and animal species including wildcats, foxes, wolves, boar and deer can be seen here, as well as 1,700 plant species – numbers exceeded only by the Great Barrier Reef in Australia and the Galapagos Archipelago.

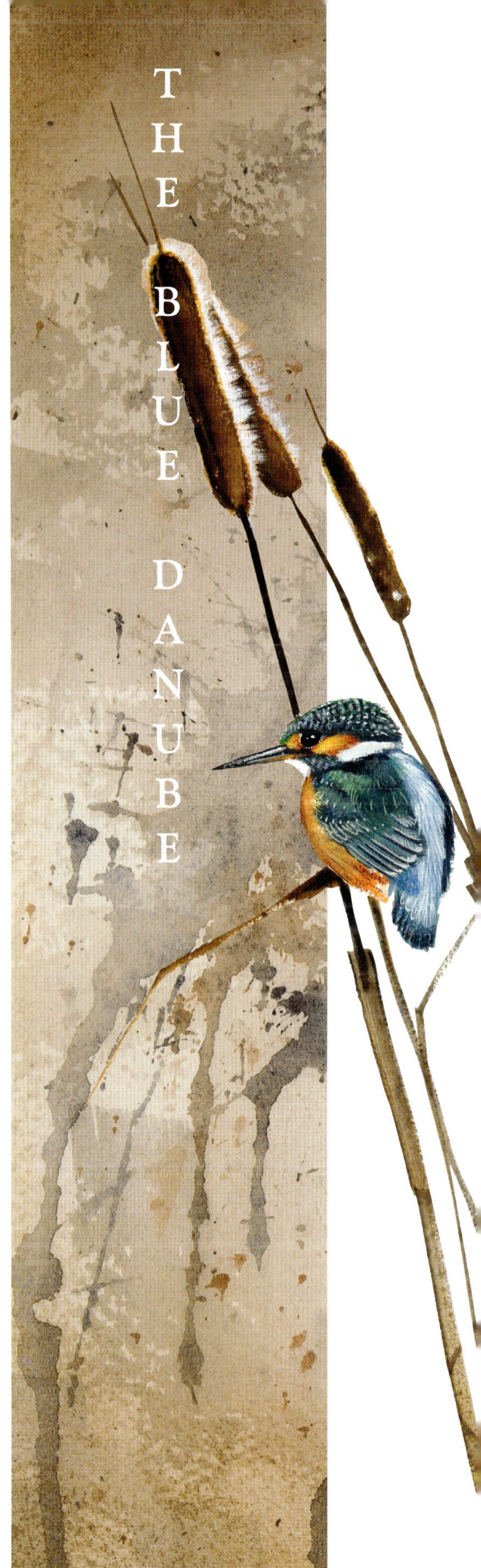

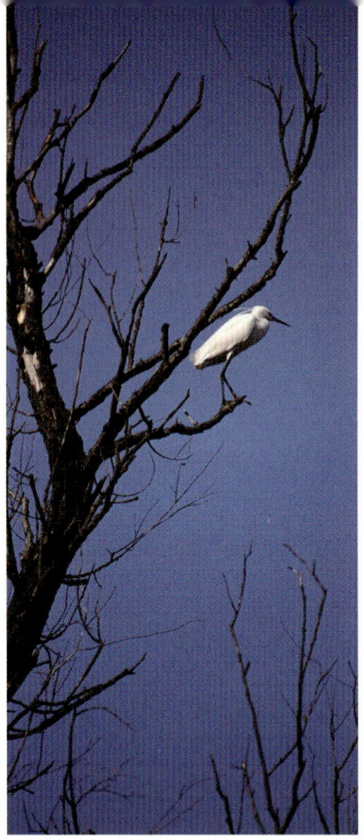

Egrets perch in the trees, their feathers snowy white against the blue of the sky

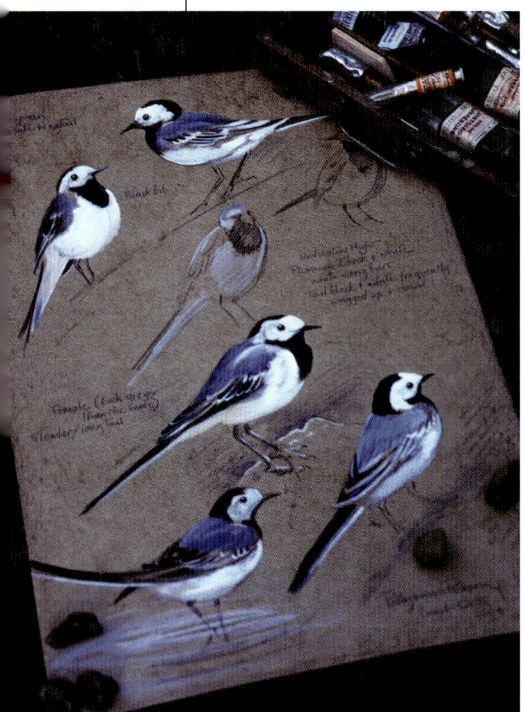

A white wagtail poses obligingly while Pollyanna sketches

Day Two

We awake to sunshine and a Blue Blue Danube. Breakfast is toast with delicate rose petal jam and a glass of yoghurt. Then, loaded down with cameras, lenses and film (me) and sketchpad, pencils and watercolours (Pollyanna) we make our way down to our boat. Pollyanna loves to paint water – she enjoys the challenge of capturing the shimmer of light reflected on the surface, the fluid shapes of the transient ripples and waves. The Danube has its own personality, the power of the constant surge of the main river balanced by the peacefulness and tranquillity of the slow-flowing pools. Egrets perch in the trees, their feathers snowy white against the blue of the sky. It is when the sky is this colour and is reflected in the water that the river earns its aura of romance. Pollyanna is constantly observing the way in which the light strikes the water, how the reflections of the clouds and trees are disturbed by the movements of the great-crested grebes as they bob on the surface, then abruptly vanish as they dive beneath the ripples.

We see numerous hobbies, and a marsh harrier. Out in the open broads we spot both red-necked and black-necked grebes, while close to the banks a white wagtail poses obligingly while Pollyanna completes a sheet of sketches. We watch as iridescent kingfishers flash past our boat, and flocks of terns and gulls take flight as we glide towards them. At Pollyanna's exhibitions she is often asked how she manages to sketch quickly moving wildlife and, in particular, birds. In fact her training began at art college, where she and her fellow students learnt to draw a moving model in life class. Recently, psychologists and experts on human behaviour have calculated that it takes ten thousand hours of practice to become an expert in almost anything. Musicians, artists, athletes, actors, writers, chess players – all may have some natural talent – but it is now thought that those at the very top of their chosen field have got there purely through practice and persistence. The magical figure of ten thousand hours has been regularly shown to be the amount of time required to gain mastery of your chosen skill. Once this level of expertise has been achieved through hours and hours of dedicated work it is possible to produce consistently high-level results. An American

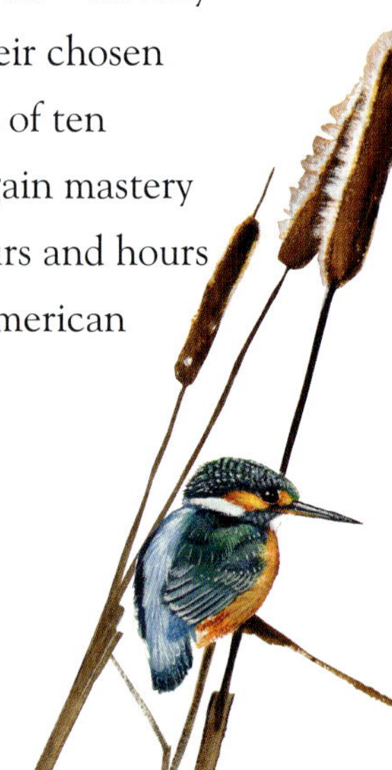

martial arts expert, George Leonard, aptly describes this as: 'The mysterious process during which what is at first difficult becomes progressively easier and more pleasurable through practice. Practice and persistence, driven by passion, is the formula for mastery'. To achieve success as an artist natural talent is not enough – you also have to have the drive, willpower and dedication to devote your life to developing your skill.

This is certainly true of Pollyanna. At the time of making this journey she has been working as a freelance artist for over thirty-five years, following five years at art college. In a typical week she works in the studio every single day, starting at 8.00 am, usually working through until around 6.00 pm – with breaks for lunch and cups of tea! Art is a vocation as well as a career. On her expeditions to study endangered species she sketches constantly, and has undoubtedly served her ten thousand hour apprenticeship. These many years of dedicated practice mean that she can trace the movement of the kingfishers as they dart across the water, and capture the elegant lines of the white egrets in a few simple, seemingly effortless strokes.

In the afternoon we stop at a fishermen's island in the middle of one of the larger expanses of water – a reminder that the Danube is not just a holiday destination for bird watchers, but a living river at the heart of the livelihood and culture of the local population. Ragged nets are hanging to dry in the sun, but the fishermen themselves are out on the waterways, and the island is deserted – apart from the fishermen's cat, who rushes up to greet us, rubbing round our ankles in the hope that our boat is bringing home the catch of the day. White-tailed sea eagles soar overhead. There is an outhouse – but we don't avail ourselves of the facilities…

However, by the time we dock at the Cormoran Hotel for a late lunch the loo on the fishermen's island is starting to look appealing – we have been on the water for the best part of five hours, and I dash into the ladies like a roadrunner dipped in Tabasco, before emerging more serenely for a leisurely lunch of fried cheese.

We have arrived at the end of March, when flocks of white pelicans arrive

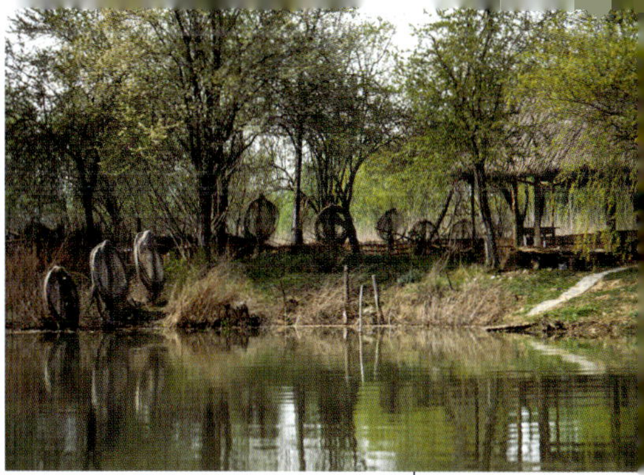

Ragged nets hanging to dry in the sun

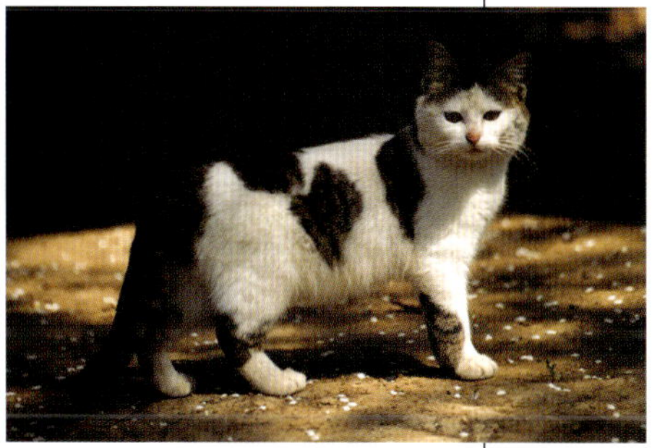

Hopeful fishermen's cat

We don't avail ourselves of the facilities…

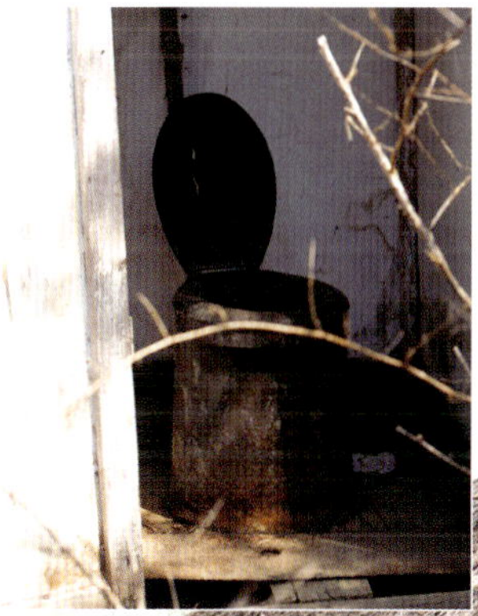

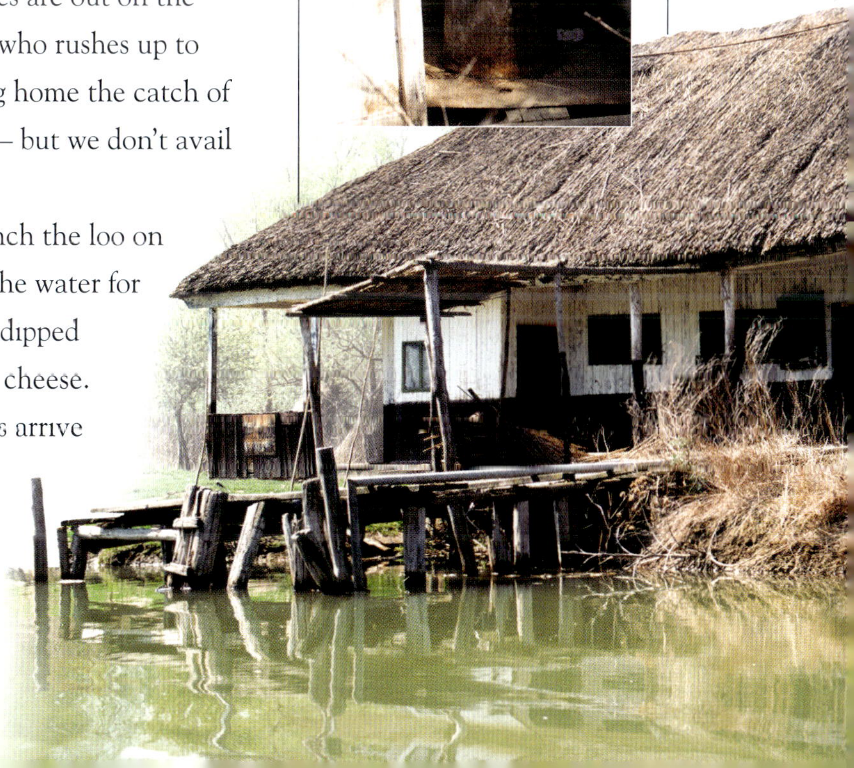

Fishermen's Island

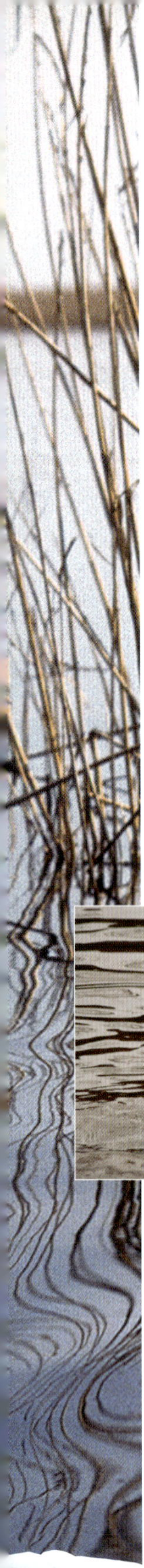

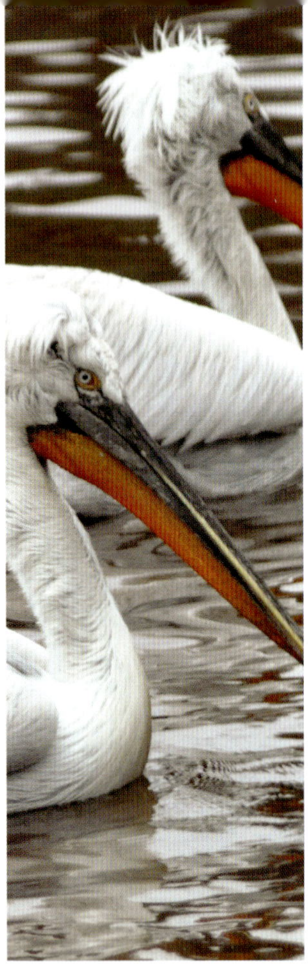

from the Nile Delta and the Red Sea to nest in this region, which becomes home to some 3,500 pairs – Europe's largest breeding population. As we turn to make our way back to the hotel a flock of large birds appears high above us – some of the first arrivals. As the day lengthens, the warmth of the sun quickly fades and my photography slows almost to a halt – after every picture I take I have to thaw my icy fingers under Colin's parka, which he has kindly lent to me to use as a blanket. We again arrive back at the hotel stiff with cold and make our way straight to the dining-room.

Tonight the tables are lavishly laid with silverware, gleaming glasses and crisp cloths with matching folded napkins. Our previously surly waitress appears to have undergone a personality transplant as she is beaming at us, and as soon as we sit down once again presents us with glasses of Tuica.

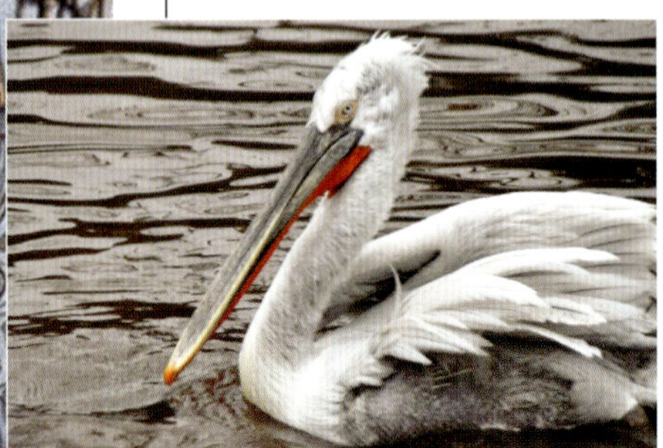

THE DANUBE DELTA IS HOME TO 3,500 PAIRS OF WHITE PELICANS WHICH NEST IN THE REGION

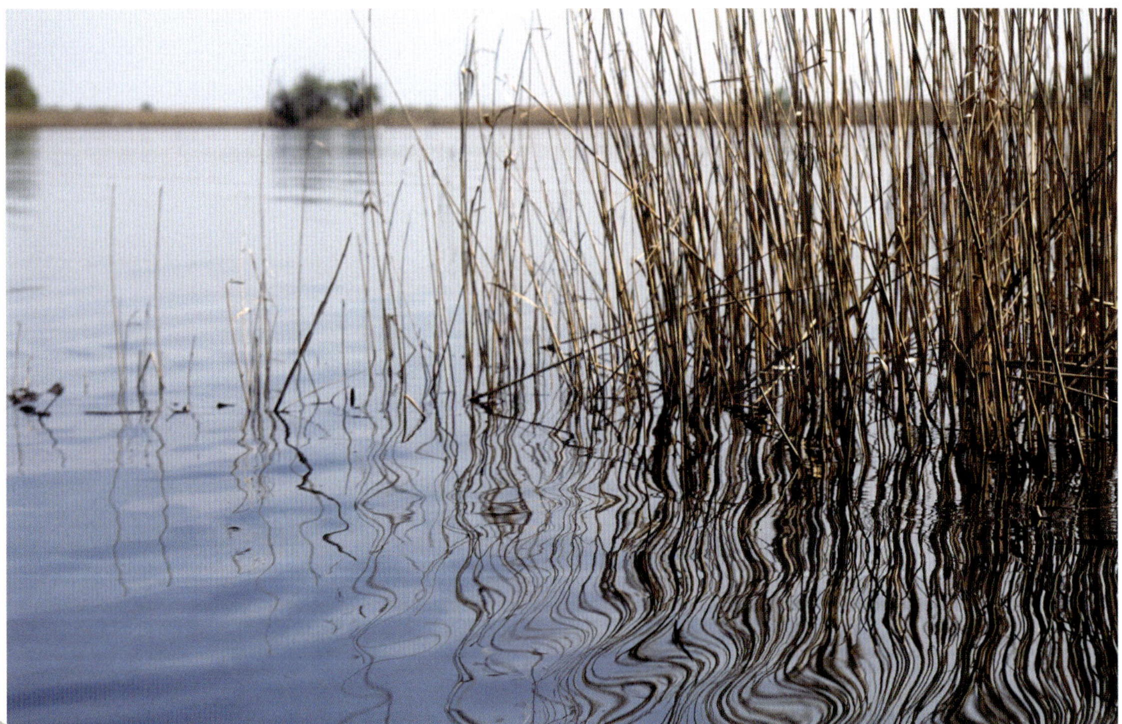

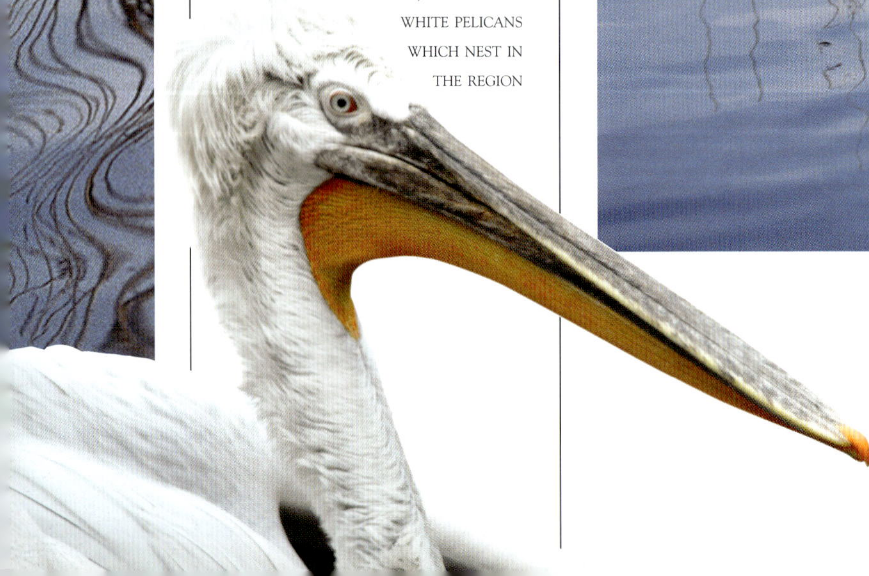

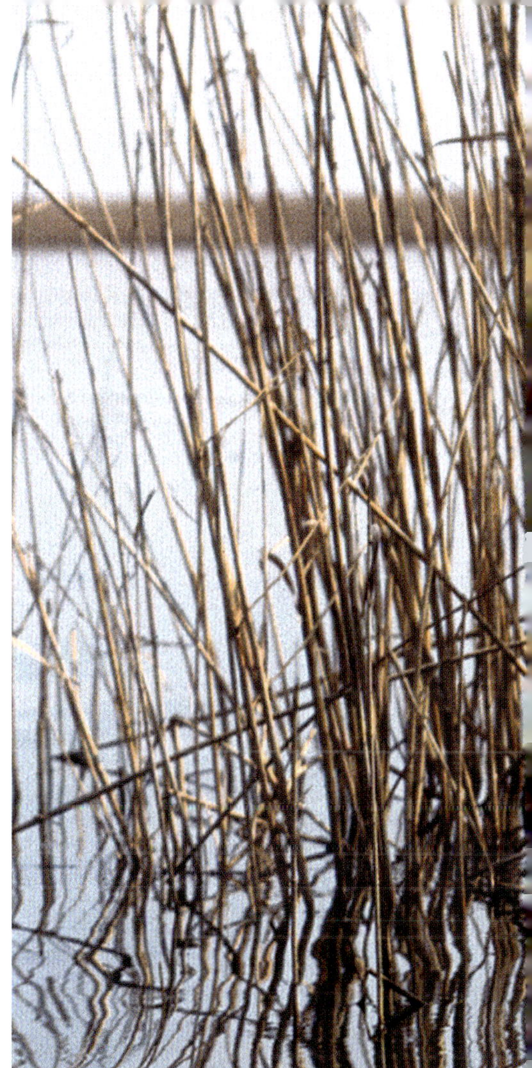

Not only do we have warming spirits to drink, the vegetable soup is on, so we place our order and sit back to sip our drinks with the Romanian toast *Noroc*. The waitress returns, and removes all the shining glassware, and before we can sully the crisp napkins she also whisks these away, replacing them with a sheet of kitchen roll each. When she returns with the menus she removes the vase of silk flowers from the centre of the table and instead of lighting the candle whips it away to a safe shelf behind the bar. Next to go is the ornate silver cutlery – when the soup arrives it comes with basic steel spoons. She also rearranges anything we dare to move on the table and constantly returns to straighten the cruet set – although she resists purloining it. Before placing the soup on the table she also removes the map Colin has brought to trace our ongoing journey and glares at his room key which is not placed to her liking. We sit around our now barren table nervously sipping our soup as she eyes us from across the room expecting that at any moment she will return to whip the tablecloth out from under our bowls like a magician at a children's party.

'It is when the sky is this colour and is reflected in the water that the river earns its aura of romance.'

For us this is an early introduction to the fact that decades of a communist regime did little to foster a service industry in the country and, even in a relatively smart three-star hotel like the Teo, some of the employees still have a tendency to view the clients as a somewhat irritating intrusion on their day, who clutter up their tidy dining-room and may even make use of some of the facilities. The fact that the more often the poor waitress returned to remove something on our table we found it the harder to stifle our giggles probably wasn't helping the relationship. At the time of our visit we felt that Romania was a country on the cusp of change, still very much in transition from its communist past to its future as a democratic member of the European Union.

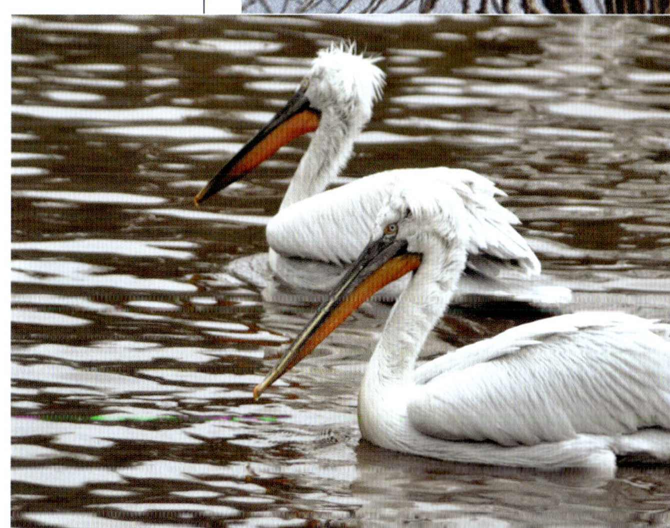

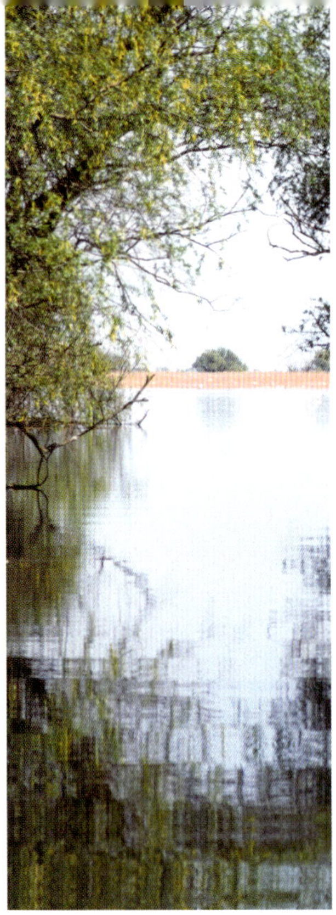

Day Three

We are up early this morning, waking with the sunrise, so take a pre-breakfast stroll south of the village towards a small archaeological site where we can see the remains of the walls of houses dating from the fourth to sixth centuries. We meander back towards the coast enjoying the tranquillity of an early morning stroll – and find we are not the only ones out in the morning air. We spot a fox – the first wild canine of our trip! – and three brown hares. There are wolves in the Delta, and of course we are keeping a careful watch for them – but it is a different and, in fact, far rarer sighting which is the highlight of our walk. We see one of Europe's rarest mammals, a raccoon dog, making its way across a field in full view in front of us. We have a wonderful clear view of this aptly named creature – although it is too distant to allow me to take any good photographs. The raccoon dog is a distant relative

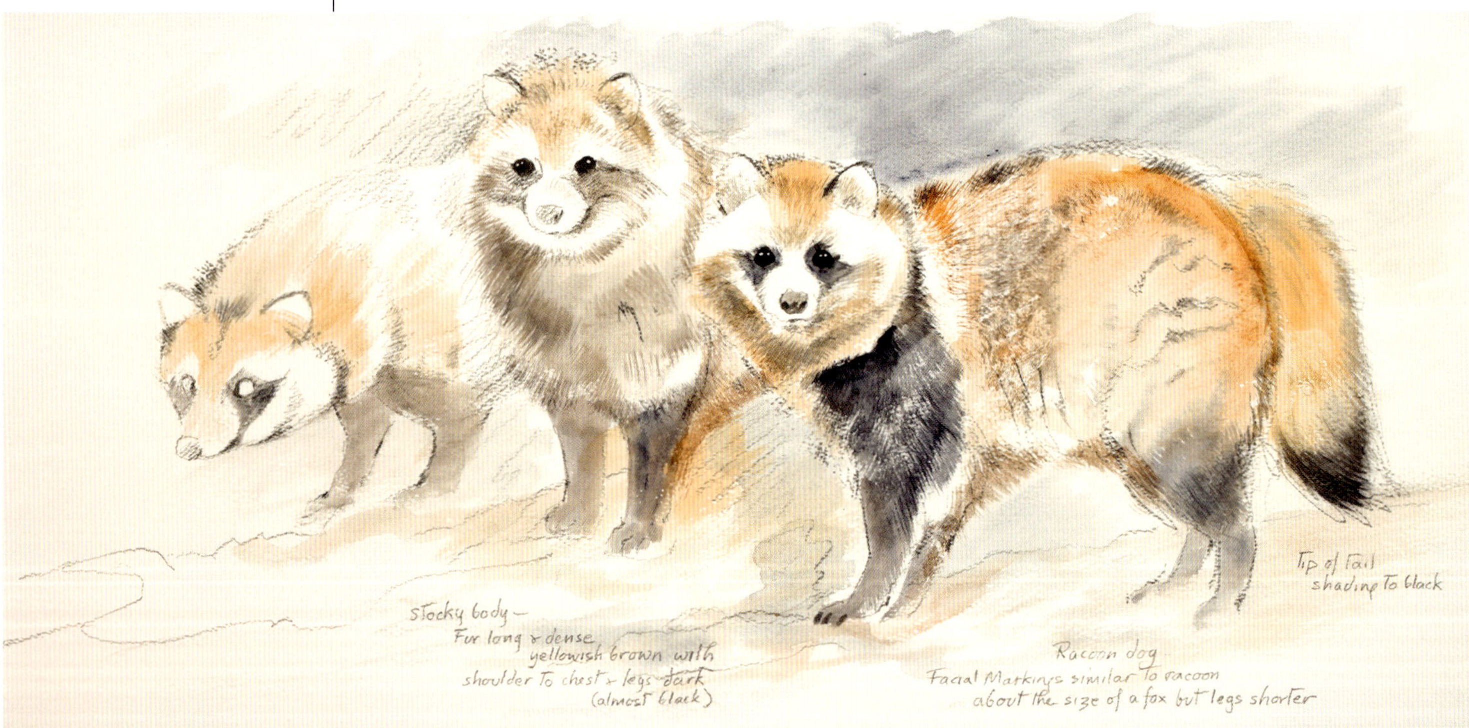

Study of a raccoon dog

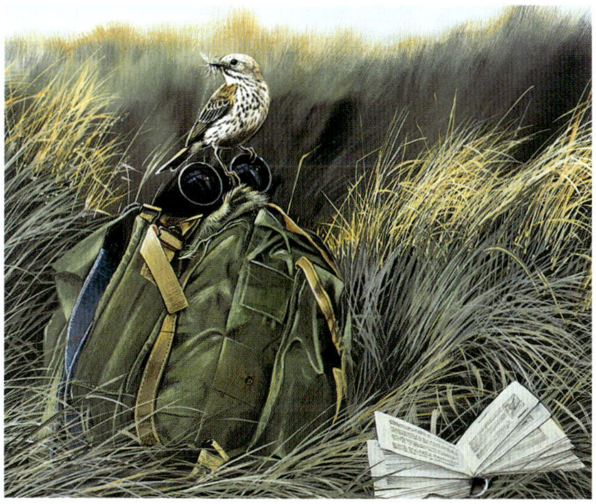

My journal becomes little more than a twitcher's list

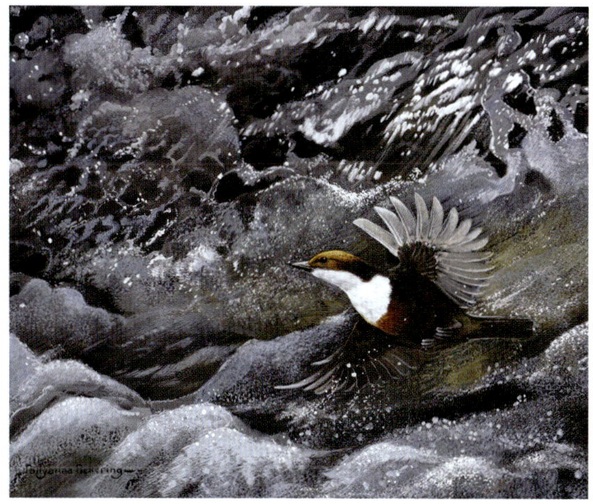

Dipper

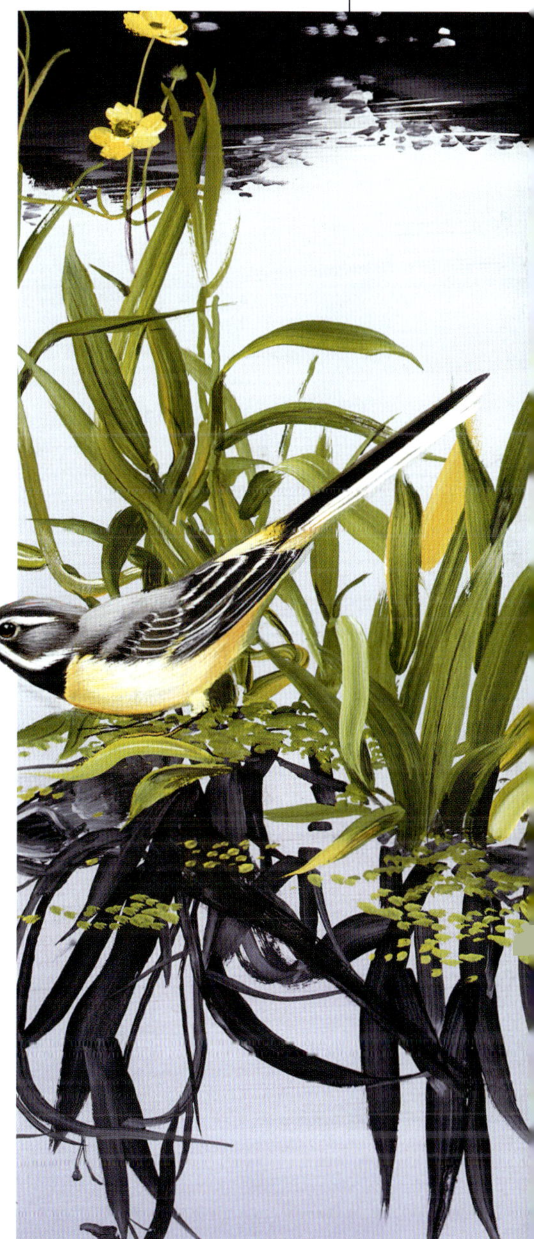

Grey wagtail

of the wolf – it is also a canine, but is considered by zoologists to be a basal canid species, which resembles ancestral forms of the family. It is named for its close resemblance to a raccoon – although it is unrelated. Originally found in east Asia, they were introduced into areas of Russia and spread into central and western Europe. Raccoon dogs are the only canids known to hibernate in winter, although in more temperate regions like the Delta they are likely to sleep only through heavy snowstorms. The Delta provides the perfect habitat for these little omnivores, who will eat fish and amphibians as well as insects, small rodents and a wide variety of plants and fruits. We are struck by just how much its markings resemble that of a raccoon – it has a perfect little black mask and a 'ruff' of longer fur around its face.

We stay in the Delta area for two further days, travelling out by boat every day in search of more birds. My journal becomes little more than a twitcher's list as I tick off more sightings: the quick glimpse of turquoise blue as a kingfisher darts past; the rook colony that had a number of pairs of red-footed falcons nesting within it; spoonbills; dippers; avocets and, my favourite, stone curlews. Although we have enjoyed our short stay in this watery sanctuary we are keen to move on to our main destination – Transylvania.

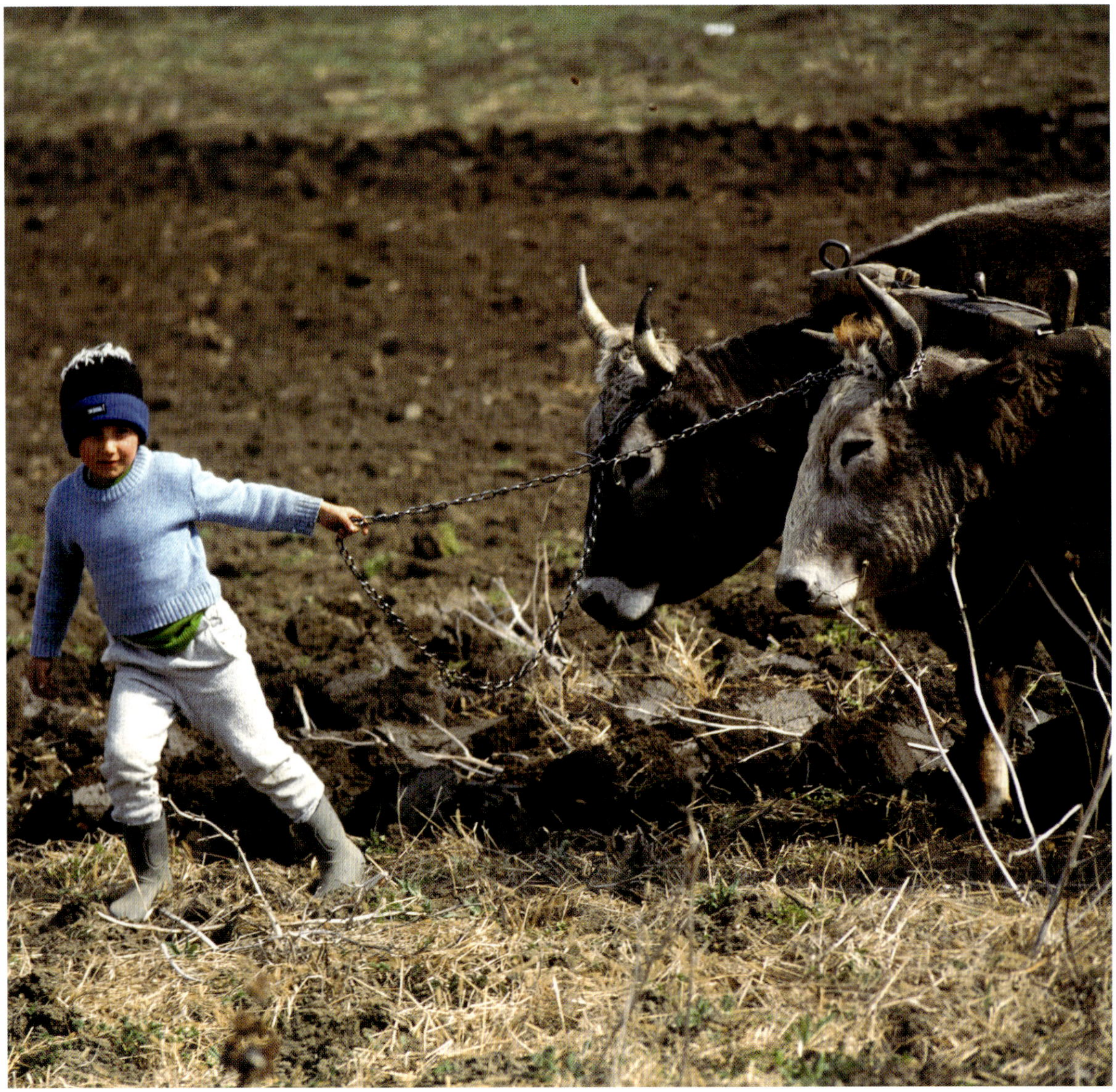

Chapter Three

'The impression I had was that we were leaving the West and entering the East; the most western of splendid bridges over the Danube, which is here of noble width and depth, took us among the traditions of Turkish rule.'

JONATHAN HARKER'S JOURNAL, *DRACULA*

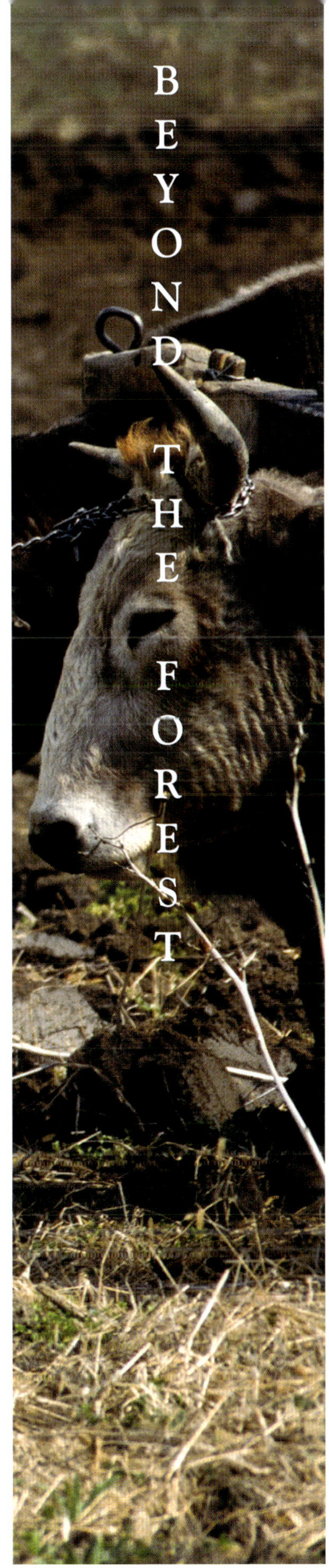

Even now many people are not sure whether Transylvania is a real place or a fictional setting for a horror story. The name itself conjures up images of a wolf-infested land of gothic castles inhabited by immortal blood-drinking fiends. The reality is a region of pastoral beauty rich in Saxon history and Hungarian culture, with unspoilt forested mountains, lakes and wildflower meadows. Time appears to have stood still in villages of prettily coloured houses with their fortified churches. Many towns and rural settlements have remained almost untouched since the Middle Ages and horse- and ox-drawn ploughs are still a common sight in the fields. The region has had a turbulent history. A series of invasions, unions, migrations and battles has left an ethnically diverse population very different from that in the rest of present-day Romania. The first recorded state that existed in this area was Dacia, populated by an Indo-European tribe, in the first century BC. After a bloody struggle in AD 106 the area was annexed to the Roman Empire. Waves of conquerors followed – the Goths, the Gepidae, the Huns and the Avars all attempted to rule. However, in the tenth century a Hungarian tribe, the Szekelys, settled the region, and it remained associated with Hungary for a thousand years up until the beginning of the First World War.

It was a time of flux and upheaval. There was a period when Transylvania became an autonomous principality, nominally under Turkish rule, but governed by Hungarian princes. In the 1600s it then came under Hapsburg rule, until the Hungarian revolt, after which it was fully absorbed into

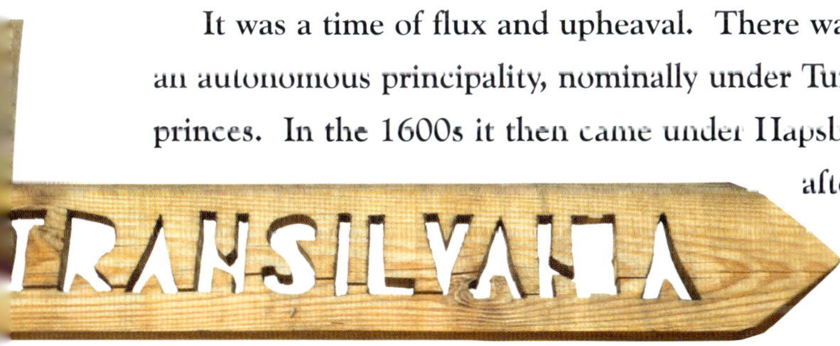

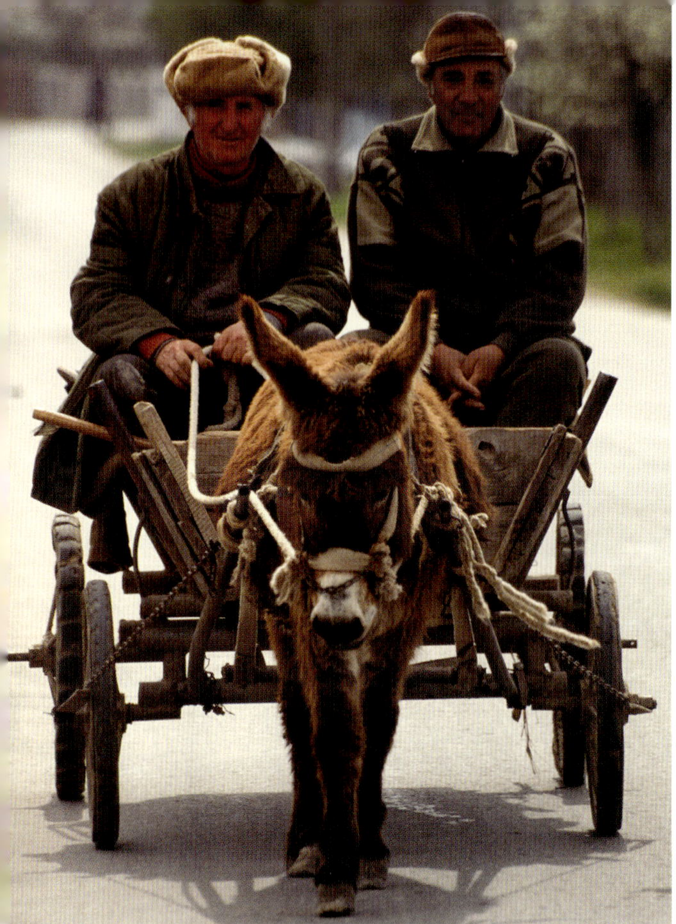

THE DONKEY CART IS STILL A COMMON FORM OF TRANSPORTATION

Hungary. Eventually the Trianon Peace Treaty that followed the First World War created a new larger Romania which included Transylvania and parts of the Hungarian plain which border it to the north and west. Disputes continue – Hungary annexed much of the region again between 1940 – 1944 – and even today there are some calls for independence.

The Hungarian influence remains strong, and although many of the descendants of the Saxon settlers have now left, their influence can be seen in the architecture and culture of the region.

The word Transylvania comes from medieval Latin and translates as 'the land beyond the forest', a name resonant with remoteness. From this we can assume it is not only in the modern imagination that the region has been viewed as a distant place, separated from the rest of the world by dense dangerous forests.

Day Six: morning

We clamber into Colin's trusty 'Roving Romania' Land Rover bright and early for the lengthy drive to the town of Brasov in the south-eastern corner of Transylvania. Colin runs a 1998 Land Rover 110 Defender – seventeen years old at the time of our visit, it has been specially adapted for travel in Romania, with improved seating and suspension to help cope with the rough potholed roads and tracks.

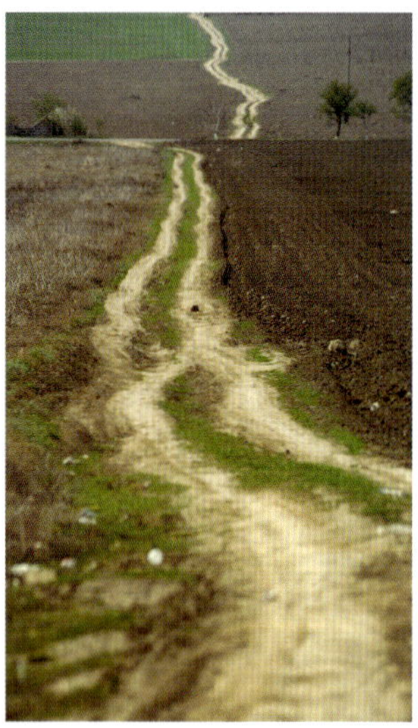

THE CORRECT CART TRACK

The roads are quiet as we drive back towards Tulcea. We stop briefly at the relay station on the hill above the village where Costica works when he is not piloting boats through the Delta. He has instructed Colin to bring us up here for the view across to the Delta. It takes ages for Colin to locate the correct cart track. Eventually we wind our way up through the vineyards only to find the Delta hidden from view by the morning haze.

We make a stop at an exposed sandy quarry, home to dozens of bee-eaters – and we also spot little owls, using the nesting holes the bee-eaters have vacated! On for a further hour and a half on narrow twisting roads which follow the southern bank of the Danube. At times the road climbs to give us views down over the thatched cottages and dense reed beds. Eventually we reach a broad stretch of the Danube, just below the industrial town of Galati, where we catch the primitive tugboat-drawn car ferry for the short crossing. This involves several false starts – they are prone to moving the ferry dock without signposting the new location. However, we eventually find it, and board. Galati is built at the point where the Delta begins – or for us ends, as we are departing! Colin drives on to the ferry at the dock in the village of I.C. Bratianu, which he tells us was formerly called 23rd August, the date of the Royal Coup in 1944. Although I understand that the desire to commemorate revolution changes with the political landscape, I like the idea of a town named after a date, if only for the hypothetically baffling conversation which could occur at parties: 'So whereabouts are you from?' '6th March. And you?'.

The ferry is quite small, and fills entirely with cars and foot passengers, so we scramble up on to the roof of the Land Rover for a more comfortable perch to enjoy the crossing and view downriver. This greatly entertains the local passengers, and many of them produce mobile phones and photograph the two strange English women sitting on the roof of a car. We climb down just before we dock to avoid being thrown into what in this industrialized area is a distinctly murky grey Danube.

On the west side of the river the landscape changes quite suddenly – we have left the flat eastern plains behind us and are driving towards the mountains through rural farmland. Colin diverts slightly from the direct route to the Carpathians to allow us to visit the Vladimiresti monastery. He has become a friend of Mother Veronica, the elderly mother superior of the community of nuns who live here – he tells us some of her extraordinary history. In the tradition of all the best fairy tales her story begins with a wicked stepmother. Her mother died when Veronica was just two years old, and her father's new wife, who wanted to gain her inheritance, locked her in a hen house. At the age of eight she escaped this

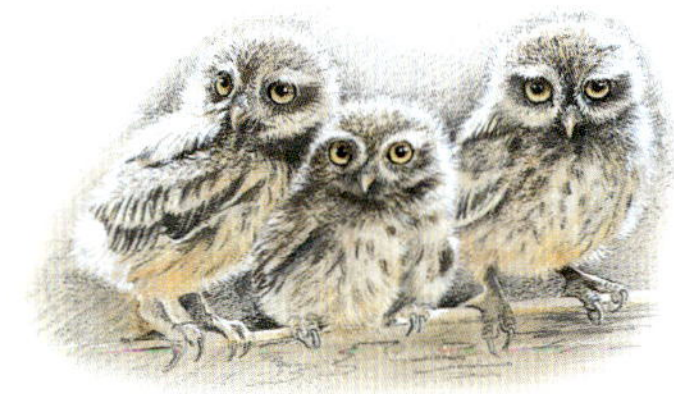

Little owls

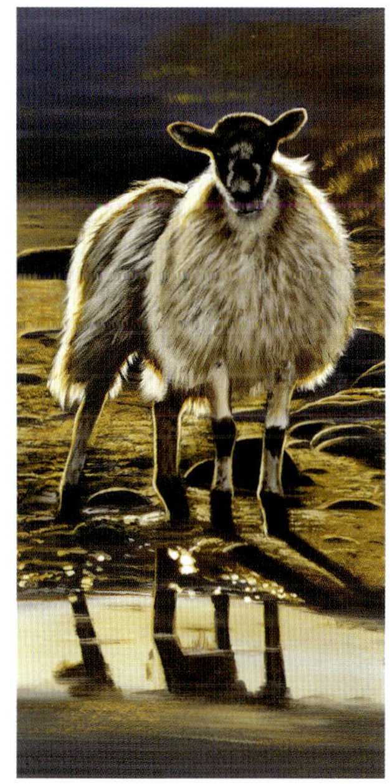

Farmers driving their sheep to pasture

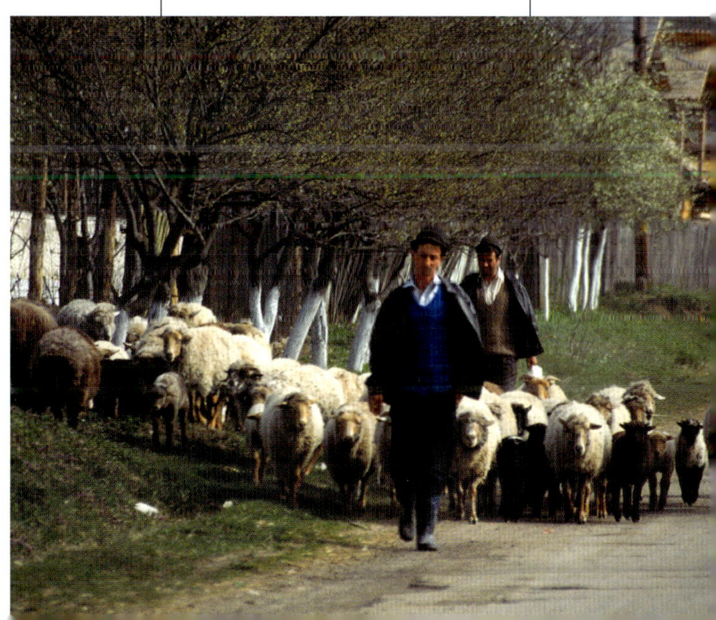

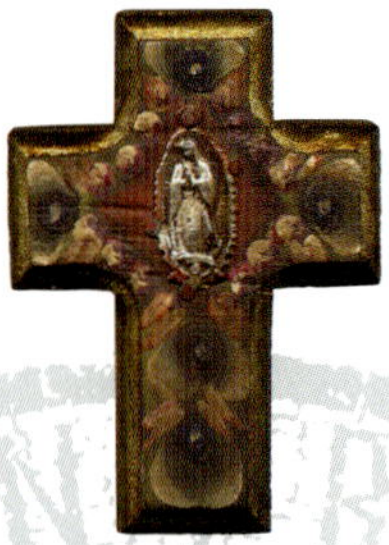

TRADITIONAL TRANSYLVANIAN BLESSING (HUNGARIAN)

Where there is faith there is love
Hol hit - ott szeretet

Where there is love there is peace
Hol szeretet - ott béke

Where there is peace there is blessing
Hol béke - ott áldás

Where there is blessing there is God
Hol áldás - ott Isten

Where there is God is all we need.
Hol Isten - ott szükseg nincen.

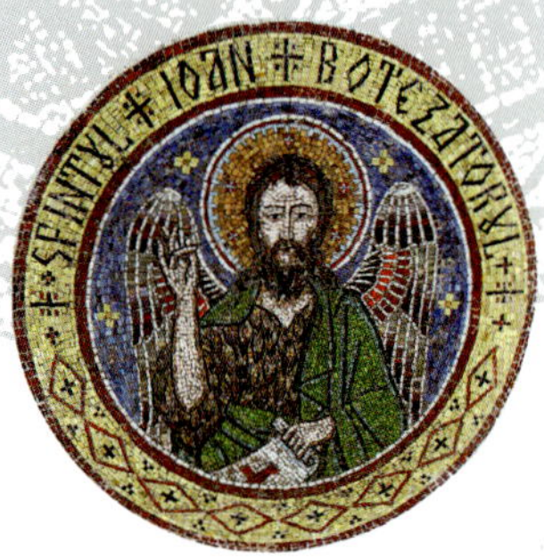

prison of neglect and ran into the nearby village where the people were amazed and horrified by the appearance of a child dressed in rags, filthy and barely able to speak. In her hand she was clutching an icon of the Virgin Mary, the last gift from her mother. The priests of the village took care of her until she was claimed by her maternal grandparents; she lived with them earning her keep by herding cattle in the pastures. At the age of fifteen she had a vision of Jesus and the Virgin Mary descending in a beam of light which became a flame when it touched the hillside. They commanded her to build a convent dedicated to Mary, where two hundred nuns would pray for the redemption of Romania. She carved a cross on the bark of the tree where God had spoken to her and the community gathered together to begin construction work that very week. In 1941 she took the veil and would eventually become Mother Superior at the new convent. In the 1950s the commune of nuns used catacombs and secret rooms excavated beneath the church to hide numerous people fleeing the communist purges. Mother Veronica, as she had then become, was herself declared an 'enemy of the people' and arrested. A year later the monastery itself was requisitioned by the communists and the remaining nuns imprisoned. After her arrest Mother Veronica was thrown into the notorious Jilava jail, where she was beaten, tortured and condemned to fifteen years hard labour. However, her sentence was later reduced to eight years and in 1964 she was released. She lived in Bucharest and, still under constant surveillance by the communists, she went to great lengths to allay suspicion that she desired to return to her former life – she even got married!

But on Christmas day 1989, just three days after the Romanian Revolution, the nearby villagers witnessed a strange spectacle – hordes of women, dressed in black, walking silently up the little track leading from the railway station to the monastery of

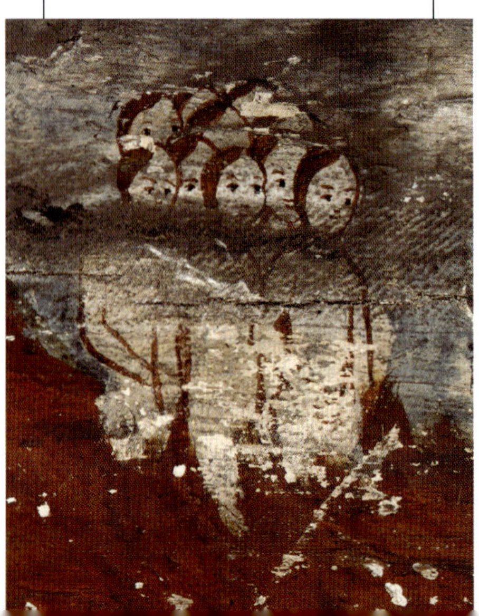

Monastery fresco

Vladimiresti, their eyes filled with tears. The remaining nuns, lead by Mother Veronica, were returning home. The monastery was in a very poor condition, but the two hundred nuns who now live in the convent have undertaken a restoration project and, by the time of our visit, the frescoes, stained-glass windows and mosaics are bright and vivid once again. Many tours of foreign countries involve visits to churches and monasteries to admire the architecture or religious art. However, knowing something of the remarkable history of both the nuns and the building somehow makes this visit more meaningful and we leave with huge respect for the devotion and bravery of this community of women.

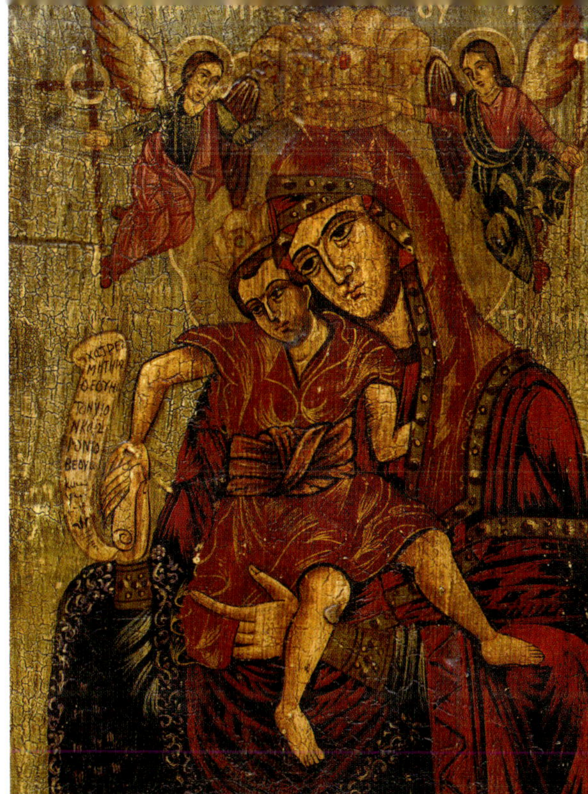

Fresco at Vladimiresti Monastery

Day Six: afternoon

We pause for lunch at an Austrian petrol station. Here we purchase a map so that we can mark the route we intend to take – and are perversely delighted by the fact that out of all the fabulously picturesque scenes of mountains, churches or villages the designers could have selected to put on the front to tempt the traveller to purchase a map of Romania, they have instead chosen a formal shot of the petrol station and, on the reverse, the identical view at night. The attached café boasts an extensive menu but, as we will learn is the case in much of the country, it is not so much a list of the food they have available as a list of food they have once heard of. The available dishes are the national staples of 'deep fried pork meat' or *ciorba de burta* – tripe soup. As vegetarians, neither of these tempt us, so we are offered deep fried *branza de burduf* – sheep's cheese with chips. (This fat laden snack will become our staple diet as we travel, with the result that we will return home skinny from hiking steep mountain trails, but with appalling acne.) The other food which will become our main staple is *mămăligă*. This is cooked by boiling water, salt and cornmeal in a special-shaped cast-iron pot called a *ceaun* or *tuci*. Historically peasant food, it was often used as a substitute for bread or even as a staple diet in the poor rural areas. It is similar in texture to Italian *polenta* but when cooked in the traditional way is much thicker and can be cut in slices like bread. Sometimes these slices of *mămăligă* are pan fried in oil and this is another

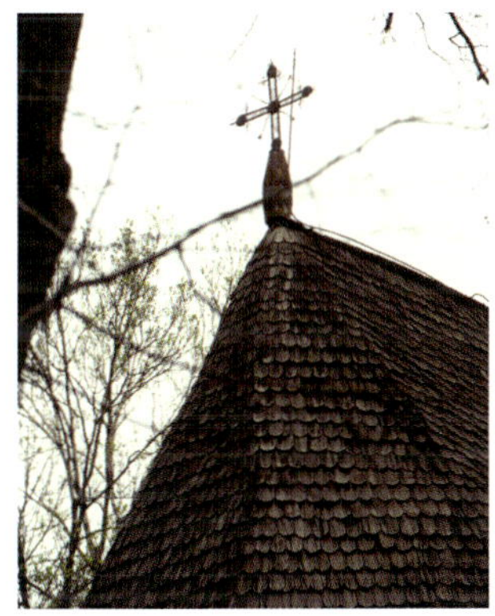

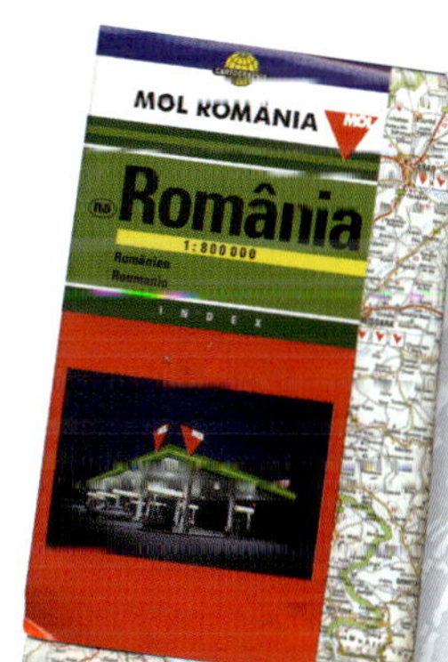

We purchase a map

Traditional Romanian Recipe
Mutton Thick Soup

INGREDIENTS

Sheep's head, tail, meat chops
2–3 carrots, parsley, onions, 1 parsnip,
green lovage, 1 borsch, 1 egg, salt
2 spoons cream, 2–3 spoons rice

METHOD

1. Boil the head, tail and meat chops in salted water.
2. When boiled, remove the meat and sieve the soup.
3. Boil the minced onion, carrot and parsley in the soup.
4. When ready add the rice.
5. Boil the borsch in separate bowl then add.
6. Put the meat in a pot and add salt to taste.
7. Mix the cream well with the yolk of the egg and some borsch and add to the pot.
8. Mince the remaining vegetables and add to the soup.
9. Cover the pot with a lid.

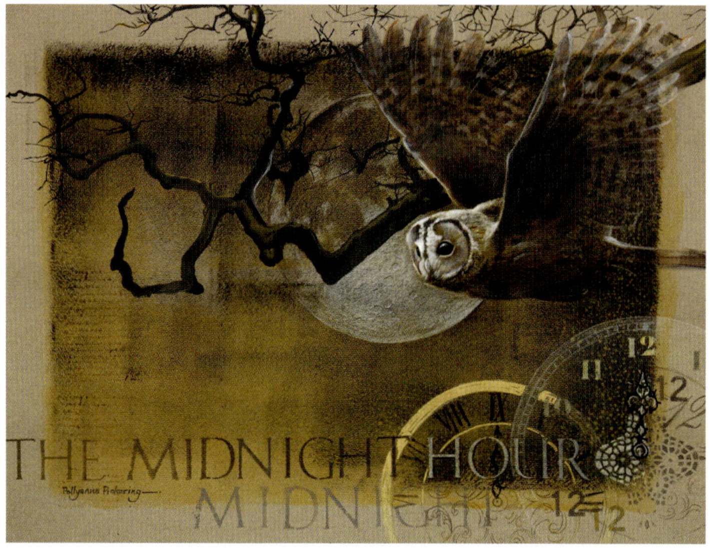

'Midnight Hour', Private collection (UK)

dish we will see a lot of as vegetarians. It is also commonly cooked to a soft mushy texture and then served with a sprinkling of salty sheep's cheese. As it has a bland flavour it can be – and usually is – served with every meal.

A further two hours' driving takes us in search of one of Romania's natural curiosities – bubbling mud pools. Anyone familiar with Bram Stoker's *Dracula* will recall descriptions of unearthly blue flames flickering in the darkness of the forests as Jonathan Harker undertakes his terrifying coach ride towards Dracula's castle. Although Bram Stoker never visited Transylvania, many of his descriptions of the countryside are both evocative and very accurate; the flames are not a figment of his vivid imagination, but a natural phenomenon. In the novel, when Jonathan Harker asks his host about the flames he has seen, the Count recounts some genuine European folklore: the blue flames appear once a year on the site of buried treasure which has never been recovered because the peasant population is too superstitious to venture out on this date. At that time in Eastern Europe Walpurgis Nacht was viewed like All-Hallows Eve, as a night when the veils which separate the realms of the living and the dead grow thin and ghosts, vampires and witches are free to roam. A villager tries to warn Jonathan Harker against travelling on this inauspicious date: *'Do you not know that tonight, when the clock strikes midnight, all the evil things in the world will have full sway?'*

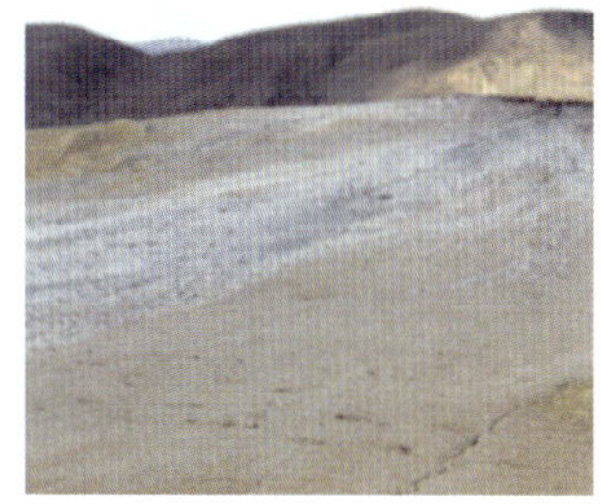

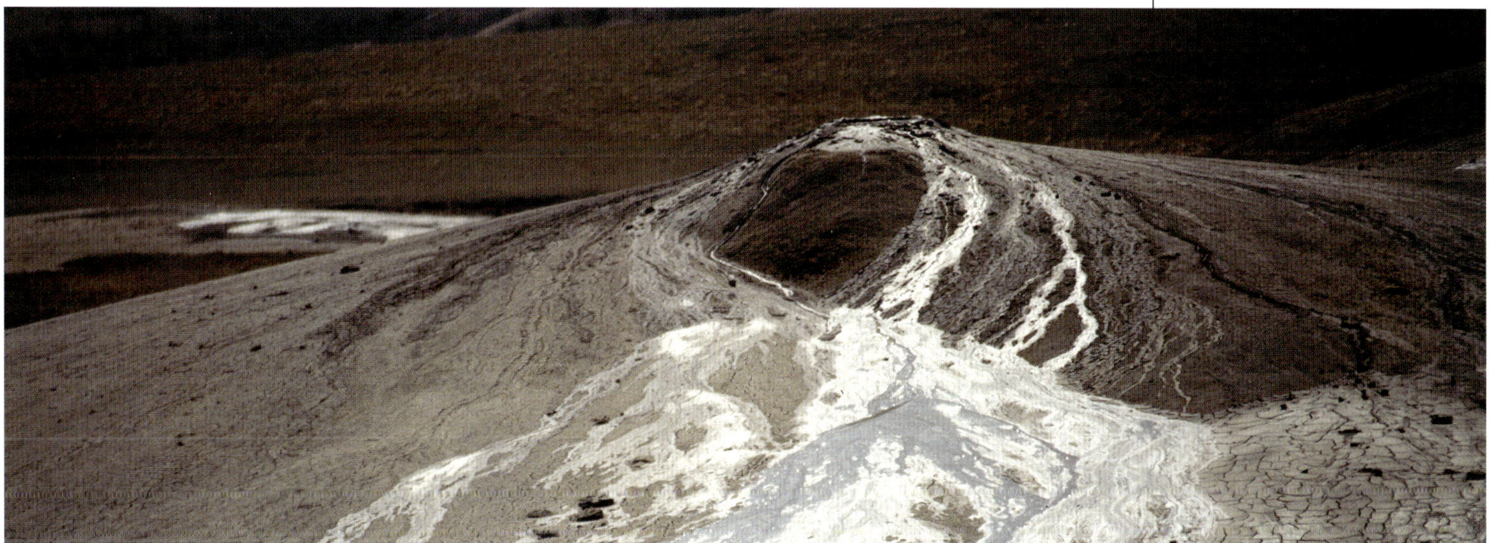

'Miniature volcanoes' formed by a build-up of sediment to create conical formations

A grey lunar landscape opens out in front of us

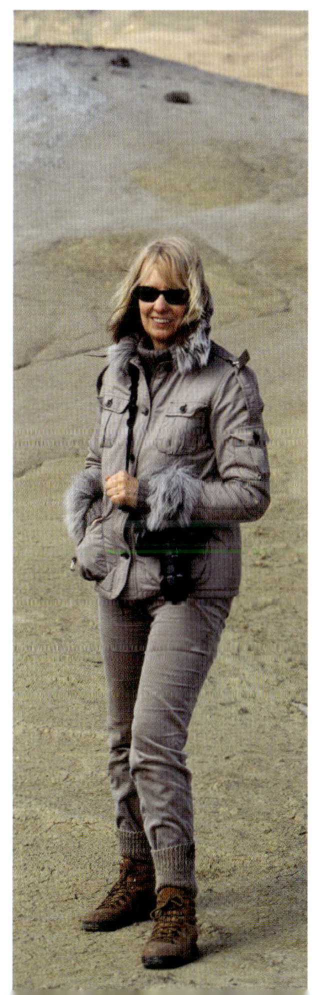

In fact the flames are not of supernatural origin. Romania is one of the European countries with a vigorous seepage of methane; this rises up through deep faults from reservoirs of pressurized natural gas and petroleum. On occasions the gas burns off with a flickering blue flame.

Our route takes us through hill country and Colin navigates the winding roads passing through many small villages until we reach Berca, where he turns off down ever increasingly narrow forestry back roads and trails searching for the bubbling mud pools of Paclele. Although Colin has visited this area several times before, it is still not an easy task finding the pools. They are not signposted and are by nature unstable and have an inconvenient tendency to move with the passage of time, with new features forming and old ones vanishing. This is an odd experience. I have spent many days and weeks in search of wildlife which is not where it is supposed to be, but never had to hunt for elusive mud. Eventually, after bumping and jolting our way down rocky tracks, a grey lunar landscape opens out in front of us. Here the natural gases seeping though the earth's crust force their

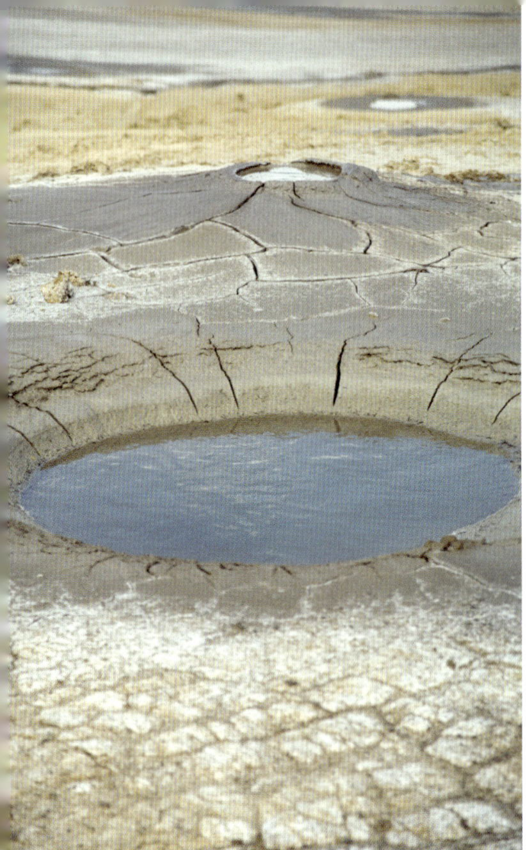

Grey lava erupts from the volcano-like formations

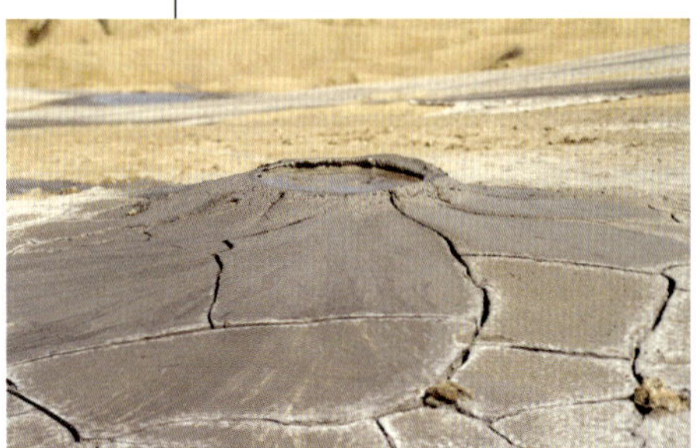

The thin crust of dried mud can cover treacherous reservoirs of quicksand

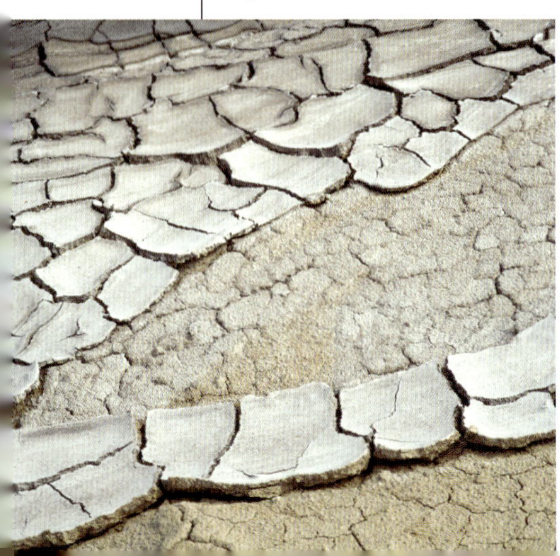

way up to ground level and mix with cold water below the surface. This creates both large and small pools of grey mud, which bubble up with a soundtrack of gloops and pops. In places the sediment builds small conical formations which look like miniature volcanoes erupting grey lava. In some areas large watery puddles bubble and splutter, elsewhere tiny holes in the earth occasionally cough up thick black mud. The mud creates a strange alien landscape, ever shifting and changing and entirely devoid of vegetation. We pick our way cautiously between the pools to take our photographs – the thin crust of dried mud can cover treacherous reservoirs of quicksand. Geophysicists describe the bubbling mud in this area as 'generally quiescent', but mention 'rare explosive episodes'.

As we drive away from the pools, Colin shares with us the story of the very first people he brought to see this phenomenon. He met three travellers at the airport: a skinny girl dressed as an Edwardian boy complete with tailcoat; a man in Bermuda shorts and an

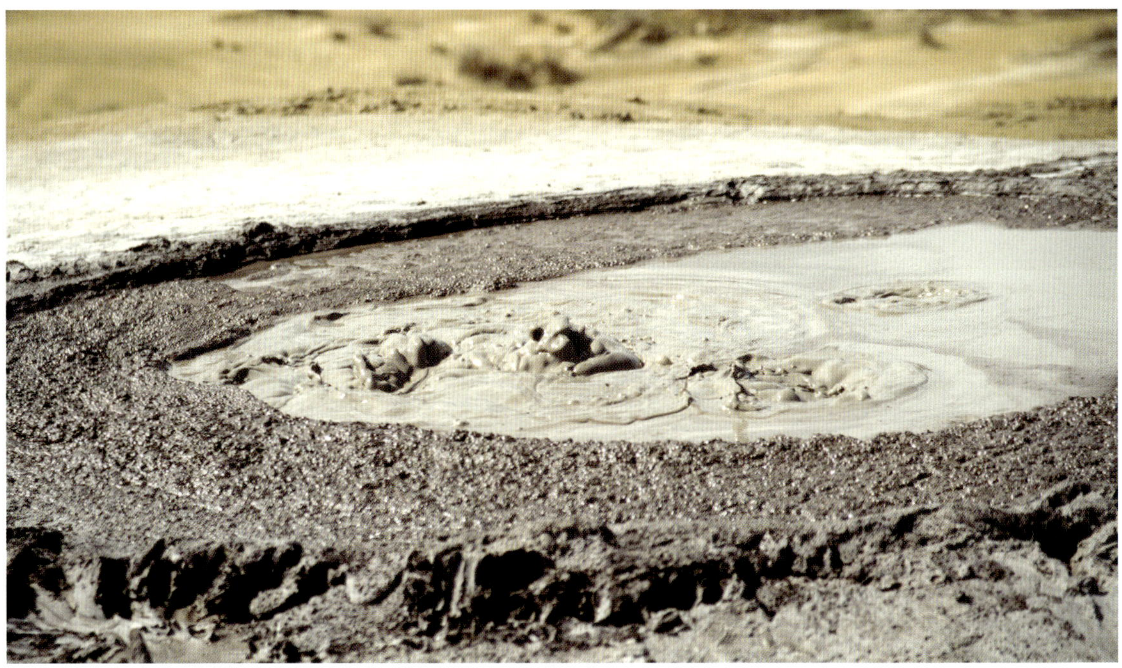

Puddles bubble and splutter

Australian hat with corks around the brim; and a large lady called Tamara, who he describes as being 'about as wide as she is tall', wearing very tight jeans over a black lycra swimsuit to further emphasize her curves. They are also accompanied by a large inflatable bat, which they perch on the dashboard of the Land Rover throughout the trip. When they reach the mud pools Colin strolls along beside Tamara, while her two companions walk some distance away. Colin is just commenting that as she is the only one who can swim she would be OK if she fell in – and on the word 'swim' she vanishes from sight. At first he thought she had simply tripped and fallen, but no, she has plunged through the crust and sunk into the mud below – her ample weight on small feet providing a fatal point loading which cracked the surface. Fortunately she has wedged cork-like in the hole she has created, preventing her from slipping through completely. Colin and her two companions eventually manage to pull her out with considerable heaving and groaning!

From here our route takes us into the Vrancea foothills. As we climb, the landscape becomes increasingly wild and otherworldly with each extra mile we drive. Eventually we turn into a valley and drive on into the woods. But these are not polite English woods with shady sun-dappled glades. These are the dark, deep medieval forests which back home exist only in our fairy tales. In these familiar stories the forest is a place of magic and danger, a place beyond which people rarely travel, where strange things might occur and stranger beings might live, home to monsters and witches. Woodland settings are common in myths and legends in countries where the natural state of the wild land is forest. As we drive through this dark tangled mass of trees I can understand why a genuine fear of travelling deep into the real forests could transform into the dangers found in the enchanted forests of folklore. It would be so easy for the inexperienced traveller to become lost and disoriented – and of course once upon a time even in England there were real bears and wolves lurking in the darkness. In the Middle Ages travellers were so fearful of walking through lonely stretches of woods that stone shelters were built specifically to protect them from attacks. The modern word 'loophole' is derived from the European term 'loup hole', or 'wolf hole', a

'Beyond the green swelling hills of the Mittel Land rose mighty slopes of forest up to the lofty steeps of the Carpathians themselves. Right and left of us they towered, with the afternoon sun falling full upon them and bringing out all the glorious colours of this beautiful range, deep blue and purple in the shadows of the peaks, green and brown where grass and rock mingled, and an endless perspective of jagged rock and pointed crags, till these were themselves lost in the distance, where the snowy peaks rose grandly.'

Jonathan Harker's Journal, *Dracula*

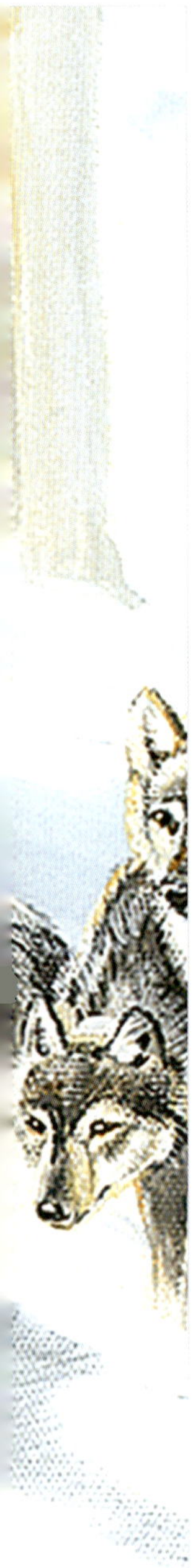

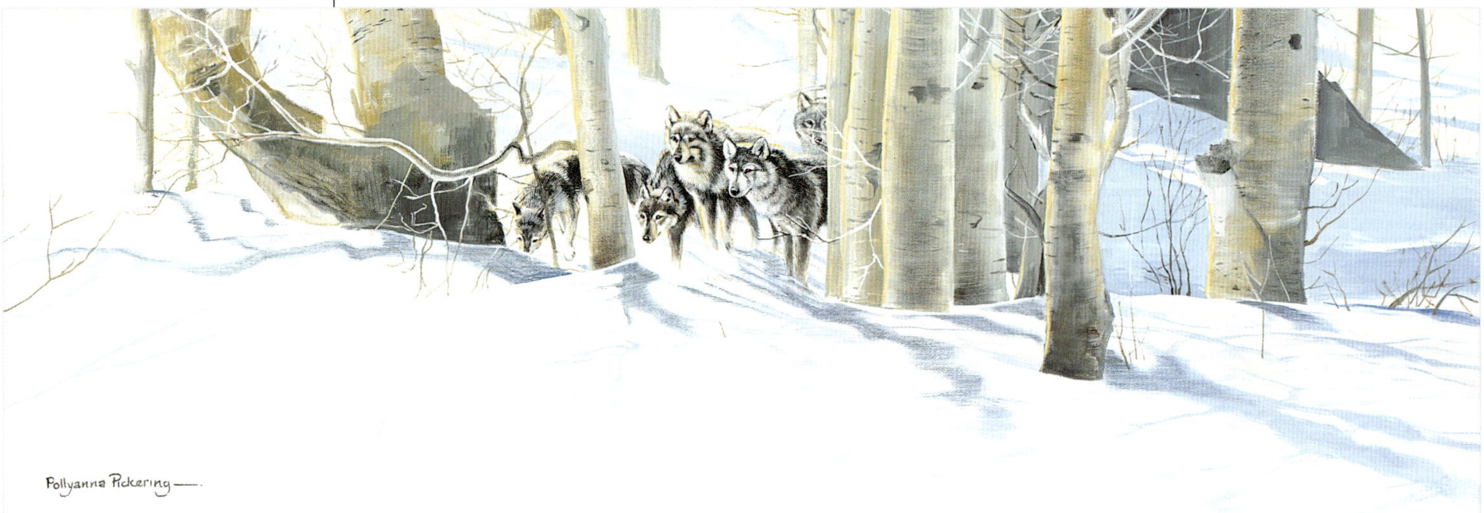

'The way is clear,
The light is good,
I have no fear,
Nor no one should.
The woods are just trees,
The trees are just wood.
No need to be afraid there –
There's something in the glade there...'

STEPHEN SONDHEIM,
INTO THE WOODS

spy hole built in the shelters through which travellers could watch for wolves.

Britain, too, was once covered by a 'wildwood' – one great uncultivated completely wild mass of bushes and trees stretching almost from coast to coast. On arrival in Britain, Julius Caesar described the whole country as 'one horrible forest'. This wildwood existed for around seven thousand years, from the end of the Ice Age until medieval times, when it was said a squirrel could cross England without touching the ground. Now we have only around five per cent of our original forest cover left – but it continues to live on in our traditional rhymes and folk tales. Here in central Europe these huge forests still exist, providing habitat for the large carnivores we have long lost.

Eventually we arrive at our accommodation for the night – a rural guest house in a very remote village, so remote they appear faintly surprised to have guests. We are given the grandly named presidential suite, which means that we have a bedroom each with a shared bathroom between.

Colin takes us out for an evening drive along the logging trails up into the mountains. We see lots of birdlife including ravens and a honey buzzard. I catch a quick sighting of a

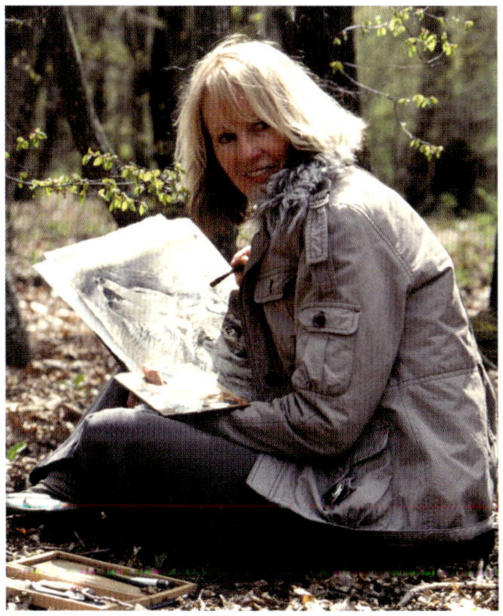

Pollyanna fills her sketch folder with impressions of wolves and the forested landscapes they inhabit

beech marten, a close relative of the pine martens which can be found in Scotland. These fast moving slender animals are the most common species of marten seen in this region. Pollyanna is already filling her sketch folder with her impressions of the forested landscapes. It is equally important for her to record the environment which wolves inhabit, as well as the wolves themselves, to enable her to complete accurate paintings when she returns to her studio. She quickly distils the complex details of the woodlands down to a series of simple brush strokes, capturing the essence of her surroundings in paint. Where the casual observer sees the forest as nothing more than a tangle of trees, for Pollyanna it is an irresistible spectrum of subtle colours, light and shadow, the shapes of the trunks and branches creating dynamic living compositions. Her paintings set in these medieval forests will capture the sense of the timelessness of this landscape through which the wolves have roamed for millennia.

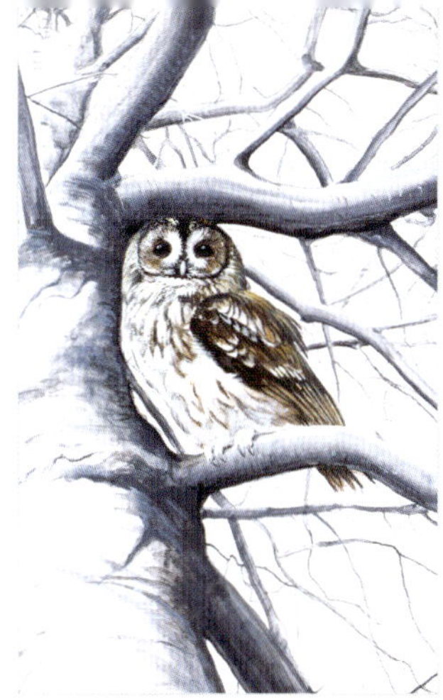

Tawny owl

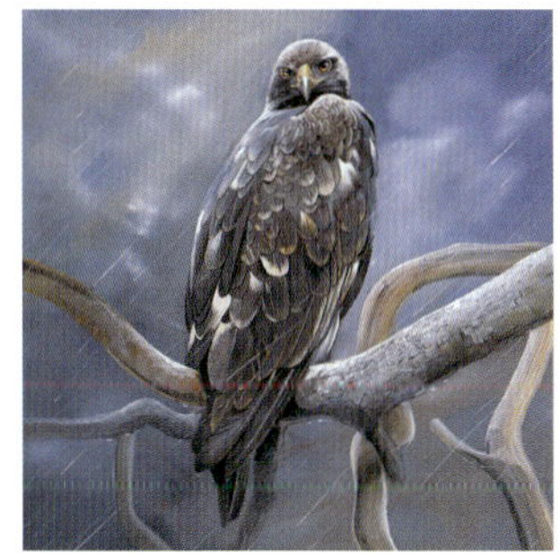

Honey buzzard

The road is a rough single track which follows and then crosses a river in the valley bottom before winding up the hillside with thick forest on either side. We keep a good look out for bears, but fail to spot any. When we turn back to the guest house, Colin declares that we are 'sensible enough' to be allowed a treat on these private roads. (Sensible? Has he really just spent the past few days in boats and Land Rovers with us?) We climb up on to the roof of the Land Rover and sit on the front clinging to the roof rack as we career back downhill into the village – much to the amazement of the two forestry workers we pass who turn and stare as a battered Land Rover hurtles wildly past them with two blonde women clinging to the roof and giggling insanely.

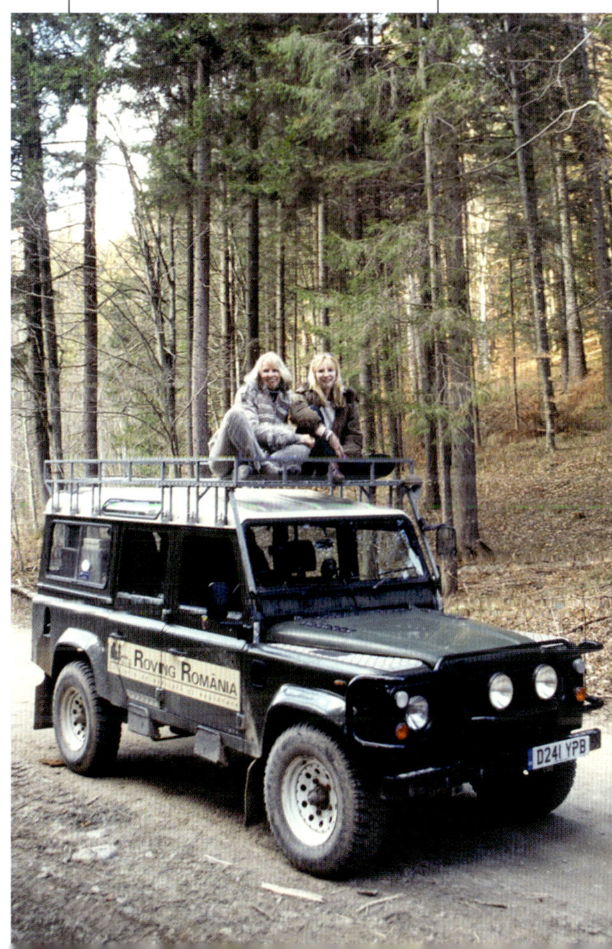

We climb on to the roof of the Land Rover

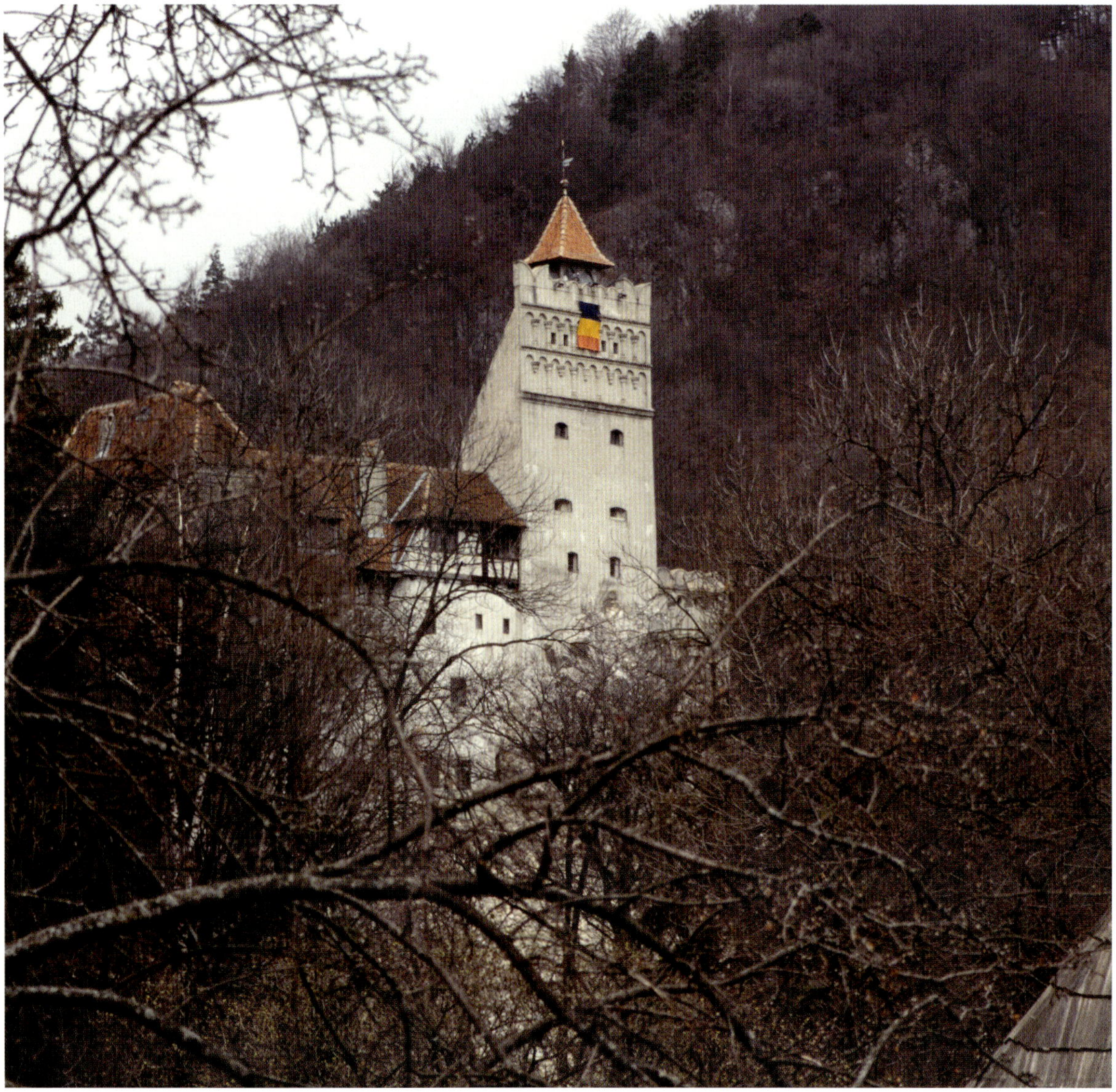

Chapter Four

THE DRACULA CONNECTION

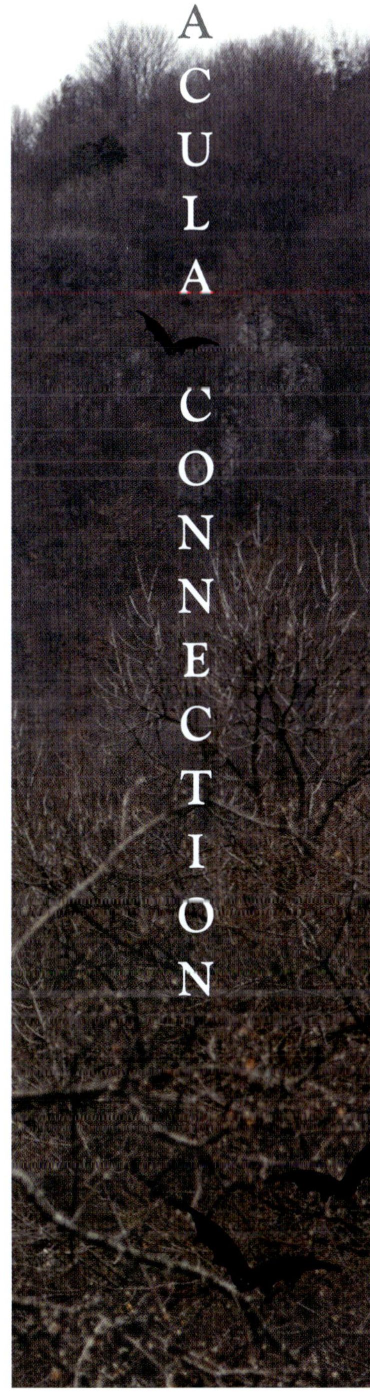

'The clear line of Dracula's castle cut the sky. For we were so deep under the hill whereon it was set that the angle of perspective of the Carpathian Mountains was far below it. We saw it in all its grandeur, perched a thousand feet on the summit of a sheer precipice, and with seemingly a great gap between it and the steep of the adjacent mountain on any side. There was something wild and uncanny about the place. We could hear the distant howling of wolves.'

MINA HARKER'S JOURNAL, *DRACULA*

Following our travels in Romania we were left with the impression that the country, and in particular the region of Transylvania, has an uneasy and somewhat ambivalent relationship with their perceived connection to Dracula. I can only imagine how galling it must be if your home country – rich in history, blessed with breathtaking landscapes and teeming with wildlife – is in the collective mind of the outside world linked only with a literary vampire created by a long dead Irish novelist who had never even visited. Adding to the problem is that although the vampiric count is a product of Bram Stoker's imagination, he borrowed the name of his anti-hero from a real life historical figure – Vlad Tepes, Prince of Wallachia.

The Prince's father took the name 'Dracul', the Romanian word for dragon, after being inducted into the Order of the Dragon, a chivalric order of the nobility created in Hungary in the late Middle Ages. The young Vlad III took the name, adding the 'a' at the end to indicate 'son of'. This historical Vlad is renowned for his tenacious resistance against the Ottoman Empire – but equally infamous for the cruel punishments he imposed on his enemies, which resulted in his other appellation – Vlad the Impaler.

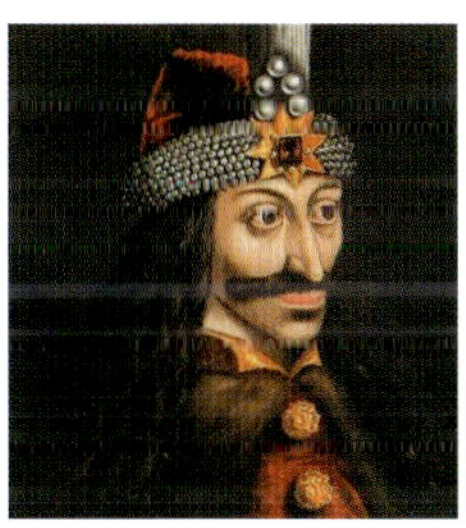

VLAD THE IMPALER

The association of Vlad Tepes with cruelty and death,

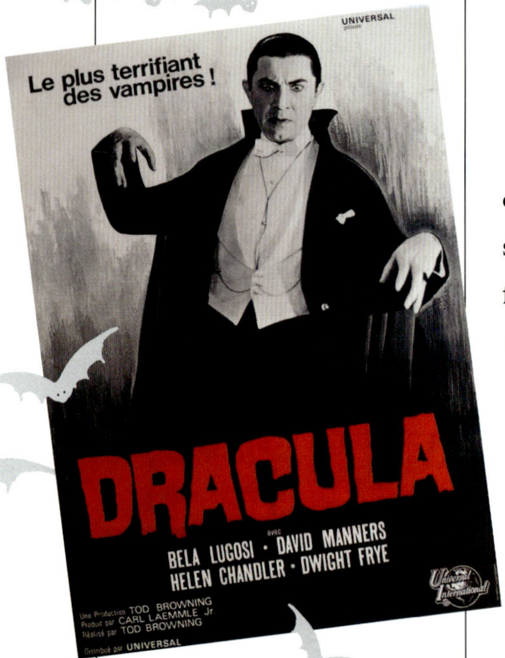

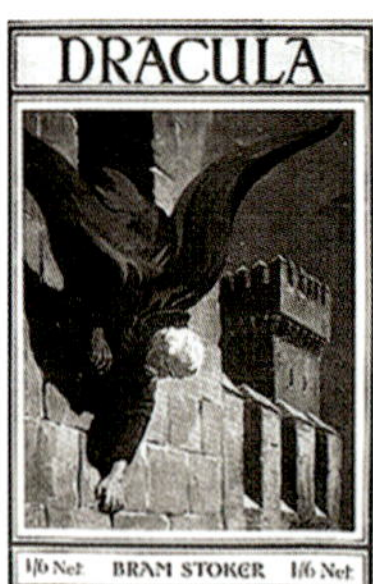

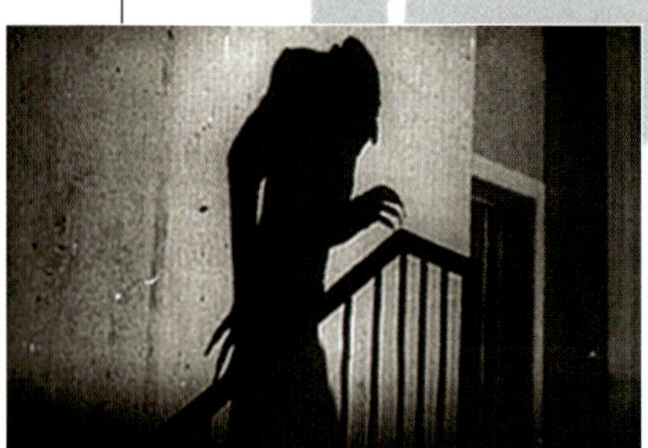

combined with the obvious link between his career as an impaler and the sharpened stakes in the Dracula story, have lead to a common assumption that Stoker based his fictional character strongly on the real life prince. But this often repeated fact appears to be largely untrue. In reality Bram Stoker had already almost completed a novel – titled *The Undead*, and featuring a Count Wampyr who had his lair in Austria – when he came across a brief paragraph about Dracula in a book he was reading (the gripping sounding *An Account of the Principalities of Wallachia and Moldavia: with various Political Observations Relating to Them*) and decided he liked the sound of the name and location. There is little to indicate he did any further research into the man, or was even aware that Vlad was a legendary impaler of his enemies, prior to the publication of the novel in 1897.

Although by all accounts the real life Vlad Tepes was thoroughly vicious and unpleasant, he is also something of a national hero in Romanian folklore and literature. He is seen as a brave fighter who stood up to both foreign and domestic enemies and is still considered one of the greatest leaders in the country's history. Understandably the linking of a national hero to an undead predator is not entirely welcome. The Romanians, therefore, spent decades downplaying any possible connection between the real and literary figures. To add to the confusion, the communist government banned both the novel and the many movie versions, and *Dracula* was not translated into Romanian until after the 1989 revolution. So while the rest of the world has been exposed to countless versions of the story, the majority of Transylvanians have never actually read the book.

This has produced something of a conflict of interest for Romania. They might not want the Dracula connection – but the rest of the world does. And Romania needs tourist revenue. This has lead to a travel industry growing up around the Count, and attractions and vampiric souvenirs aplenty attempt to capitalize on the global fascination. Plans for 'Draculaland' – a Dracula theme park – foundered due to poor planning and

opposition from conservationists, but as we followed in Jonathan Harker's fictional footsteps through a pass in the Carpathian Mountains and entered Transylvania we became ever more aware of the lingering presence of Dracula.

Day Seven

We set out early, in the grey misty hour before first light. Our drive takes us around the edge of an unfinished reservoir from where we make our way steadily uphill heading north-west to the southern edge of the wild Vrancea Mountains and into Transylvania. We enter a fairy-tale landscape of lakes and forest, of snug villages and tidy farmhouses – a corner of Europe lost in time. Winding mountain roads and passes bring us at last to Brasov, Romania's second city. In medieval times it lay at the intersection of trade routes which linked the Ottoman Empire to Western Europe and became a prosperous trading centre. The Saxons built ornate churches and houses here and then fortified the town with a massive wall, which still remains today. We stroll around the oldest part of the city – where we wander into a small Orthodox church – thick with the heavy scent of incense, fantastically painted and lined with golden icons. We buy candles to light before walking back to Piata Sfatului, the main square. Here we visit the famous Black Church – the largest gothic church in Romania. Its name derives from damage caused by the Great Fire of 1689, when flames and smoke blackened its walls. During restorations the original gothic interior enjoyed a makeover. The lofty, light space is undoubtedly beautiful, with its balconies and stained-glass windows, but it is a little austere for my taste. I prefer the crowded Orthodox churches, every space crammed with painted icons hung with jewellery, their huge layered silver-and-gold-encrusted frames making Pollyanna's exhibitions look woefully underframed.

The baroque facades of the square are interspersed with cafés and patisseries, so we call in for lunch – slices of delicious cheese pie.

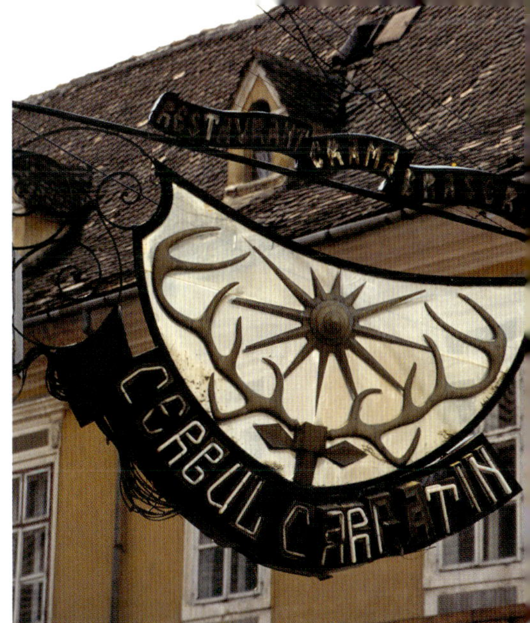

THE STREETS OF BRASOV, ROMANIA'S SECOND CITY

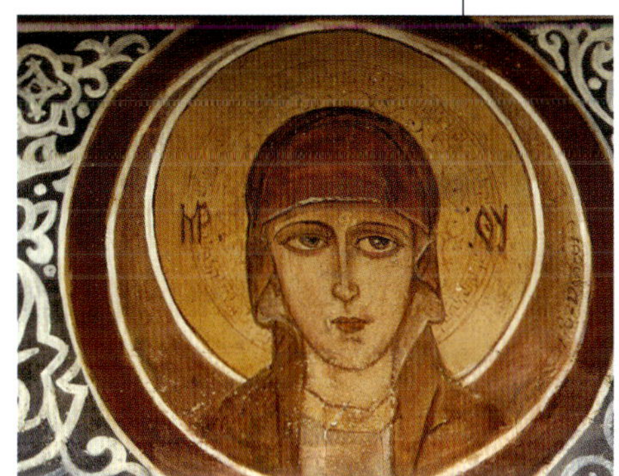

FRESCOS ADORN THE WALLS OF THE CHURCH

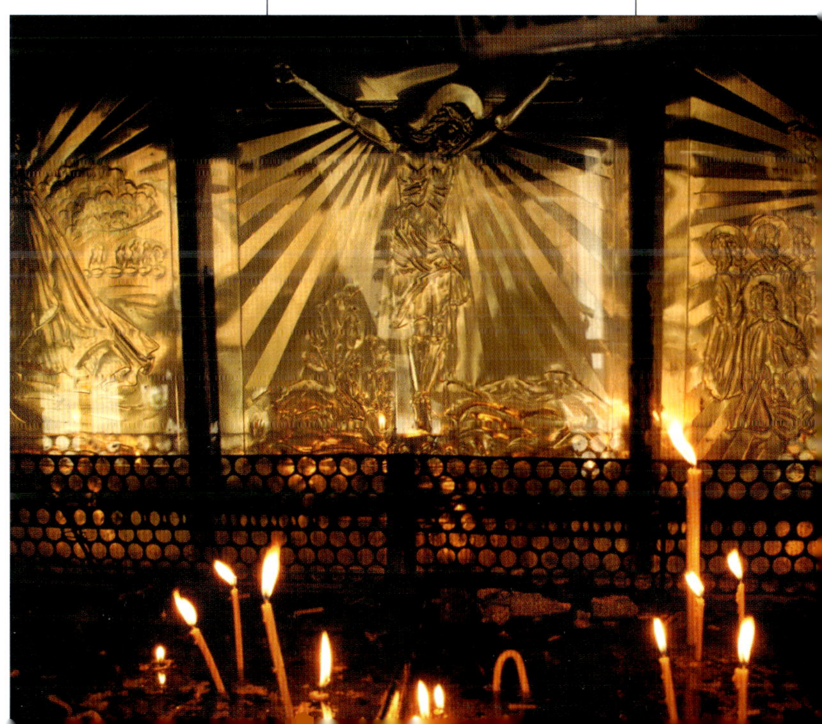

INSIDE THE ORTHODOX CHURCH GOLDEN ICONS GLEAM IN THE CANDLELIGHT

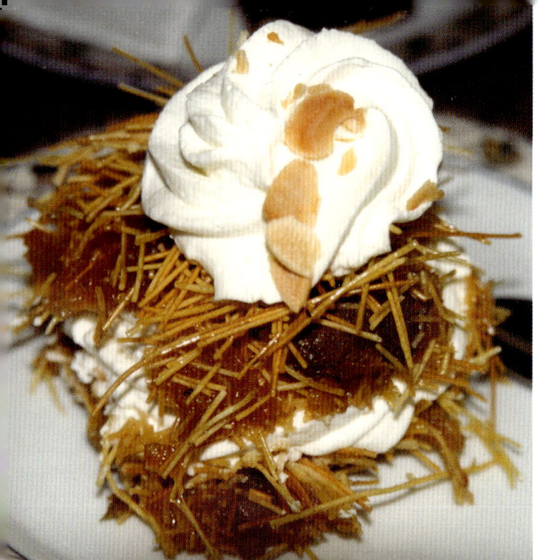

We get carried away and order four luscious cream cakes – one shaped like a delicate bird's nest

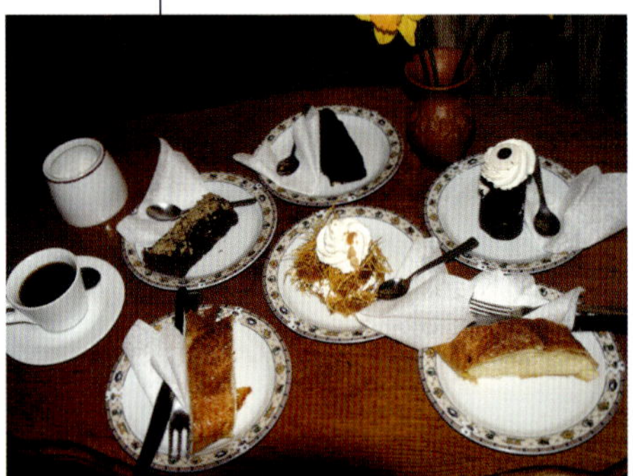

Fabulously rich looking chocolate truffles

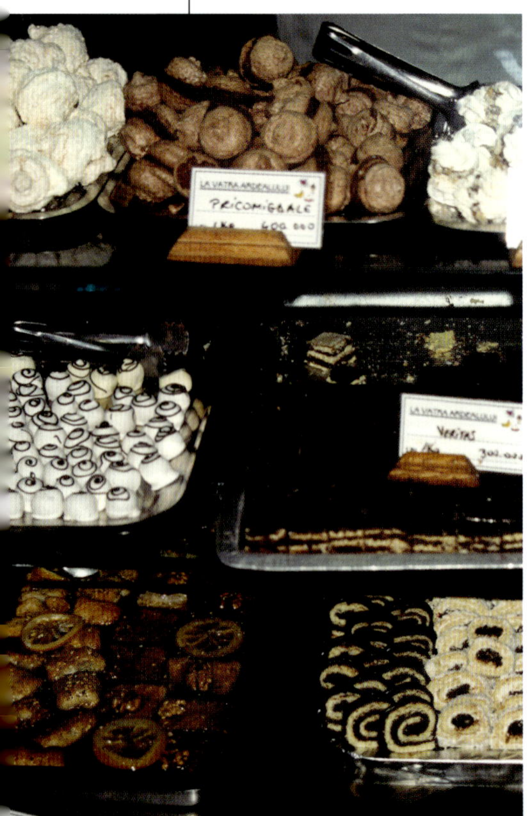

We get carried away and order four luscious cream cakes – one shaped like a delicate bird's nest – to enjoy with a coffee. Eventually, admitting defeat, we end up having to ask for two of them to be wrapped up to take away – so throw in some fabulously rich looking chocolate truffles as well.

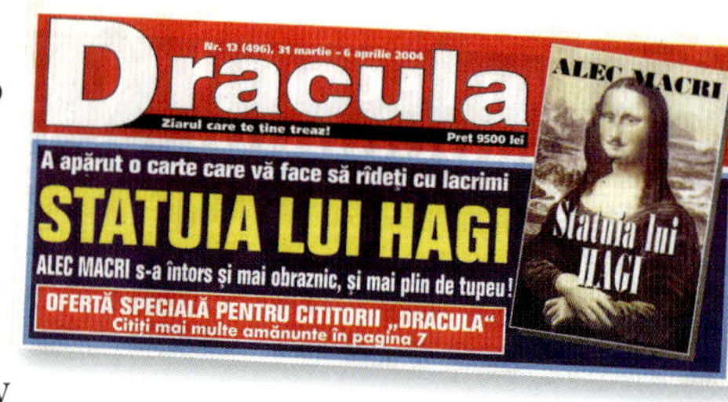

Dracula
Nr. 13 (496), 31 martie – 6 aprilie 2004
Ziarul care te ține treaz!
Preț 9500 lei
A apărut o carte care vă face să rîdeți cu lacrimi
STATUIA LUI HAGI
ALEC MACRI s-a întors și mai obraznic, și mai plin de tupeu!
OFERTĂ SPECIALĂ PENTRU CITITORII „DRACULA"
Citiți mai multe amănunte în pagina 7

Colin needs to make a few stops in the town to sort out paperwork, so we take the opportunity to wander the narrow streets. We splash out around 30p for a bottle of Dracula beer and pick up both the *Transylvanian Express* newspaper and the trashy tabloid *Dracula*. Brasov is also the last convenient place to change money and stock up on essentials. We visit a bank to convert more of our euro into lei and for the first time in my life I find myself in possession of a single bank note worth a million. Of course at this point a million lei is worth around £5.00 – in the aftermath of the revolution the value of the currency evaporated overnight – but it still looks fairly spectacular. We find a fabulous second-hand bookshop, where Pollyanna falls for an ancient illustrated German prayer book, which weighs approximately five tons (OK that may be a slight exaggeration, but I still think it will take up our entire luggage allowance on the flight home). I am drawn to a witchy looking Hungarian text dating from 1865. We make our purchases. On returning to the Land Rover I take out my book for closer examination. In the daylight the cover looks worryingly as though it is stitched from skin. Of some kind. As I open it a small spooky cloud of dust rises from the pages. When we next alight from the Land Rover, Pollyanna peers at me and asks: 'What are those two red marks on your neck?' I laugh, thinking she is joking. She isn't. I have two tiny red dots fang distance apart on the side of my neck. I can only imagine that I have been bitten by an insect with a vampire fetish who enjoys playing practical jokes on unwary tourists.

'What are those two red marks on your neck?'

From Brasov it is only a forty-minute drive to Bran Castle, which has become the centre of the Dracula tourism industry. Back in the 1970s the communist government decided it was time it began to capitalise on its infamous connection and it declared that this impressive looking castle with whitewashed towers and turrets was Bram Stoker's inspiration for Dracula's lair. In fact there is nothing to suggest Stoker had ever heard of it – but it did meet a few of the essential criteria which inspired this claim. Its impressive location high on a cliff face may well have been a factor, but most likely its close location to the hotels of Brasov, and the fact it could be reached on relatively good roads from the airport were the major deciding points. Some sources suggest that the real life Vlad Tepes might have sought refuge in the castle for a few nights while fleeing a Turkish attack, but this is the only tenuous link to the legend. The castle is actually a beautifully preserved museum. Originally built under Hungarian rule, a castle fortress on this site was first used in 1378 as defence against the Ottoman invasions. Later it became a customs post on this high mountain pass between

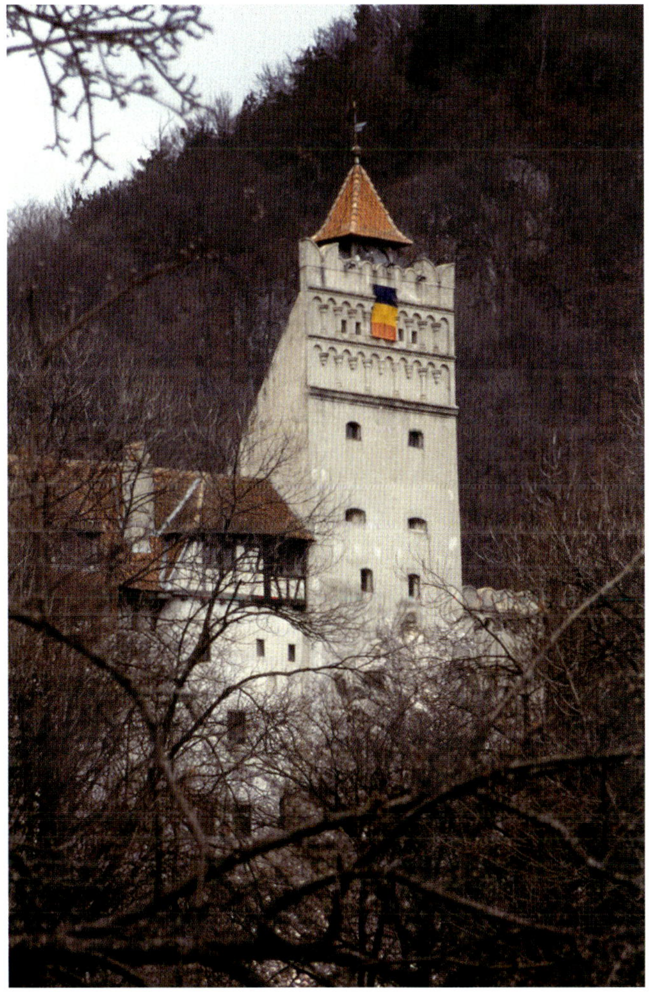

Bran Castle

'...a vast ruined castle, from whose tall black windows came no ray of light, and whose broken battlements showed a jagged line against the sky...'

Jonathan Harker's Journal, Dracula

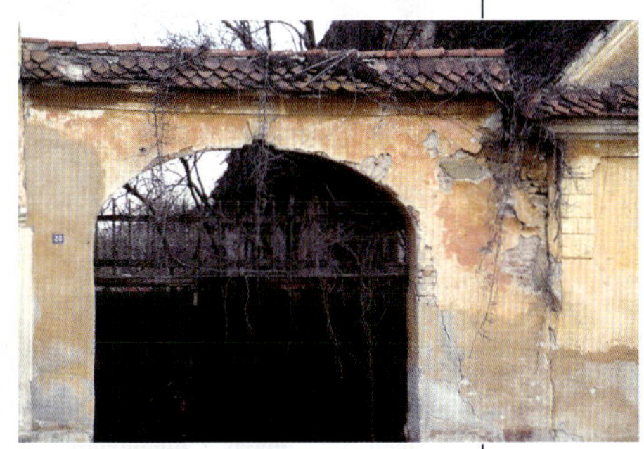

The castle walls

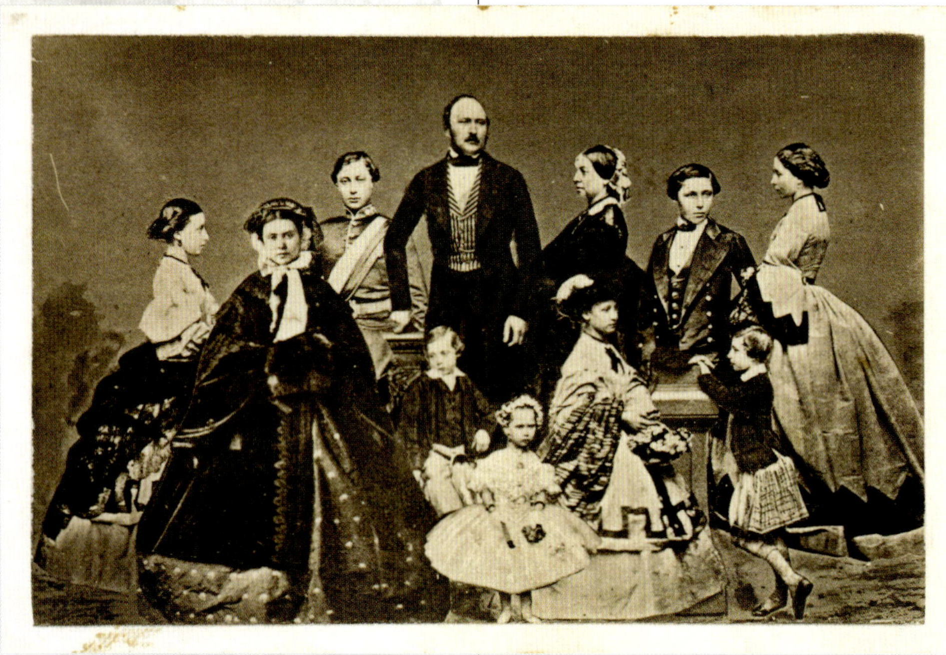

Queen Victoria, Prince Albert and their children

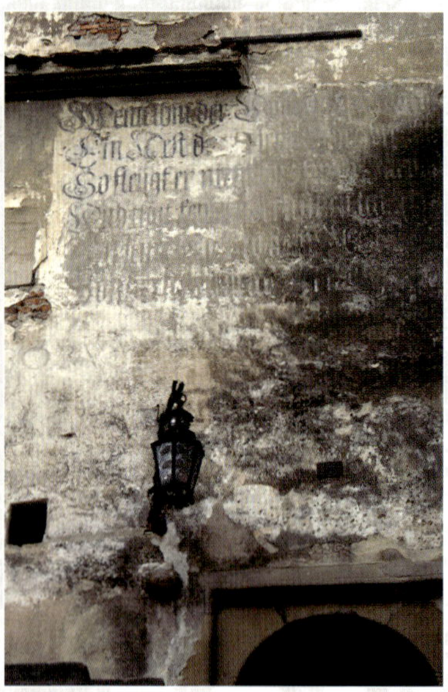

Transylvania and Wallachia. By 1920 the castle had become a royal residence.

It was the principal home of Queen Kestine Marie, the grand-daughter of Queen Victoria. Born in Kent, she became Queen of Romania on her marriage to King Ferdinand. The castle is primarily decorated with artifacts from her time, including traditional Romanian furniture and the tapestries that she collected to highlight the crafts and skills of the nation. The red-tiled roofs surround a pretty Mediterranean looking courtyard, and there is nothing sinister or spooky about the place at all. None the less this doesn't seem to discourage the tour guides who are leading coach loads of Danish schoolchildren and Japanese tourists round the castle. The only words I recognise from the guides' spiel are the frequent repetitions of 'Dracula' and 'vampires' – although in fairness with my lack of knowledge of either language these could be preceded by 'has no connection with' and 'you won't find any'. Even my guide book doggedly tries to create a more sinister atmosphere with references to a lost labyrinth of secret underground passages beneath the fountain in the courtyard – the perfect location for a vampire's lair. But under the blue skies and bright sunshine the castle stubbornly resists any attempt to transform it into the '*vast ruined castle, from whose tall black windows came no ray of light, and whose broken battlements showed a jagged line against the sky*' of Stoker's novel. This hasn't stopped an entire Dracula industry springing up in the car park, where Colin is waiting for us with the Land Rover. There is a themed coffee bar and a market full of a bizarre mix of local crafts and plastic tat with fangs.

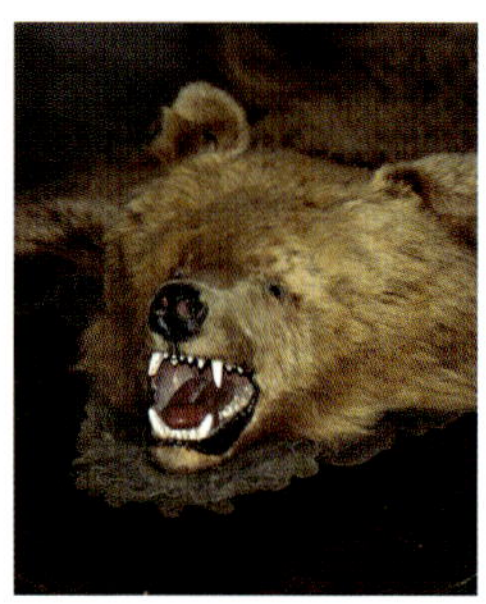

Alas, this bear no longer roams the forests

We continue our drive towards Zarnesti, stopping en route at a petrol station to use the facilities. The ladies loo is locked, so we use the gents, much to the immense shock of the Transylvanian gentleman who shoots out of the rest room like a vampire faced with a bowl of garlic soup when he sees us. When Colin sent us a pre-trip email he did warn us: *'Public toilets in Romania are rare, and can leave much to be desired in the way of hygiene. A remote bush or tree is often preferable'.* We, however, are more concerned by the next sentence: *'Please give me plenty of notice if either option is required en route.'* This implies we need to let Colin know if we need to use the loo before we even know ourselves.

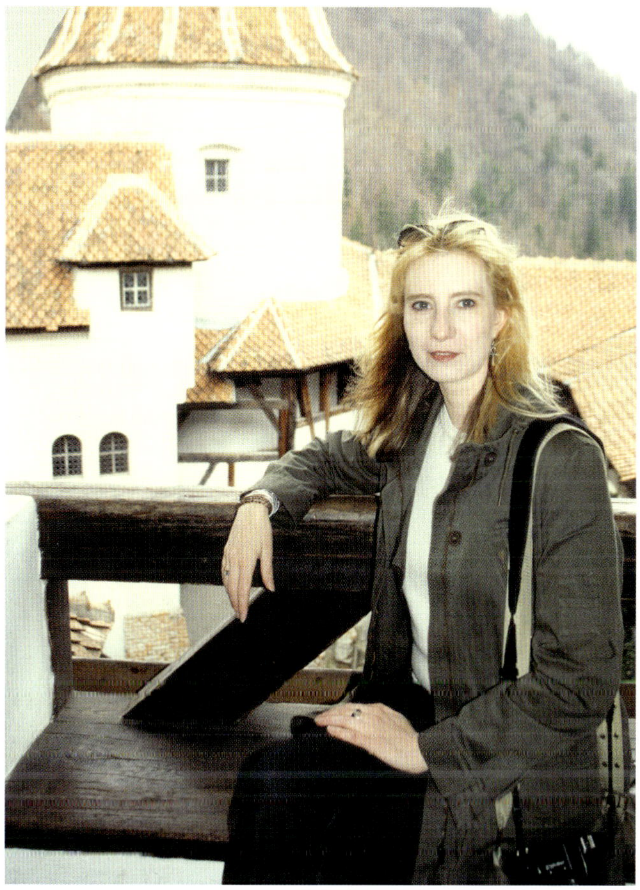

Anna-Louise at Bran Castle

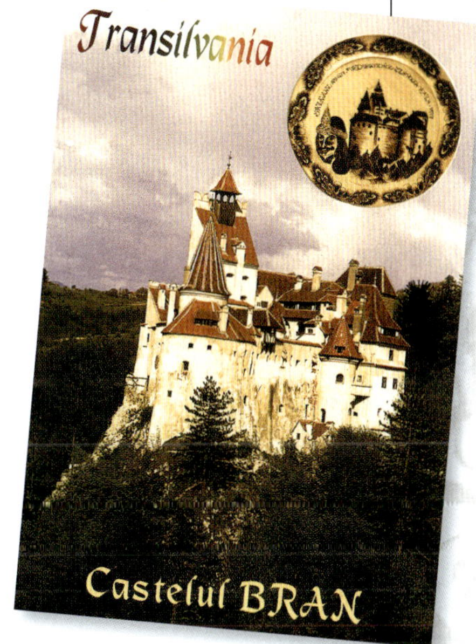

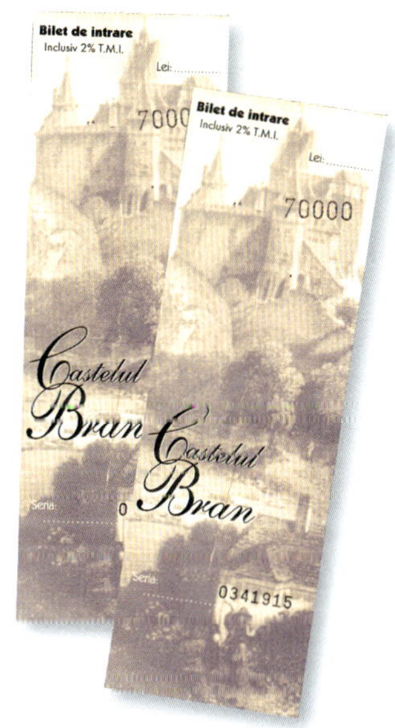

By nightfall we have arrived at our accommodation, the Pension Elena – known locally as 'Gigi Popa's'. It is run by Gigi and Elena, a delightful and welcoming couple who immediately press glasses of fiercely warming Tuica into our hands when we arrive. The guesthouse was established in 1998 in cooperation with the CLCP to provide a base for ecotourism.

Elena prepares a tasty meal of chunky vegetable soup with potatoes and mushrooms followed by delicious little home-made cakes. The food in this region has its roots in spicy Hungarian recipes, rather than the blander Romanian fare.

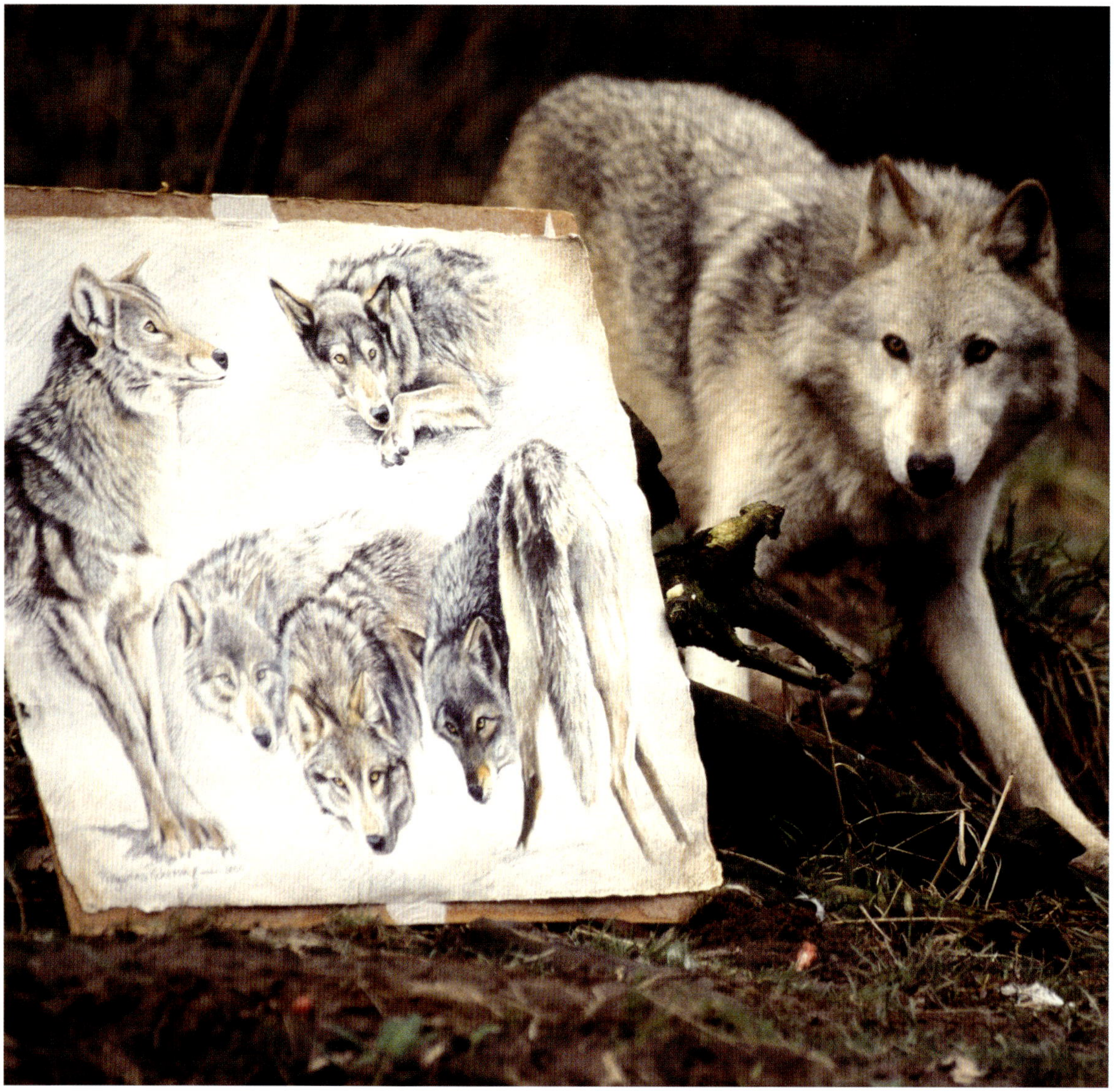

Chapter Five

THE WOLVES OF ZARNESTI

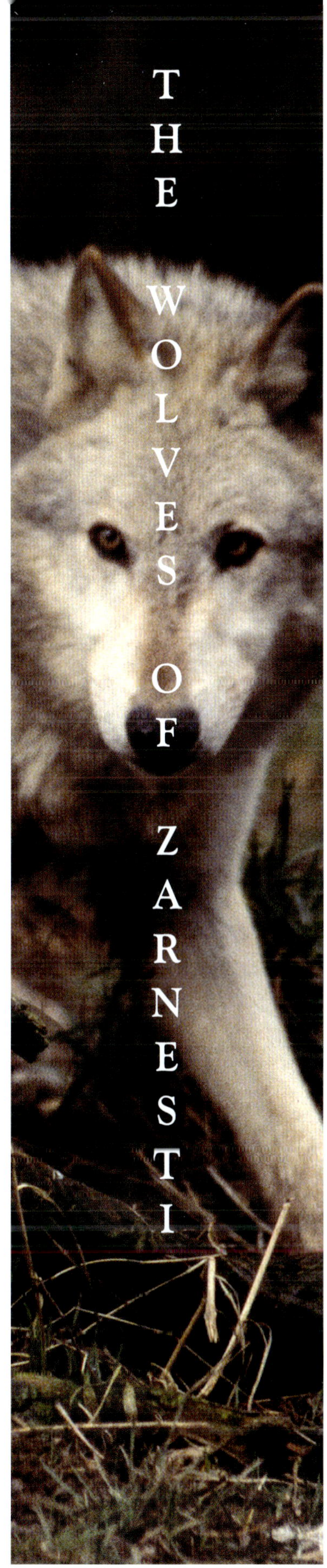

'I understand that the Count comes from a wolf country...'

SEWARD'S DIARY, *DRACULA*

If Transylvania is inextricably linked with Dracula, then the Count himself is inextricably linked with wolves. Travelling to meet his host, Jonathan Harker finds himself surrounded by a wolf pack: *'I saw around us a ring of wolves, with white teeth and lolling red tongues, with long, sinewy limbs and shaggy hair. They were a hundred times more terrible in the grim silence which held them than even when they howled'*. Dracula himself has the power not only to control the wild pack, *'He could control the wolves, as he did, by only holding up his hand'*, but could also transform himself into one: *'If he wished he could go either as man, or wolf, or bat, or in some other way'* – a neat trick if you can do it! Throughout the book, the wolves are a sinister presence, linked to the supernatural, always threatening. Writing about them in this way Bram Stoker taps into our primeval fairy-tale fear of the Big Bad Wolf, lurking in the depths of the forest. He uses them to help build a frightening environment for his hero to explore – a place where howls rend the air and it is unsafe to wander far from the beaten path. Although he is using our inbuilt fear of wolves for dramatic effect, it is a reflection of the attitude of most of western Europe to wolves over the last few centuries.

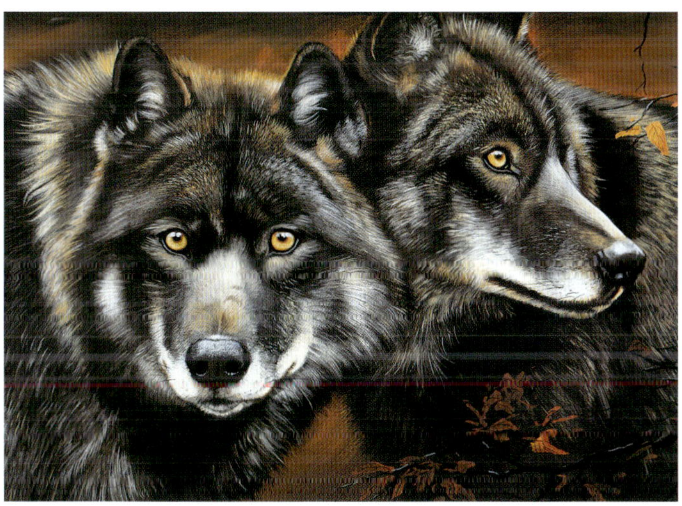

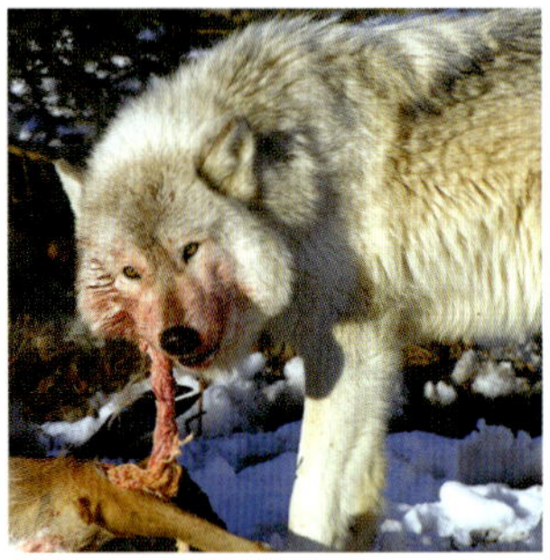

SHEPHERDS HAVE NO LOVE FOR THE WOLVES WHICH PREY ON THEIR FLOCKS

'Be quiet, or the wolf will come and take you!'

In the cultures of the earliest human societies wolves were usually admired and the legends surrounding them were positive. Wolves were widespread across Europe until the Middle Ages, when the human population increased and began to compete with the wolves for food. Human settlements grew as the land was cultivated. People kept livestock on which the wolves preyed, the image of the Big Bad Wolf appeared and these large carnivores were vigorously persecuted. Aside from humans, wolves were once the most widespread species of mammal on earth, but by the mid 1800s the wolf population had been decimated; they were completely extinct in many countries where they had once roamed.

During the 1990s, public interest in saving the remaining wolves resulted in the World Wide Fund for Nature working with experts from seventeen countries to create a wolf action plan to ensure their continued survival. They began to enjoy some limited protection, with the result that populations have largely stabilised, and in some areas have increased slightly.

Unfortunately, in many parts of rural Romania, the attitude to wolves has not changed a great deal since the Middle Ages. The shepherds have no love for the wolves which prey on their flocks of sheep. Grandmothers still tell their grandchildren that the wolves will eat them up if they go into the forests alone, and mothers commonly cajole their offspring to behave saying: 'Be quiet, or the wolf will come and take you!'. These prevailing views do not create an easy environment for the conservationists who are attempting to protect the remaining wolf packs. Fortunately there are people who are prepared to work in these difficult circumstances to ensure the survival of wild wolves.

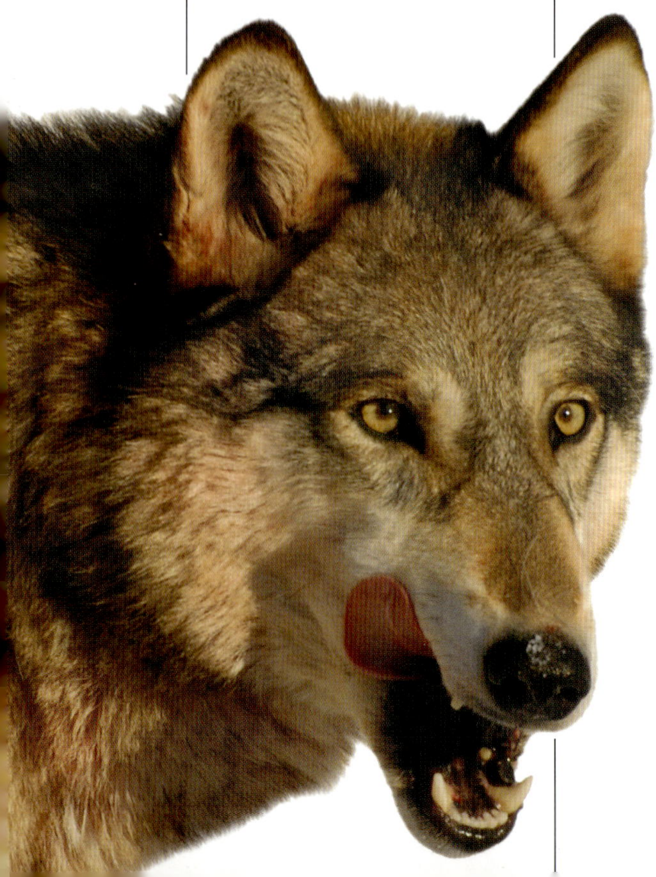

It is believed that around forty per cent of the wolves remaining in the wild in Europe are found in Romania, primarily in the isolated region surrounded by the Carpathian Mountains. In 1993 the German biologist and wildlife researcher Christoph Promberger initiated the Carpathian Large Carnivore

Project (CLCP) to research the lives of wolves in this area. Other large carnivores, such as bears and lynx, which share the wolves' habitat were also part of this study.

The project was set up to run for ten years and had ended the year before our visit, in 2003. The overall goal of the programme was to establish a community-based conservation of large carnivores and their habitat in a model region in the southern Carpathians through an integrated management approach. It was largely successful in achieving its aim. The CLCP helped in the setting up of the Craiului National Park, worked to ensure the legal and effective practical protection of wolves and, most importantly, helped to educate and inform the human population in ways they could co-exist peacefully with the wolves.

We are heading for the small town of Zarnesti, at the north-east end of the spectacular twenty-five kilometre-long limestone ridge of Piatra Craiului. A former industrial town, it was badly hit by the recession of the 1990s and the paper and armaments factories which used to employ about eighty per cent of the town's workforce are now effectively closed. But we have been drawn here because Zarnesti was the birthplace of the Large Carnivore Project.

Carpathian Large Carnivore Project

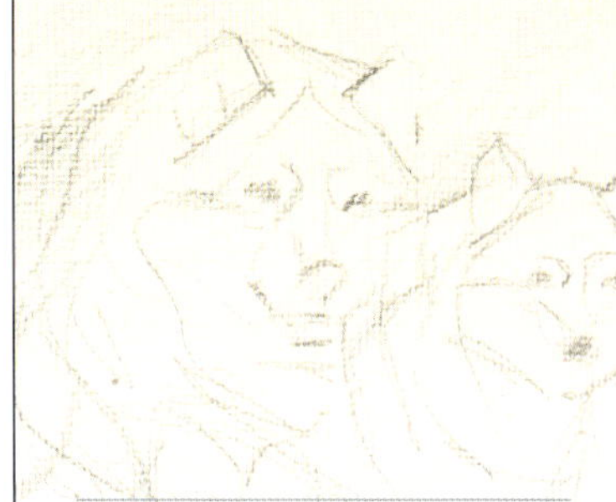

'It is believed that around forty per cent of the wolves remaining in the wild in Europe are found in Romania, primarily in the isolated regions surrounded by the Carpathian Mountains.'

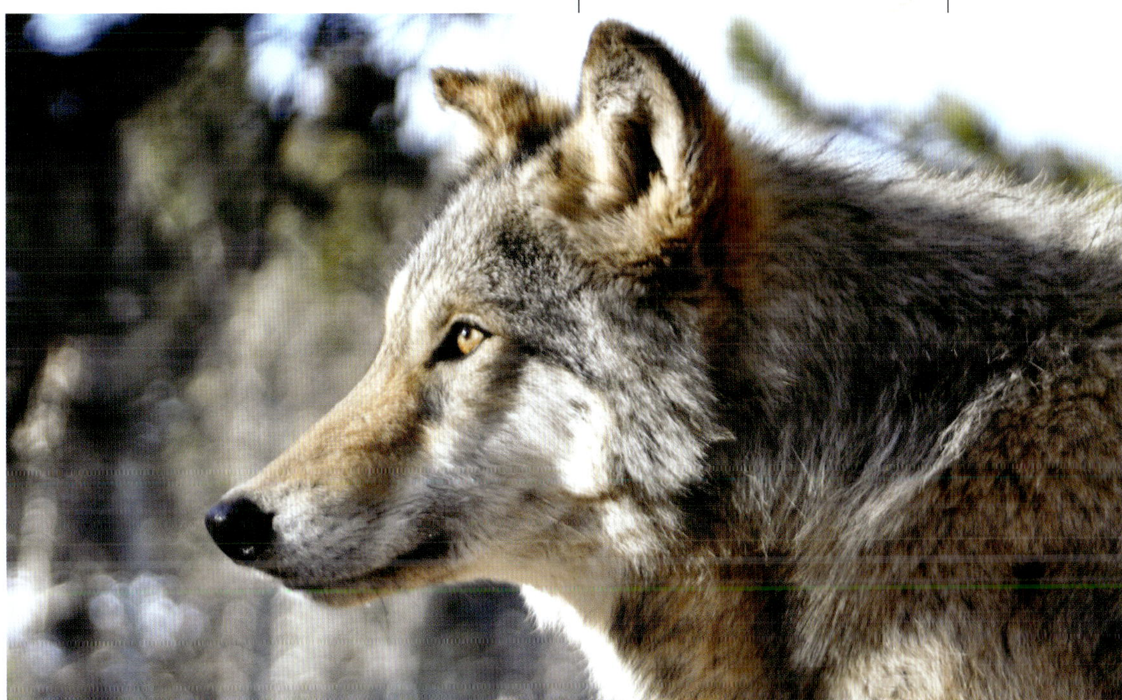

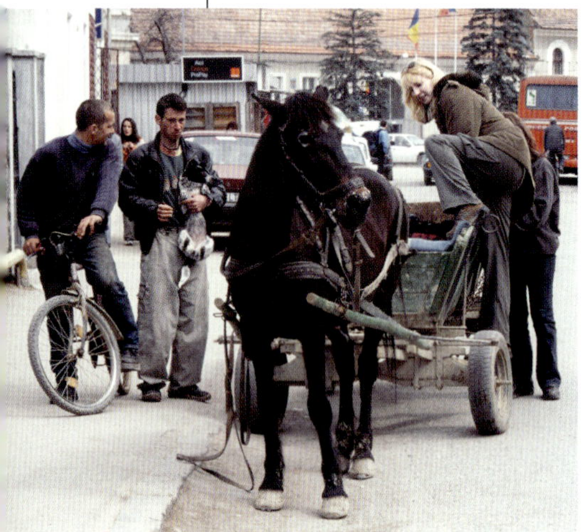

We elect to travel by the more environmentally friendly option of a traditional horse-drawn cart

A bone-rattling ride...

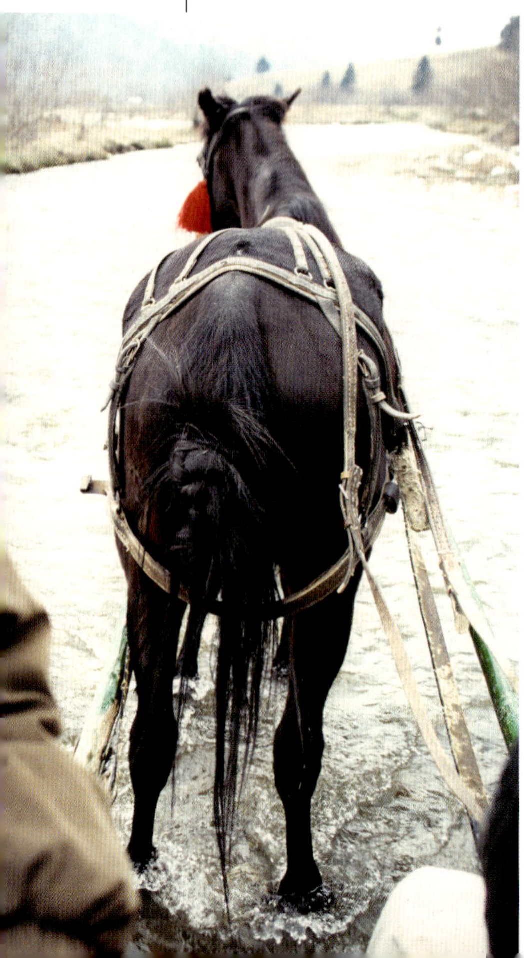

Day Eight

After breakfast, we meet Simona Buretea. As the public awareness officer for the Carpathian Large Carnivore Project (CLCP) she is responsible for media contacts, the school programme and 'anything else necessary in the office'. Simona is taking us to the CLCP's cabana to meet her two most important colleagues.

Shortly after establishing the project, Christoph rescued two young wolf pups named Crai and Poiana from a Romanian fur farm. Just six weeks old, they were reared by Simona and her team of volunteers. Although they grew into healthy adults, they could not be released into the wild because of the difficulties of integrating them into the complex social structure of a wild wolf pack. However, they now live in a vast enclosure – a huge fenced area of forested hillside.

Rather than drive our four-wheel drive vehicle up into the beautiful habitat of the Barsa Valley, we elect to travel by the more environmentally friendly option of a traditional horse-drawn cart – which also provides an extra source of income for the farming families. We arrive to be told that the horse belonging to their usual driver is suffering from an injury and won't be able to pull the cart for a few weeks. We meet the substitute cart driver who, I can't help but notice, has a powerful aura of plum brandy surrounding him. Can you be breathalysed while in charge of a horse and cart?

We jolt our way into the national park, accompanied by Simona and her cute little dog Lucky; she explains more about the CLCP and its work during the six-kilometre journey. The project operated in four fields: research, conservation and management, rural development and public awareness. In the research programme scientists studied the basic biology of wolves, bears and lynx and their relationship with local people in terms of livestock issues, economic importance and attitudes towards the carnivores. During the programme seventeen wolves, three lynx and twelve bears were radio-collared and economic aspects of livestock breeding, ecotourism and trophy hunting were closely monitored. An ecotourism programme was developed in the area around the national park and local politics was influenced in favour of conservation. At the time of our visit the CLCP is considering a new project – building an education and arts centre in Zarnesti. Simona invited Pollyanna to be the first artist to exhibit in the gallery once it is completed.

Eventually, after a bone-rattling ride, we reach our destination. We can see the little wooden cabana beside a huge enclosure stretching up the mountainside. We have been warned that the wolves are not always cooperative and may not emerge from the thick undergrowth in their large wooded enclosure. However, as soon as Simona walks up to the fence, the two wolves come racing down the mountainside to greet her – possibly lured by the very strong smell of the cheese we stopped to buy en route, apparently it is their favourite treat. They rub their backs along the netting for Simona to pet them. When Simona calls Poiana, the name sounds so similar to 'Pollyanna', both canid and artist come running! Once the greetings are over, and the cheese is finished, the wolves stroll back into the trees.

We scramble up the sheer slope alongside the edge of the pen to a half-way point where I can take photographs without the fencing blocking my view, thus providing a wonderful opportunity to see the wolves in their natural forest environment. I then slither back down, merrily shredding my newly purchased combat trousers and parka on brambles, thorns and sharp rocks.

The best wolf watching vantage point is from the roof of the cabana – we can reach it by scrambling up a ladder and across a small bridge – where we perch on the apex of the roof to observe the wolves. Now eight years old, Crai and Poiana are in their prime. Wolves show the most variation in colour of any furred mammal species. They range from the pure white of Arctic wolves through to jet black. White wolves remain white throughout their lives, but the coats of other wolves can change colour over the years. Crai and Poiana have beautiful thick creamy salt and pepper speckled grey fur. Their muzzles and legs are paler than the rest of their coats. A wolf's coat comprises an outer layer of thick coarse 'guard hairs' which provide protection from the elements and a soft under fur that traps air to

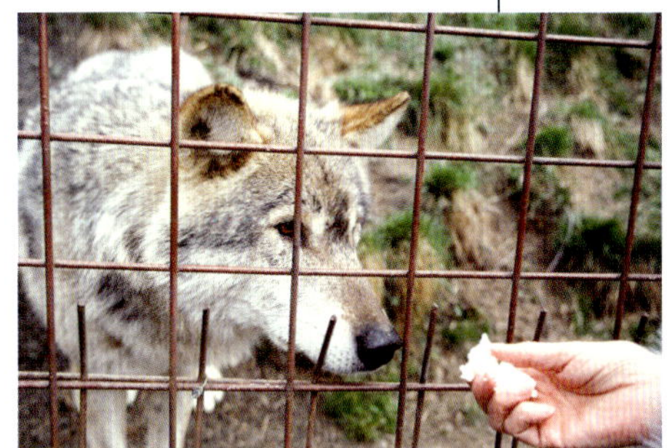

The wolves are lured by the very strong smell of the cheese, their favourite treat

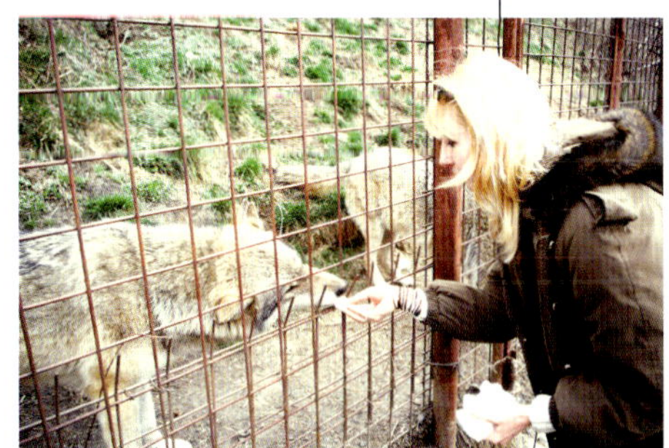

The best wolf watching vantage point is from the roof of the cabana

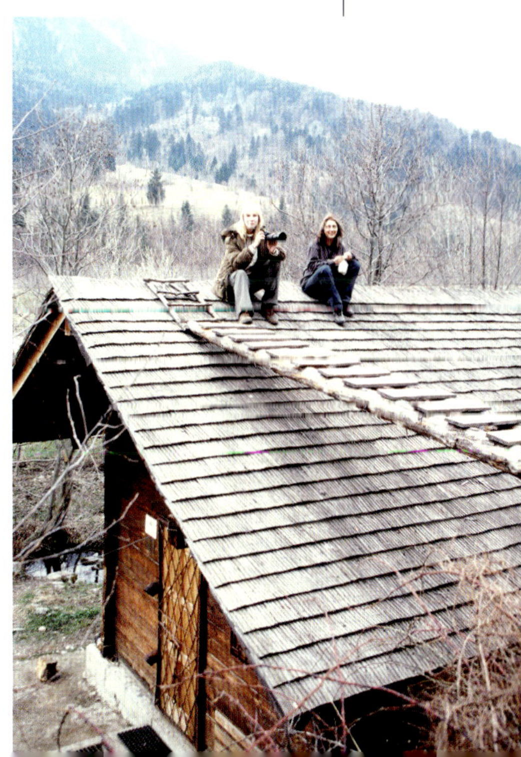

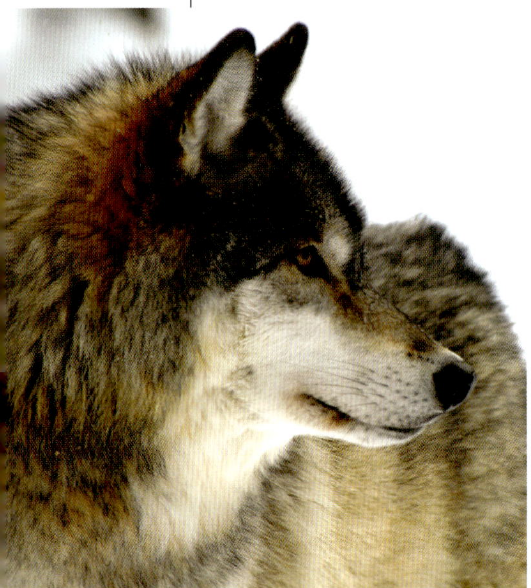

THE WOLVES' COATS ARE IN BEAUTIFUL CONDITION

provide insulation. Wolves generally moult during the summer and then grow a new coat through the winter – so at the time of our visit they are in beautiful condition, with thick lush fur. Their tails have dark tips of black fur and like a domestic dog, you can judge a wolf's mood by its tail. Although they rarely wag their tails, the tail of a cheerful wolf will be carried high while a despondent wolf really will have his tail between his legs. They have golden eyes which seem to shine with intelligence. Simona tells us that when the wolves arrived as tiny pups their eyes were blue, gradually turning to a deep amber as they matured.

A newborn wolf pup weighs approximately half a kilogram. They have dark brown fur, tiny ears and their eyes are closed. In the wild pups usually stay in the den for around five weeks, during which time their mother suckles them and keeps them warm. Research has shown that wolf pups which suckle from the teats closest to the centre – where the milk is richest – will grow to become the most dominant members of the pack, while those at the outside will be submissive. By the time they are a year old wolves will have reached their full adult size. There is wide variation in the relative size of the male and female

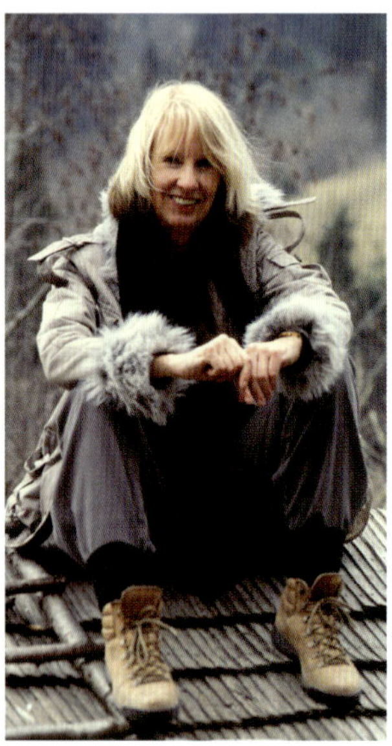

WE ARE ABLE TO SIT...

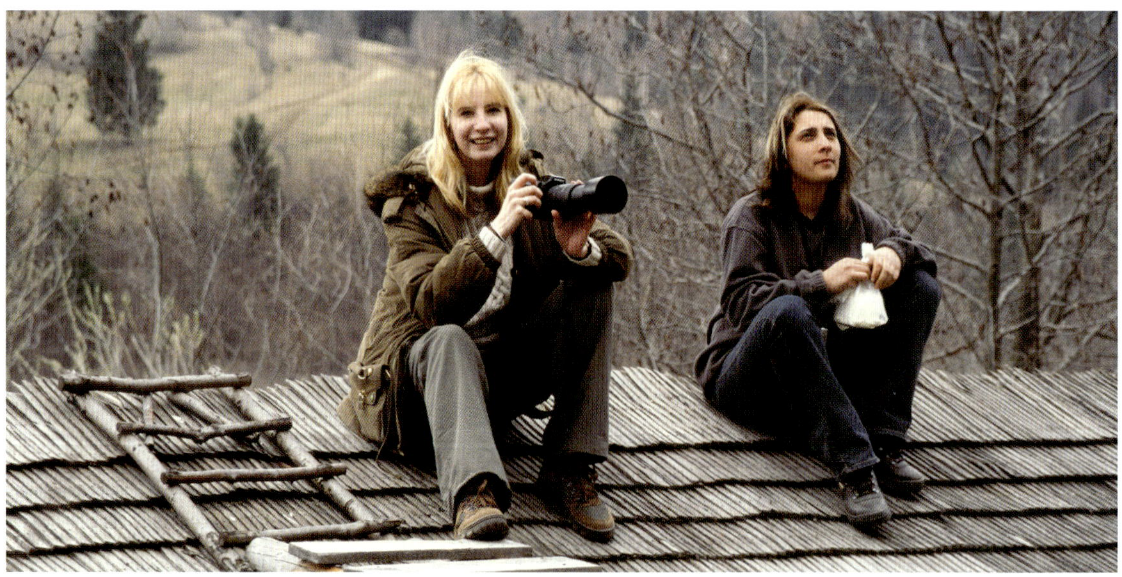

...AND WATCH THE WOLVES FOR THE WHOLE DAY, WITHOUT LEAVING OUR LOFTY PERCH

wolf populations found throughout the world, but generally the males are around twenty per cent larger than the females, measuring between one and a half and three metres nose to tail.

We are able to sit and watch the wolves for the whole day – we made some cheese and tomato sandwiches early in the morning and enjoy them as a lunchtime picnic without leaving our lofty perch.

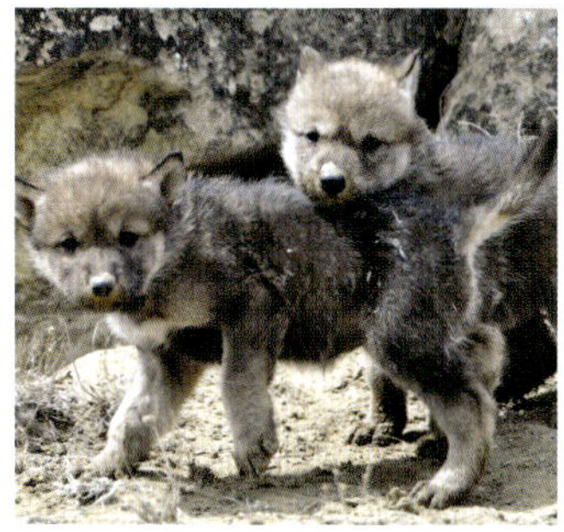

WOLF PUPS EXPLORE THEIR SURROUNDINGS

This provides the perfect opportunity for Pollyanna to sit and sketch. Having had a classical art education Pollyanna always makes sketches and studies on location – my photographs will be used for her illustrated talks, and to accompany magazine articles about the journey. For Pollyanna my photographs are merely a 'back up'. She never paints directly from one photograph, although she may occasionally use them for reference detail. Alongside her line drawings, Pollyanna also makes field notes – these act as a reminder of the colours and textures she has observed – and adds annotations on the behaviour of the wolves. All of this information is then incorporated into her completed paintings. She firmly believes that sketching from life is the only way an artist can capture the true essence of a bird or animal; this is the best way to observe and understand the shape, form and movement of her subjects.

Day Nine

At the end of another day sketching and photographing the wolves we return to Zarnesti late in the afternoon. Pollyanna has been invited to visit a local high school to talk to some of the children about her work as a wildlife artist, especially with wolves and wolf conservation. The students are particularly fascinated by the fact that she has travelled all the way to Zarnesti to sketch Crai and Poiana – and she explains the importance of sketching from life, as well as experiencing the natural environment that the wild wolves inhabit.

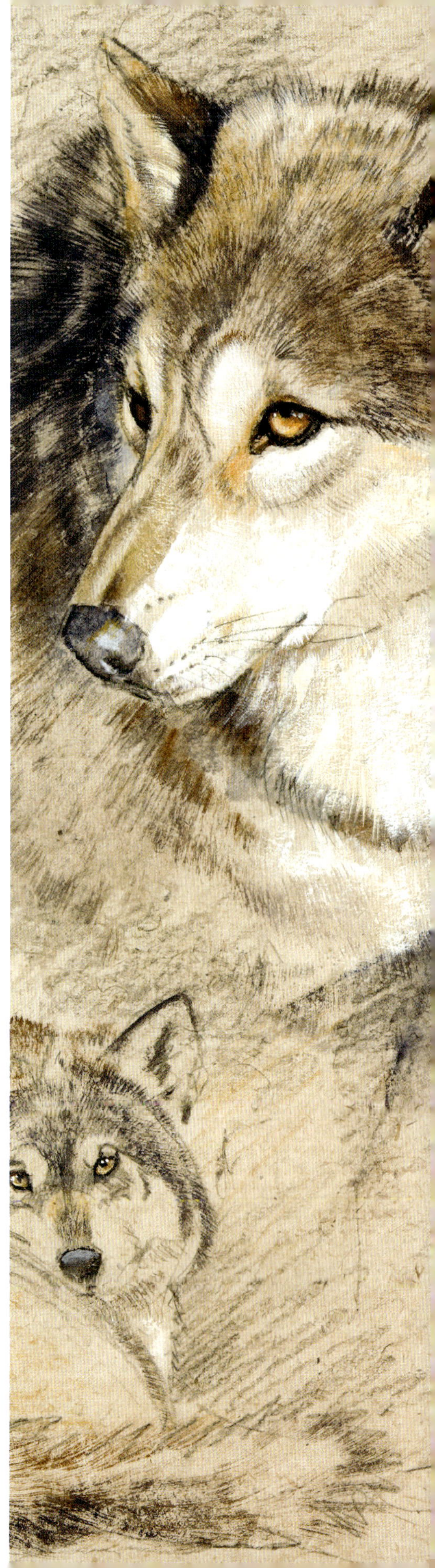

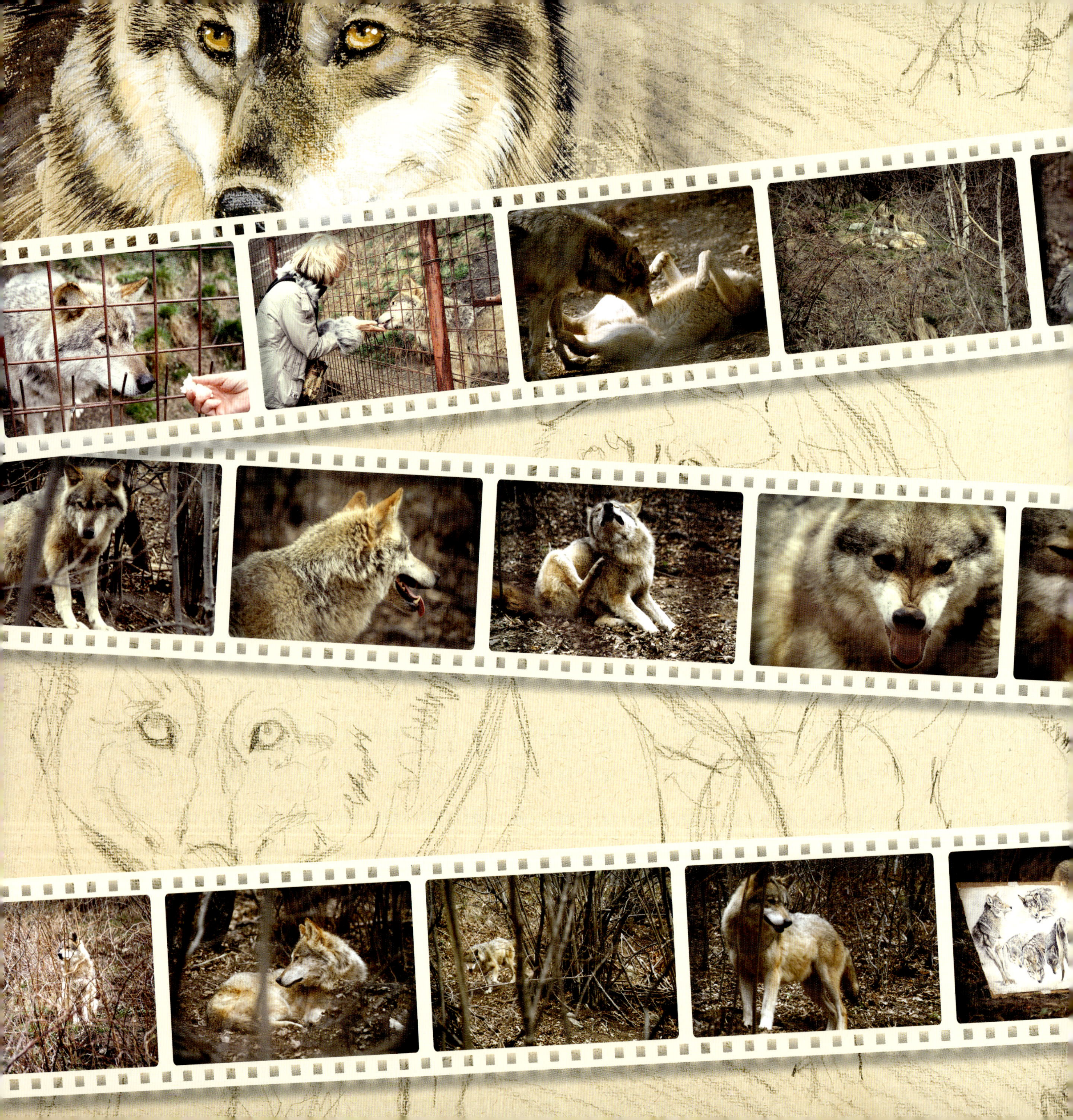

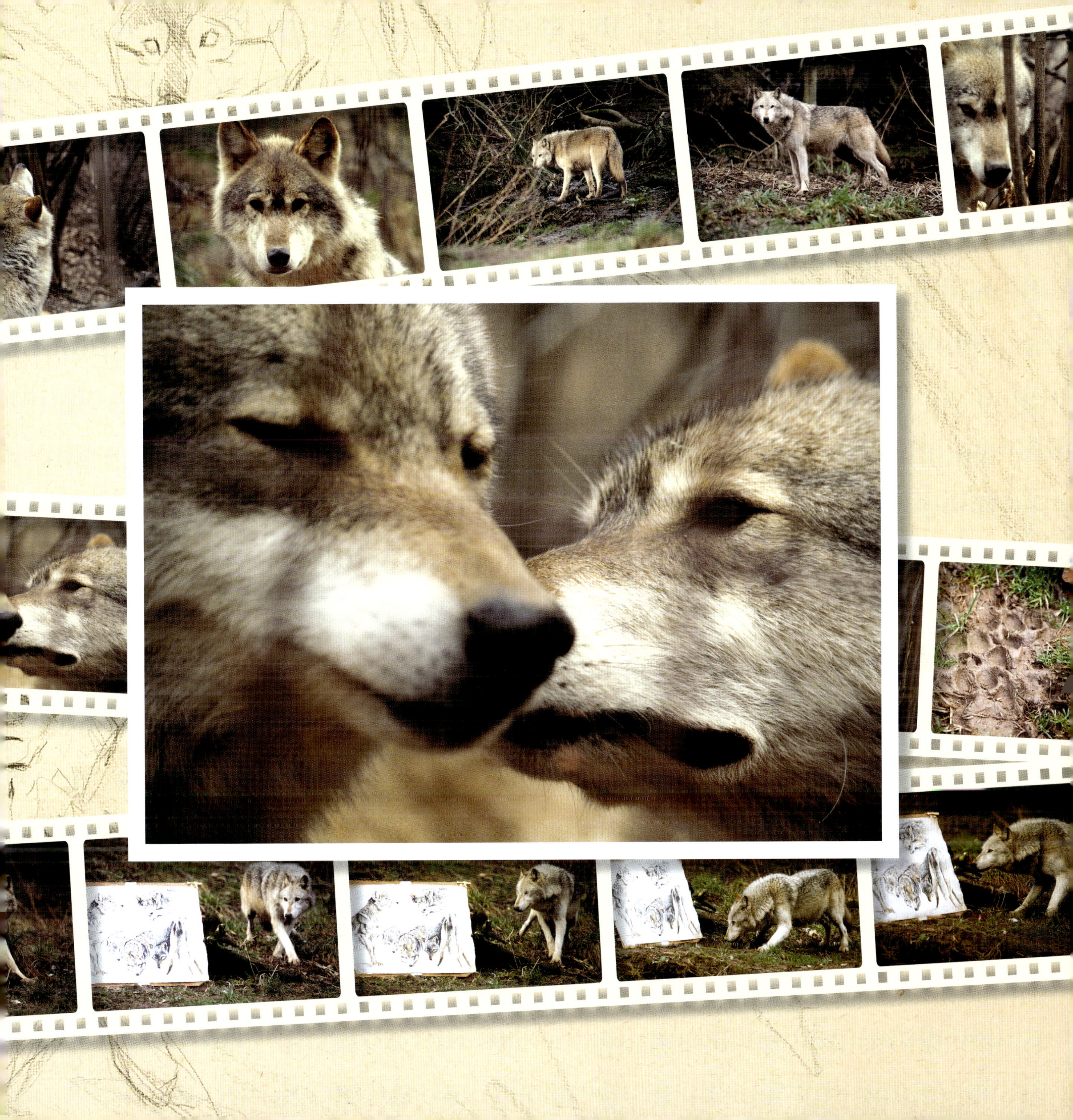

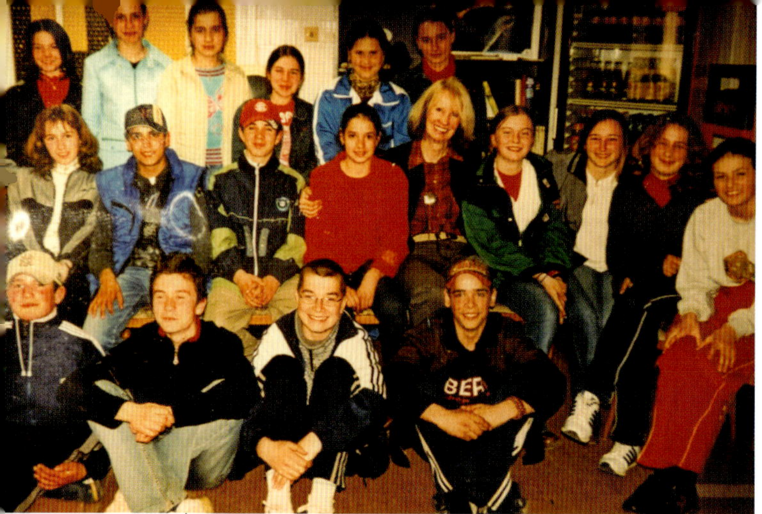

Pollyanna has been invited to visit a local high school to talk to some of the children about her work as a wildlife artist, especially with wolves and wolf conservation

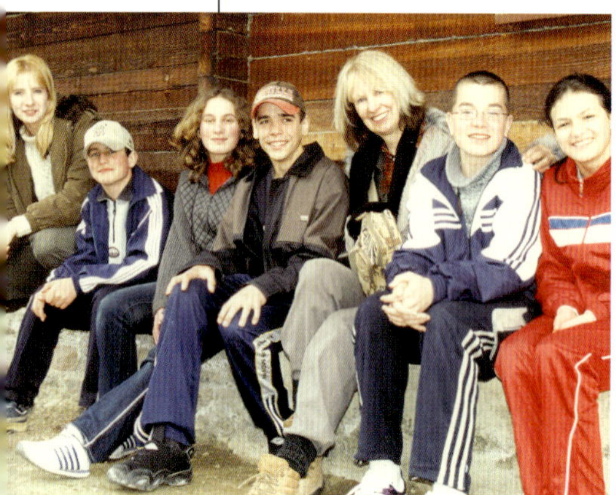

Five of the children from the *Mujarel* magazine team: Alex, Alex, Andrea, Andrea and Elvis

Anna-Louise with children from the local school

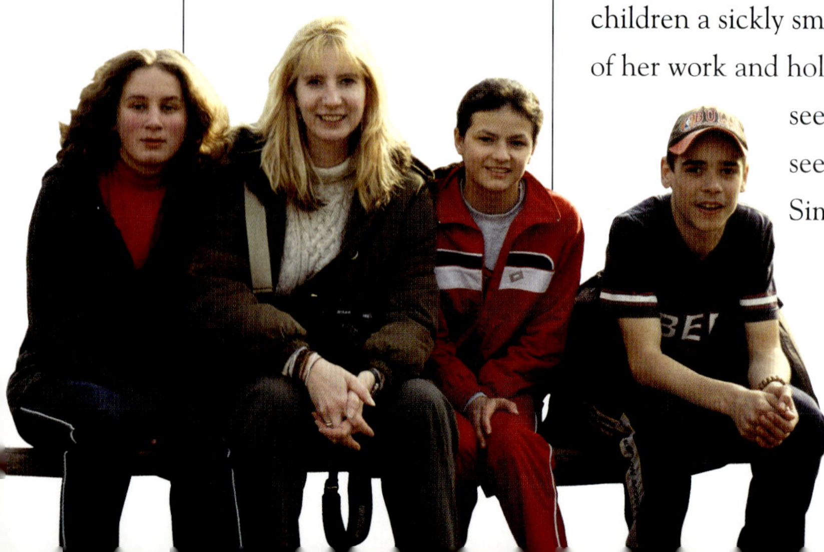

Simona has worked a great deal with schools in this region, helping to educate the children about the richness of the wildlife which surrounds them – and the importance of protecting it. Much of her work with the children is to help combat the age-old fear of wolves which has been handed down through the generations. This fear is rooted in stories about wolf attacks that are told repeatedly until they become part of the folklore or rural legends. She has set the students the task of investigating the stories of wolf attacks on people that circulate in their own villages. The students are asked to interview their friends and relatives, discovering where they have heard the stories and, if possible, tracing the stories back to the source. They then present a report on their findings to the community. In the report that these students gave to their neighbours in the nearby village of Pestera, they confirmed that they were not able to verify a single one of the local wolf-attack stories. Simona told us, 'Some members of the audience were initially reluctant to hear that message, but as the discussion progressed, the villagers eventually acknowledged that their fear of wolves was unjustified and that in fact they had more problems with bears'.

The group of children are perched on rigid benches, still wrapped up in their winter coats and hats, watching Pollyanna with great interest. A faded map of the world hangs crookedly on the wall, depicting nations that have not existed for half a century. There is a slight hitch when the school's dusty slide projector refuses to come to life, and we are unable to show the selection of slides we have compiled from our various working trips to study wildlife. Pollyanna gives the assembled children a sickly smile – and improvises, handing round cards of her work and holding up pictures in our books for them to see. The children are all learning English and seem to understand the talk very well, although Simona helps by translating more complex

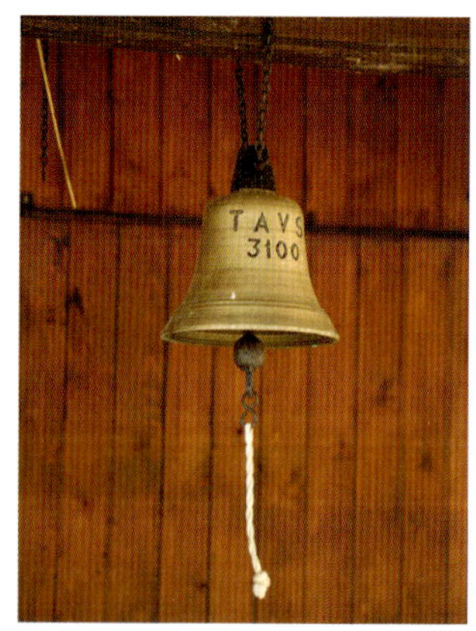

The school bell

sentences. Initially very shy about asking questions in a second language, we are soon bombarded with queries about Pollyanna's painting and our travels. One of the boys, Alex, is especially interested in art and Pollyanna presents him with one of her wolf sketches. At the end of the evening several of the children give us paintings and drawings which they have completed. Alex presents Pollyanna with a truly beautiful painting in the style of the traditional icons which we plan to have framed to display in her gallery at home.

A group of the children to whom we have been talking also collaborate to produce an in-school environmental magazine called *Mujarel*. They ask if they can meet Pollyanna the following day to interview her for the publication. She is happy to oblige.

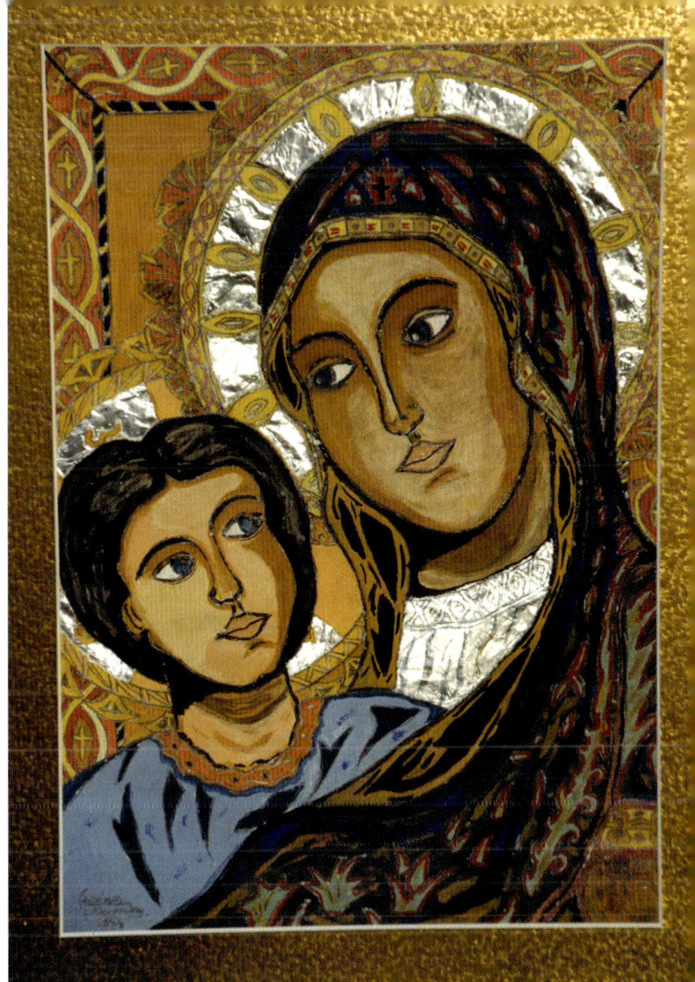

Alex's beautiful icon-style painting – a gift to Pollyanna

Day Ten

After spending the morning wolf-watching on the roof, we clamber down to eat our sandwiches at ground level. Colin arrives with five of the children from the *Mujarel* magazine team, who introduce themselves as Alex, Alex, Andrea, Andrea and Elvis. Simona has suggested that during the afternoon we walk up the mountainside to visit Coltul Chillii, a monastic shrine hidden in rocks. The shrine is in an area rich in birds and other wildlife, as well as being fascinating in its own right. We suggest to the children that they come with us, and we chat as we walk. However, Simona's idea of a walk turns out to be our idea of an endurance hike, up an incredibly steep area of mountainside. We are at almost 5,000 feet above sea level, and Pollyanna and I are both struggling with the altitude, becoming breathless quite quickly. After an hour's walk we find we are scrambling up an almost vertical patch of

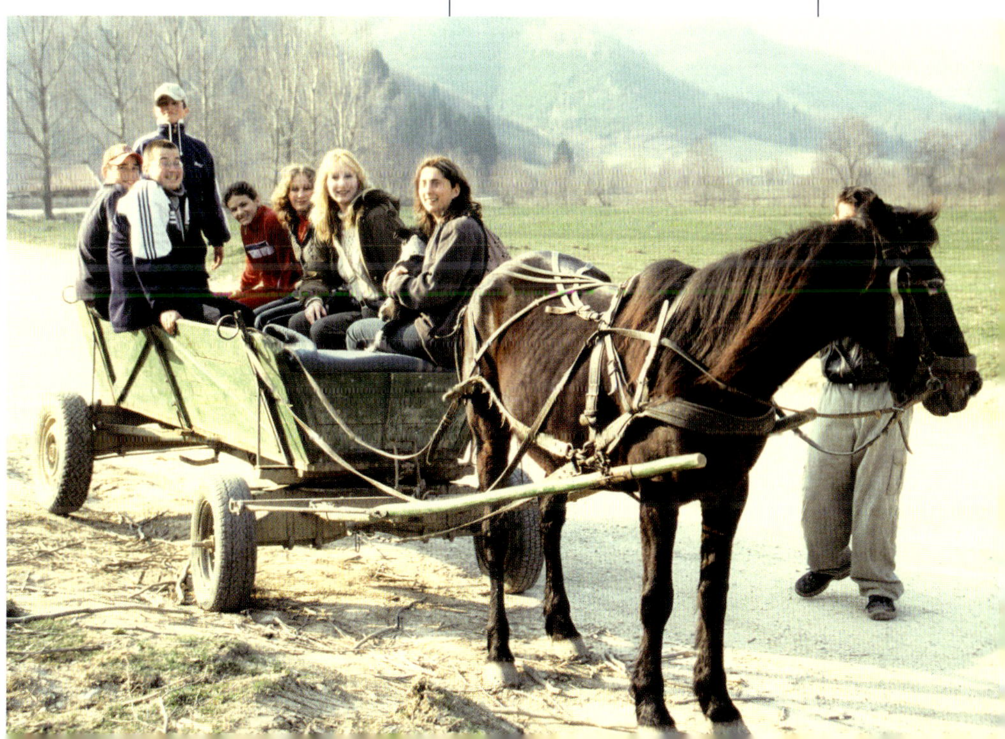

Pollyanna, Simona and the writers of *Mujarel* travel to the Barsa Valley

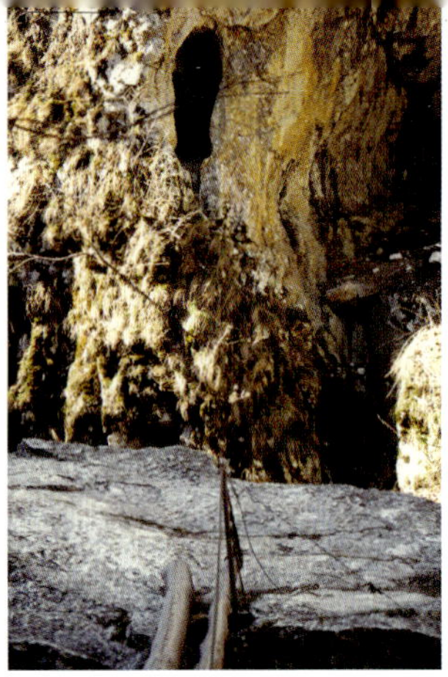

Coltul Chillii, a monastic shrine hidden in rocks

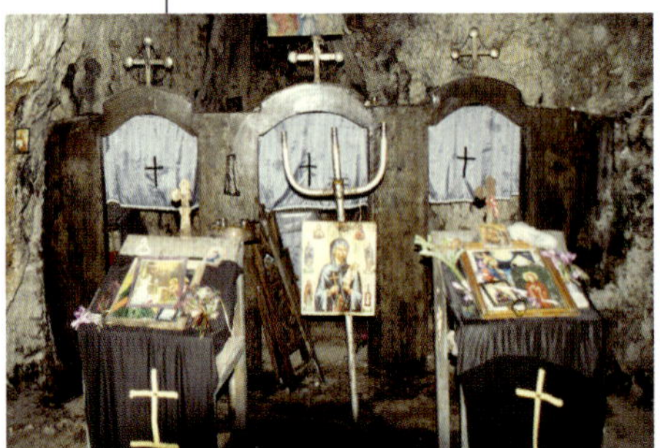

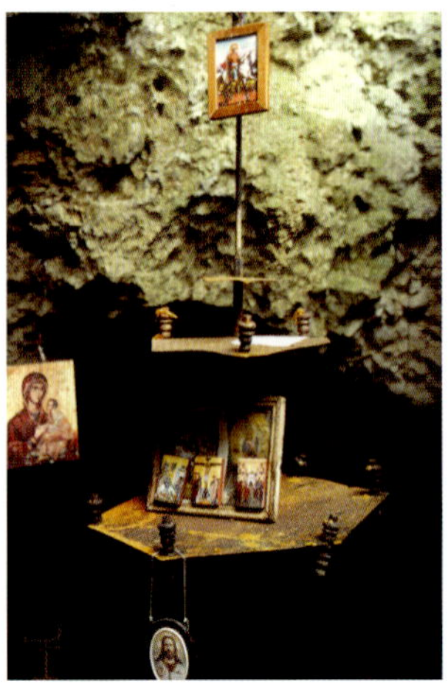

Relics and icons are suspended from the ceiling

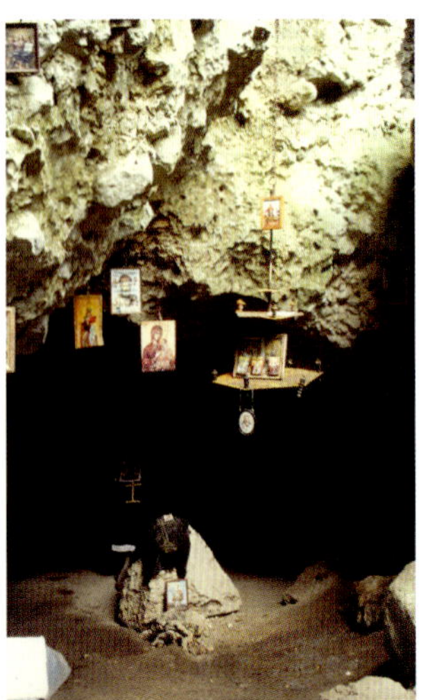

woodland, scrabbling for handholds and footholds where we can. This means that it is not the ideal situation for the children to interview us – we are puce and gasping. When Elvis asks Pollyanna, 'How did you start out as an artist?' her reply begins, 'Well… gasp, gasp… when I was… gasp… a small… gasp… child…' We decide that it would be better that they refrain from asking any more questions until we are going back down the hill. Having grown up at these altitudes the children are scampering around the hillside like mountain goats.

Eventually we reach our destination – a holy cave sunk into the rock. It contains two altars and an iconistery. The rough walls of the cave are hung with pictures and icons, ornate gilded frames glowing in the gloom of the interior. More relics and icons are suspended from the ceiling. It is wonderfully atmospheric, well worth the struggle up the steep slope. Going downhill is, of course, much easier and we practically saunter back, while

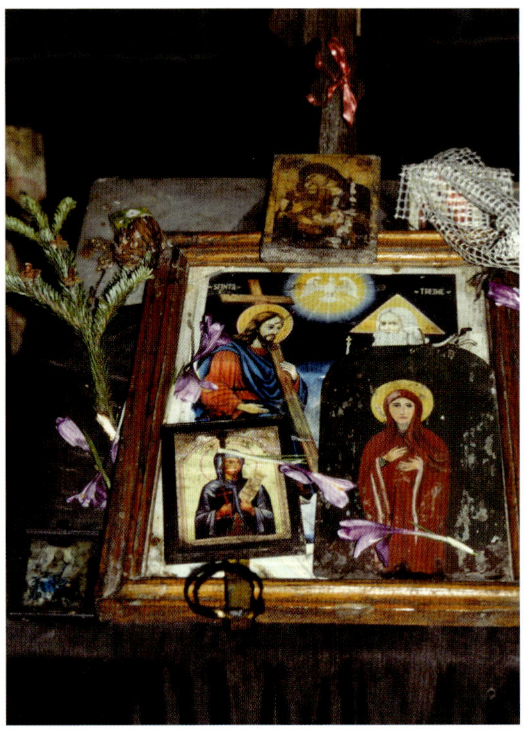

the kids race ahead of us – the interview is eventually concluded sitting on benches near to the wolves' enclosure. Simona sweetly expresses the hope that we won't have 'blizzards' on our feet after the long walk.

We return to the school in the early evening to show the children the slides they missed out on the previous day, on another ancient projector which Colin has miraculously dug out of a storage cupboard. Elena makes us another lovely dinner, accompanied once again by glasses of industrial strength brandy. At least that should kill off any germs I may have picked up working around the wolf enclosure. Although, if I do get sick today, no doubt it will be due to the fact that when Simona's dog Lucky leapt up to greet me extremely enthusiastically this morning, I did get a little dog tongue in my mouth. Blueurggghhh.

After our meal we stroll round the corner to the small CLCP shop, which is run by Simona's mother, who is delightful. She also introduces us to her father, who kisses our hands in the traditional Romanian manner. We simper girlishly, before purchasing nearly everything in the shop – which sells mostly local crafts. We buy a cute hand-knitted teddy, as well as CLCP bags and hiking socks, decorated with images of wolves and bears. We crawl back to the pension, very tired after our afternoon's hike, and ooze into bed – only to discover the walls are tissue thin, and I can hear every word of the mobile phone conversation another English guest is having in the room next door – including the latest cricket scores!

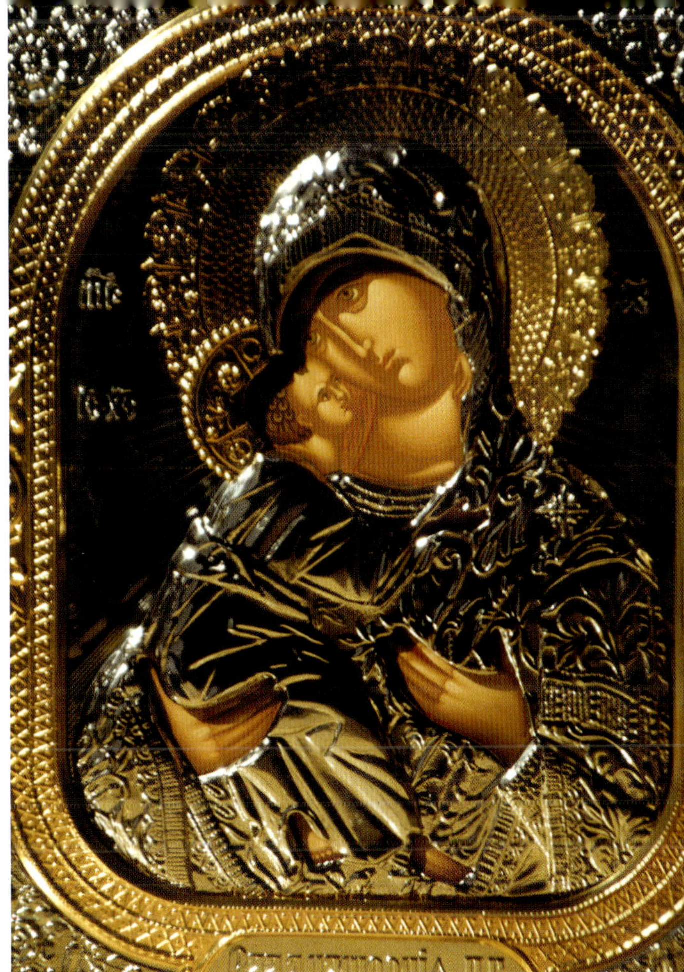

Icons with ornate gilded frames line the interior

Day Twelve

Our last day with Crai and Poiana. Once again we stop off at a grocery shop to purchase the ripest smelliest cheese we can find. As soon as we arrive Crai and Poiana catch the scent and come bounding up to the fence. Scientists have estimated that the wolf's sense of smell is up to one hundred times more sensitive than that of a human, but frankly they don't need exceptionally keen noses to locate the source of this smell – I think I will be carrying it round with me for weeks. We are allowed carefully to give the wolves pieces of cheese from our fingers. They take them from us very gently – which is a blessing. Adult wolves have forty-two highly specialized teeth – ten more than humans. The canine teeth, or fangs, can be two and a half inches long and are used for puncturing skin and gripping their prey. The incisors are used for nipping small pieces of meat; the carnassial teeth are like knives which they use to sheer flesh away from bones. Wolves use their powerful molars for grinding and crushing their food.

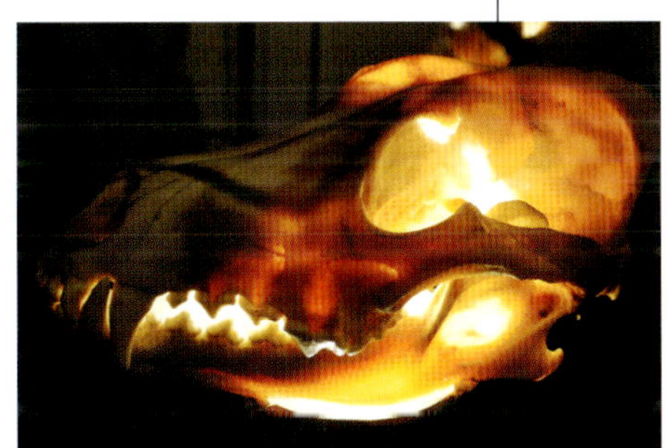

Adult wolves have forty-two highly specialized teeth

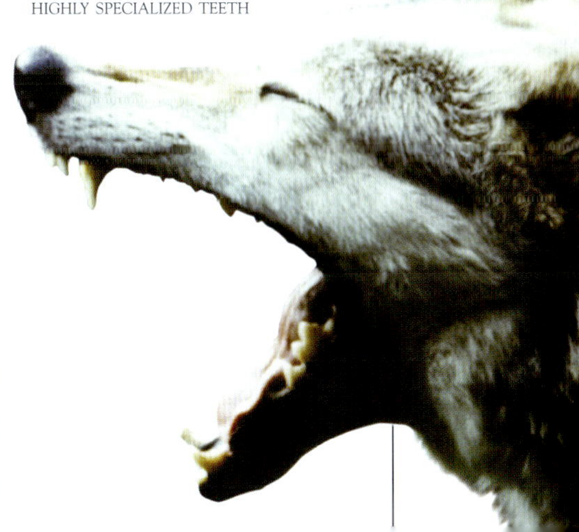

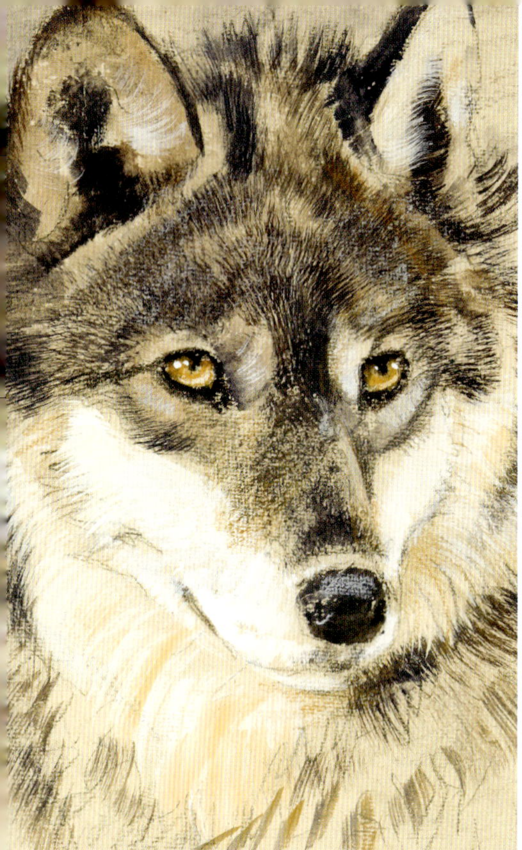

Crai shows a rather unnerving interest in the fake fur trim around the cuffs of Pollyanna's jacket. We are unsure whether he thinks it is another wolf and his interest is romantic – or he has identified it as a small tasty mammal. Either way it does not bode too well for the survival of the jacket. Poiana, meanwhile, is fascinated by the movement of Pollyanna's pencil across the sketchpad as she moves closer to the netting to begin work.

Separated by just one layer of wire mesh, Pollyanna sits and stares directly into the yellow eyes of the wolf – and the wolf stares steadily back. She picks up her pencil, selects a sheet of handmade paper from her folder and wonders how to record that luminous gaze – how to capture the soul of the wolf. Quickly she begins to sketch using a soft 6B pencil – shading against the fur where the sun highlights the edges. Using the basic watercolour palette which she carries with her on all her working trips, she makes a simple wash of raw umber, running indigo into the wet paint. Then using Payne's grey on a dry brush she picks out the dark flecks in the fur to create texture and, with white gouache, softened to a creamy shade with a hint of yellow ochre, she dry brushes some highlights where the sunlight catches the wolf's soft coat. For the eyes she begins with burnt sienna – deepened at the top with burnt umber, and then a brush of pale ochre where the sun passes through the lens, with a highlight of white to bring to life the wealth of intelligence she can sense in the golden gaze. She concentrates once again on working into the texture of the fur – but when she next glances up the wolf has melted silently into the cover of the trees like a ghost. Even these relatively quick studies will enable her to catch the essence of this enthralling animal when she returns to her studio. She aims to make her wildlife portraits as accurate as possible so that the viewer can feel they are right there beside her, just a few feet away from the gaze of the wolf. This sense of reality has become the hallmark of her work.

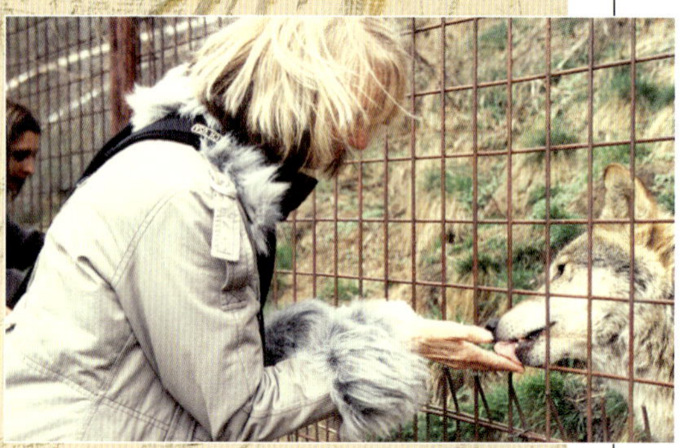

WE ARE ALLOWED CAREFULLY TO GIVE THE WOLVES PIECES OF CHEESE FROM OUR FINGERS

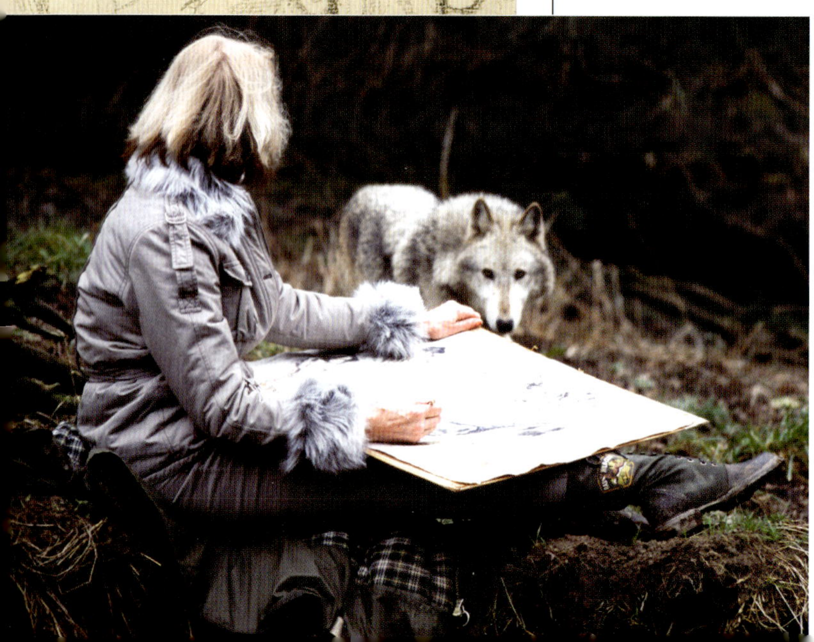

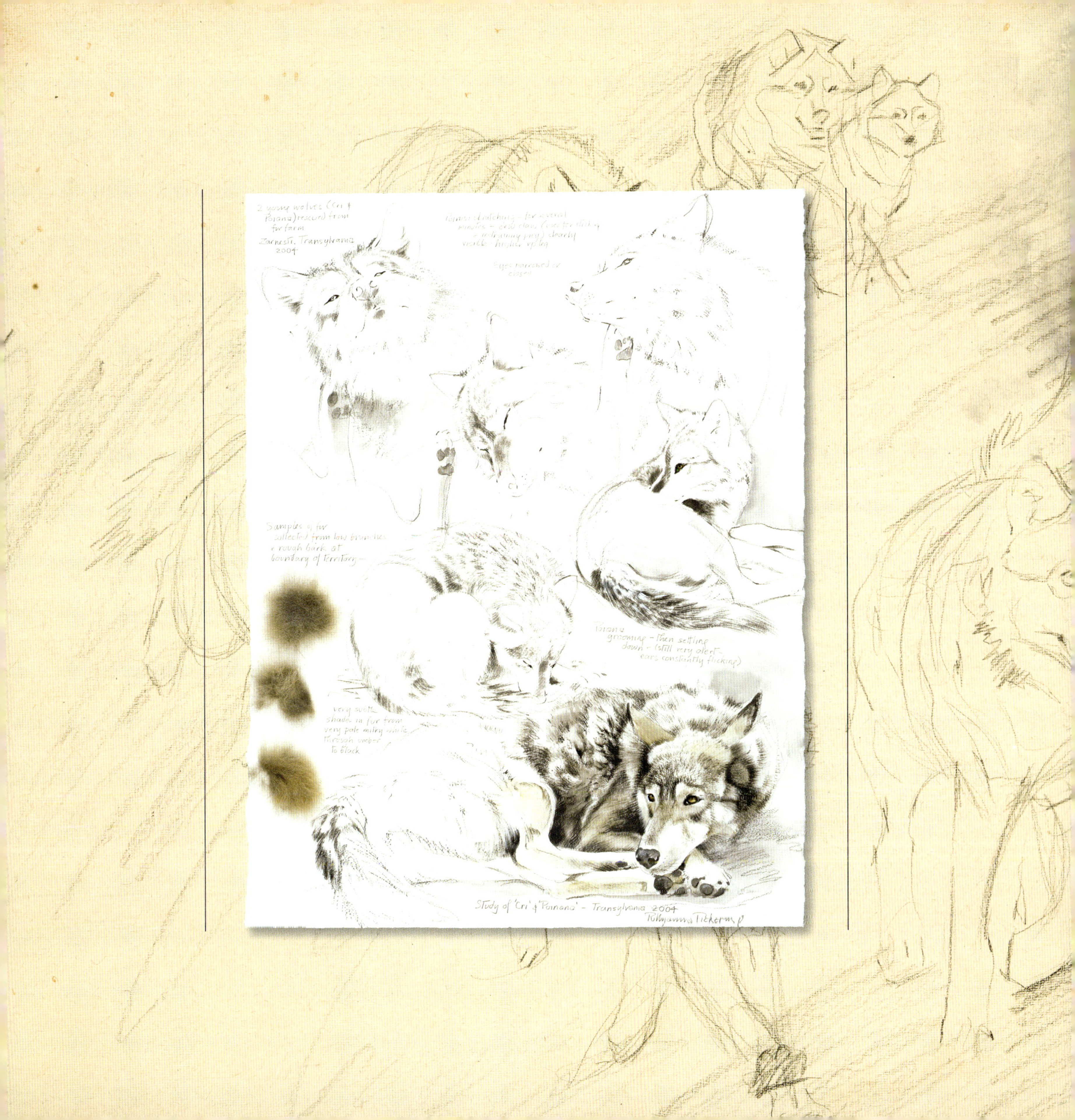
2 young wolves (Cri &
Poiana) rescued from
fur farm
Zarnesti, Transylvania
2004
Eyes narrowed or
closed
Samples of fur
collected from low branches
& rough bark at
boundary of territory –
Poiana
grooming – then settling
down – (still very alert –
ears constantly flicking)
very subtle
shades in fur from
very pale milky white
through umber
to black
Study of 'Cri' & 'Poinana' – Transylvania 2004

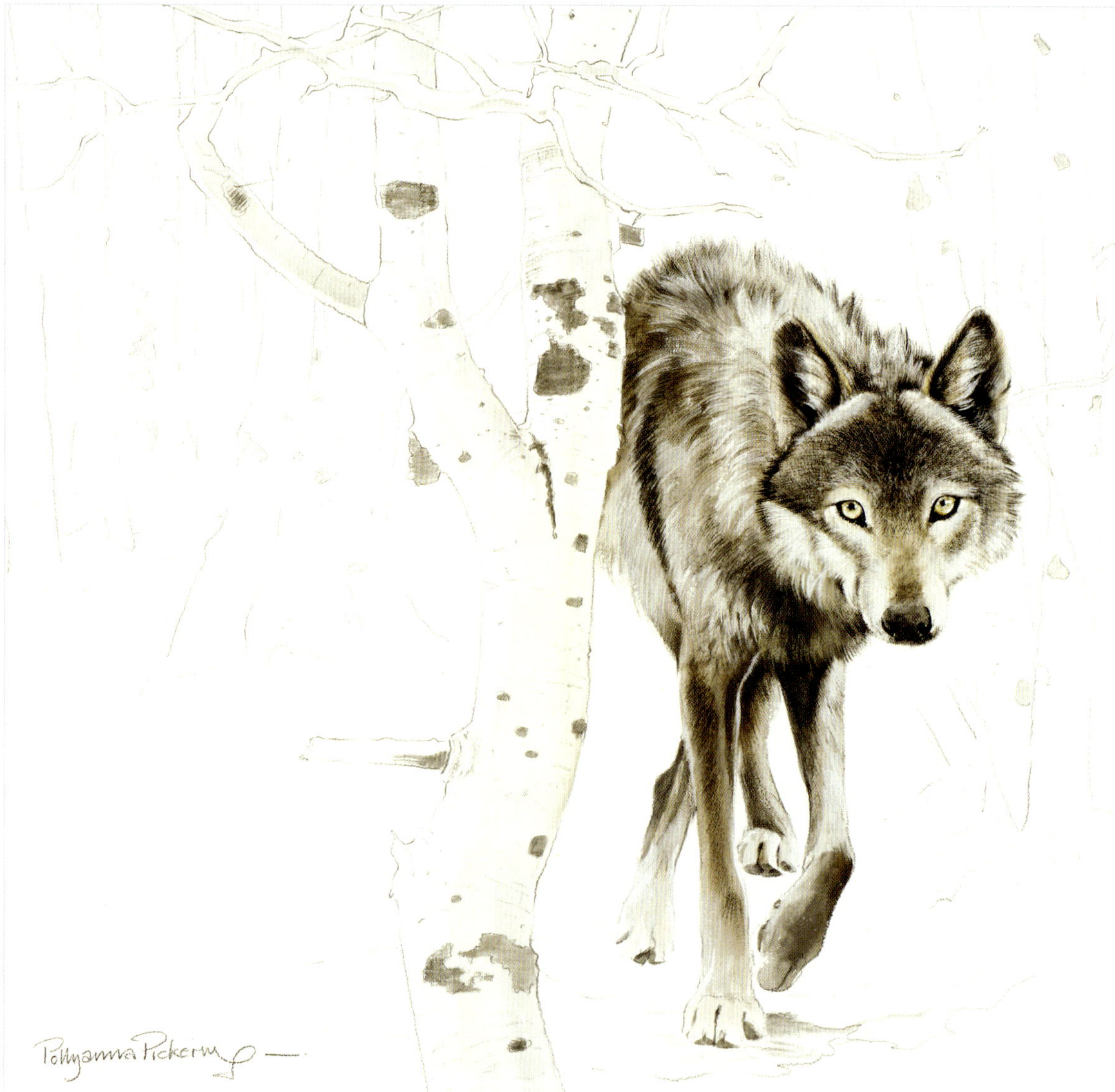

Chapter Six

Where Shepherds Watch

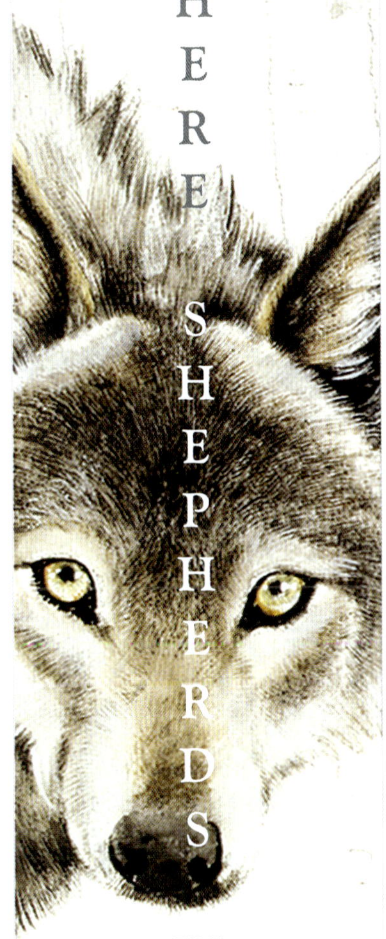

'There seem to be endless streams running down the mountains...'

JONATHAN HARKER'S JOURNAL, *DRACULA*

Before we leave Zarnesti we want to take the opportunity to hike through the forests around Piatra Craiului. This is prime wolf habitat. However, our chances of seeing a wolf in the wild are depressingly slim. In parts of the world where wolves have had little contact with humans, or in areas where they are heavily protected (or even fed!) they show very little natural fear. However, in areas where they have been persecuted, they quickly learn that humans pose a threat and are best avoided. Wolves do coexist with humans in some of the more remote Romanian villages, passing along the roads at night, or foraging on the outskirts – but their activities occur under the cover of darkness when the chances of direct contact are low and they are rarely observed. Here in Transylvania, the main human / wolf conflict occurs where shepherds graze their sheep on the mountain pastures. The wolves are naturally attracted to the flocks of sheep as a plentiful food supply. Historically the shepherds would kill the wolves to protect their livestock and, as a consequence, the wolves of Transylvania have become secretive and stealthy, preferring to remain unseen in the shadows. Despite this our best chance of seeing wolves remains to trek up to the pasture lands, the areas the wolves are drawn to by the presence of the flocks.

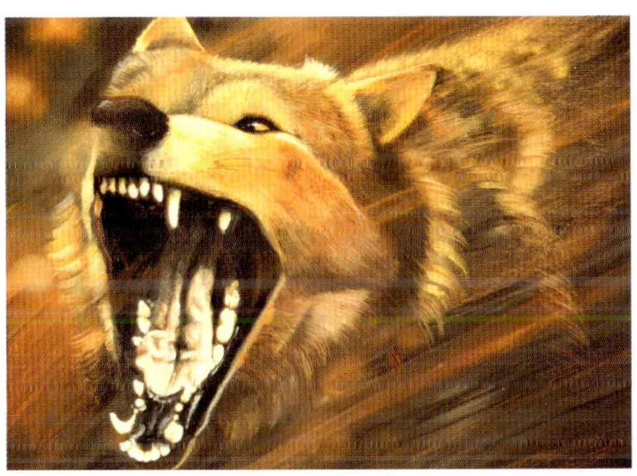

The wolves have naturally learned to be afraid of us – but as we are heading unarmed into their wild home, I wonder if we should fear them? Although there are no documented cases of humans being killed by wolves in

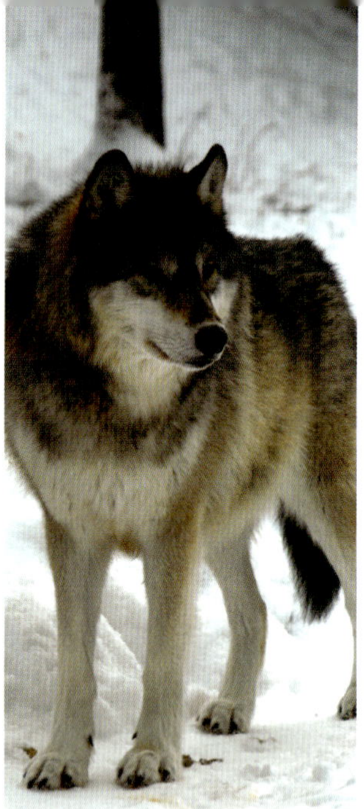

North America or Canada, wolf attacks have certainly occurred over the centuries in Europe and Northern Asia. Researchers have studied church and court records, and archives of correspondence dating back to the fifteenth century to identify occurrences of wolves killing humans. As late as the 1900s, papers and magazines were printing lurid first-hand accounts of wolf attacks, under headlines like 'The Grey Scourge'. An article published in *The Wide World* magazine in 1910 described the Transylvanian wolf as: '*Harmless and cowardly in summer, when he hunts alone and there is plenty to eat, but a deadly menace in winter, when hunger gnaws at his vitals and he joins his grey-coated fellows in concerted attacks upon hapless travellers – such is the dreaded wolf of the Carpathian forests*'. The article goes on to describe a '*tale of escape from these four-footed fiends of winter*' when a police sergeant is returning on horseback from his weekly patrol through some of the outlying villages. He describes a daring escape from a pack of '*gaunt, red-eyed brutes waiting their chance to pull me down to death*' – which he estimates at nearly two hundred. Despite these lurid accounts – which show every sign of enthusiastic exaggeration for a thrill-seeking audience – there are very few authenticated records of attacks by wolves. Not only rare, such attacks seem to occur only under unusual circumstances. Like most animals, wolves will be aggressive when defending pups, or when cornered and unable to run. Certainly in modern times confirmed attacks appear to involve either tame wolves, wolf-dog hybrids or rabid animals.

In fact if a wolf should spy us as we walk through his territory he is most likely to slope away unseen into the forests.

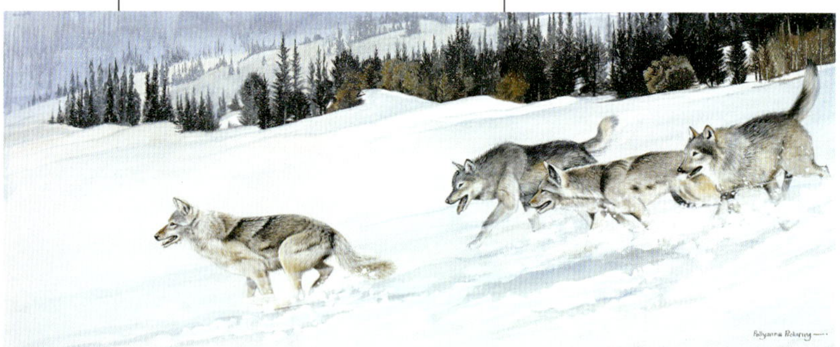

'The Pack'

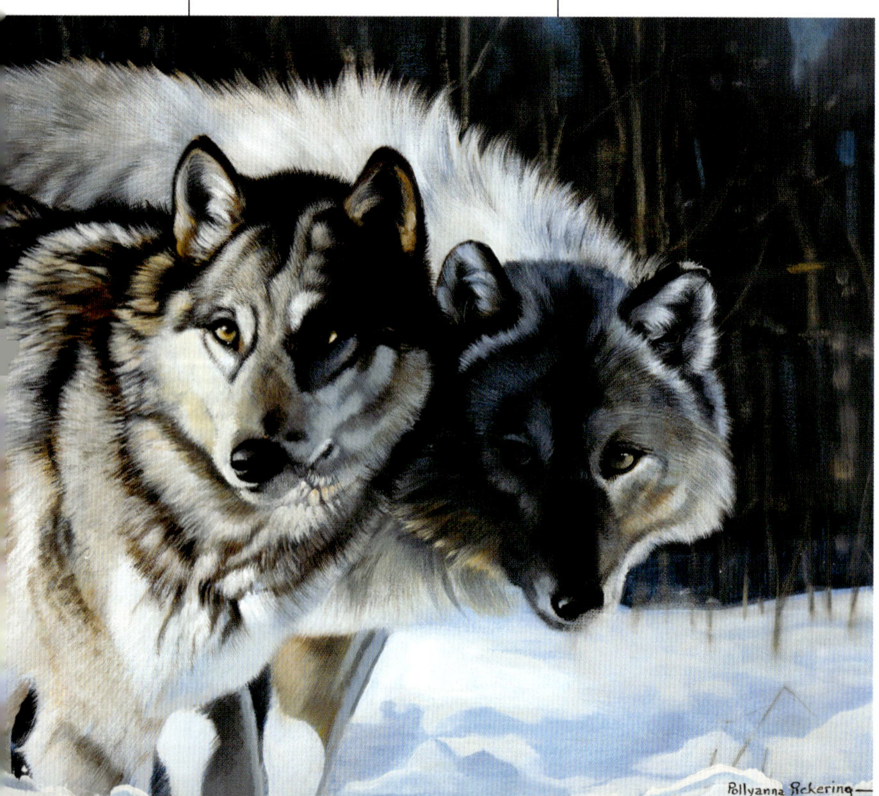

Day Thirteen

Woken by the dawn chorus. Tempted to open the window and throw a bucket of water over it.

Breakfast with Colin at 8.00 am – and then we meet Dan Marin, who worked with the Carpathian Large Carnivore Project (CLCP) for three years. Dan has lived near Zarnesti all his life. We discover that he left school at fourteen and started work on a production line in an ammunition factory where he worked for sixteen years, assembling bomb and weapon pieces. When the factory closed following the revolution he took a wildlife-guiding course and has been working in the Piatra Craiului National Park area and the Danube Delta ever since. When the communist regime collapsed all the major industries in this area – including a fire-brick factory, a paper mill and the ammunitions factory where Dan worked – went bankrupt. This was in some ways good news for the environment – the air and water are clean, the plant and animal life abundant – but the unemployment rate remains very high and life very difficult for many people who found themselves without work almost overnight.

We grab our bags and cameras and head out. Our horse and cart is AWOL – apparently yesterday's driver went to work last night (in a distillery?) and has not yet returned. So Colin drives the three of us up to the bottom of our trail in the mountains. We head off following Dan up a clearly signed easy path through beautiful scenery, already keeping our eyes open for any wildlife. The first thing Dan points out to us is a clump of edelweiss growing on the side of a cliff – which he has spotted through his binoculars.

We pass a beautiful church, high on the wooded slope. In this corner of Romania religious paintings are not confined to the interior of churches and monasteries, but also adorn the outside walls. We have seen many of these intensely coloured, almost psychedelic buildings, as we have travelled, but are a little surprised to see this one so far away from any towns or villages. In fact this 'ancient' church is brand new. As we approach we see that the scale is slightly 'off' – it is smaller than would be expected – and workers are still erecting

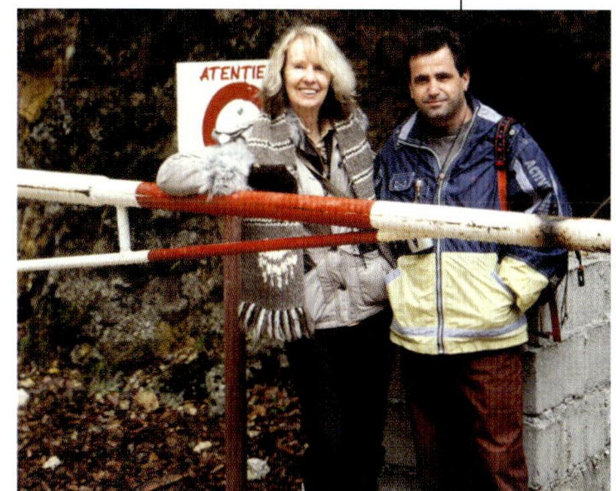

Dan Marin, who worked with the Carpathian Large Carnivore Project (CLCP) for three years

The mountain trail

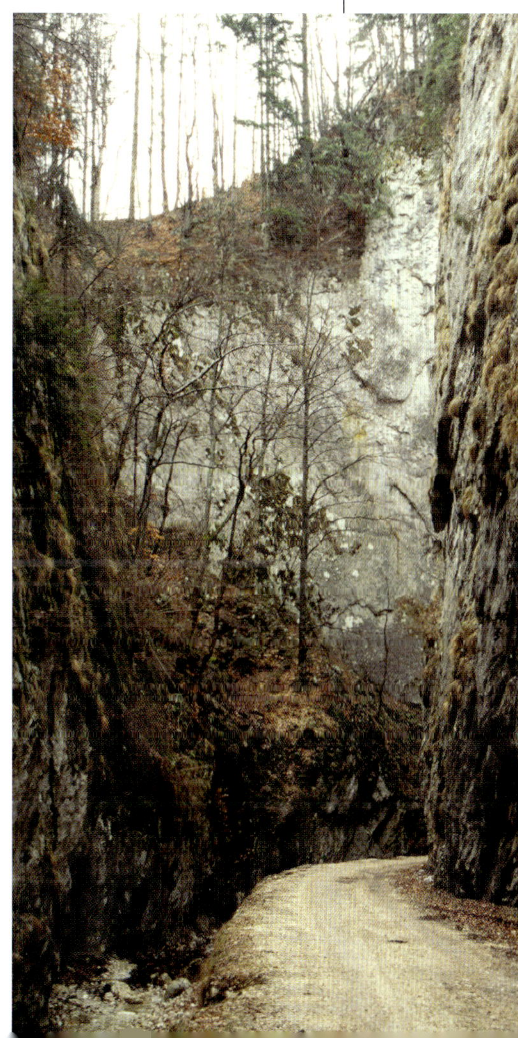

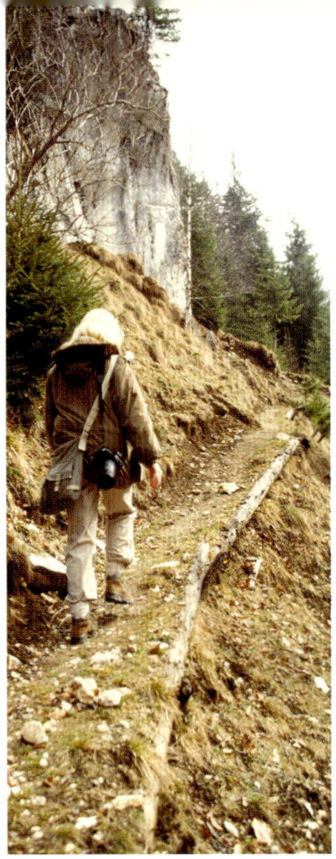

A TRAIL USED BY THE SHEPHERDS

'Our boots are millimetres from a terrifyingly precipitous drop'

a thin plaster-board wall. The church is being constructed for a movie which will begin filming in this location over the coming week. The unspoilt landscape, fairy-tale villages and turreted medieval castles – along with the availability of temporary labour created by the high unemployment in this region – have made Transylvania an increasingly popular destination for film makers.

All is going well with our walk, until Dan points out a narrow groove in the steep wooded mountainside, and tells us that this is a trail used by the shepherds. We nod and smile – until he tells us that it is our route as well. We follow him on to the barely discernible track through the undergrowth – which unlike the broad easy path we have left winds sharply up the mountainside – and begin to trek upwards on an ever increasing gradient. For one memorable stretch our boots are just millimetres from a terrifyingly precipitous drop down the cliff side. We continue onwards and upwards, pausing only to examine the wild flowers or sip from our water bottles. As we climb higher, Dan leads us 'off trail' on the edge of a steep valley which runs between two lovely clearly defined paths. I am looking at them longingly, but Dan dismisses them as being for 'tourists and walkers' – as presumably he has decided we don't fit into either category we find ourselves clambering over fallen branches and rocks, while ducking low-hanging branches and trying not to twist an ankle in any of the holes and burrows of the uneven territory. However, Dan is quite right to bring us to this area – we find bear tracks in the snow and claw marks on the trees – and to our great excitement fresh beech marten droppings, although on this occasion they remain elusive. As we are scrambling up one especially difficult part of the 'non trail' I have fallen a little way behind – and am sure I hear a low growl coming from the dense cluster of rocks and fallen trees above me. When I catch up with Pollyanna and Dan we all stand quietly and listen out – but don't hear another sound. I am actually quite relieved that a bear doesn't appear – although we are in almost no danger from the packs of wolves which live in the forest, bears can be quite dangerous. Although I would of course love to see a wild bear, I feel a little

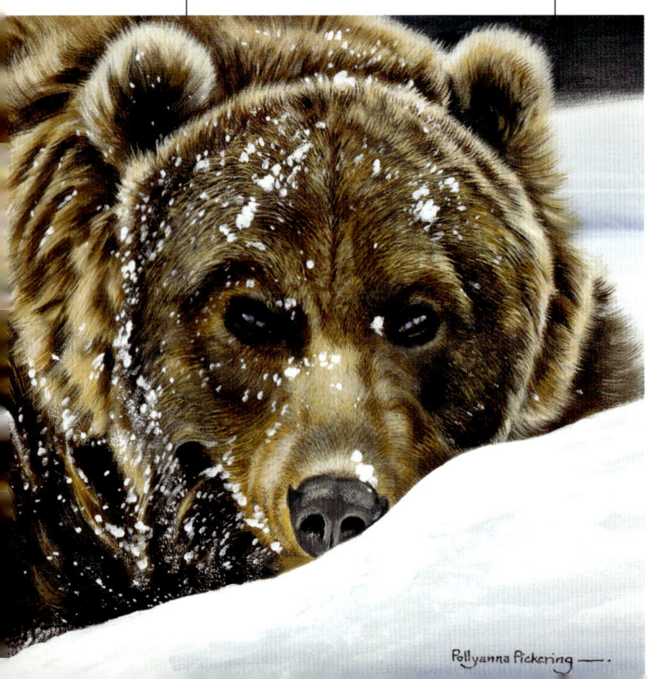

HALF OF EUROPE'S BROWN BEARS LIVE IN ROMANIA

vulnerable on foot in such dense forest.

Romania is home to about half of Europe's brown bear population. Attacks on humans are rare but they do occur. In 2007 a group of American hikers was attacked in these mountains – a thirty-one-year-old woman was mauled to death and two were injured. The following year a German tourist was badly injured when a bear slashed open his tent in search of food. Bears have learnt that tents pitched in the mountains can contain food – and the local population never sleep out under canvas, preferring the protection of huts or stone shelters. Most attacks on humans are caused by people surprising a bear which then reacts aggressively through fear – or because people either provoke them, or even – incredibly – attempt to hand feed them. Dan advises us that in this area the best course of action is to lie down and pretend to be dead, as there is then a good chance that the bear will just pass by. The biggest mistake is to follow our natural instinct to run. Bears are much, much faster.

This undisturbed area may make for tough hiking, but it is perfect habitat for

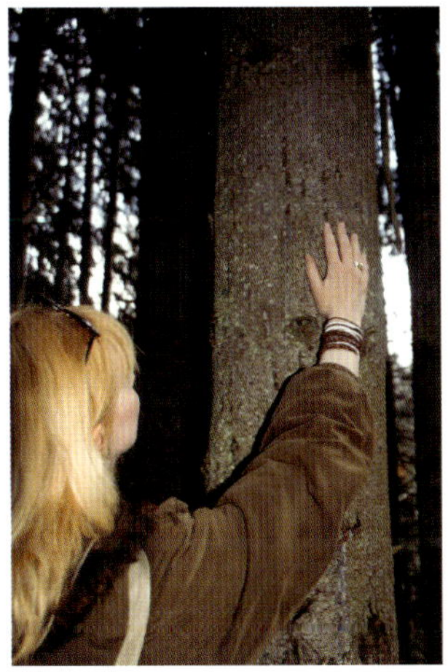

Bear claw marks are clearly visible on the tree trunks

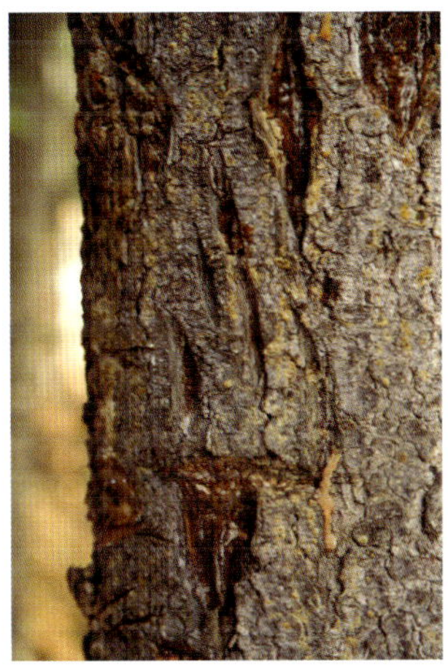

'I hear a low growl coming from the dense cluster of rocks'

HOW TO SURVIVE A BEAR ATTACK

1 Carry bear pepper spray

Experts recommend that hikers in bear country carry with them a highly concentrated capsaicin spray that creates a large cloud. Although it doesn't cause any permanent injury, it will usually stop a bear in its tracks.

2 Don't run

When you run, the bear thinks you're prey and will continue chasing you. Stand your ground. Make a noise and wave your arms around to make yourself appear as large as possible. Don't imagine you can out run a bear – they can reach speeds of 30 mph. Never block a bear's possible escape route – and never place yourself between a mother and her cubs.

3 Drop to the ground

Put your face down and cover the back of your neck with your hands, elbows spread out to prevent the bear from rolling you over. If you don't have pepper spray or the bear continues to charge even after the spray, this is your best remaining defence.

4 Play dead

Brown bears will usually stop attacking when they feel there's no longer a threat. Once the bear has finished tossing you around and leaves, continue to play dead. In North America, grizzly bears are known for waiting around to see if their victim will get back up!

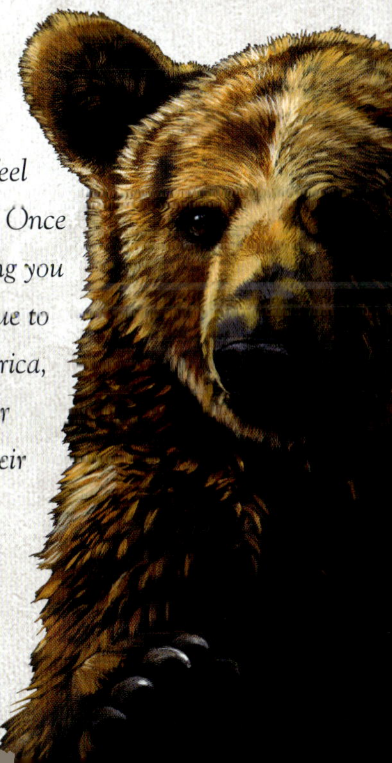

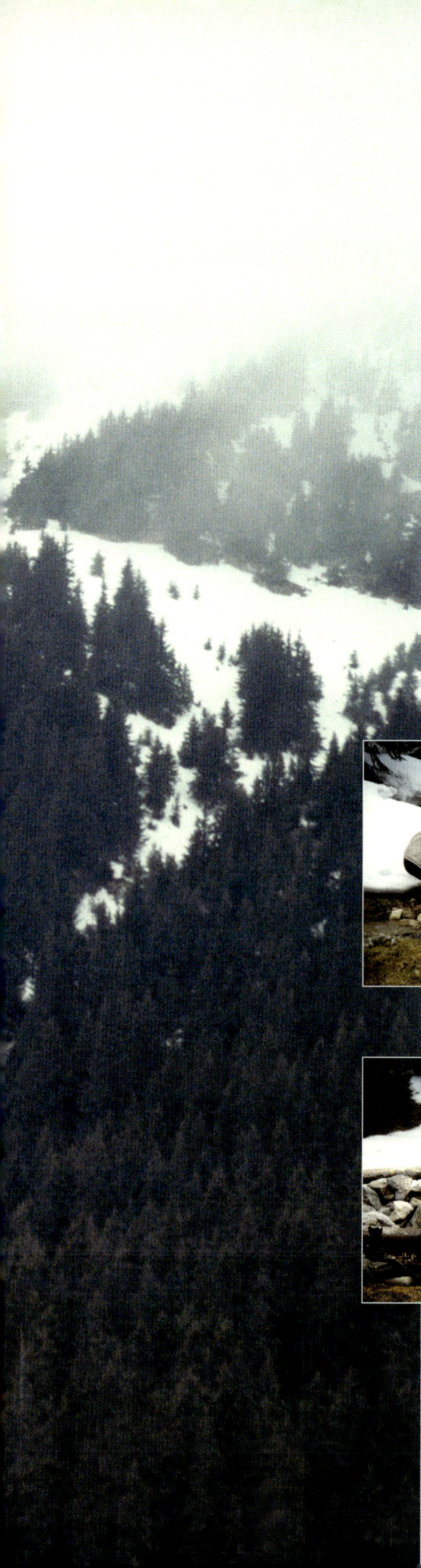

lynx, bears and, of course, wolves. We pause in a small clearing and Pollyanna fills her water bottle from a mountain stream – up here the water is clear and unpolluted.

Eventually we emerge from the dense forest into an Alpine meadow. For the past couple of hours we have only been able to see a short distance into the heavily wooded undergrowth. Suddenly we have an uninterrupted view across the mountains – and I am amazed to find we have emerged on a level with the snow-covered, mist-shrouded mountain peaks which until now we have been gazing up at from the valleys below. We have been crunching through (and sinking knee deep into) snow for some time and also battling our way around and climbing over fallen trees and debris brought down by the heavy snowfall – but I hadn't appreciated just how far above the snow line we had now climbed.

We stop in the meadow to enjoy our sandwiches and fruit with these beautiful views spreading in front of us. Pollyanna also sits and outlines the scene in her sketch folder. She especially enjoys painting snowscapes – distant snowy peaks behind white patches of snow on the lower slopes create abstract patterns which will inspire the composition of her larger works back in the studio.

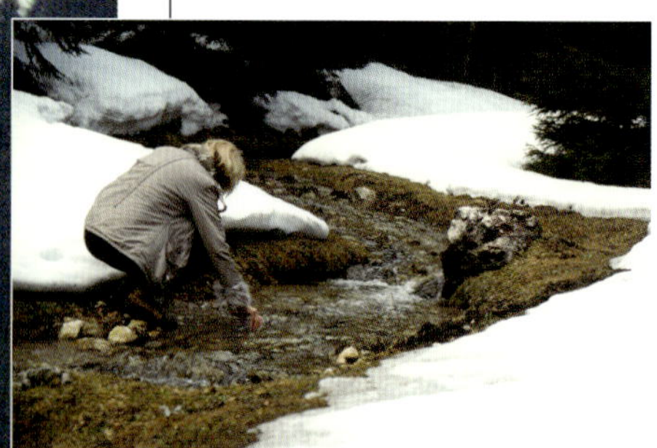

POLLYANNA FILLS HER WATER BOTTLE FROM A MOUNTAIN STREAM

In the spring the shepherds will bring their flocks back to the high pastures and will live in the tiny shepherd's huts where they also prepare and store the sheep's cheese throughout the summer. Each family typically owns between five to ten sheep. Throughout this season the head shepherd oversees an intense period of milking during which each animal is milked three times a day. Each family then receives six kilos of a semi-soft cheese wrapped in bark as well as one and a half kilos of urdu – similar to ricotta in texture – all of which is theirs to eat or sell.

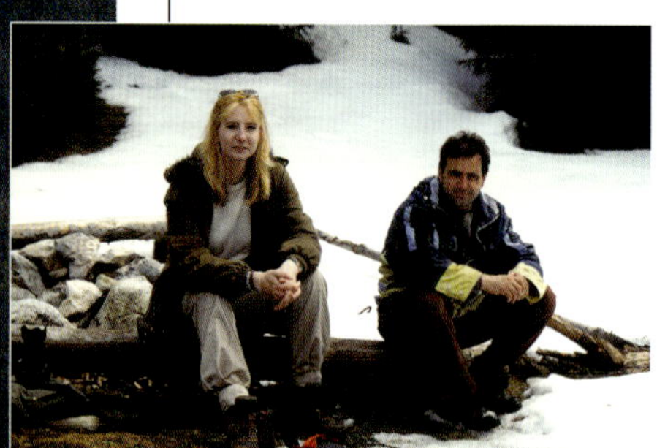

ANNA-LOUISE AND DAN.
WE HAVE EMERGED ON A LEVEL WITH THE SNOW-COVERED, MIST-SHROUDED MOUNTAIN PEAKS

Although it is too early in the year for the sheep to return to these meadows, the first Alpine flowers are beginning to

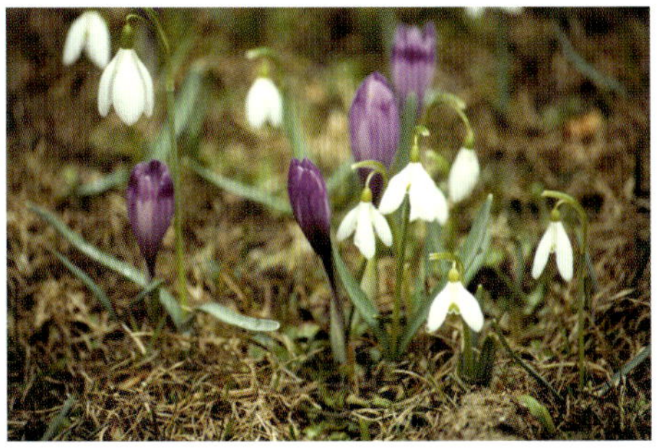

Snowdrops and wild purple crocus

show through – delicate snowdrops mingling with wild purple crocus. With these first signs of spring the shepherds will start to think about moving their flocks back to higher pastures, and we meet one of the shepherds who has come to check the condition of his accommodation following the long winter. The little wooden hut will provide shelter for the shepherd and his family throughout the summer – but the winter storms have caused quite a bit of damage to the structure and it is time to make repairs. With Dan translating for us, we learn that a tiny annexe on the side of the hut will be the children's bedroom – several siblings will share what appears to be an impossibly tiny space.

Wolves are a very real threat to the sheep here and the shepherds have to be constantly vigilant. In his memoirs of this region novelist Gregor Von Rezzori wrote: '*The poetic gentleness of the flowery slopes was all too deceptive in obscuring the wildness of the deep forests*'. Sitting in the sunshine it is easy to forget the wolves watching from the shadows.

The CLCP has worked closely with the shepherds to discover humane ways in which they can protect their livestock from the wolves. Wolves have been a protected species in Romania since 1996 and it is illegal to kill them – but some shepherds, keen to protect their flocks from these hungry predators, still do so. The CLCP provides education and training to farmers in methods of preventing losses to wolves including using portable electric fencing which surround the sheep pens at night and native mountain shepherd dogs to guard the livestock.

Dan tells us how ingenious the dogs are. In this region some of the shepherds tie long sticks under the necks of the dogs so that they cannot run fast enough to chase prey – but

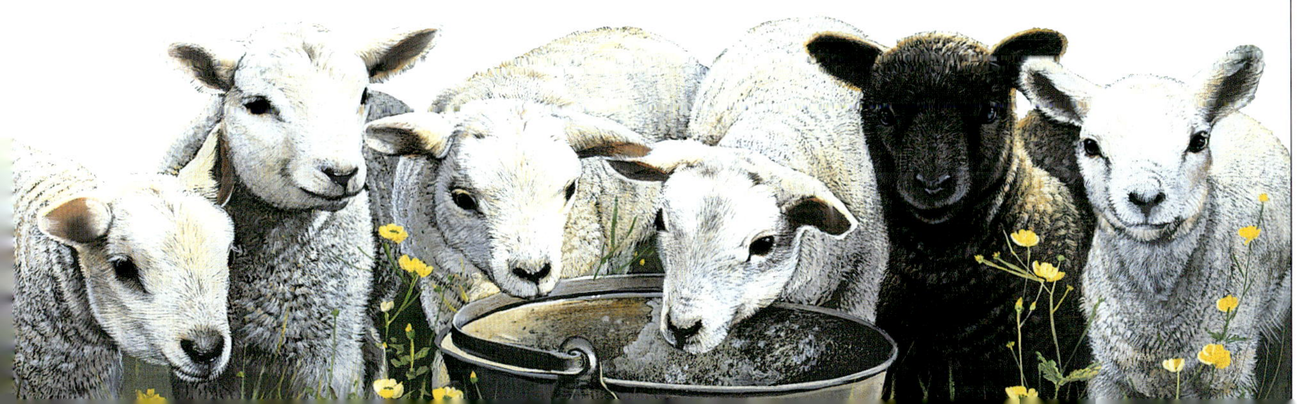

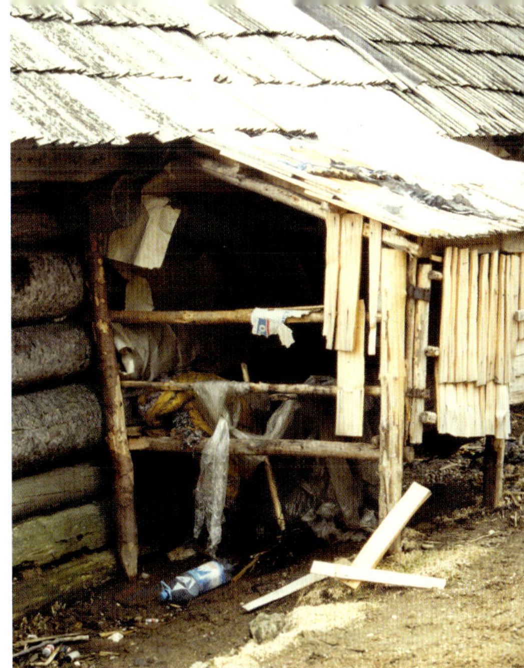

A tiny annexe on the side of the hut will be the children's bedroom

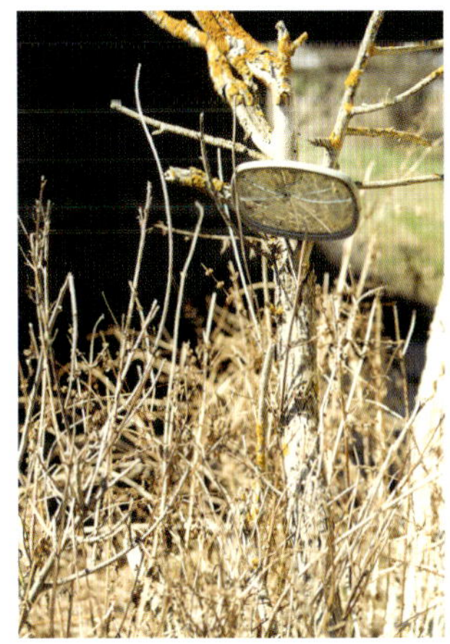

Above: shepherd shaving mirror
Below: fresh water supply

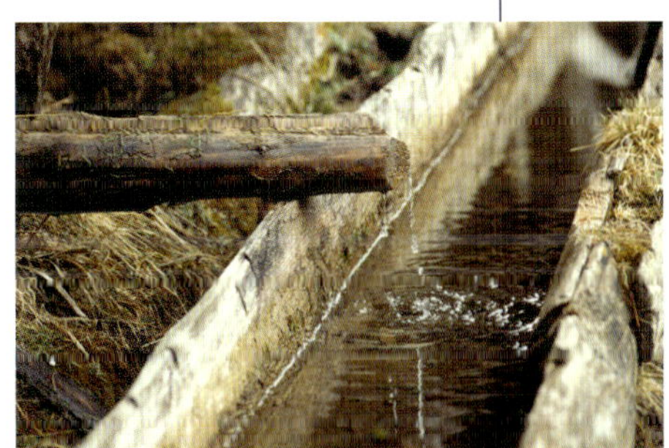

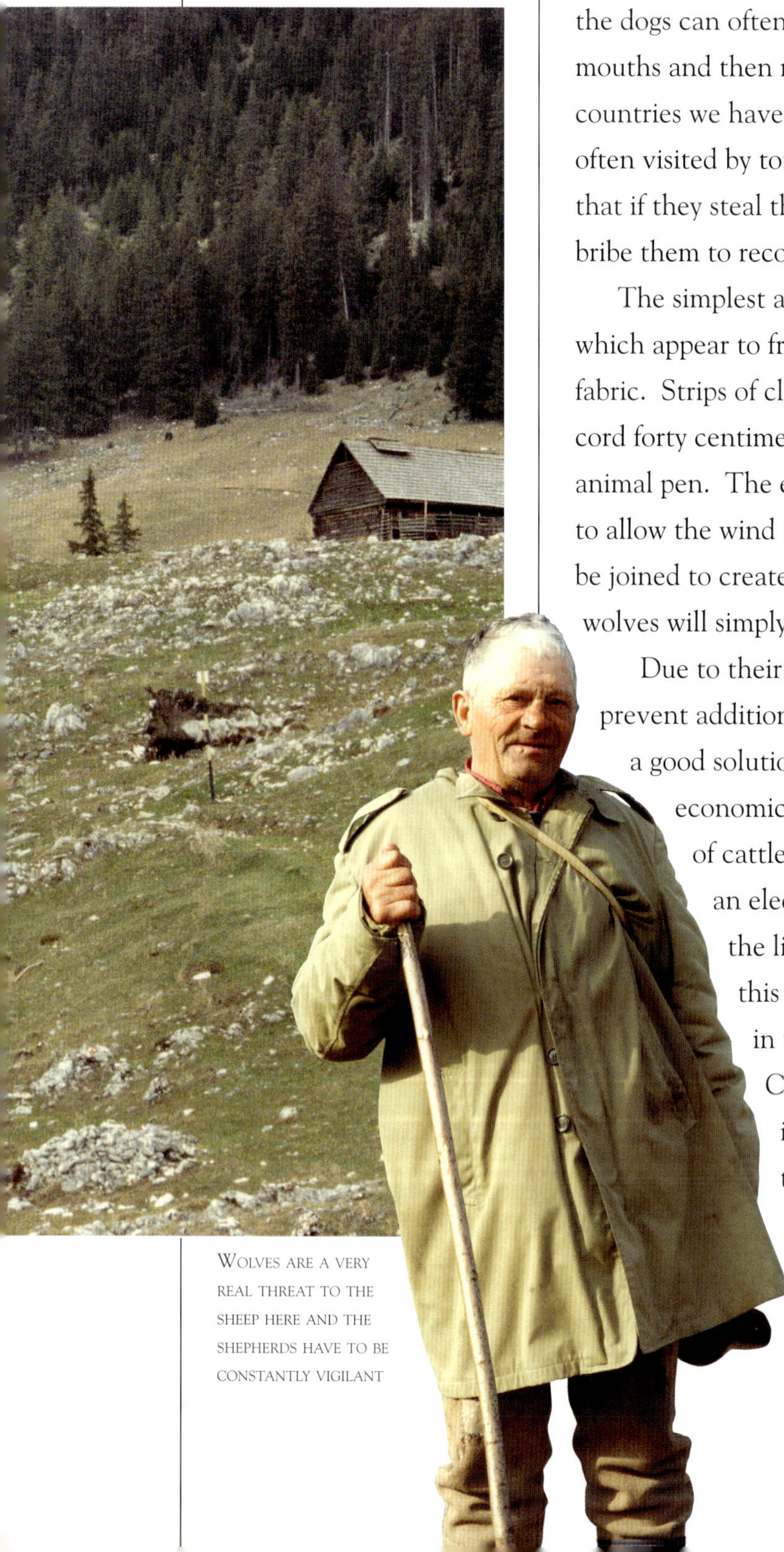

Wolves are a very real threat to the sheep here and the shepherds have to be constantly vigilant

the dogs can often be observed twisting around until they can hold the sticks in their mouths and then run freely. We are reminded of the cleverness of animals in many of the countries we have visited – like the monkeys living in a ruined temple in India which is often visited by tourists. The monkeys used to run up and steal food, but have now realized that if they steal the tourists' cameras instead they receive twice as much food, as people bribe them to recover their cameras!

The simplest and cheapest method of livestock protection involves the use of 'fladry', which appear to frighten or at least deter wolves. These are flags, made from thin, red fabric. Strips of cloth fifty-five centimetres long and ten centimetres wide are sewn on to a cord forty centimetres apart. The fladry are then installed on sticks around the pasture or animal pen. The ends of the red strips hang about twenty centimetres above the ground to allow the wind to move them constantly. The beginning and end of the fladry must be joined to create a rectangle or circle. If they are placed in a straight line along a forest wolves will simply follow the line to find its end.

Due to their ease of installation, and relatively low cost, fladry are very useful to prevent additional damage when a wolf attack has just occurred. Fladry can also be a good solution when a flock consists of just a few sheep or goats and there is no economic reason to introduce a guard dog. This method is also used for big herds of cattle or large sheep flocks, and is especially effective when combined with an electric fence and/or dogs. Fladry are likely to be more efficient when the livestock pen is moved every few days to a new part of the pasture, as this prevents the wolves from getting used to it. This practice is common in the Carpathian Mountains and helps to protect and fertilize the grass. Consequently this region has the largest variety of wild flowers in Europe, including many species of wild orchids which have vanished on the rest of the continent. Christoph Promberger, director of the CLCP, is convinced that 'People and wolves can live together,' and is confident that the

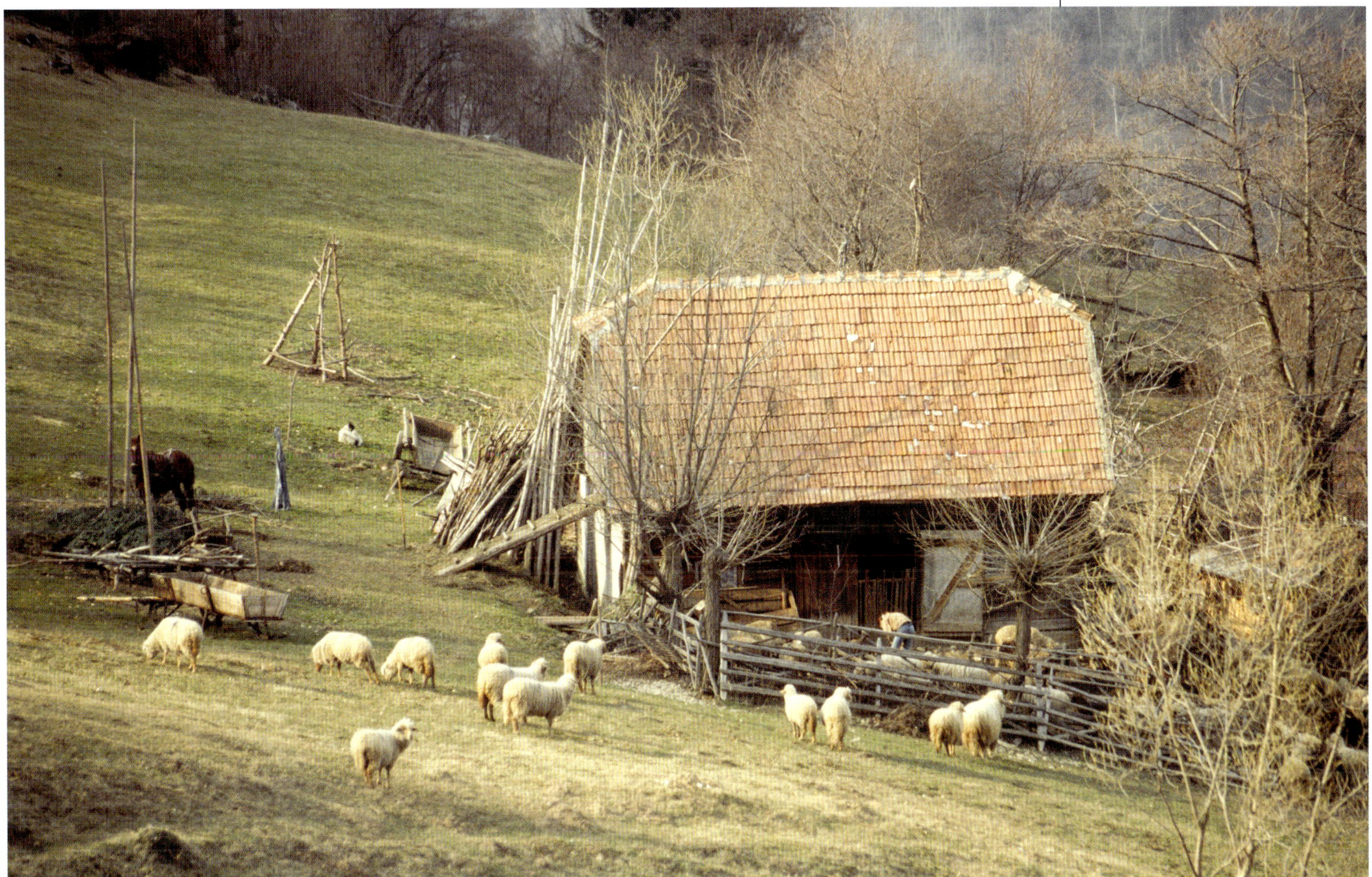

SHEEP GRAZE CLOSE TO THE SAFETY OF THE BARN

work of his organisation has helped this to happen: 'When I first came here, a lot of people felt like they were living in the Middle Ages because there were so many wolves running around. They felt they had to change this in order to join the West. I said "No! You have to keep them here." There was no sense of pride about this. Now people see wolves as a heritage that Western countries have lost.'

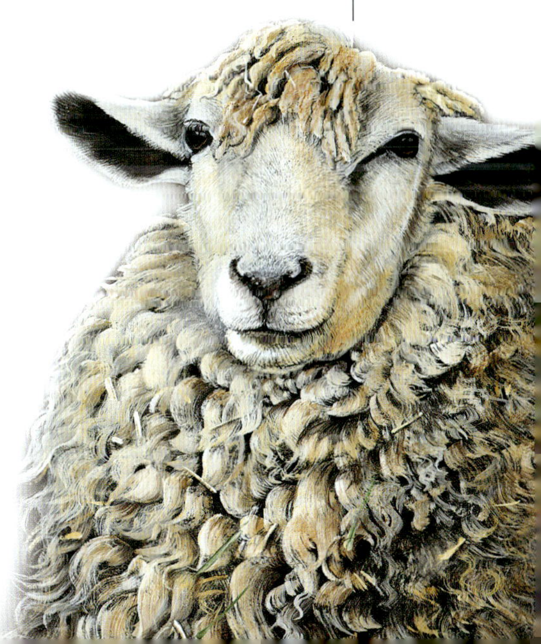

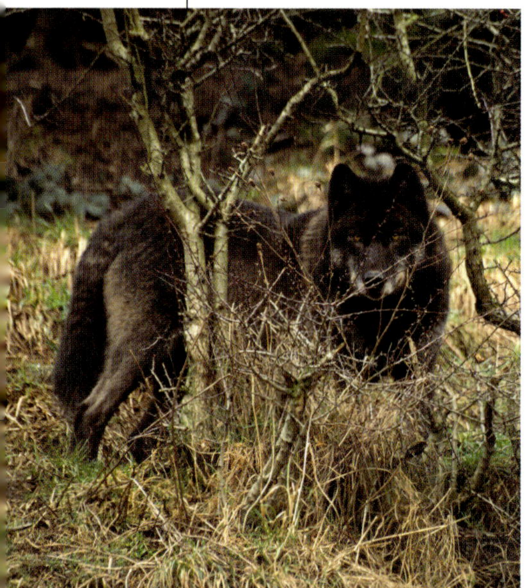

WOLVES WILL WATCH THE FLOCK OF SHEEP FOR SEVERAL DAYS BEFORE MAKING THEIR MOVE

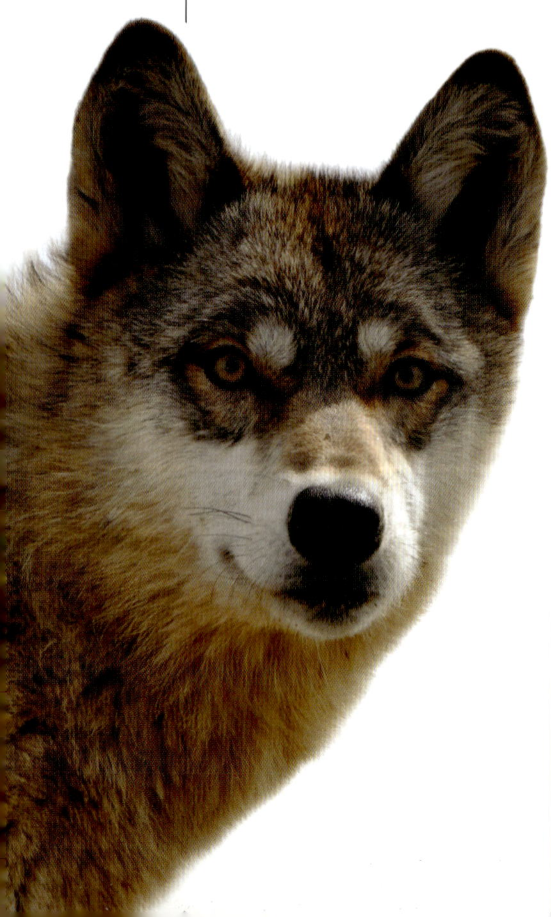

WOLVES AND PREDATION

While sketching Crai and Poiana Pollyanna could see the intelligence in their yellow eyes. And it is this deep intelligence which makes wolves such effective hunters. Hunting as a pack allows them to take large animals – deer in the Transylvanian forests, elk and moose in North America. By killing large animals they need to hunt less often than if they were feeding on small game – giving a far better ratio of energy expended to calories consumed. They are not generally opportunistic hunters – rather they will often observe their prey for days before attacking. They will watch a herd of deer, observing signs which mark an animal as suitable prey – often a weakness such as a limp. The oldest and youngest animals are usually the most vulnerable to attack. The shepherds tending their flocks in the pastures know that a wolf would often be present for several days watching the sheep.

It is always the alpha female who selects the prey and the pack members feed on different parts of the animal according to rank and need. The alpha pair will eat the rich high-quality meat of the internal organs, along with some muscle meat from the legs or rump. Beta animals will feed almost exclusively on this meat, while the lower ranking members of the pack eat the remaining meat along with the vegetable matter from the stomachs of their herbivorous prey.

Shaun Ellis, the wolf expert who lived as part of a pack of wild wolves in North America, believes that the wolves have an amazing ability to eat precisely the food they need for the individual requirements of their bodies. He once observed the pack feeding on a putrid carcass 'so rotten it was practically liquefied' – the wolves had worms and knew that eating the tainted meat would flush them out of their systems.

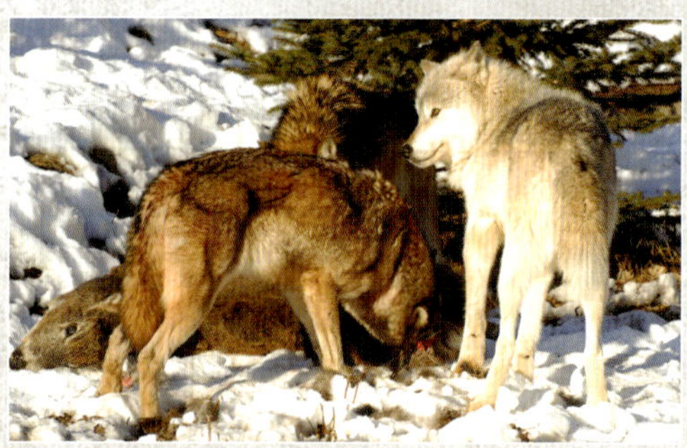

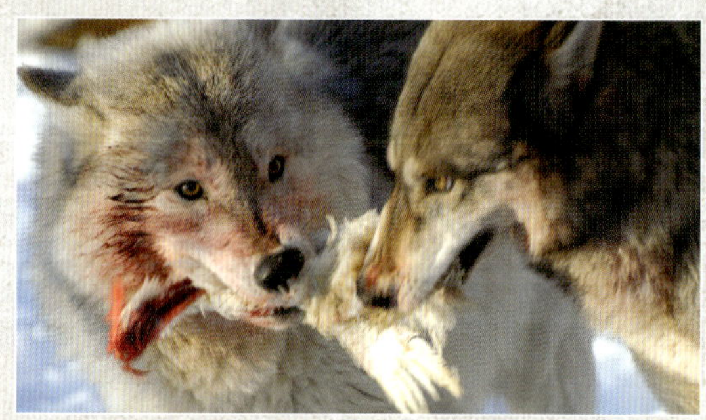

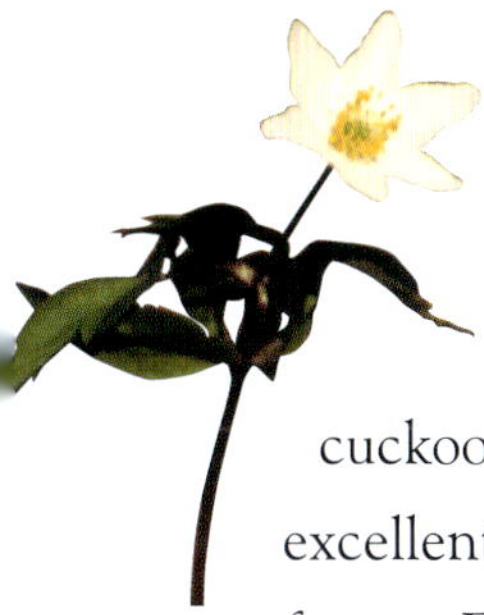

Finally we begin to head down – initially still off trail – until the snow and ice make the steep slope impassable and we cut back to rejoin a marked path. We have spotted a variety of bird life through the day – black woodpeckers, a cuckoo, a ring ouzel and just one mammal, a tiny vole. However, as well as being an excellent wildlife spotter, Dan is very knowledgeable about the flora and fauna of the beech forest. Forestry in this area is strictly controlled. He explains that loggers use a natural regeneration system, cutting older trees in small patches, while leaving younger trees to grow over a thirty-year period. The forest floor is rich with plants that Transylvanians still use today as traditional medicine. He points to plantain leaves, which are wrapped around wounds as bandages to speed the healing process, and colt's foot which is turned into an infusion to help cure lung diseases. He tells us that foxglove helps the heart pump harder and that the elder leaf makes a fizzy juice that acts as an anti-inflammatory for arthritis and rheumatism.

Eventually we emerge on to the road at the base of the mountains. Dan dashes off to find our transport. Just as we reach this point, heavy drops of rain start to fall. We have been very lucky not to get completely drenched during our hike. As we are nearly back at the base of the mountain, Dan decides to share with us the fact that the weather forecast for the day was extremely bad – even threatening thunderstorms. He tells a story of taking some Dutch ornithologists up into the mountains when a storm broke directly overhead; laughing merrily he describes huge trees crashing to the ground all around them. They found a deep creek and ran back down until they reached the road we have now joined. We find a small seat with a roof which appears to have been designed just for us to wait until our horse and cart arrive. This evening we have two horses, one of which is very skittish. It bolts at the sight of a car, another cart, a child on a bicycle, a plough, a blade of grass blowing at the side of the road. Somehow we cling on and make it back to pension Elena, a little soggy, but otherwise unharmed. We retreat to the room to change into dry and less muddy clothing.

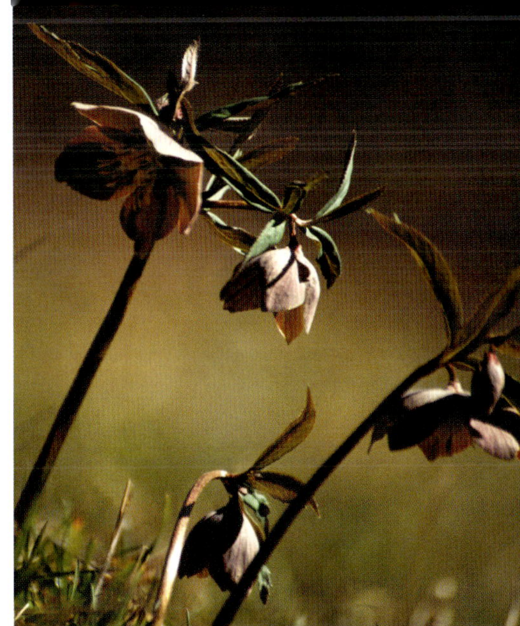

Hellebores

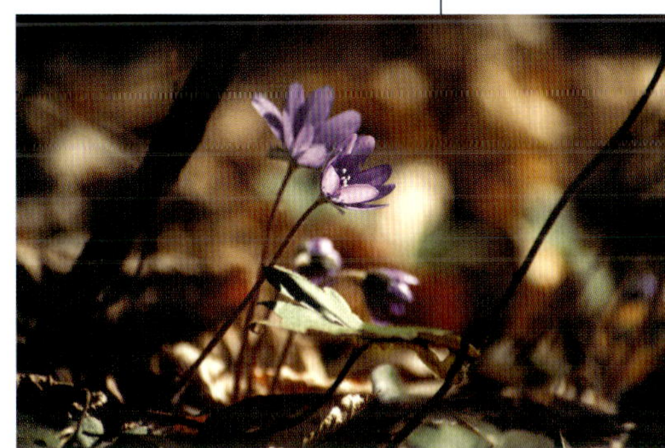

Liverwort

Wood anemone

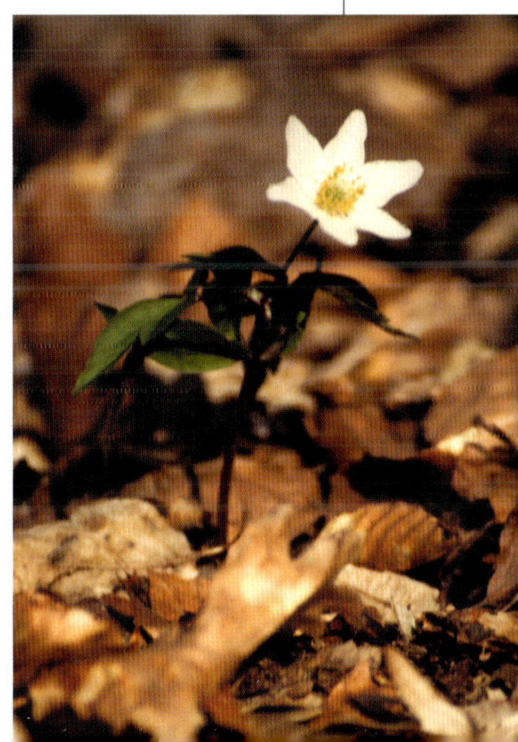

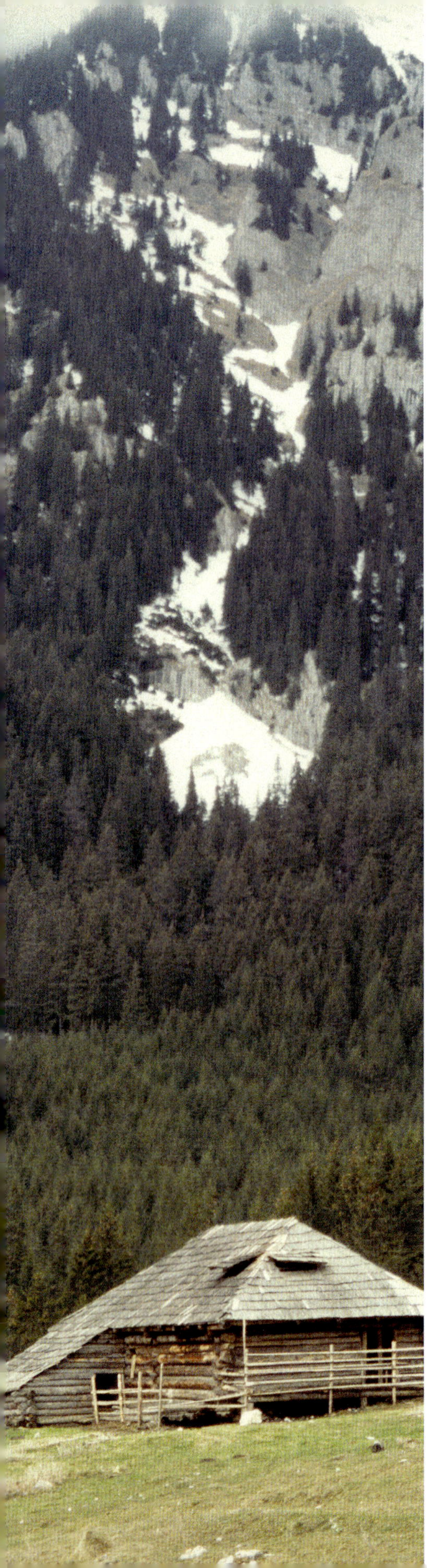

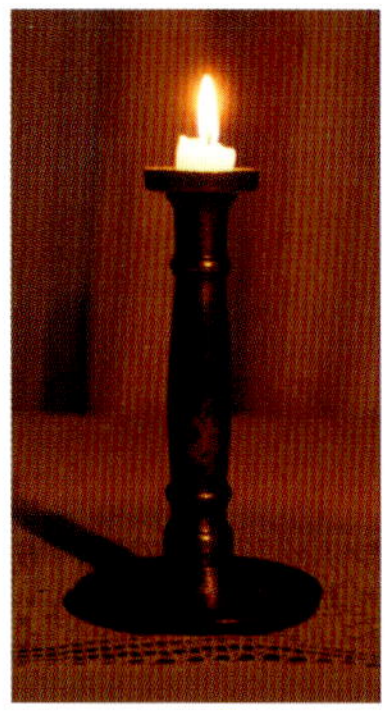

I venture down to test out some of my phrase-book Romanian in the kitchen, by asking for water to make tea. *'Pot aveh apa pentru Chai?'* I stammer, waving a tea bag in the air to help convey my desire. A flask is quickly produced along with two cups, so either the Romanian or the mime has worked. However, when I return to the room to make tea we find the water is barely tepid. Note to self: learn the Romanian word for HOT. Nonetheless, I am English so cannot go too many hours without a cup of tea – which I sip while sitting at a little wooden desk to write my journal. The desk is in front of a window so when I look up from the page I am gazing across a quintessentially Romanian view of higgledy piggledy red roofs with snow-capped mountains rising behind.

After dinner Dan joins us to show a very interesting documentary about the conflict and balance between human populations and wild predators. When it is over he attempts to engage us in an in-depth intellectual conversation about ecotourism/conservation/government corruption – but after our day's hiking followed by Elena's delicious dinner the only thing we are sinking deeply into is sleep, and we regretfully make our excuses and crawl up to bed.

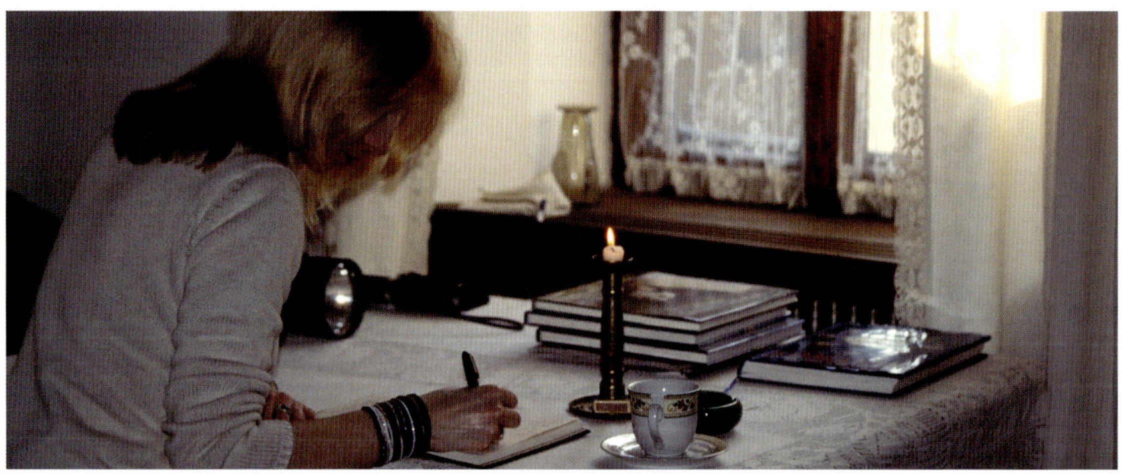

Anna-Louise writing her journal

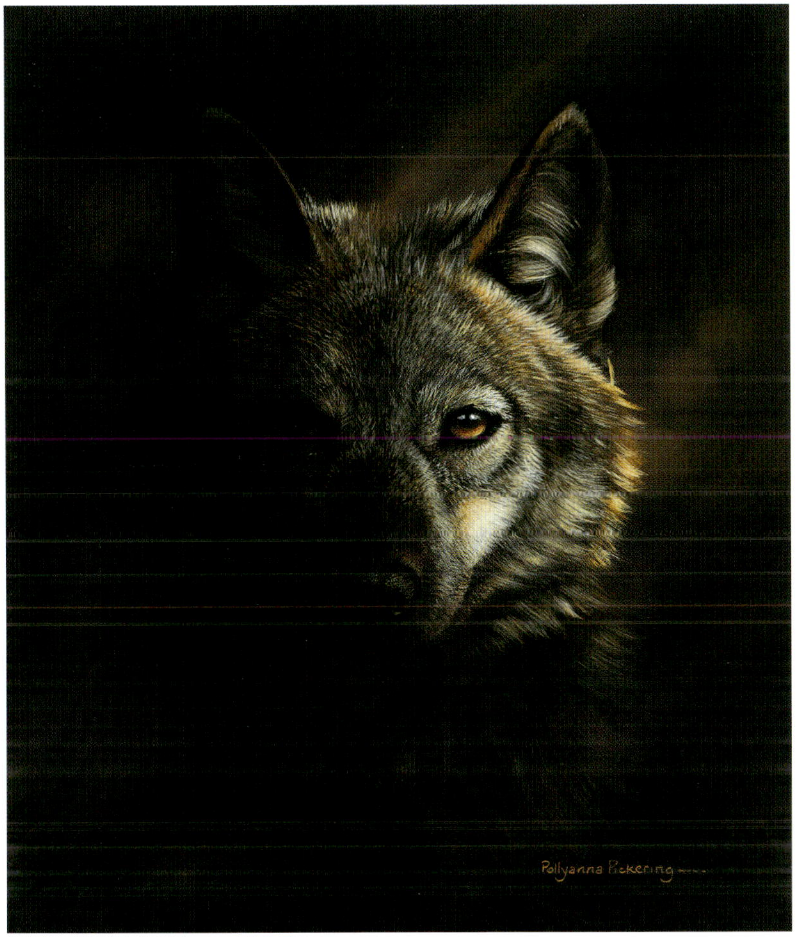

'Into the woods, each time you go,
There's more to learn of what you know.
Into the woods, but not too slow -
Into the woods, it's nearing midnight -
Into the woods to mind the wolf...'

STEPHEN SONDHEIM

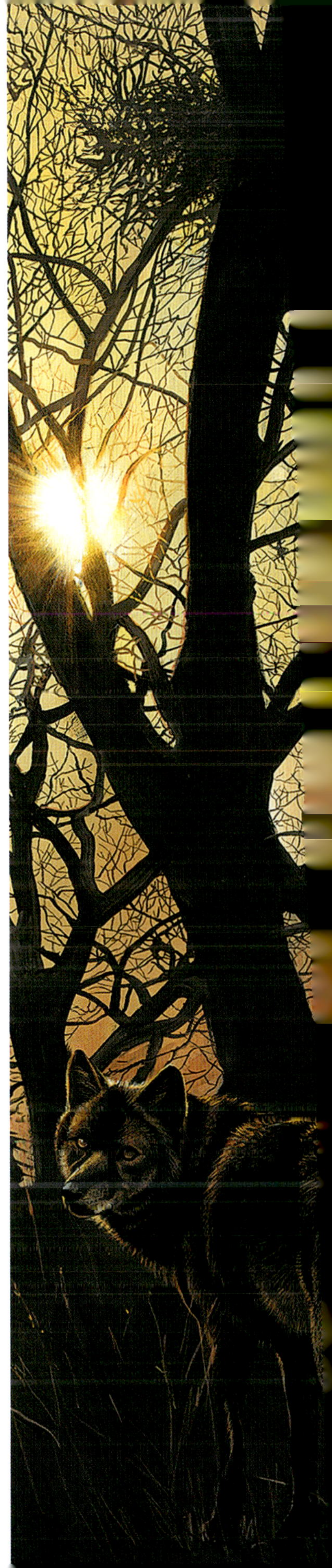

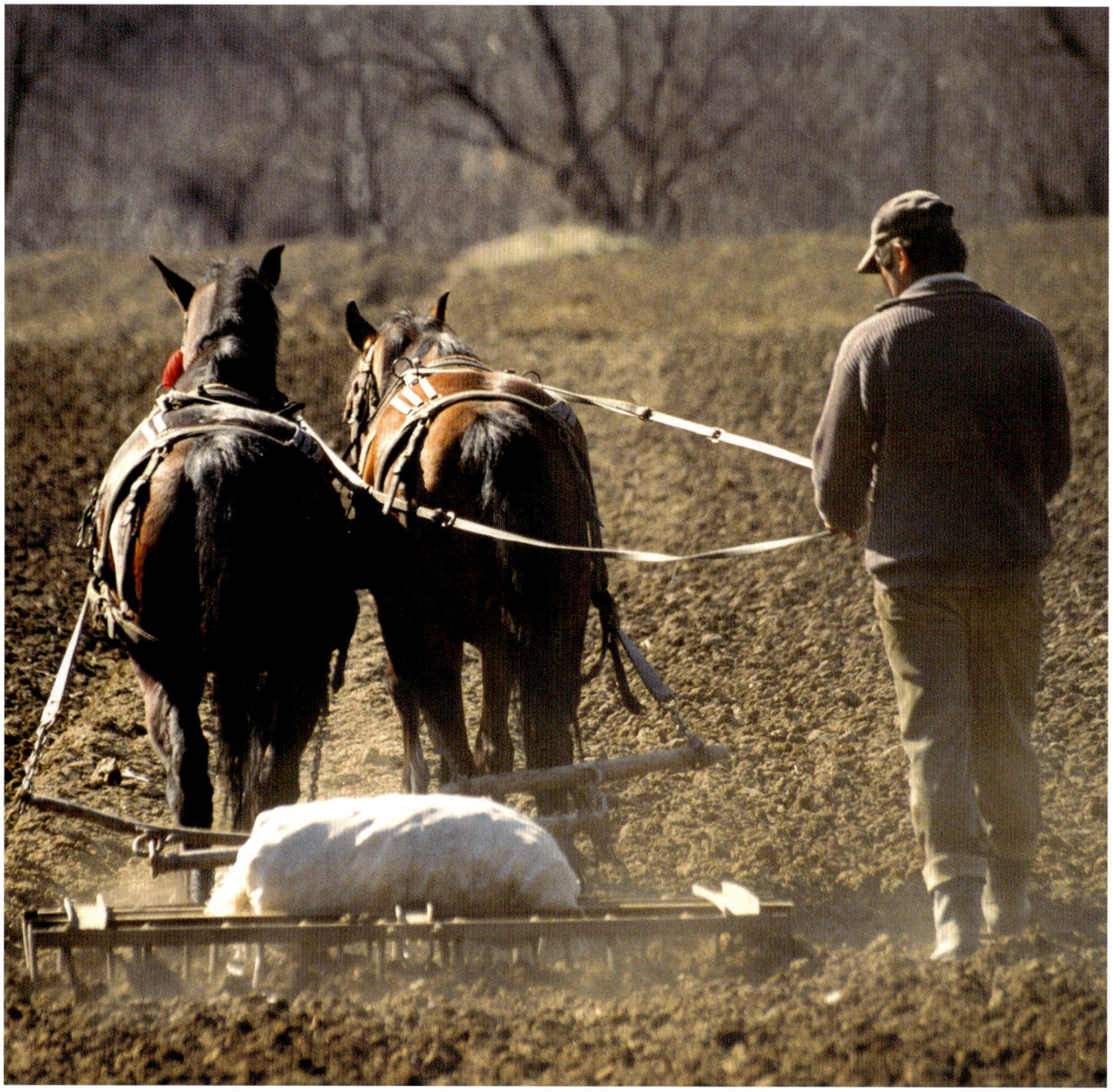

Chapter Seven

A Legendary Piper

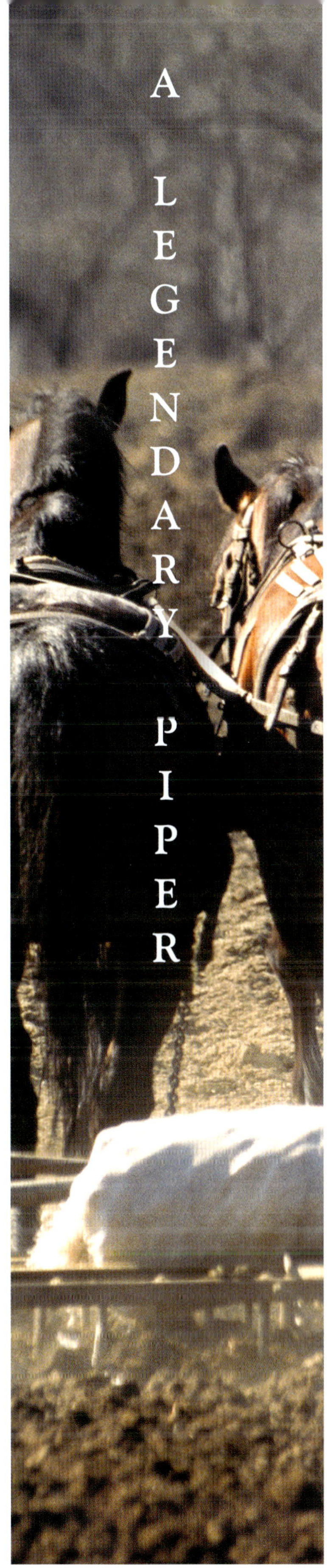

'The old man motioned me in with his right hand, with a courtly gesture saying in excellent English, but with a strange intonation: – "Welcome to my house! Enter freely and of your own will!"'

Jonathan Harker's Journal, *Dracula*

Ecotourism is a growing industry which encourages responsible travel to destinations where wildlife and cultural heritage are the main attractions. It aims as far as possible to be low impact and generally remains small scale. Successful projects not only educate the traveller, but provide funds for conservation and benefit the economy of local communities. Since Romania opened its doors to the outside world a few individuals have seen the potential for using small scale tourism to help preserve the relatively untouched mountainous habitat. Prior to our journey we made contact with one ecotourism pioneer, Count Tibor Kalnoky. (In my earliest phone calls with his office they appeared to be referring to him by the fabulously gothic name of Count Igor, so you can imagine my disappointment when the first emails arrived, and I discovered his first name was in fact Tibor!)

A two-year restoration project has lead to the opening of the first guest houses in the grounds of the castle and a few pioneering guests are now being welcomed. The project is still in its infancy and we are part of the 'test period', while the Count and his staff discover what visitors to the area are most interested in seeing and how best to accommodate them.

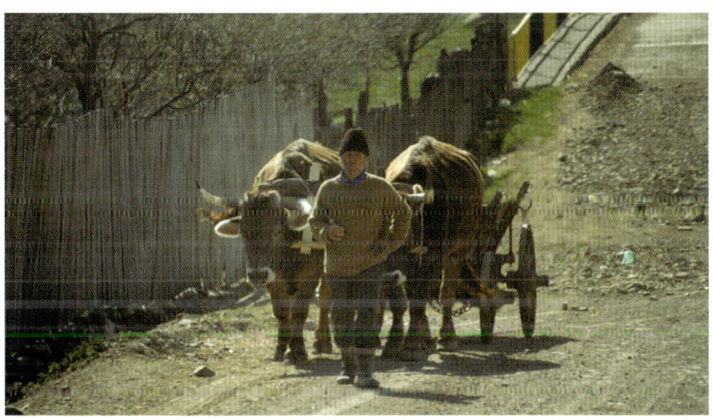

Ox drawn carts are common forms of transportation

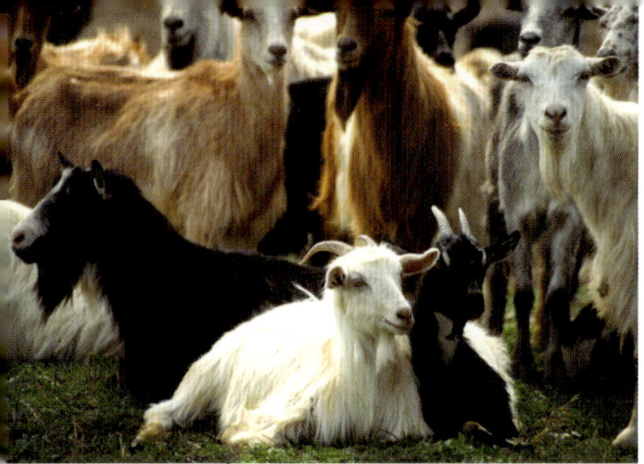

Goat herd

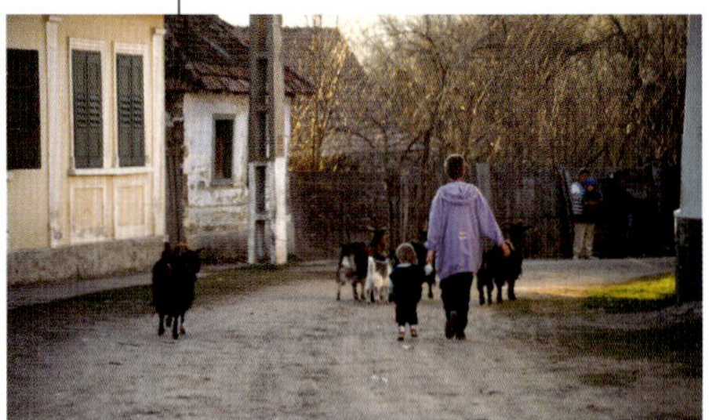

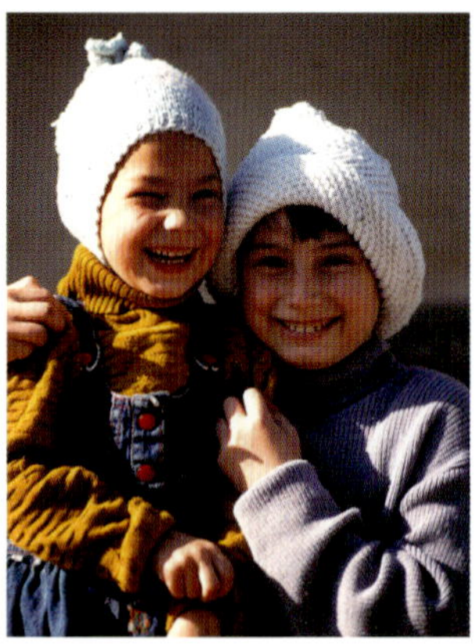

Scenes from a Saxon village

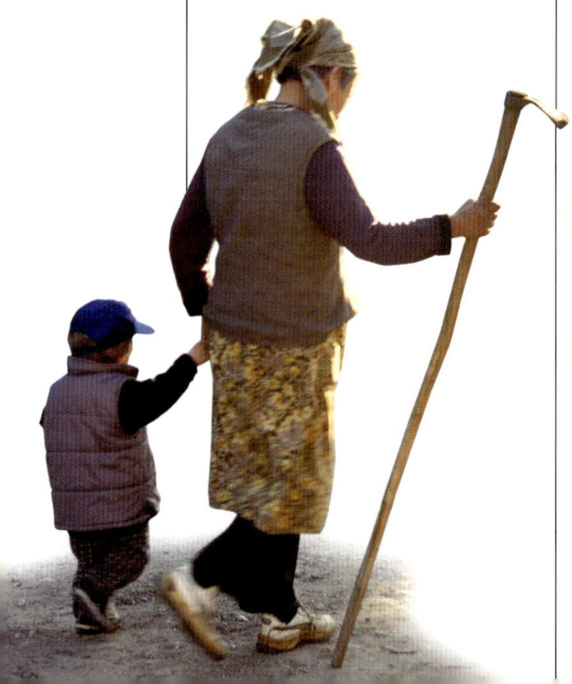

Day Fourteen

We bid farewell to Gigi and Elena and climb into the trusty Roving Romania Land Rover for our journey to Count Kalnoky's estate. I cannot help but notice we are now heading through a pass in the Carpathian Mountains to a Szekler village to stay with a Transylvanian Count we have not yet met. Sounds like it might make a good plot for a novel. I have, of course, purchased an old copy of Bram Stoker's infamous book to bring with me – and finally began reading the first chapter yesterday evening.

We are now very much in Saxon country. There is very little traffic on the roads, we are slowed only by cattle wandering into the road, and the many horse-drawn carts which clatter through the villages. The fields are dotted with pointed hay ricks, black-skirted women sweep the roadsides, while dark eyed children stare warily at us as we pass. The houses in the fairy-tale villages sit end-on to the street, painted in a rich variety of ochre, pink, green and blue, with fretted eaves and shingled roofs. They have cobbled courtyards and large vegetable patches. Large timber-frame barns stand at the rear end of the courtyards, and beyond these we can see more vegetable plots and orchards. Pear trees are planted on either side of the streets. Arable and pasture land extends from the backs of the villages up to the woodland behind. Cows, horses, pigs and chickens live in and out of a ramshackle assortment of barns, stables, sties and coops. As in much of medieval Europe, the Saxon communities divided their farming lands into strips. A family might own a number of separate strips, some distance apart from each other, on which different crops could be grown. Unlike the rest of Europe, this method of strip farming remains very much alive in these villages. Each village we pass through has its own distinct character, but there is a certain look which defines their 'Saxon-ness': fortified churches, the architecture of the houses, barns and streetscape, and the centuries-old agricultural methods that survive to this day.

The men in this region seem to share a fondness for luxuriant facial hair and many of the older women are dressed in traditional costume. We feel more than ever as though we are driving into a film set ready for a remake of *Dracula*.

We had read a little about the troubled history of Romania prior to our journey, and knew of Ceauşescu's campaign to 'systemize' the rural areas. This programme of urban planning consisted largely of the demolition and reconstruction of existing villages, towns, and cities – including the destruction of historic churches and monasteries. The original plan was to bring the advantages of the modern age to the Romanian countryside with hundreds of villages slated to become industrial centres.

As part of this plan smaller villages were deemed 'irrational' and listed for forced removal of the population and physical destruction. New buildings had to be at least two stories high, so people could not build small houses. Courtyards could not be larger than two hundred and fifty square meters and private agricultural plots were banned from within the villages, effectively destroying at a stroke the traditional self-sufficient lifestyle of the majority of the rural population. We are therefore relieved to find that this region at least has remained relatively unscathed from this state organized vandalism, which is now almost universally agreed to have been a disaster for Romania and indeed a major contributing factor to the violent fall of the Ceauşescu regime during the revolution. Here we barely see a modern building – and if any of the land was converted to large scale intensive farming, there is no sign of it now.

We finally arrive in Miklósvár as the sun is setting. It is an ancient village surrounding the Count's residence, and our first impression is that we are driving into a medieval painting. On either side of the dusty unpaved earth track we are following there are fields being farmed by hand, or with traditional horse-drawn ploughs. We imagine that the fields around our home in England would have looked like this in the Middle Ages, before the introduction of intensive farming. We learn later there isn't a single tractor in the village, and the majority of the residents survive on subsistence agriculture and a barter economy.

We drive straight to the main house where we are met by Robert Zeitz

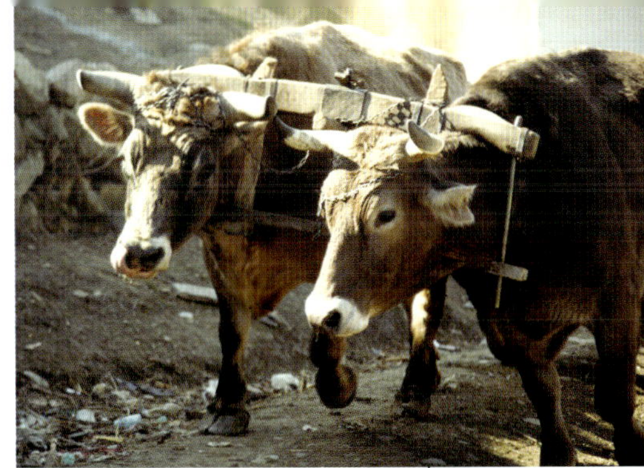

Farmers depend on oxen...

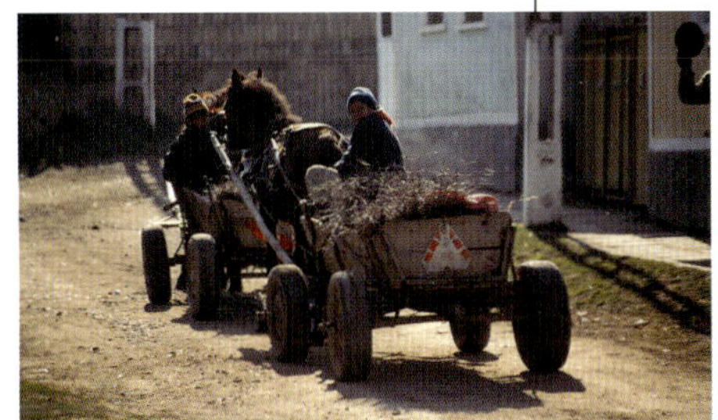

...and horses

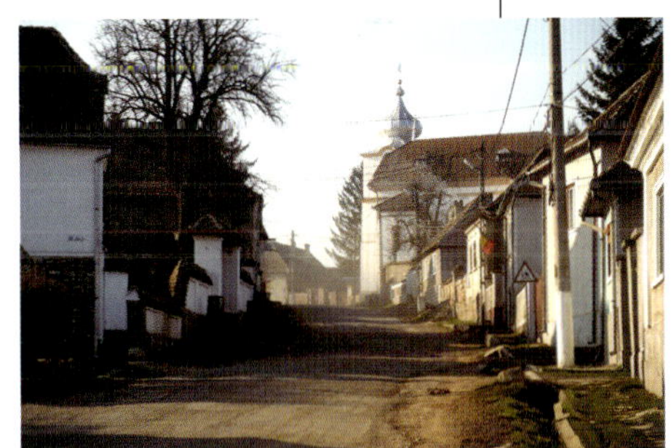

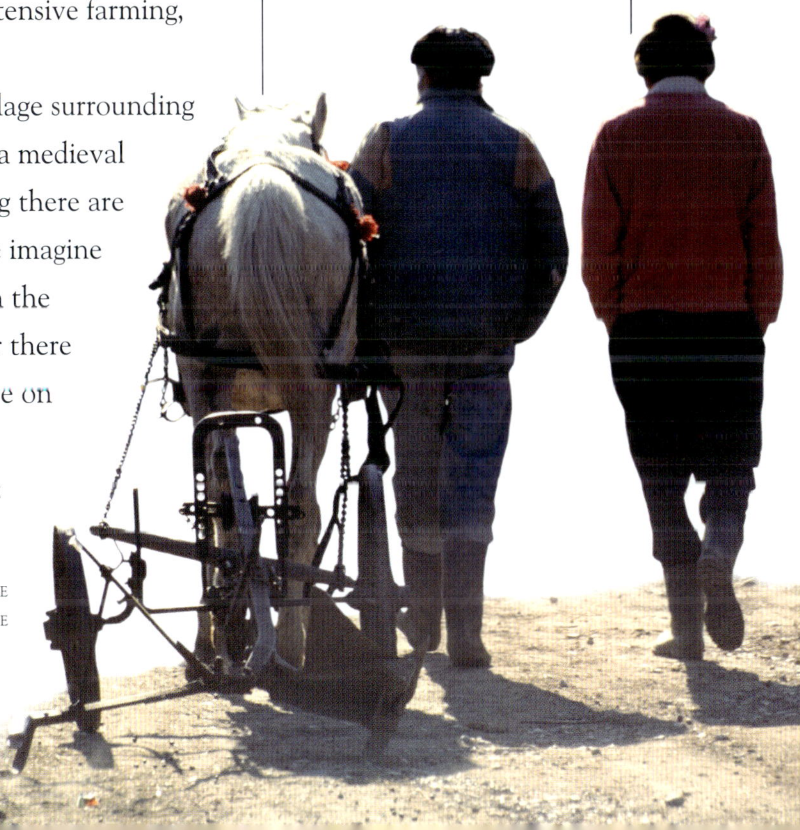

Horse-drawn ploughs are commonplace

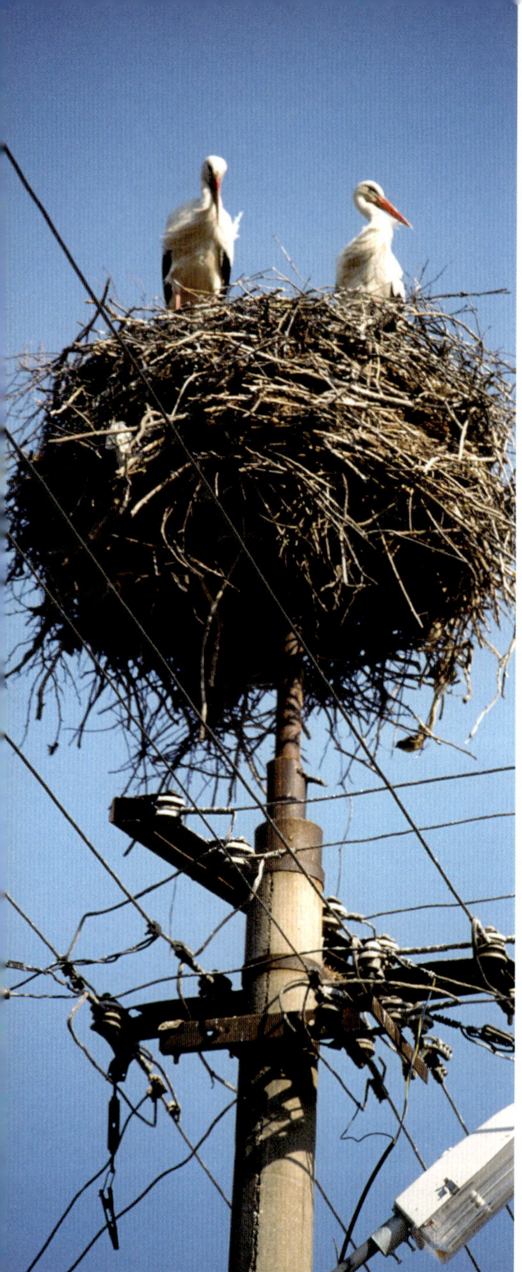

We are charmed by the untidy structures created by the storks. Tiny sparrows also make them their home

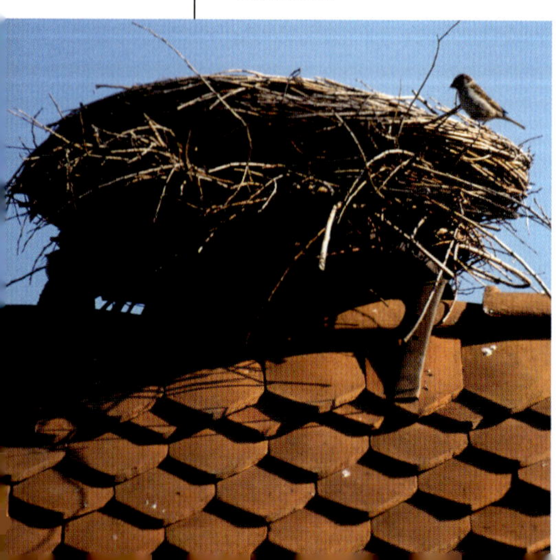

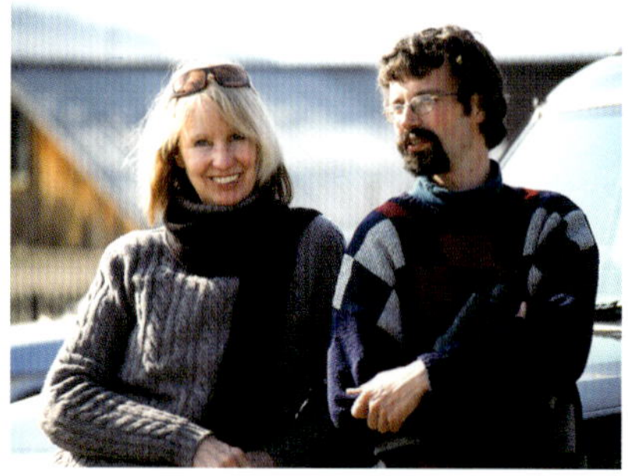

Pollyanna and Robert Zeitz

– a naturalist making a survey of the endangered flora and fauna of the region with a view to persuading the government to declare it a protected area. He greets us with a glass of brandy – giving us the choice between 'fierce or smooth' – and we down it with the traditional Romanian toast of *Noroc*. We can tell immediately that he is incredibly knowledgeable about the region's wildlife – and he speaks absolutely fluent English! Robert is recognised as one of the top two ornithologists in the country – and one of its leading raptor experts – so we feel very fortunate to have the opportunity to be in his company while he acts as our guide over the coming days. We bid farewell to Colin, who is returning to Bucharest, and thank him for all his wonderful organisation during the trip.

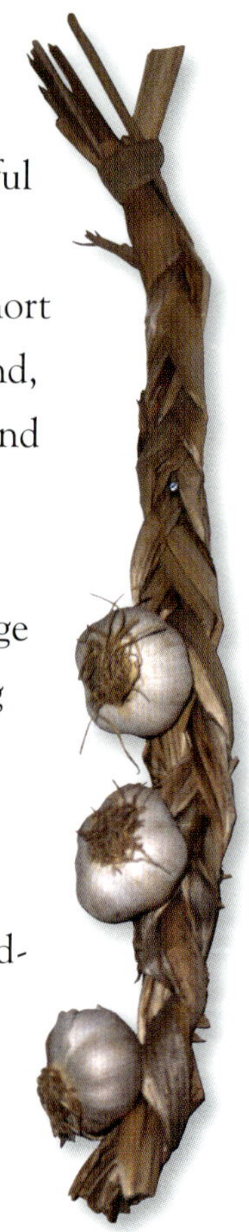

Robert shows us to our accommodation, which is in Miklósvár village, just a short walk from the Count's main residence. The houses here date back to the 1800s and, although they have been recently restored, they retain all of their original charm and character.

Our guesthouse is situated within spacious gardens, overlooked by the storks nesting on nearby rooftops. We are charmed to see that within the sides of the huge untidy structures created by the storks tiny sparrows also make their home, nesting within a nest. Sometimes they will perch momentarily on the top of the larger nest, looking for all the world as though they have built the massive construction themselves.

The interior of our cottage is filled with huge gothic antique furniture and hand-embroidered textiles, including heavy lace curtains at the windows. We also have a modern en suite bathroom, and even bathrobes hanging in the wardrobe. There is a bottle of the ubiquitous Tuica on the table 'to help you sleep and ignore the things which bite'. We also can't help but notice the crucifix hanging over our beds. And the string of garlic nailed above the door...

We walk back up to the main building where we enjoy another glass of Tuica and Josef introduces us to some more guests on the estate who will be joining us for dinner. Jack is a lawyer from San Francisco, accompanied by Rose, a young woman we initially take to be his daughter. We have been planning our journey to Romania for months and arranged to stay at Count Kalnoky's estate in this remote mountainous location after weeks of emails and conversations. Jack and Rose, however, are only in Romania because they were refused entry to their original destination, South Africa, because Rose's passport had a tear on one page. They were immediately bundled back on the next return flight to London, where they had taken a connecting flight. At this point they sensibly decided that as they couldn't spend their planned two weeks in South Africa they would make the most of the situation and explore part of Europe. Having packed a suitcase full of clothes suitable for a luxurious holiday lounging in South Africa's famous Sun City resort, they scan a rack of guide books at the airport for a suitable destination – and apparently randomly decide on mountainous chilly Transylvania. They duly book flights, hire a car at the airport and, having seen a mere mention of the Kalnoky estate in their guide book, set out to drive through the mountains. By some miracle having located the village they knocked on the door – much to the surprise of the staff! I am actually very impressed by the fact they have managed to find the place. Having to negotiate the Romanian roads and drivers, then follow the map purchased at the airport, while travelling through tiny villages where few people speak English to help with directions if required, they not only found Miklósvár, but also the lodge which does not have any signs or nameboards, or any indication at all that it has facilities to accommodate guests. As the ecotourism business is both very new and quite specialist, and the region they are in is pretty remote, the estate was not expecting to attract passing trade.

However, a room was quickly prepared for them and they are now looking forward to a hearty meal. Jack appears delighted to be in Transylvania, but Rose is very quiet and looks

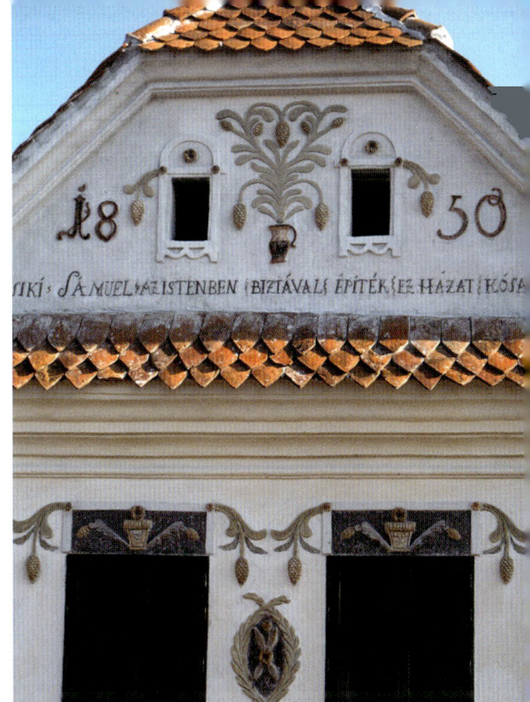

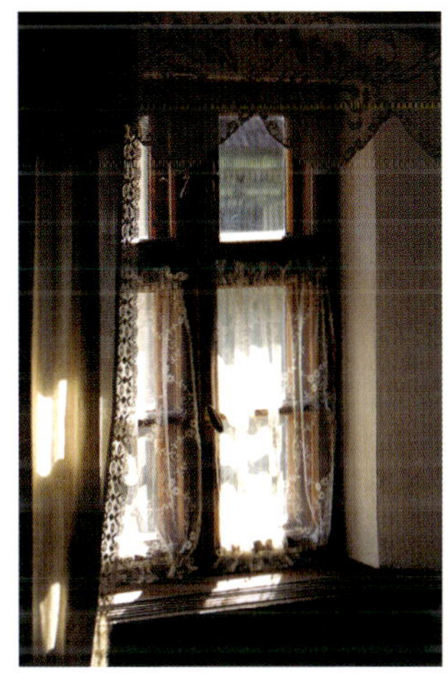

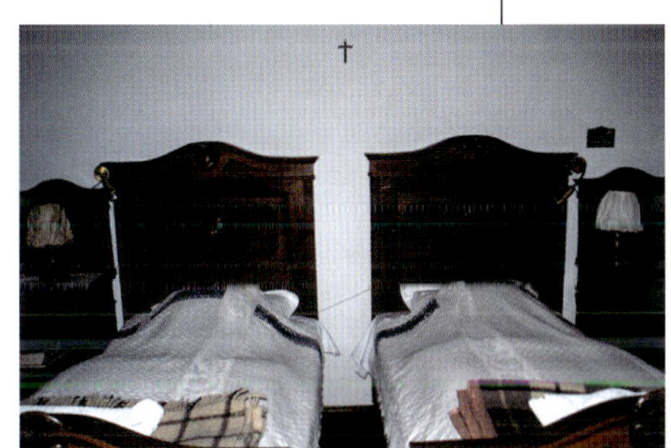

Our guesthouse in Miklósvár village

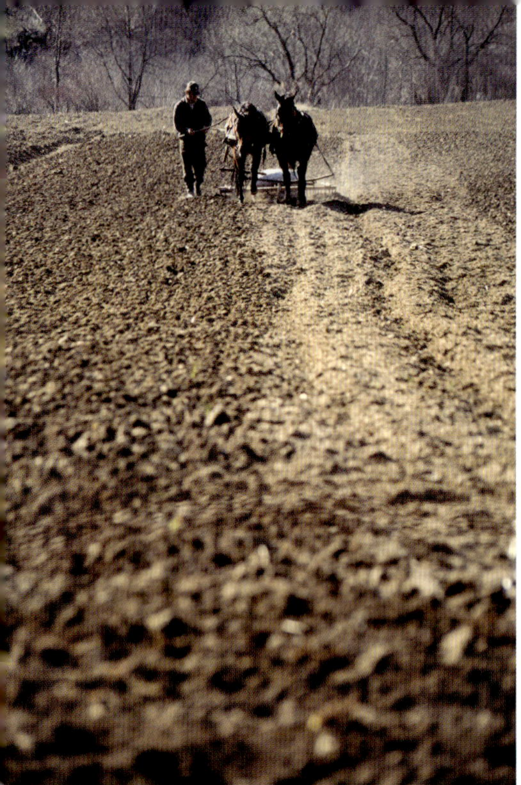

Horse power

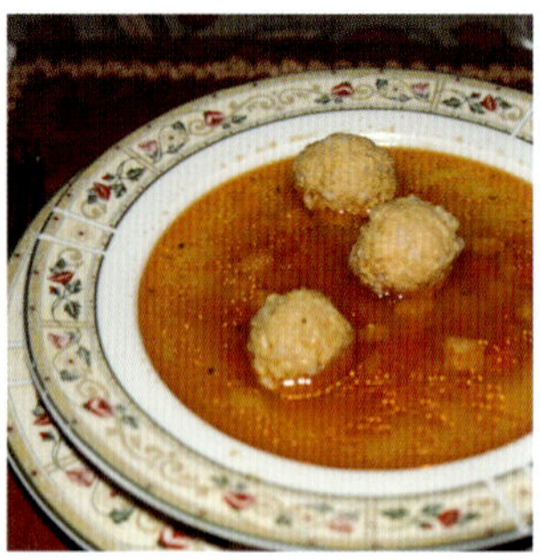

A substantial soup

very cold in a thin cardigan – she tells me her case is mainly full of bikinis and sun lotion, with a few T-shirts and pairs of shorts.

We dine on soup full of huge chunks of carrots and turnips – all locally grown and organic. Jack and Rose then have some round balls of minced meat, while we settle for the *mămăligă* with goats cheese washed down with a glass of local Hungarian wine, followed by delicious fresh crepes with homemade jam. Jack and Rose are very affectionate with each other throughout the meal; by the end we are assuming that despite the considerable age gap – he must be in his late fifties while she is 'in school' – she is *not* his daughter (or certainly we are very strongly hoping that she isn't!). They don't have any plans for the next day (or indeed any plans for the rest of the trip as far as we can tell!) so we invite them to join the two of us and Robert on our planned hike – which will take us into caves at the base of a deep gorge where we are hoping to find bats. Jack seems quite keen, but Rose far less so – understandably, as while we have packed sturdy hiking boots, she has flip-flops and high-heeled sandals. The poor girl was looking forward to reclining on a lounger in the sun, not trekking into bat-infested caves.

We leave them discussing the matter and return to our lodge, where I sit writing my journal, listening to the dogs barking and howling in the village streets. I then settle into bed to read the next few pages of *Dracula* – and read that on his arrival at Count Dracula's castle Jonathan Harker dines on *mămăligă* before retiring to his room where he sits writing his journal while a dog howls beneath his window...

Of course his journey through the mountains featured considerably more in the way of howling wolf packs, mysterious blue flames and peasants giving him muttered and threatening warnings than our pleasant day's drive.

I put the book aside and start browsing through some of the books and leaflets about the region which have been left on the table for the enjoyment of guests. The first line of one of them informs me: '*The population of this region is of solid stock, warlike and well armed*'. Always the first thing you want to read in a guide book. Luckily as I read on I discover that

COMMUNICATION

Wolves have a complex range of vocalisation and their howls are just one part of their 'language'. When a wolf howls it throws its head back, eyes closed, muzzle angled to the sky, and mouth slightly open. To us it is an eerie, spine-chilling sound, especially when heard echoing through the mountains in the dead of night. But wolves howl for many reasons, and in many ways. Each wolf has a unique howl, as recognisable as a human voice. The howl contains a great deal of information: to summon back a missing member of their own pack; to warn of danger; or to confirm that all is well. A low deep howl is a warning to defend territory from other packs – and sometimes when threatened some of the wolves will disguise their voices and yelp in between howls to make the pack appear bigger than it really is. It has been estimated that in open tundra, wolves can hear each other howl from as far away as 16km – although we humans can only hear the eerily beautiful sounds of the wolf chorus from around 2km away. From the age of 3 weeks pups will attempt to join in with the adults howling – although their howls are short and high pitched. Howling is used to communicate over long distance; face to face they communicate in whines, yelps and whimpers, with growls, snarls and barks used in more aggressive situations.

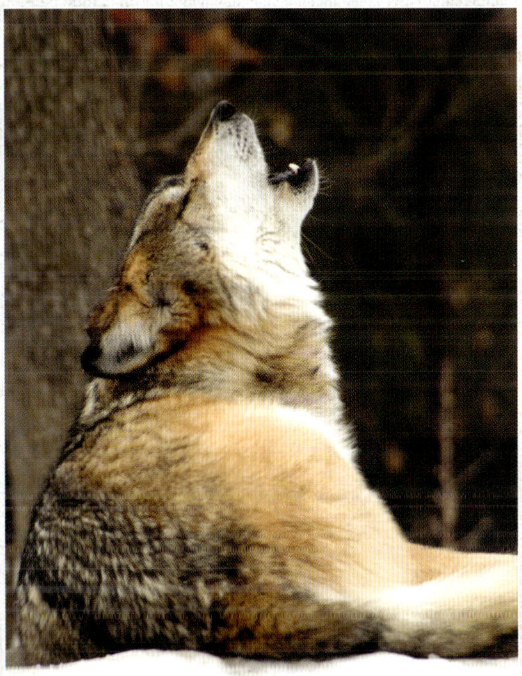

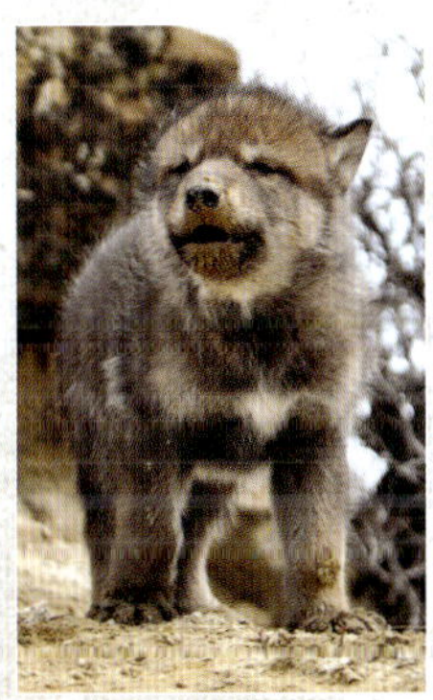

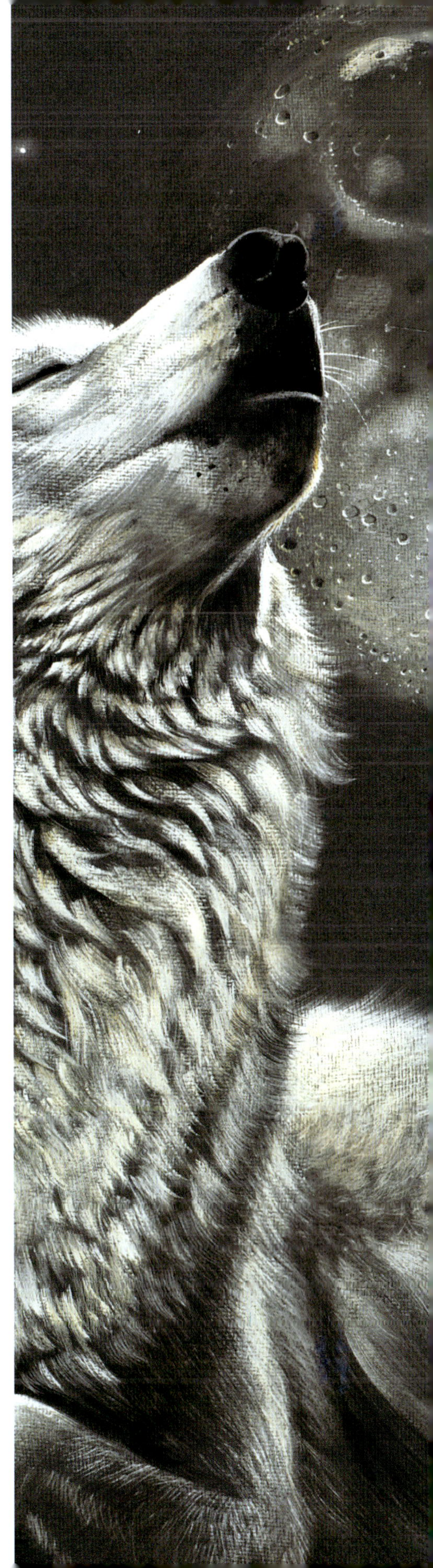

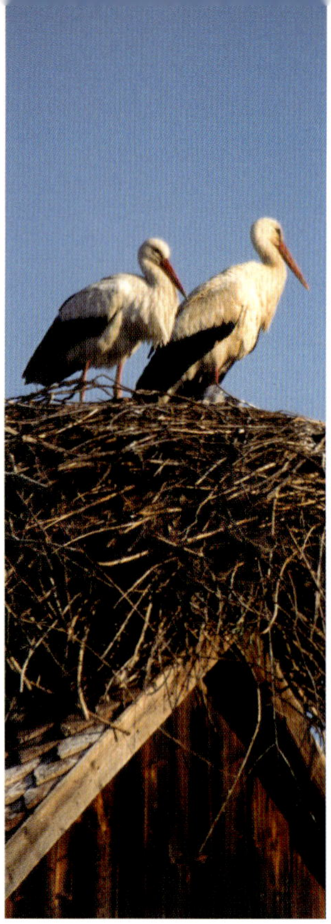

Storks build messy nests perched precariously on roofs

this description of the Transylvanians was in fact written by Miklos Olah in the 1560s, so I switch out the light reasonably confident that I will survive the night.

Day Fifteen

Having passed a peaceful night listening to village dogs barking and howling at the moon, Pollyanna arises bright and early at 5.30 am. It emerges this is because she misread her watch and thought it was in fact 6.30 am – so I doze for another hour or so, before trying my luck with the plumbing. The copper pipes clank and groan, before gushing around four inches of yellowish water into the bath. It is piping hot – but it also has a strongly metallic smell. Drawn straight from mountain springs it has a high mineral content – drinking the soup last night I could taste the iron in the water. (Robert will tell us later that as a child he would dig for spring water if he became thirsty while exploring in the woods – and if he then dug a second hole just metres away up would bubble water from an entirely different source

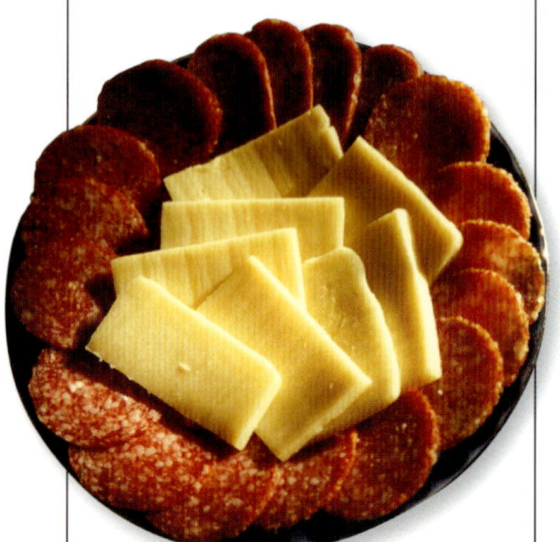

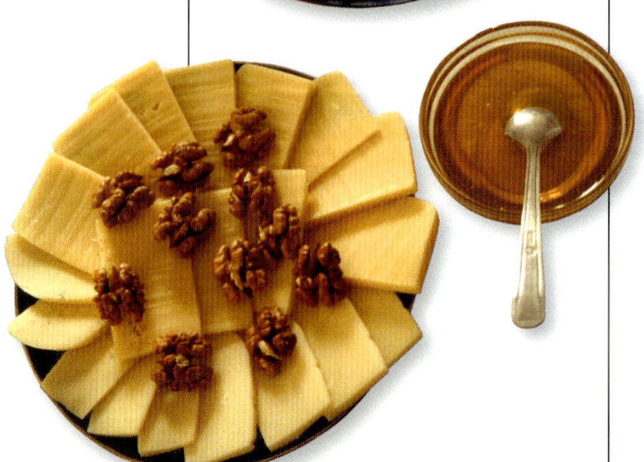

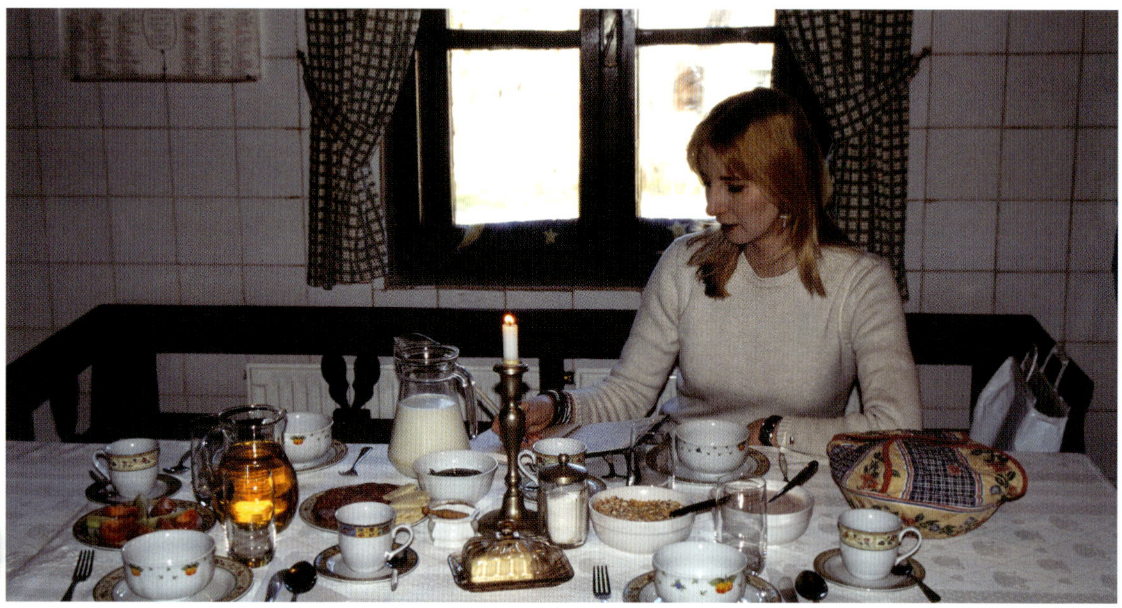

We eat breakfast in the kitchen – seeds and nuts moistened with apple juice and delicious yoghurt produced by local farms

with a different mineral composition and entirely different taste.)

Dry and dress speedily as the bathroom is not too warm – but the weather outside is glorious. Clear blue skies and bright sunshine. We eat breakfast in the kitchen – seeds and nuts moistened with apple juice and delicious yoghurt produced at the local farms. We stroll back to our guest house to pick up my cameras and Pollyanna's sketching equipment for the day; as we walk we greet the villagers we pass with a cheery *Buna!* – hello in Romanian. But rather than returning our greeting with a smile, they simply stare at us – and draw back into doorways gathering their children into their skirts.

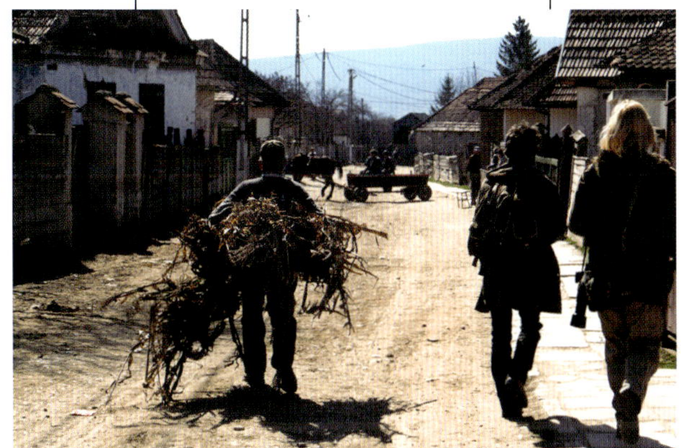

Pollyanna and Robert explore the village

Robert tells us that for decades during communist rule no foreigners set foot in the village. However, this isn't the problem. What we haven't taken into account is that we are now in an entirely Hungarian settlement and the people here don't speak any Romanian. The Szeklers who settled in this region are believed to be the descendants of the Huns from Attila who settled here in the Middle Ages. Robert teaches us a new greeting: *Yo Regat*. He also tells us that we need have no fear of attack by 'well armed' locals – there is almost no crime in this tiny community. In fact the only reported crime in recent years was the theft of one chicken and, of course, the gypsies were blamed. The village is too small to have a resident policeman, but there is a night watchman 'who walks the streets – when he is not asleep'. Each man in the village takes it in turn to be the watchman, starting in the south and working north – each person will do around three nights a year.

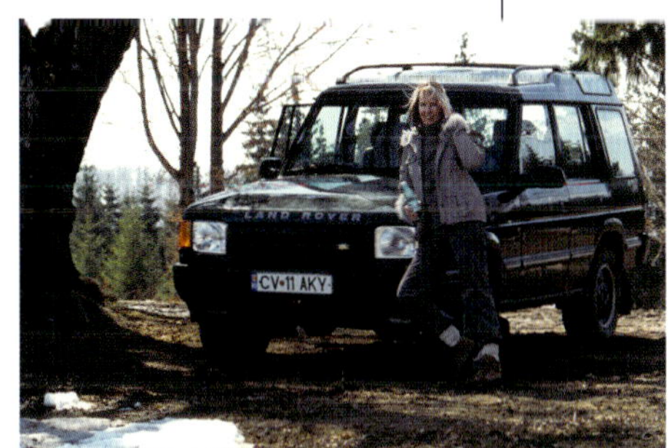

The Count's Land Rover – complete with family crest on the door

Robert and Josef, the driver for the guests, pick us up in the Count's Land Rover – complete with a crest on the door. We rattle and jolt through the village stopping every few hundred yards to photograph a stork's nest or a horse and cart. In this area the carts still far outnumber the motor vehicles, and the only other 'traffic' on the streets is the stray dogs who lulled us to sleep – they stop

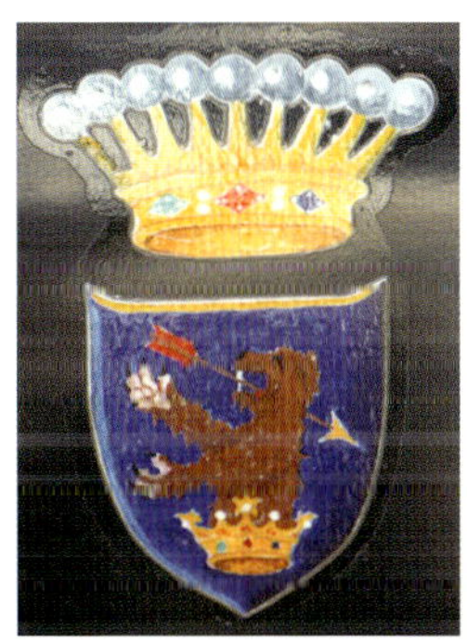

The Kalnoky family crest

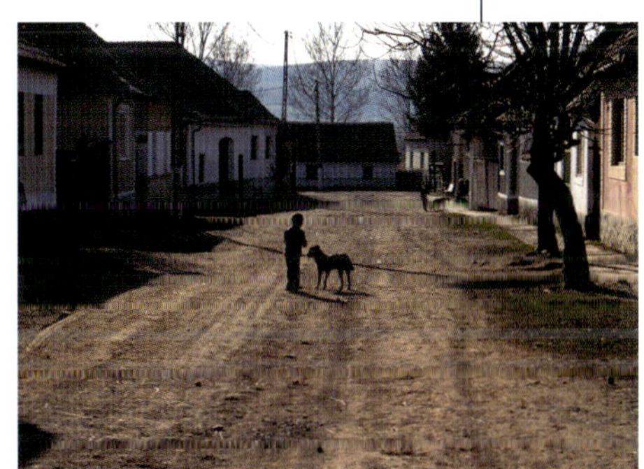

Miklósvár

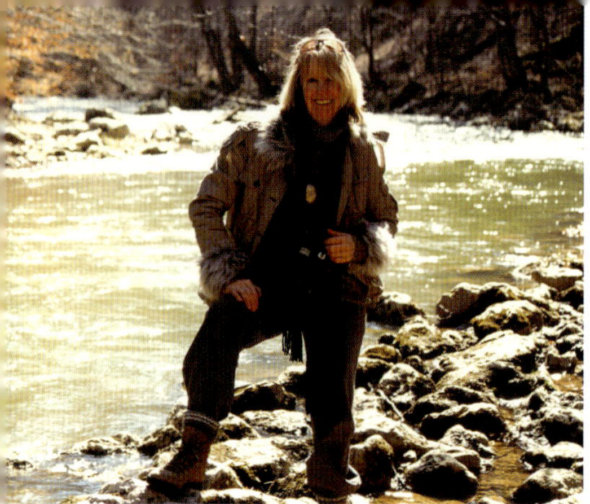

Pollyanna by the Vargyas river

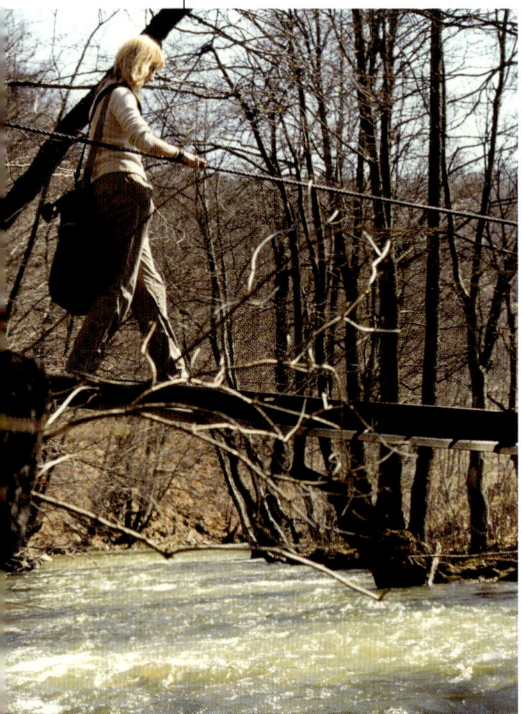

The first bridge

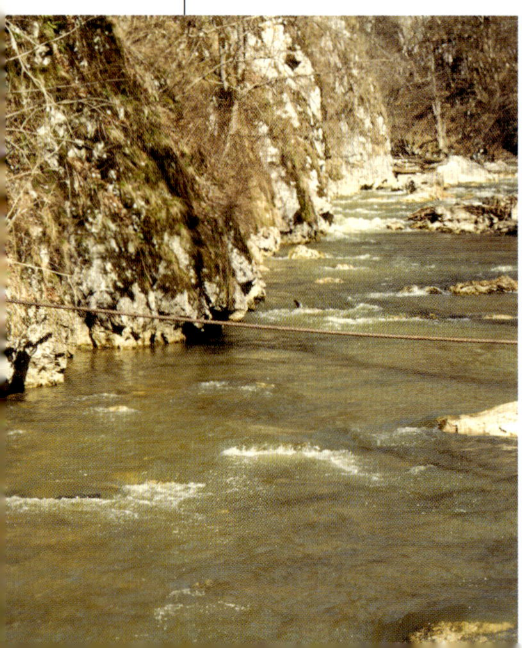

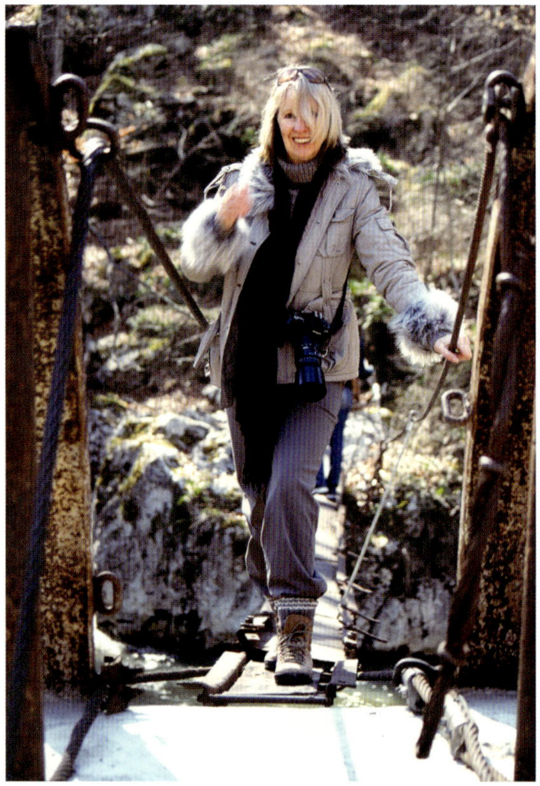

and watch as we pass. The dogs come in a multitude of colours and shapes but nearly all have very short legs, as though the great-grandfather of every street dog was an itinerant (and randy) corgi.

We turn off the dusty roads on to cart tracks, and finally off road altogether into the Vargyas (rhymes with gorgeous) gorge. (Also home to the Vargyas river and Vargyas caves.) From here we start to hike out towards the cave system. The hike is beautiful as we follow the course of the river. At times the path is wide and smooth and easy, but suddenly we are scrambling for footholds on narrow ridges of rock and branches. Here and there the river is running so high with melting snow the normal path is flooded and we are forced on to higher rough and rocky ground. We have to cross the river three times, each time on a progressively more alarming rickety bridge. The first is a narrow suspension bridge reached by a spindly iron ladder. Some of the iron panels which make up the floor of the bridge are missing, but it feels relatively stable. At the far end we have to climb on to the ladder going down by swinging under and round a huge overhanging branch unsupported by either bridge or ladder for an unnerving grab. The second bridge is similar, but this time only half of the suspension structure remains, so Robert and Josef hold the remaining iron 'rope' taught so that we can use it for balance as we edge across the remaining metalwork which is listing

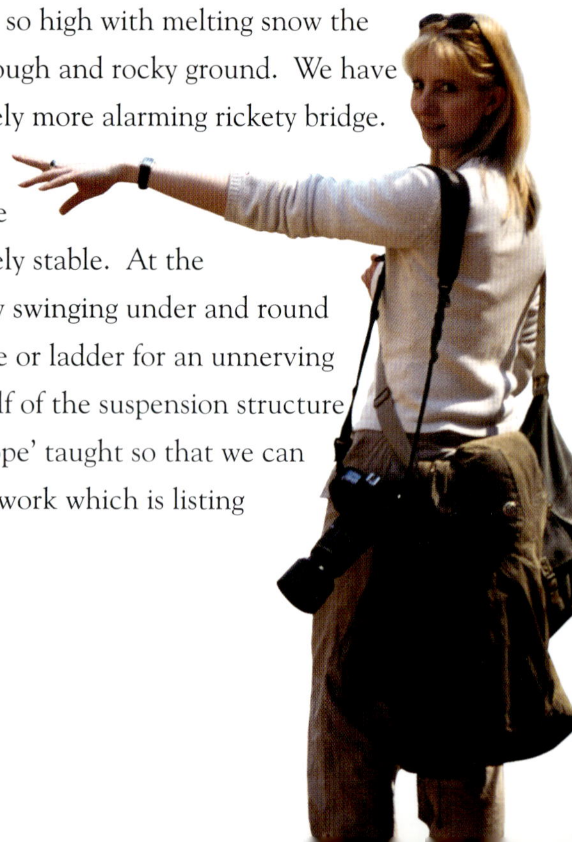

Only one iron rope remains of this bridge!

and tilting beneath our feet. All along they have been muttering about bridge number three and its potential 'crossability'. They are cheered by its condition when they arrive; we are considerably less so as two tree trunks have been carelessly lashed together with one iron 'rope' above to hold for support, while the swollen river rushes below. Robert attempts to reassure me that if I should fall in I won't be badly hurt because the water is barely two metres deep at this time of year. I am less than reassured. This is snow melt and the water is freezing – I think hypothermia would set in within seconds if I so much as dipped my little fingernail in. We cross safely, with much photographing of the event – and of course we also photograph the river, the trees, otter tracks (but no otters).

Finally we reach the entrance to the cave system. There are seven hundred caves in this limestone gorge, the deepest lie some seventeen kilometres below ground. The largest of them is reputed to be the cave from which the Pied Piper emerged followed by the missing children of Hamelin – and according to legend thus established the Saxon communities in Transylvania. In fact this is a romantic explanation for what seemed to many who travelled through this region in days gone by to be an inexplicable presence of blond blue-eyed German speaking people who followed ancient customs but were hundreds of miles from the German border. The

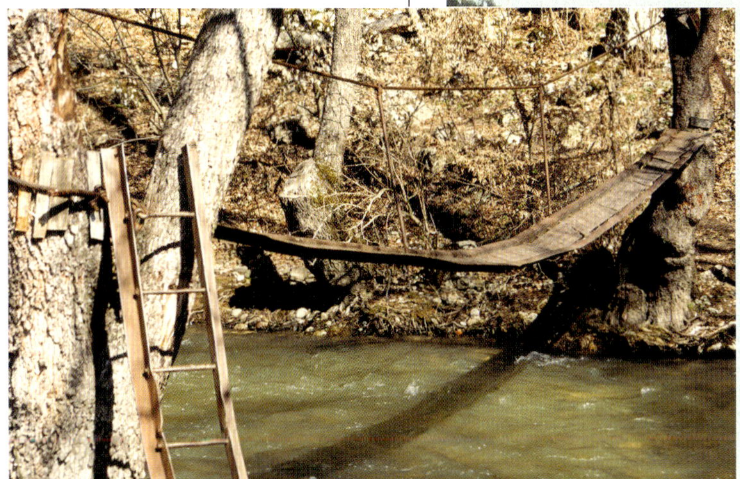

The second bridge

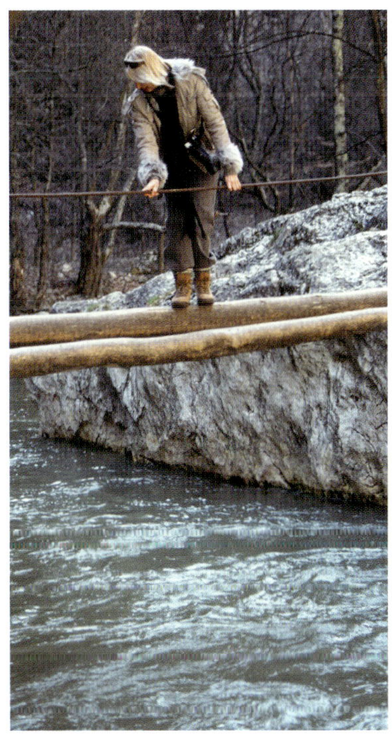

The third bridge

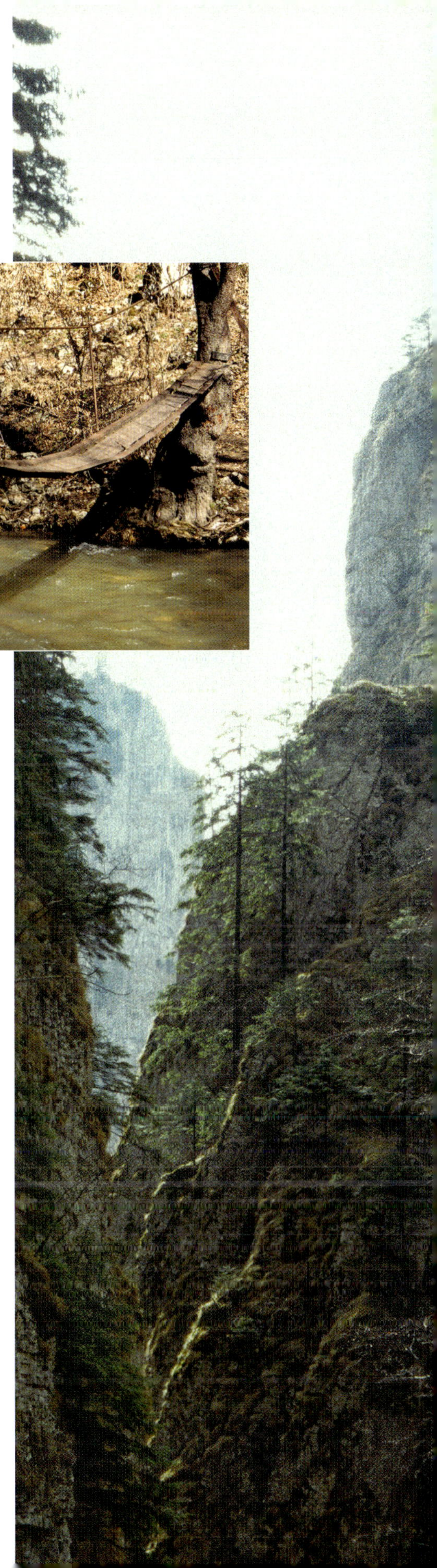

Vargyas gorge

'In Transylvania there's a tribe
Of alien people that ascribe
The outlandish ways and dress
On which their neighbours lay such stress,
To their fathers and mothers having risen
Out of some subterraneous prison
Into which they were trepanned
Long time ago in a mighty band
Out of Hamelin town in Brunswick land,
But how or why, they don't understand.'

Robert Browning

more prosaic reality is that the fortified towns and villages of Transylvania were established in the twelth century by settlers from the Moselle region of Germany, drawn here by favourable market rights offered by the Hungarian rulers who wanted their help in guarding the mountain passes against invaders. They were referred to locally as Saxons.

Whatever the origins of their arrival, many of the Saxons whose architecture and traditions have left such a mark on this region were repatriated during the communist rule and their villages left deserted. They had coexisted in harmony with the Szekler villagers for centuries, but the abandoned villages are now being taken over by bands of gypsies – who are not coexisting quite so harmoniously. We will come up against widespread prejudice and condemnation of the gypsy communities throughout our travels – as we have already seen they are immediately blamed for any crime or trouble. Sometimes this is not without some basis in fact. Driving to Miklósvár, Colin was careful to avoid certain areas which had been settled by gypsies as they often lay tyre-slashing materials across the tracks so they can ambush the stricken cars. However, Robert tells us that there is one gypsy family who have settled very well in Miklósvár, where they share in the farm work and live without tension – apparently a rarity. They have been nicknamed the ravens – (big raven, little raven, etc). Everyone in the village has an affectionate nickname –

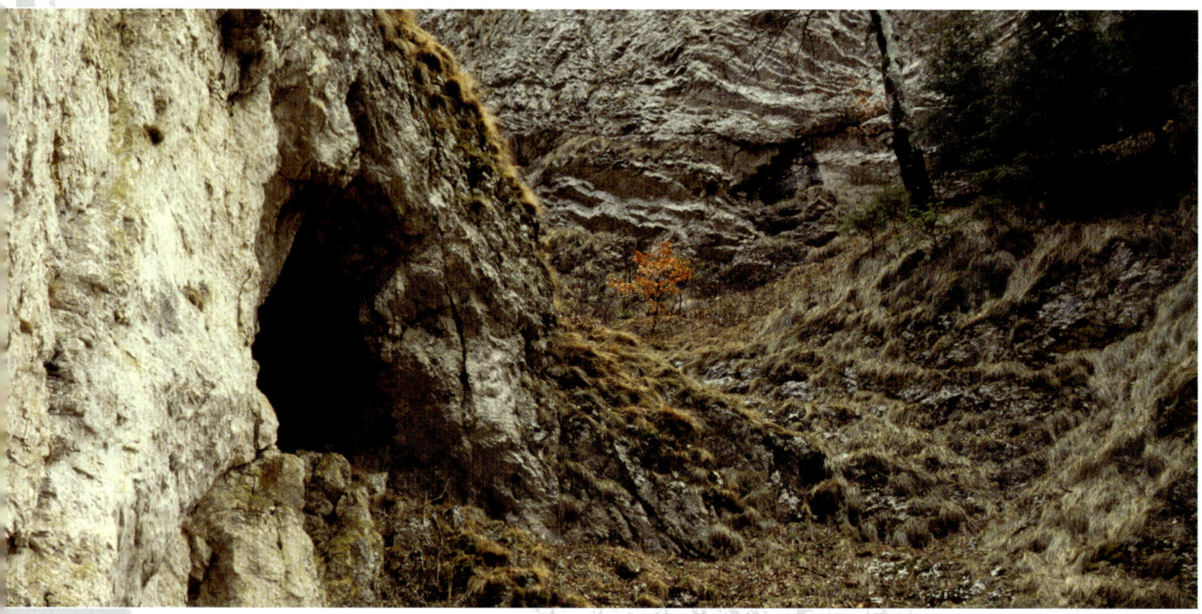

The largest of seven hundred caves in the Vargyas gorge

including a farm worker named Dino, as he only has four fingers on his right hand, and therefore leaves a handprint like a dinosaur track.

A second local legend tells of the story of the leader of a group of villagers who took refuge in the caves when the Turks invaded. (This much is certainly true – the remains of a wall they built at the cave entrance can still be clearly seen, as can the altar stone used by the priest.) Both the villagers and the Turkish forces ran out of supplies, so the Transylvanians baked the last of their flour mixed with ashes into a huge loaf which they placed at the entrance to the cave to fool the Turks into thinking they had enough food to last for months. The Turks turned and retreated, at which the leader of the villagers apparently climbed to the top of the high rocky outcrop to make sure it was a true retreat and not a trap – and seconds after confirming the Turks had truly left he fell to his death. He was hailed as a hero and the rock still bears his name.

We make our way into the myth filled caves. They are pleasingly high roofed for most of the initial section, and we only have to crouch low to pass through a couple of small passageways. It is very slippery underfoot, with water constantly dripping from the ceiling. Almost as soon as we have entered the caves we spot our first bat – a tiny pigmy bat entirely wrapped up in his wings, wintering in the caves. As we continue we find huge roosts of horseshoe bats, huddled together for warmth on the high ceilings of the caves. One or two are roosting separately, a little lower down the cave walls. The bats will hibernate in here for most of the winter in a state of inactivity characterised by slower breathing and a lower metabolic rate. Their body temperature will drop by a few degrees. We spot one bat covered in tiny droplets of

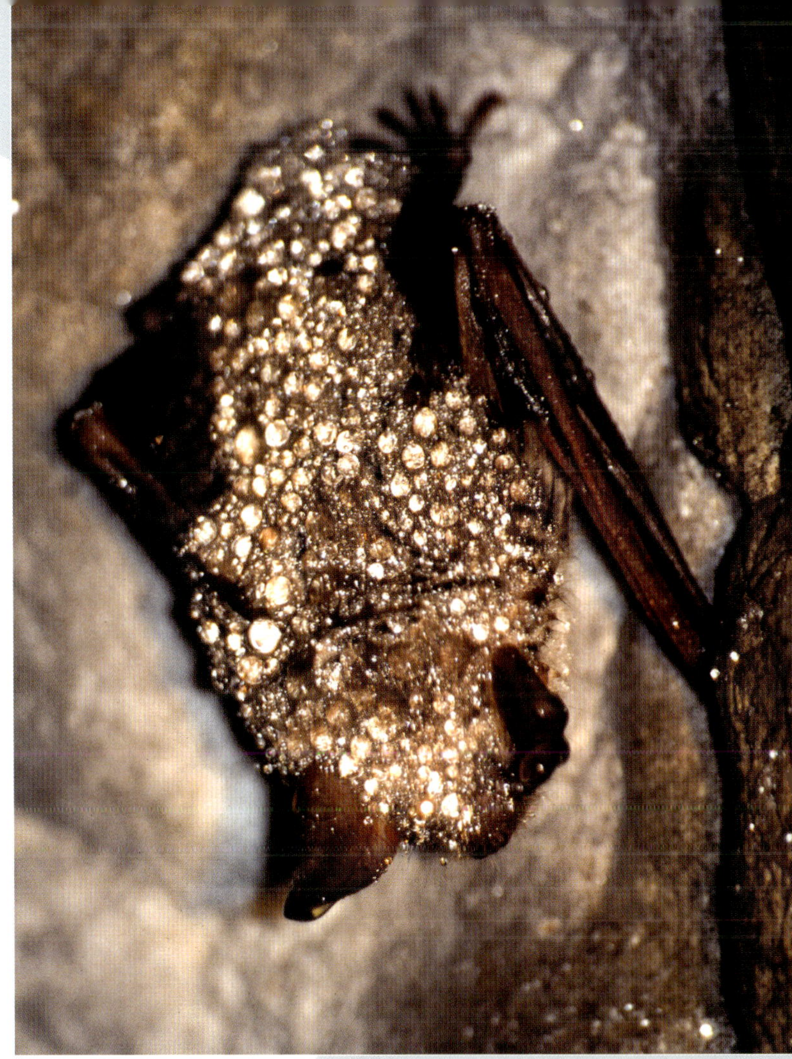

A BAT APPEARS TO BE ENCRUSTED IN DIAMONDS AS WATER DROPLETS SPARKLE IN THE TORCH LIGHT

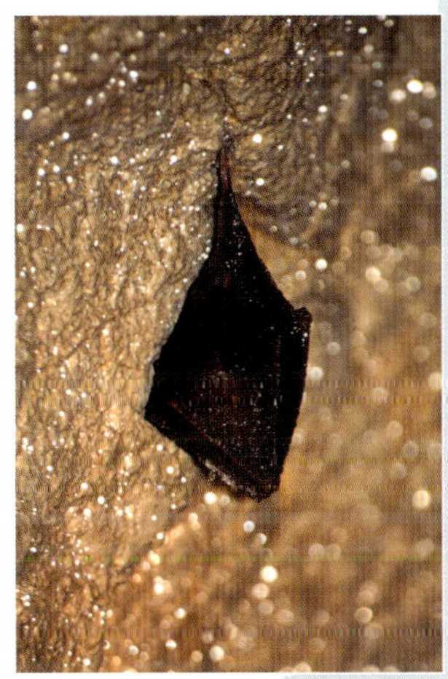

A PIGMY BAT ENCASED IN ITS WINGS

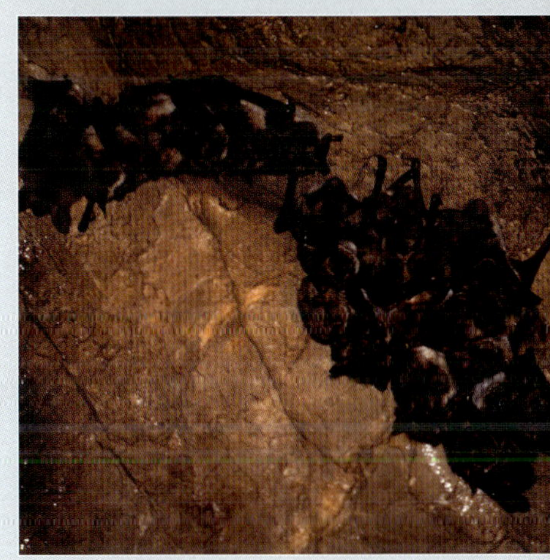

DEEPER INTO THE CAVE SYSTEM WE COME TO ROOSTS OF LONGWING BATS

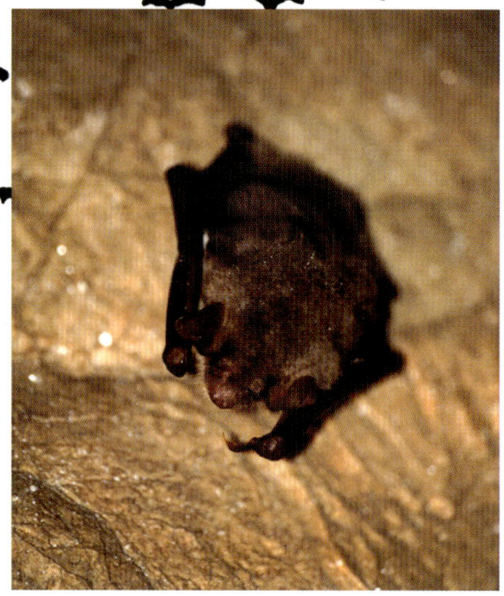

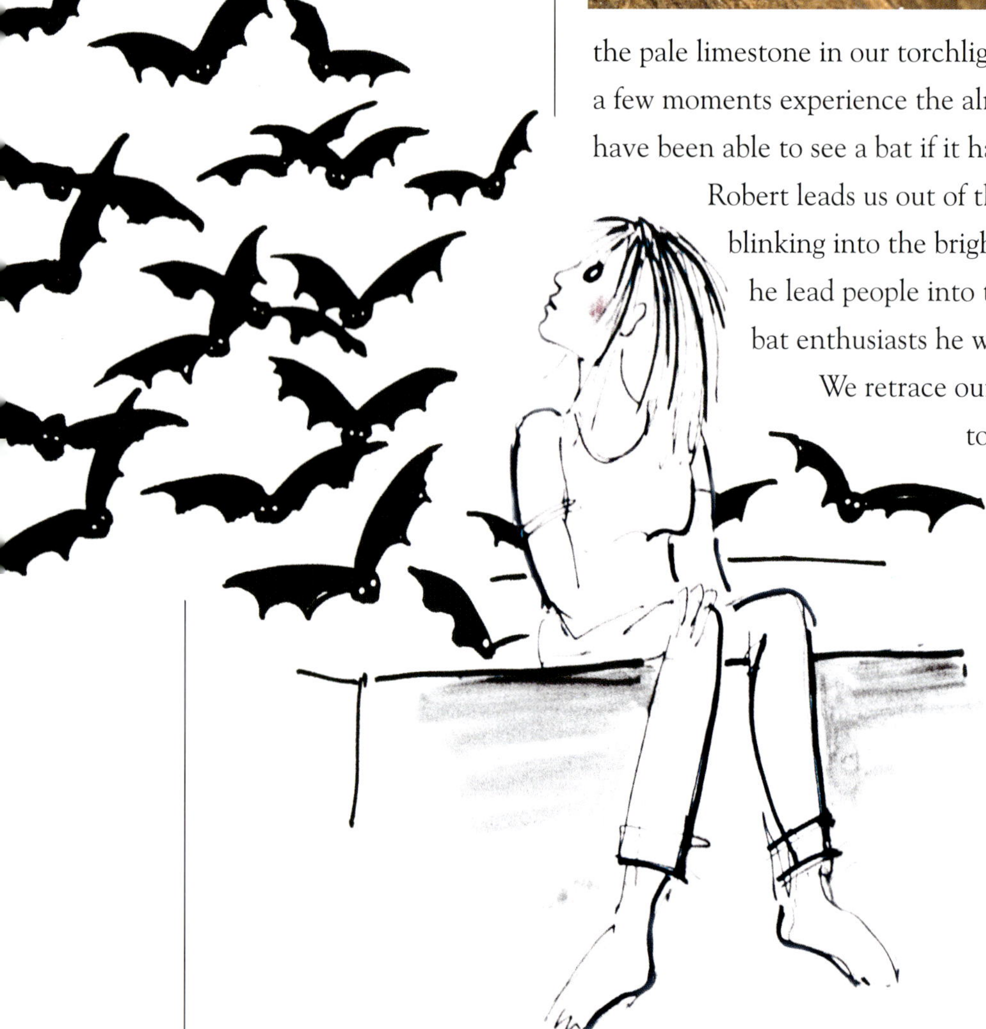

water which sparkle and shimmer in the torch light like diamonds. It is like finding a tiny bejewelled treasure in the dark depths – and one of the most remarkable sights I have seen in all the years we have been travelling to search for wildlife.

Deeper into the cave system we come to roosts of longwing bats; one or two of these take flight as we enter swooping down round our heads. Most hang upside down, their white fronts making them look like inverted penguins hanging from the roof. They have perfect little pointy bat ears which cast silhouettes on the pale limestone in our torchlight. Deep underground we switch off our torches and for a few moments experience the almost unimaginable denseness of total darkness. I wouldn't have been able to see a bat if it had been hanging from my fringe.

Robert leads us out of the caves by a slightly different route and when we emerge blinking into the bright sunlight once again he admits to us that the second time he lead people into the caves he got lost – but decided not to mention it to the bat enthusiasts he was guiding!

We retrace our steps and recross our bridges to the vehicle and drive back to the lodge – taking a slight detour to see Count Kalnoky's castle. It was originally a hunting lodge for his great-great-great-great-grandfather and is first mentioned in records from the twelfth century. The castle is in the first stages of renovation, following its semi-destruction under the communists. Since beginning work they have discovered the remains of the walls of even earlier structures beneath the foundations.

In the evening we are expecting to join Count Kalnoky for dinner – a date arranged while we were back in England. However, we arrive at the lodge to find the Count has sent his apologies – he will not be able to join us, as he has been delayed by other pressing matters. As well as being the owner of three fledgling tourist destinations, and overseeing the renovations of all his ancestral properties, the Count is also a diplomat, so we are not surprised that he has many claims on his time. We are told he will try to join us for dinner the following evening instead, but this is a pattern which will be repeated throughout our stay: we will arrive at the lodge to be told 'the Count will not be joining you' until we begin to think he is either fictional, or will not appear in daylight.

However, this evening we still have the pleasure of Jack and Rose's company, who, despite finding they were able to resist the lure of a Transylvanian bat cave, join us for a delicious meal of bean soup, followed by freshly caught trout with crispy potatoes and a couscous-like side dish. Although the cuisine in Romania may not be world renowned, the home cooking from the castle kitchens is absolutely wonderful. The meals are based upon traditional local Transylvanian recipes and cooked using vegetables produced in the kitchen garden or within the village. Milk products and meat come from animals bred and reared in Miklósvár. None the less we spend much of the meal putting the fear of God (or more specifically of diarrhoea and projectile vomiting) into 'Humbert and Lolita' who up until this point have been merrily drinking the tap water and eating the salad throughout their stay in Romania. Here in the mountains, where the water is drawn straight from the earth, they have probably been quite safe, but in the city they were taking their lives in their hands.

As we stroll back through the dusty streets the batteries in our torches fade – they have had heavy use in the caves today! There is no street lighting in Miklósvár so when we reach our lodging we are able to see only by the light of the full moon.

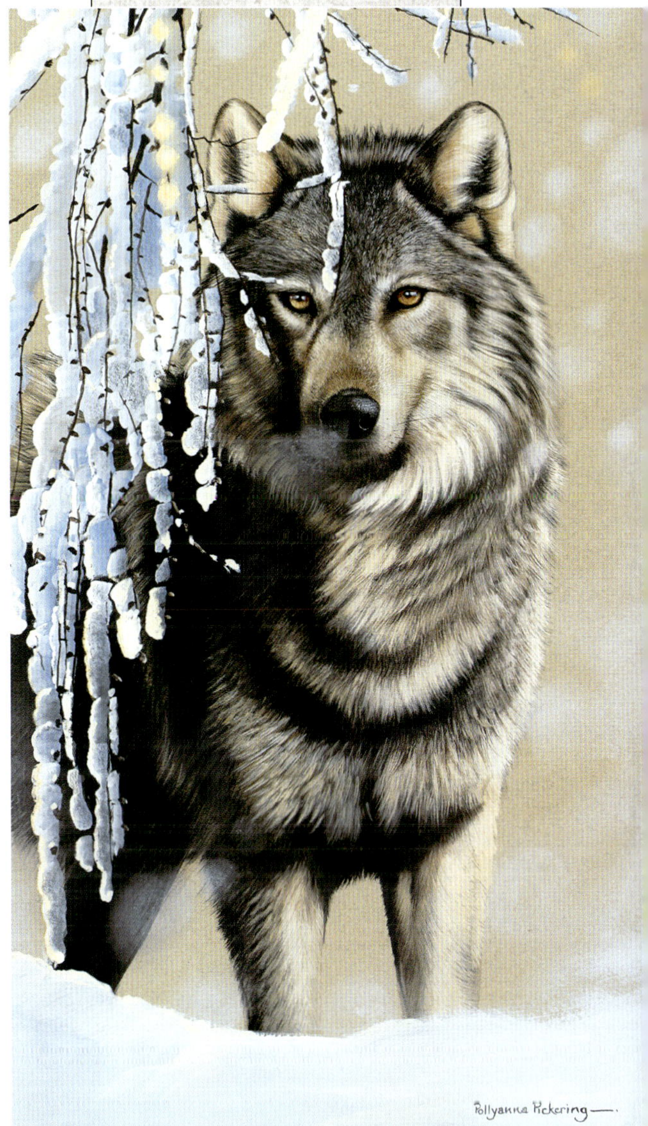

'...And a Deep Cold Descended Upon the Land'
Private collection

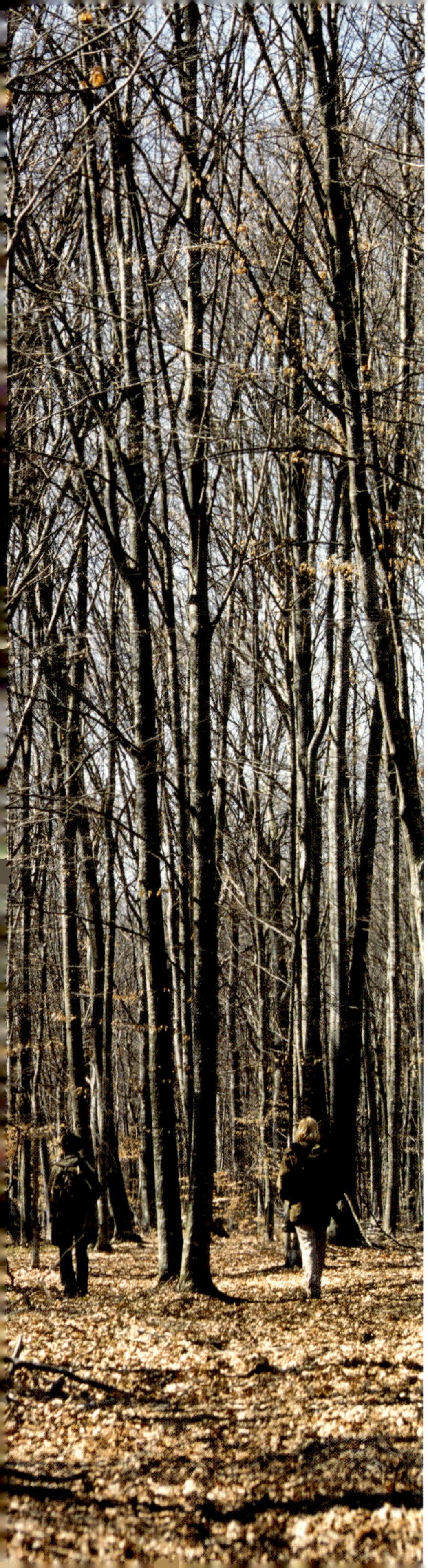

Day Sixteen

Room freezing. Immerse self in steaming bath, then hop out and get dried in three seconds flat. Can still see self in mirror, so assume the garlic over the door is continuing to work as a vampire repellent. In the kitchens for breakfast we learn that Jack and Rose have departed early that morning – perhaps to find a warmer part of Europe in which to enjoy the rest of their unplanned vacation. Robert and Josef collect us after breakfast and Josef drives us into the heart of an ancient valley for our day's hiking.

The Land Rover has to forge the river several times to get us to a suitable point to begin our hike, and from here we set off – upwards of course – to the crest of hills surrounding the valley. We are walking through ancient oak woods, empty acorn cups crunching beneath our hiking boots. A breathless clamber up is followed by much easier walking along the ridge, with beautiful views of the Carpathians in the distance. Robert points out the first of the wildflowers, and many birds – we spot wagtails, coal tits, jays and a middle-spotted woodpecker. He identifies the calls of wrynecks, finches and many more. We see small lizards scurrying away beneath branches and some salamander tadpoles in a puddle. Robert bounds up eagerly to every puddle we pass in search of yellow-bellied frogs – the prey of the lesser-spotted eagle – but fails to find any, although we do find big clumps of frogspawn. The weather is glorious – bright, sunny and warm with an occasional cool breeze.

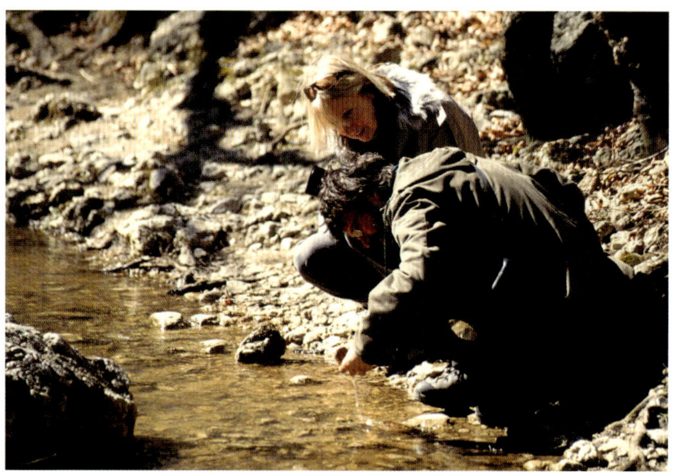

In search of the elusive yellow-bellied frog

We stop for a picnic lunch of hardboiled eggs, raw peppers, bread and butter with homemade tangy apricot jam. Convinced of the purity of the water in this

Walking through ancient woodlands

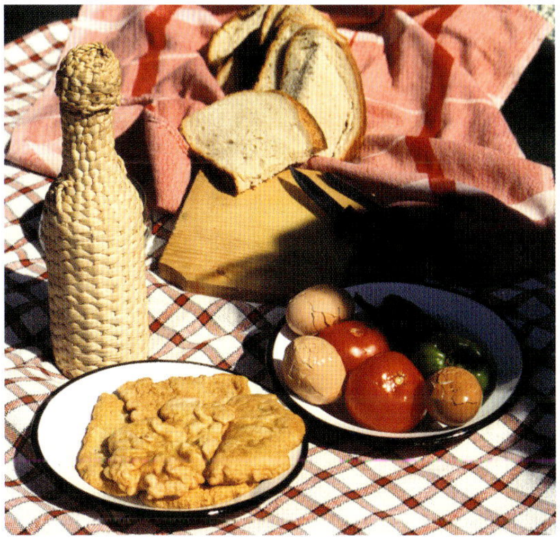

Picnic lunch

region I throw caution to the wind and eat a TOMATO! Yesterday I was venturing across collapsing bridges and now this – is there no end to my intrepid risk-taking recklessness? The kitchens have sent a bottle of brandy for us, so we knock back shots in chunky little glasses – I am sure it is powerful enough to kill off anything noxious I may have unwittingly eaten. While we are eating, a horse-drawn cart loaded with logs (from fallen trees) trundles picturesquely past us. Robert greets them and Pollyanna sketches while they pause to make conversation. Although she rarely paints portraits her sketch folders and travel journals are often filled with simple line drawings of the people she sees on her travels. Shortly afterwards Robert starts muttering about going to work in Outer Mongolia – I think he is starting to find rural Transylvania a little overpopulated. After all we have seen three people and two horses during our six and a half hour hike. After the horse and cart have passed, a younger horse trots by in their tracks – too young to work she is never the less following the cart.

We wend our way back down the wooded hillside – I can now tell the difference between beech and hornbeam – and eventually rejoin a 'cow' road which leads us back to the village and our lodgings. This well-worn track is used daily by a parade of pale golden cattle. At 8.00 pm each evening we join the villagers outside their gates to watch as a long line of cows appears on this track and processes into the village where individual beasts then peel off into the family courtyards. The first evening we witnessed this spectacle we assumed that a cowherd would be following the cows down the track. In fact the cows are self-trained and every morning they take themselves up to the grazing lands, and every

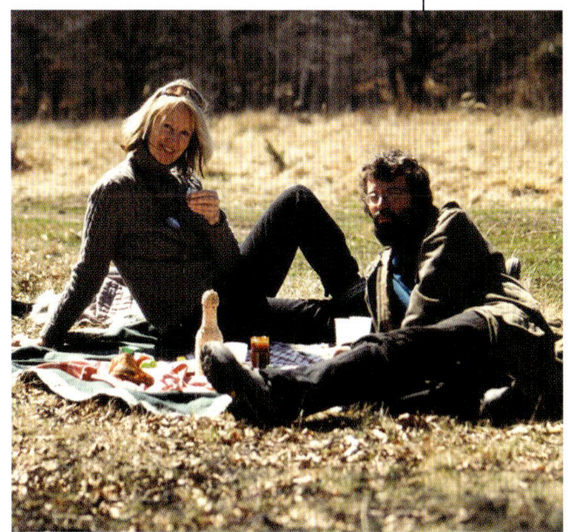

Al fresco lunch in the sunshine

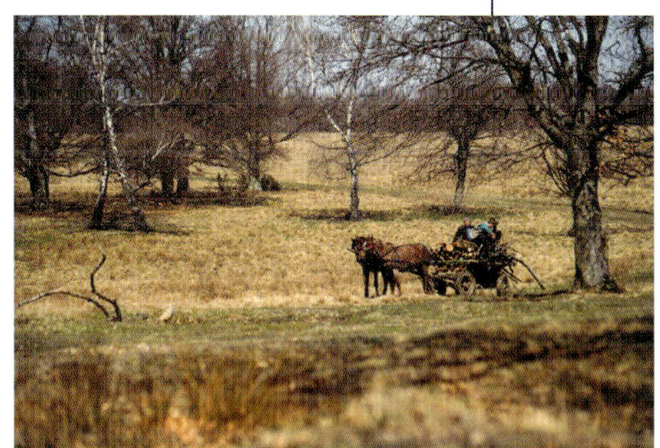

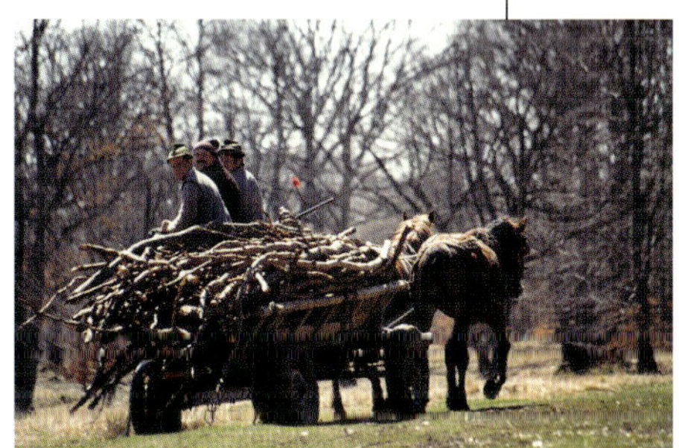

A horse-drawn cart trundles past as we eat our lunch

A strong shot of the local brandy to kill off any bugs

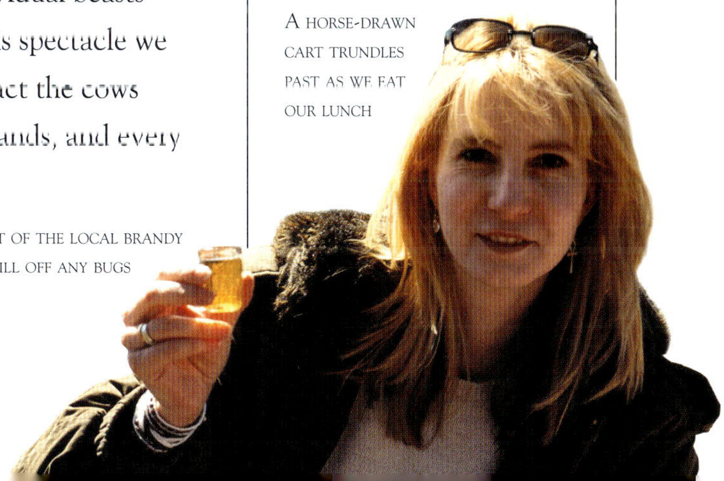

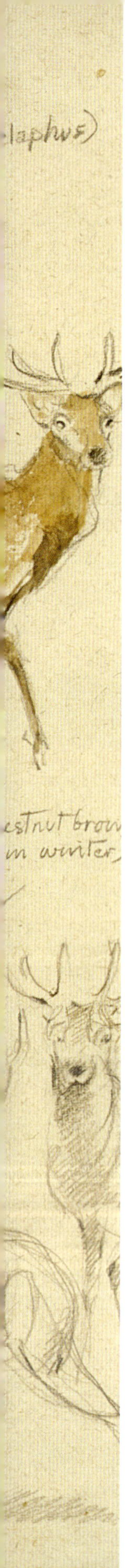

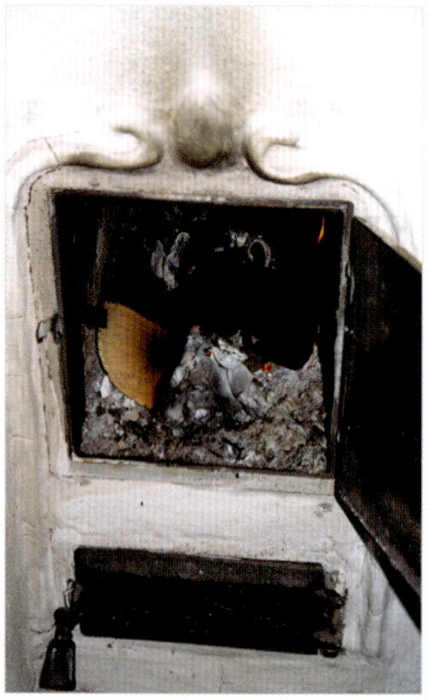

A TRADITIONAL ROMANIAN WOOD-FIRED STOVE

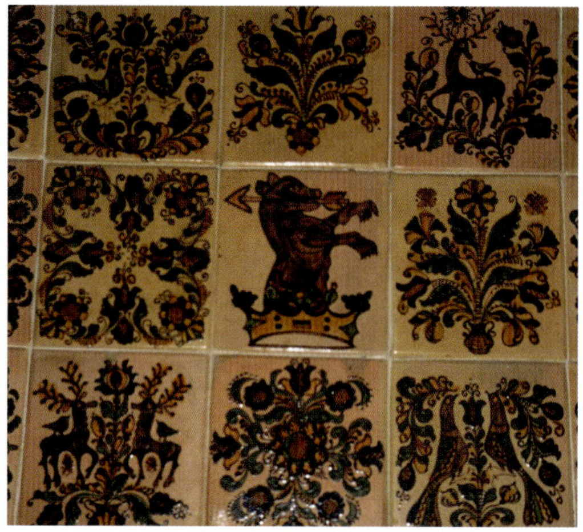

A SOBA IS ALWAYS COVERED WITH GLAZED CERAMIC TILES

evening at 8.00 pm they return. So each night the villagers stand beside their gates, waiting until the cows come home.

As we walk back through the village streets we pass the general store. The shop is shut so we peer in at the dusty window display of soap powder, chenille dressing gowns, cigarettes and a grapefruit. Next door is the bar, which is open – and full of men relaxing after a hard day's work in the fields. Peering in we can see the walls are decorated with a fading Cindy Crawford calendar and a poster of JR from the American soap *Dallas*. As we turn down the side road which leads to our guest house I glance back and see that all the guys from the bar are standing crowded in the doorway to watch us walk through the village! As we have been walking a cool breeze has picked up and when we enter our accommodation we find the fire has not been lit. Our room features a traditional Romanian wood-fired stove, or soba, in the corner. Used to heat most rural homes in Romania a soba is always covered with glazed ceramic tiles which are made with open channels on the inside. The heat is stored in these and the stove radiates warmth for hours after it has been lit – so usually our room is toasty warm. The stove is fed with logs through a door on the outside wall. However, it is now very cold, so I wrap in layers adopting the traditional style of an ageing peasant woman, while Pollyanna boils the kettle and sits huddled over it as the water heats.

The weather in these mountains is always changeable and in spring there seem to be constant fluctuations in temperature. In twenty-four hours we have been from warm air and blue skies to snow and icy temperatures. It drops down to freezing overnight and we wake to frost followed by a warm morning and a hot blue sky at noon. Thunder storms can boil up in minutes.

For the rest of our time in Miklósvár this will be the pattern of our days. Josef and Robert will collect us in the morning and we will spend our days hiking – constantly watching for wolves, but seeing wild boar, red deer and roe deer darting between trees. Wolves in this part of Europe prey on animals ranging from large ungulates like these deer, though to small rodents. Like urban foxes in Britain some wolves even subsist on food scavenged from bins, having adapted to life on the edges of human habitation. The wolves, like the bears, have become predominantly nocturnal to avoid contact with humans. Robert shows us the lesser-spotted eagle nests where he hopes to ring the youngsters later in the year. Pollyanna's sketch books continue to fill with her impressions of this wildlife-rich region with its flower meadows, woodland pastures, forested gorges and mountains.

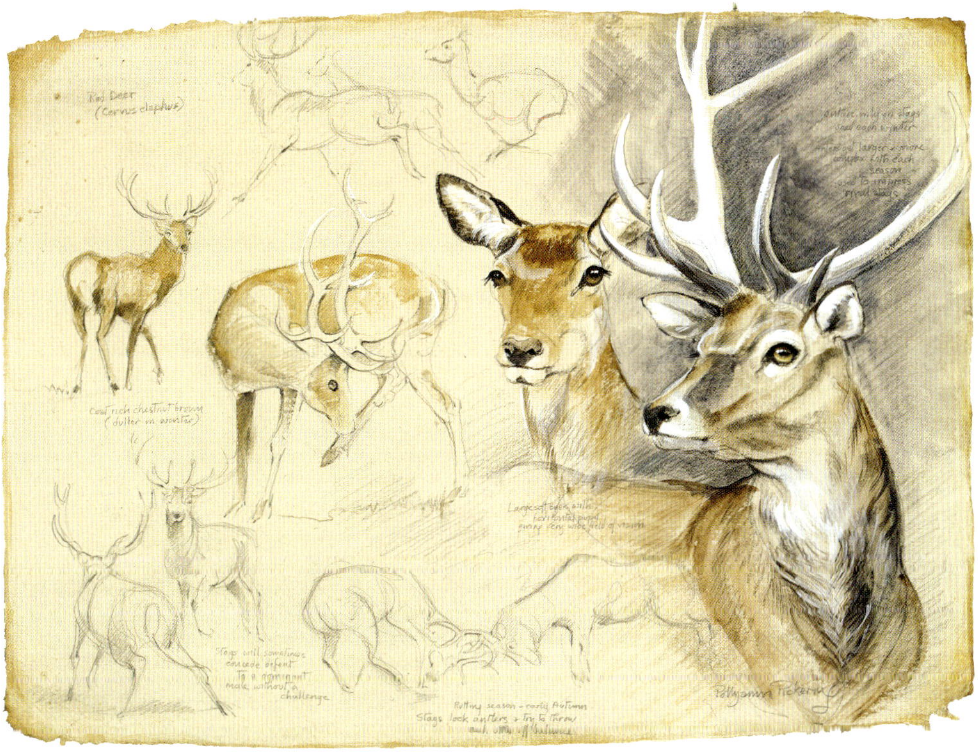

Pollyanna's sketches of red deer

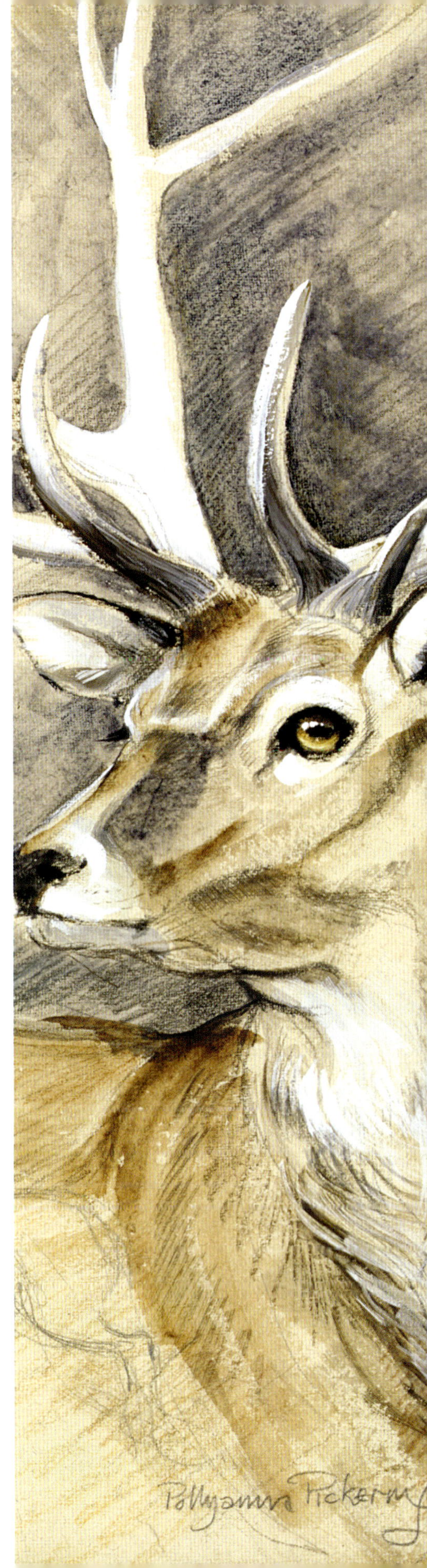

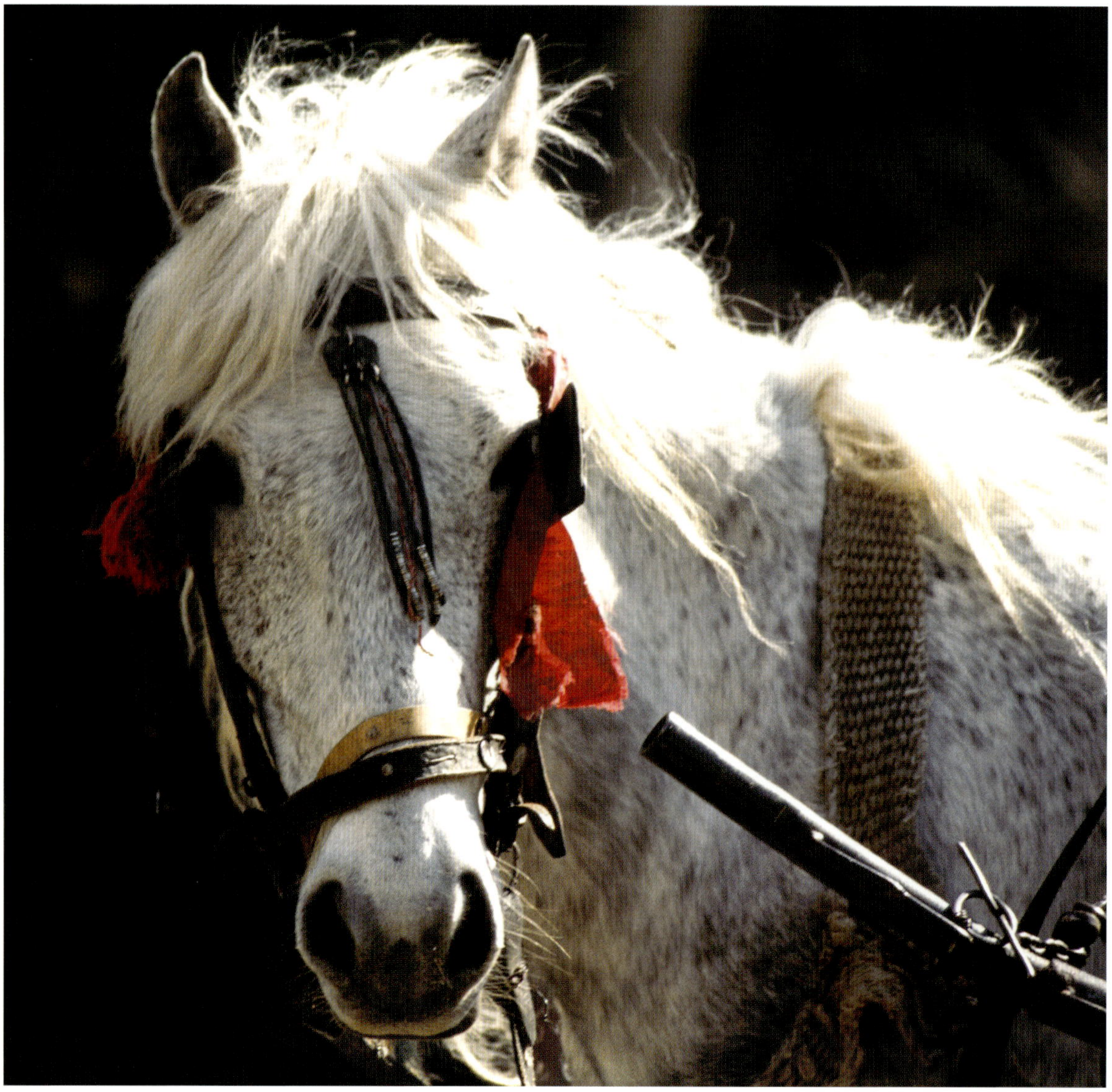

Chapter Eight

CART TROUBLE

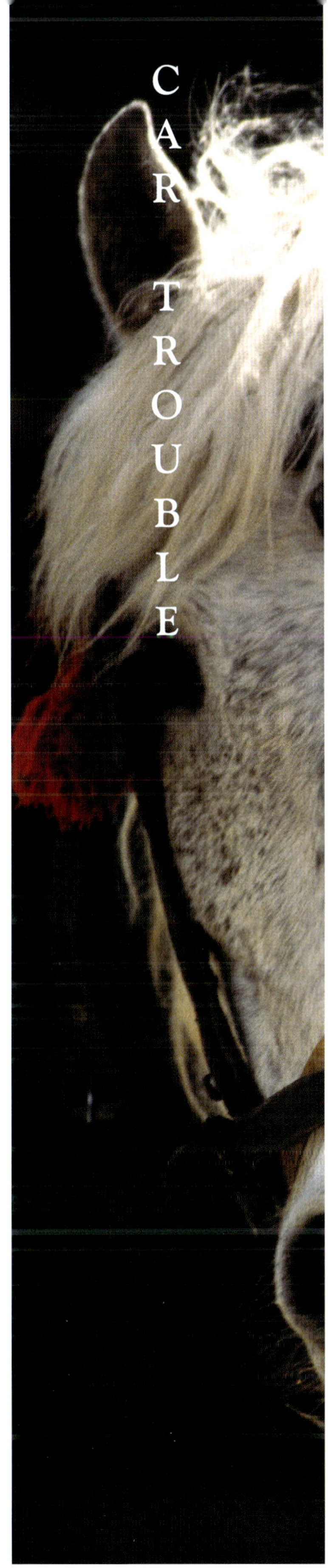

'I read that every known superstition in the world is gathered into the horseshoe of the Carpathians, as if it were the centre of some sort of imaginative whirlpool; if so my stay may be very interesting.'

JONATHAN HARKER'S JOURNAL, *DRACULA*

During our first two weeks in Romania we had been told time and time again that the Transylvanians are not a superstitious people, and that they have no belief in vampires – it is a misconception of the country foisted on them by the *Dracula* connection. However, for a non-superstitious population, we cannot help but notice they have one or two interesting quirks. Every horse and donkey we see pulling a cart has two red tassels on either side of its bridle, which we learn are there to 'ward off the evil eye'. A lot of people, even in the larger towns and cities, persist in keeping small balls of cotton wool soaked in garlic in their ears in the firm belief this will keep illness at bay. Here a black cat crossing your path is seen as bad luck – fortunately this can be warded off by taking three steps backwards then three forward again before continuing on your journey.

Vampiric legends and beliefs are spread throughout the world, and appear to have been with us since the days of primitive man, tapping into our twin fears of death and darkness. For all their protests Romanian folklore is no exception – one traditional tale begins with the words: *'There was a time when vampires were as common as blades of grass or berries in a pail... And they never kept still but wandered round at night among the people.'* There are lingering beliefs

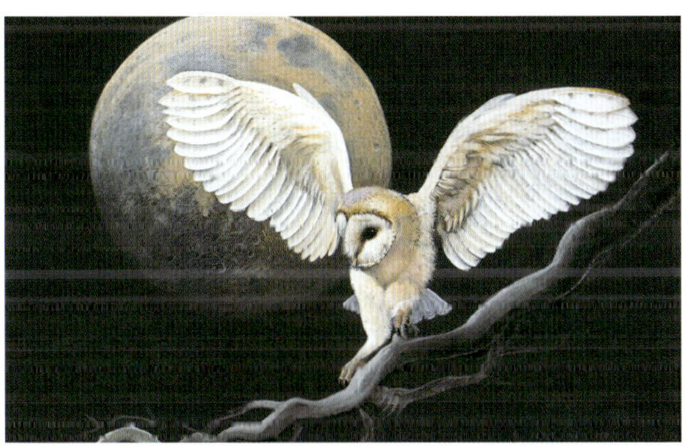

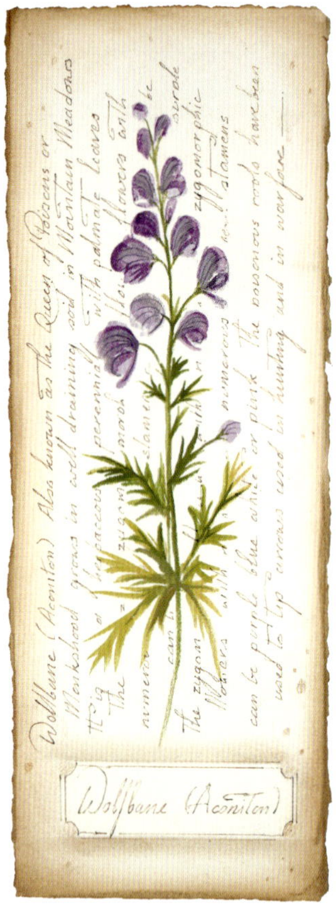

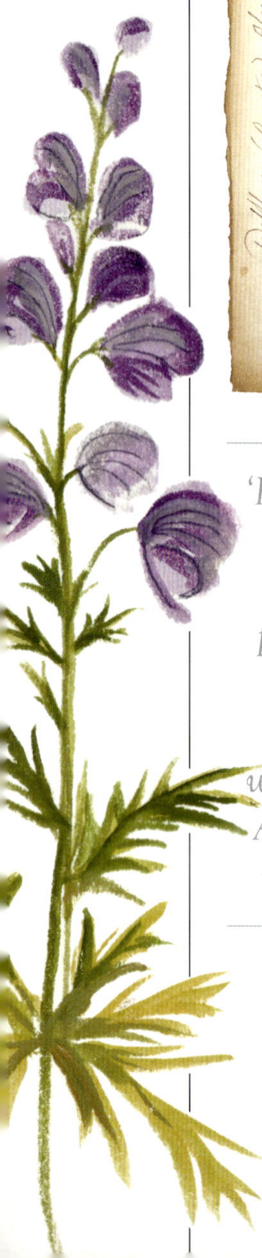

'Even a man who
is pure in heart
And says his
prayers by night
May become a
wolf when the
wolfsbane blooms
And the autumn
moon is bright.'

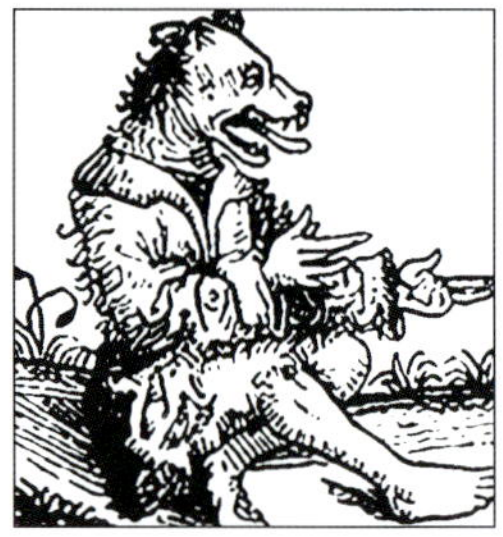

that those who commit suicide become vampires – and if seven siblings are born of the same sex the seventh will have a little tail and grow up to be a vampire. The traditional mythology recounts tales of the 'Varcolac', an extraordinary type of vampire. They can be recognized by their pale faces and dry skin, and by the deep sleep into which they fall when they travel to the heavens to swallow the sun or moon causing an eclipse. They can control the clouds and when they wash their dry skin rain will soon fall. The legends hold that during a period of drought wealthy landowners and nobles would make all the peasants on their estate wash in the river in the hope that there might be a vampire among them, thus bringing welcome rain.

Superstitions have also attached themselves to the wolves we are seeking, in particular the belief in werewolves – humans who can transform into wolves during a full moon. It is thought that this belief has its roots in the psychotic condition of lycanthropy, the sufferers of which drop to all fours, howl, develop an appetite for raw meat and exhibit violent tendencies. During medieval times werewolf beliefs were widespread in European and Baltic countries, and many otherwise inexplicable deaths or incidences of violence were blamed on these mythical creatures. An ancient gypsy rhyme outlines the fear of transformation: ***'Even a man who is pure in heart and says his prayers by night, may become a wolf when the wolfsbane blooms and the autumn moon is bright'***.

Outlaws and bandits played on these superstitions by sometimes wearing wolf skins to further terrify their victims. In the Middle Ages it was widely believed that if a horse stepped in a wolf print it would be crippled and that meeting the gaze of a wolf could cause blindness.

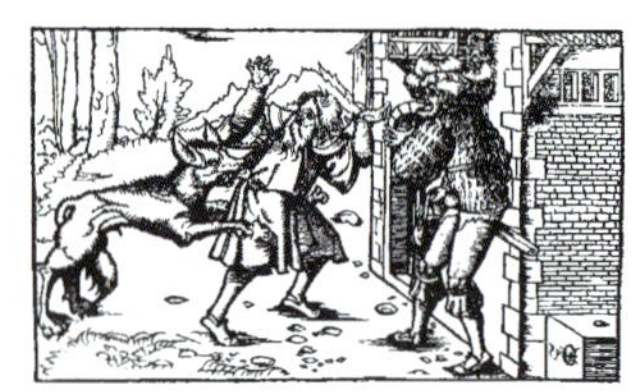

Medieval etchings depicting werewolves

Unfortunately these historically negative views of wolves

have probably played a role in their persecution and destruction. Efforts to eradicate these predators who haunted our nightmares, while at the same time preying on our livestock and hunting the same food source, began in earnest in classical times with the result that only a few pockets of wolves now remain in western Europe. These same Europeans brought their fear and loathing of wolves to the new world when they settled the Americas and pushed them to the edge of extinction in mainland United States. Fortunately attitudes are gradually changing and conservationists are working hard to protect and preserve the remaining wolf populations. Packs have been reintroduced into areas of America where they were previously extinct and wolves are now a protected species in most European countries.

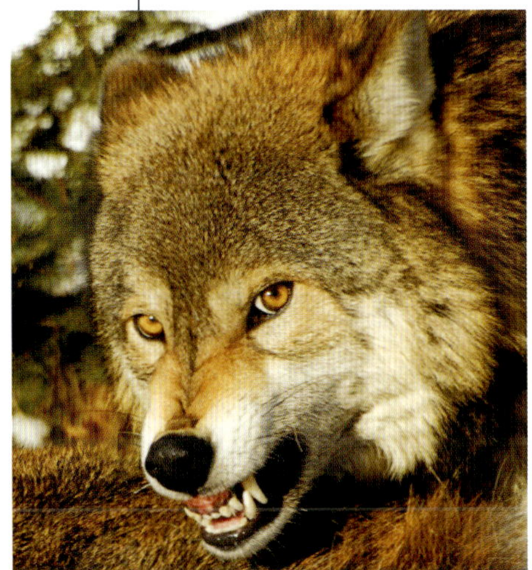

Wolves haunted our nightmares in the Middle Ages

Day Twenty-three

We are almost coming to the end of our stay in Miklósvár. It has felt like we have not only travelled geographically to arrive here, but also stepped out of our own time for a brief period. In the word's of the Count, his estate 'Preserves values and a state of mind still largely unaffected by modern consumer society. It has remained under the spell of a centuries old harmony between the human community and nature itself'.

Walking to breakfast we see lots of elderly ladies sweeping the dirt road in front of their houses with twig brooms. It is Sunday morning, so we assume this must be a weekly ritual. Robert is a little late to collect us in the estate Land Rover – some of his entomologist colleagues have just arrived in the village to carry out an insect survey in the region. He takes random phone calls throughout the day: 'No I don't have a map of my garden'; 'no you can't use my car, I have the keys with me' and so on. As we drive through the neighbouring village of Capeni, Robert points out the immaculate pink and white school building. The director of the school is also a former colleague of Robert's – and an expert on spiders. He is apparently quite obsessed with his subject and has so far discovered four new species in

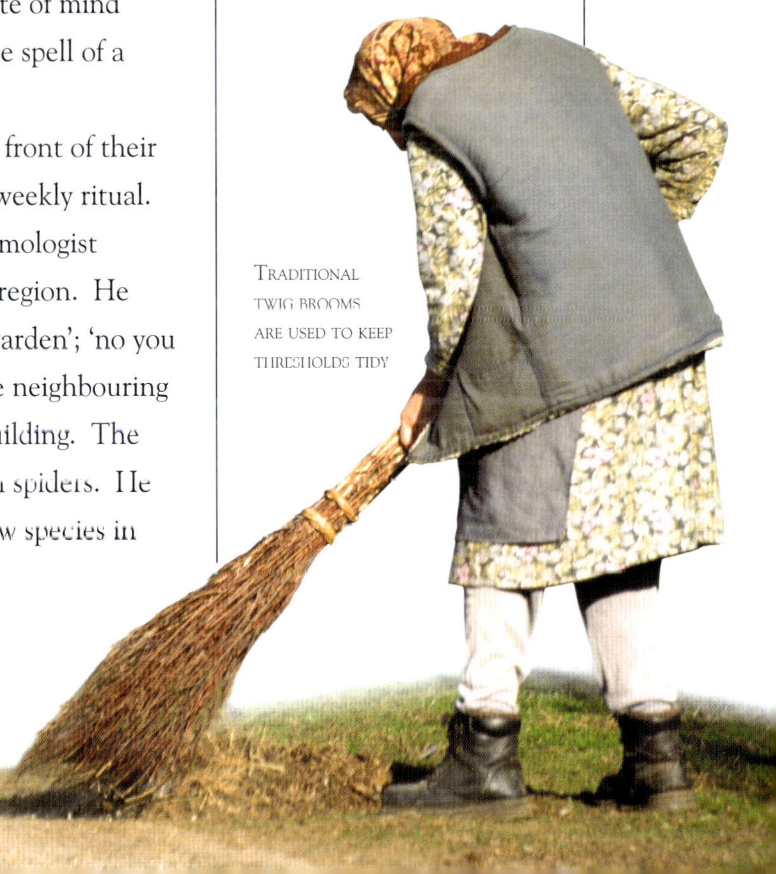

Traditional twig brooms are used to keep thresholds tidy

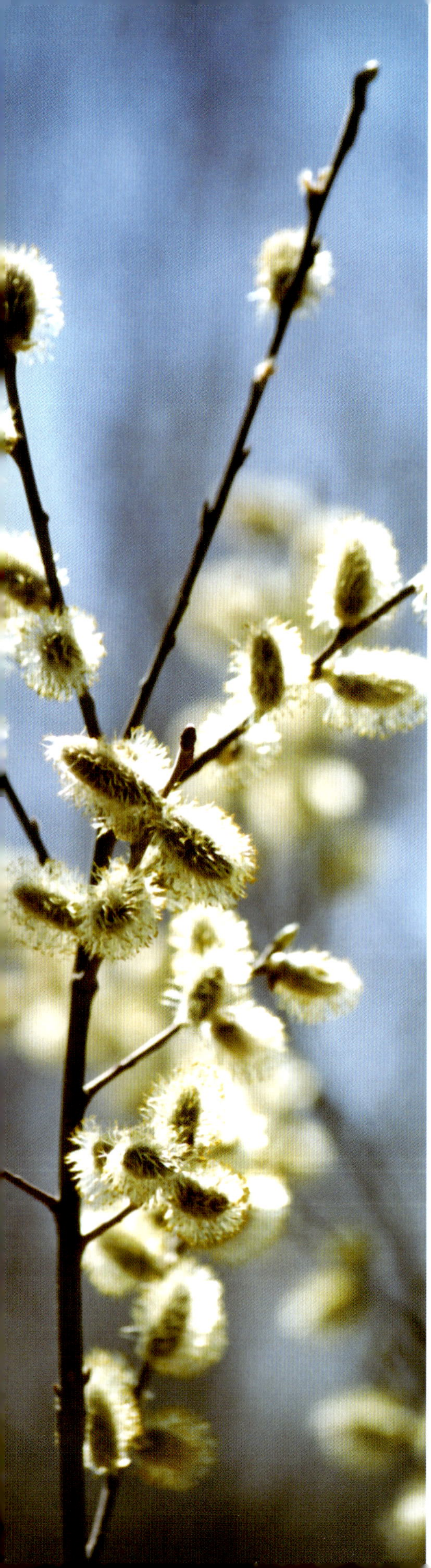

the village – including one lurking in a neighbour's wine cellar. Apparently he is more than a little eccentric. He has waist-length hair and a full flowing beard and insists his pupils listen to Romanian heavy metal music during his biology lessons. At the last prize giving he greeted all the smartly dressed parents wearing jeans and a T-shirt emblazoned with a giant cannabis leaf.

The village streets are fairly quiet, it is the last Sunday before Easter, which is celebrated here as *Douminica Floriilor* – Sunday of the Flowers. We see a few of the villagers dressed in their Sunday Best, carrying small bunches of flowers down the street and, in one of the villages we pass through, branches of pussy willow. These will be blessed and then both children and cattle touched with them so that they will grow and bloom as strong as the willow tree. The holy branches are then taken back to people's homes where they will be placed near icons or above the door and used throughout the year as medicine or for protection against natural disasters and for averting storms. It is believed that people who eat three catkins will not suffer from sore throats through the coming year.

Sunday of the Flowers celebrates the end of Lent – here seen as a time of purification. It is said that if the frogs can be heard croaking today the coming summer will be warm and sunny. In some villages in Transylvania rituals are also performed at the beginning of Lent to charm evil spirits away from houses. On Saint Martyr's Day the elderly residents stamp the ground with their feet to bury the winter cold deep underground. (At least their feet will be warm!) Traditionally as night falls on Sunday of the Flowers girls place a mirror and the blouse they will wear the next day under a pear tree, so that the sun will rise above them. These objects are then used for love and health spells. Few people here see any conflict in the belief in a Christian God alongside the continuing practise of these timeless pagan traditions, and magic and witchcraft remain part of everyday life in many of the rural villages.

As we leave the houses behind we climb into the hills, dip back down into a valley, then head higher once again into the mountains. The roads become ever more excessively

Pussy willow

windy and curvy with dramatic hairpin bends as we climb above the snow line. We make a brief stop so that I can photograph the view from our new vantage point – when Robert hears an odd noise from the engine. He pops the bonnet open and we find the engine has spectacularly overheated. (And speaking as someone who has heard the words 'I think the engine has overheated' in many languages, on many different continents and in a wide range of ramshackle jeeps, trucks and four-wheel drives, I could be said to be something of an expert.) In this instance steam is billowing out and as Robert loosens the radiator cap to release the pressure boiling water spurts and spews out volcanically – and keeps coming. Eventually he prizes the entire cap off and the contents boil and bubble out, pouring on to the road below. As we are only a few kilometres from our destination, Robert decides to drive on; when we stop for the day the engine will cool down and he can then refill the radiator.

We are not the only ones having car trouble this morning. Robert tells us that when Josef delivered the Land Rover for Robert to use today he forgot to bring the keys for his Dacia, which was parked in the castle courtyard. Robert handed over the keys to his own Dacia and the car started perfectly. The fourth vehicle this set of keys has successfully started!

We arrive at Lacul Sfanta Ana. The only volcanic lake in central Europe, it is a famous local beauty spot and also a nature conservation area. The lake itself sits alongside the Mohos peat bog in a double volcanic crater. All sorts of legends and traditions have built up around this body of water. Although it is actually filled exclusively with rain and melting snow, with an average depth of just four meters, there is a persistent belief that it has a subterranean link to the ocean.

On the day of Saint Anna (26th July), the lake is the site of a big procession and a Roman Catholic mass. The centre of this event is the lake-side chapel dedicated to the saint, rumoured to have been built on the site of pagan sacrifices. Robert tells us the church is currently occupied by a hermit.

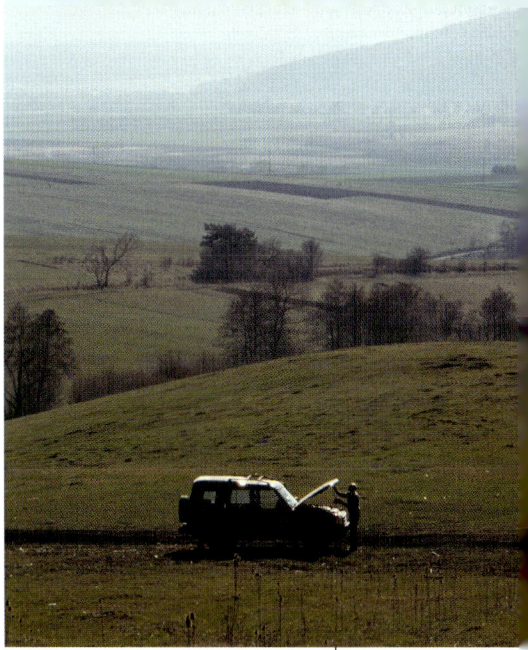

Despite the cold conditions the engine has overheated!

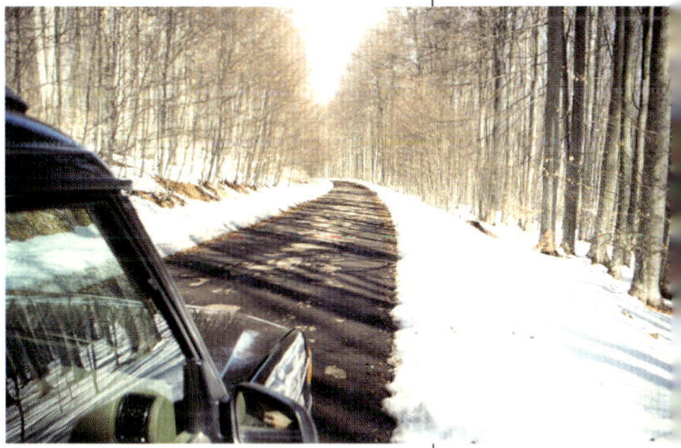

We climb above the snowline

Lacul Sfanta Ana: the only volcanic lake in central Europe

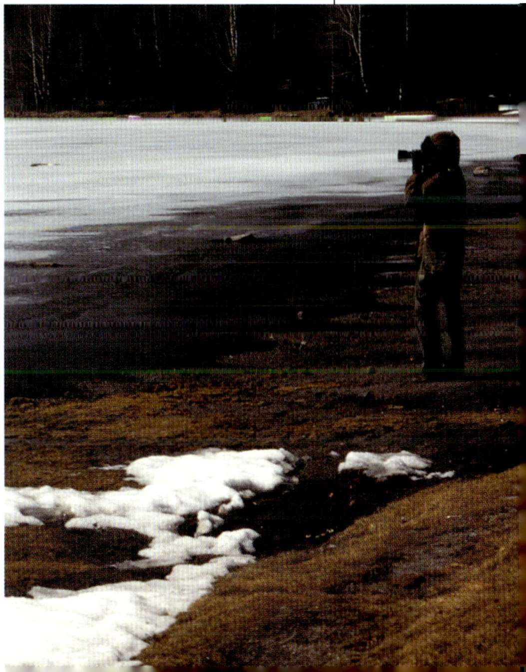

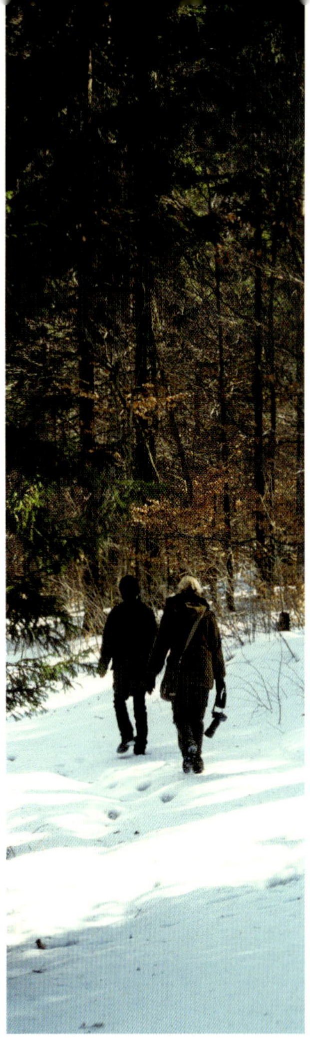

Visiting the region when the snow is on the ground means it is much easier to trace the path left by some of the native wildlife

We see our first wolf tracks

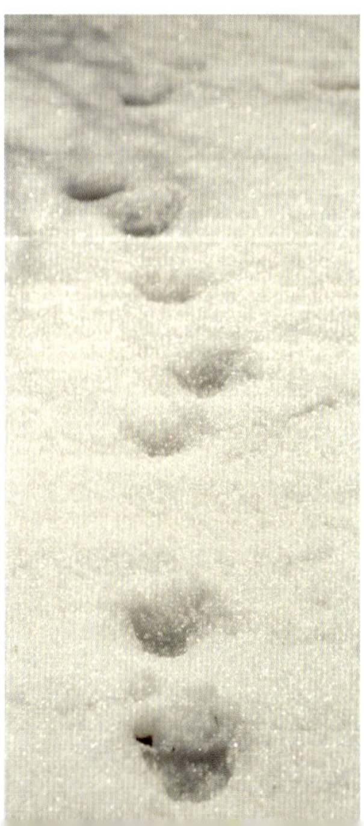

He also tells us one of the more popular local myths. Long ago fairies lived in the water of the lake. At night they used to come out and dance, but their revels were disturbed by the sound of the bell from the nearby chapel. They offered a large reward to a young man if he would steal the bell, but instead of paying him they drowned him in the blue green waters. Apparently this proved no obstacle to him marrying one of the fairies who had left the water to dance with the villagers one fateful night – she kissed her human dancing partner and lost her magic powers. After many adventures the young couple retrieved the bell from the fairy king at the bottom of the lake. When the bell rang once again on dry land the water-dwelling fairies were turned into fish and the fairy king into a very ugly toad, which is how we see them today.

We walk down towards the water – which gives the illusion of being somehow closer than it is. The lake is frozen and the forested slopes thick with snow. We set out to walk around the lake. Reaching the far side we come across a set of wolf tracks, clear in the fresh snow.

Robert tells us that wolf prints are similar in shape to dog prints, but they are larger than those made by all but the very largest dogs. A wolf's paws are configured to walk on snow thus making the prints wider due to the animal's weight being concentrated in a small area. The wolf's front paws are larger than the hind paws and the toes more spread out. This helps the wolf to cling to slippery boulders in rough terrain, and is another clear indication of a wolf print. When a wolf is trotting through the snow the hind foot will land in the print made by the front foot on the same side. Dogs place their hind foot beside, rather than inside, the front footprint. When travelling through deeper snow as a pack the wolves run in single file landing in the tracks created by the leading wolves, reducing the energy required to run. Their long legs mean they have no trouble moving through a thirty-centimetre depth of snow – but above forty-three centimetres it will reach their chests, slowing them down. Wolves are physically very well adapted to travelling seemingly tirelessly over long distances. Their normal walking speed has been estimated at an average of eight kilometres an hour – around twice that of humans – and a wolf has been recorded travelling ninety-six kilometres within twenty-four hours.

Robert can also estimate the approximate size of the wolves who made these tracks. He tells us that one stride length is roughly equal to the distance between hip and shoulder. Wolves tend to walk in a straight line while dogs meander and their trails zig-zag. It is thrilling to think that within hours of our hike around these trails a wolf was here, making his way purposefully along the lake side.

We only see the tracks of one wolf, but in this region they typically form packs of around six animals. Each family group is dominated by the alpha male and female. The rest of the pack is comprised of juveniles and occasionally a 'demoted' alpha who has been ousted by a younger stronger wolf. They hunt communally, share their food, and all take part in caring for and raising the pups.

While we are walking back we come to a small perpendicular track in the woodlands. A ranger drives down specially to tell us that there is a great spot for photographing the whole lake from a little way further up the hillside. He also informs us that if we sniff the small holes in the banks of the lake, we can smell ammonia from below the volcano. If the smell is strong the weather will be bad. We stick our noses in and sniff, but with no basis for comparison are unsure of the forecast. (When planning this journey I never imagined it would lead me to this point – sniffing underneath a Transylvanian volcano in an attempt to predict the weather.)

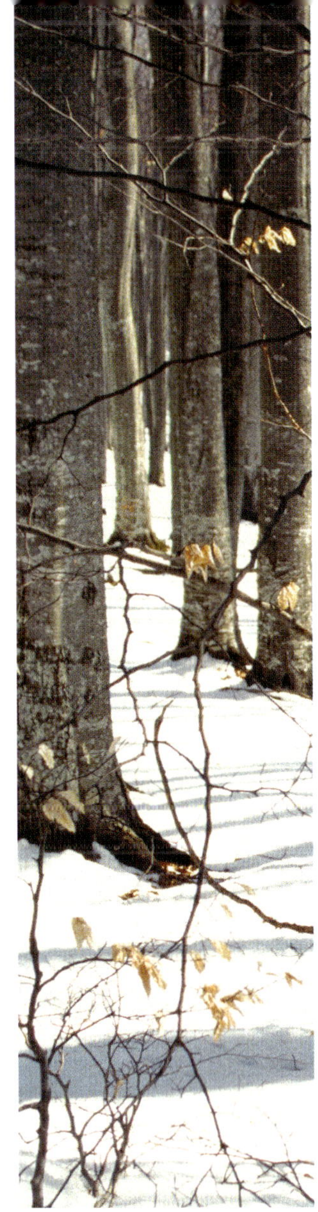

Wolves are physically well adapted to snowy terrain

Robert tells us there are also sulphur caves close to the lake – they are used as a natural 'sauna' by the locals. In the cave nearest to the entrance the sulphur forms a gas cloud from the floor to approximately waist height. If you sit on the bench provided you will start to feel warmth as the blood vessels begin to dilate. 'Of course,' he adds, 'if you stick your head below the surface you will die'.

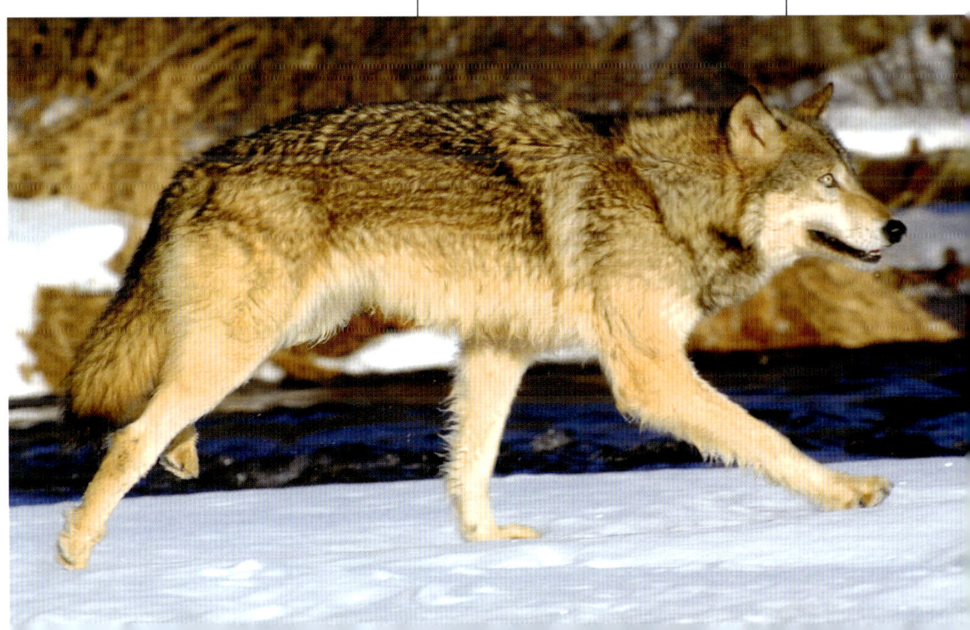

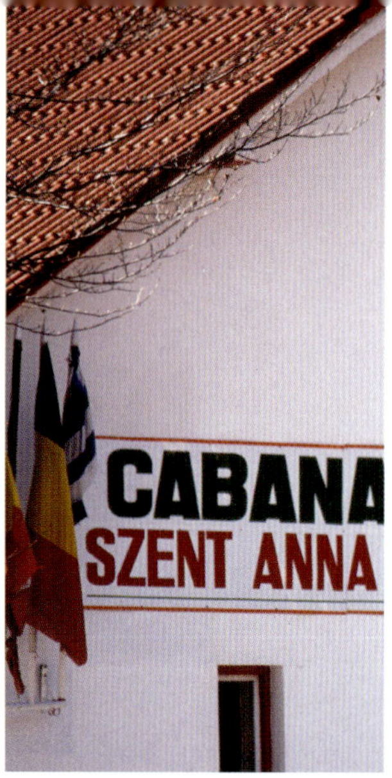

We enjoy lunch at this typical Transylvanian restaurant

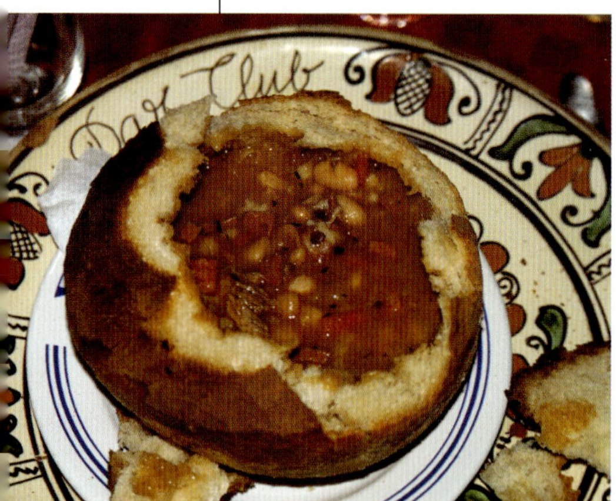

Bean soup served in a 'bowl' made from bread

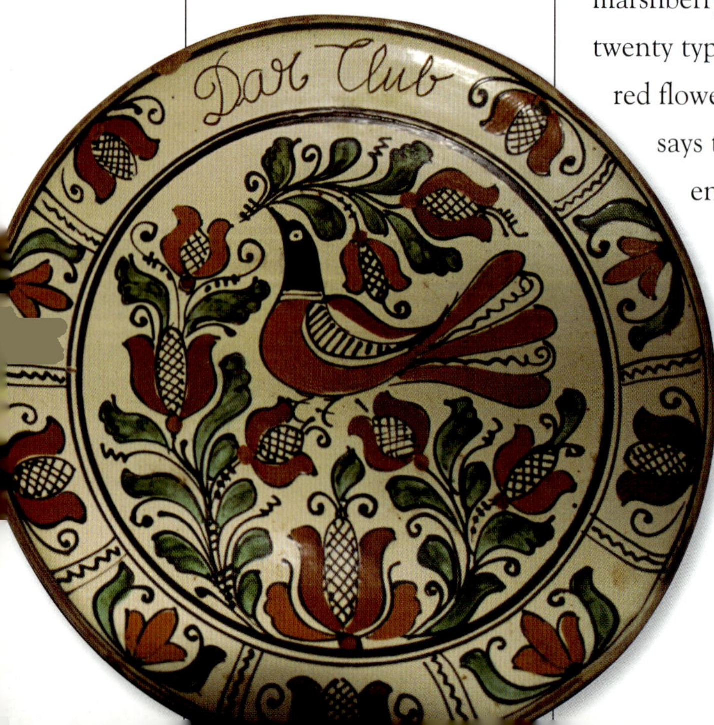

Personally, I prefer my saunas a little less life threatening. The interior of the caves are completely filled with sulphur; walking far into the caves used to be a favoured method of committing suicide.

We scramble up the steep slope to the point where we can see the promised view – and are also rewarded with a sighting of bear tracks. Visiting the region when the snow is on the ground means it is much easier to trace the path left by some of the native wildlife. Robert suggests that if we carry on higher up the slope, which is little short of vertical at this point, it will be a short cut. Luckily he is right. We emerge breathless at the top. From this point we can easily cross to the second crater where the Mohos peat bog is located.

It is known that Lacul Sfanta Ana has been here for more than six centuries, as there is a reference to it in a 1349 charter. But naturalists say it will only be here for another two centuries as it is gradually turning into a swamp. In 1870 its depth was twelve and a half meters, now its deepest point is just over eight metres. The lake is getting ever shallower as sediment piles up from below.

Looking across at Mohos, we are seeing the future of Lacul Sfanta Ana. Once a lake three times as big as Sfanta Ana, Mohos atrophied around 3,500 years ago. This means that so many plants were growing in the lake – especially sphagnum moss – that it became a bog. The peat bog is a botanist's paradise. Home to round-leaved moor-grass bog berry, marshberry, bog-rosemary and – non botanists try to contain your excitement – more than twenty types of sphagnaceae. There are three different species of insect-eating sundew here: red flowers that look quite harmless but are able to catch passing mosquitoes. Robert says that he would have loved to have shown us these rare carnivorous plants, but the entire surface of the bog is covered in snow and they are 'hibernating'.

Back at the car Robert grabs a water bottle, which he fills from a nearby well, and chances refilling the radiator. We cruise down the hill (Robert: 'I don't really need brakes') to a very typical Transylvanian restaurant where we sample their house speciality – a hearty rich bean soup served in a bowl and with a

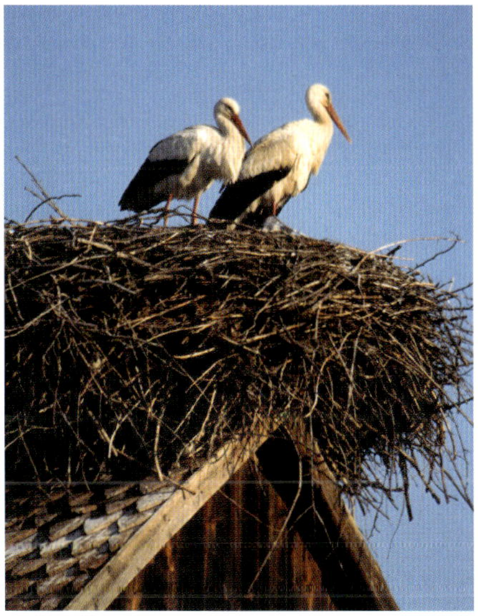

spoon both made out of bread. This is an ingenious method of saving on the washing up – when you have finished the meal you have also eaten your tableware. The ceilings are decorated with wild boar and bear skins and there is a stuffed Ural owl gracing the walls, giving us the impression that we are eating in a medieval Transylvanian hunting lodge. All the air of tradition and history is somewhat marred by the very loud and less than traditional music which veers between anodyne Euro-pop, and hardcore hip-hop (blow my whistle bitch).

We career back down the hairpin bends in a miraculously working car (offer quiet thanks to Saint Anna). Robert veers idly from side to side as he spots distant birds, nests, flowers, lynx tracks, and even bees at the side of the road. We all gaze hopefully at a distant speck that may or may not be a lesser-spotted eagle. Constantly peering out of the windows, Robert's gaze seems to be everywhere except on the road ahead of us – fortunately there is almost no other traffic and he always manages to swerve back to the correct side of the track if another vehicle approaches.

Back at our guest house we take advantage of the still-warm late afternoon sun to sit out in the garden – me to catch up on my journal and Pollyanna to add detail to the sketches she has been making throughout the day. The world's largest bumble bee buzzes around us. (Quick! Alert the entomologists!) It is about the size of a blue tit and even seems to have its own nest box. In the next door garden an elderly man is hoeing the soil, while his wife plants potatoes in the furrows he is creating and a stork peers down at them from its nest on the chimney pot.

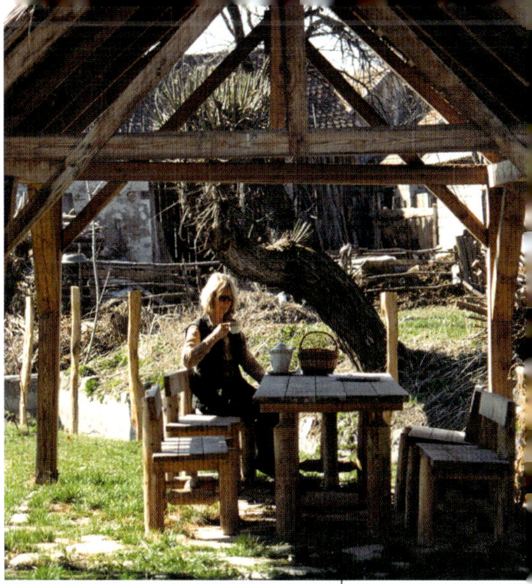

Time to enjoy a cup of tea in the late afternoon sun

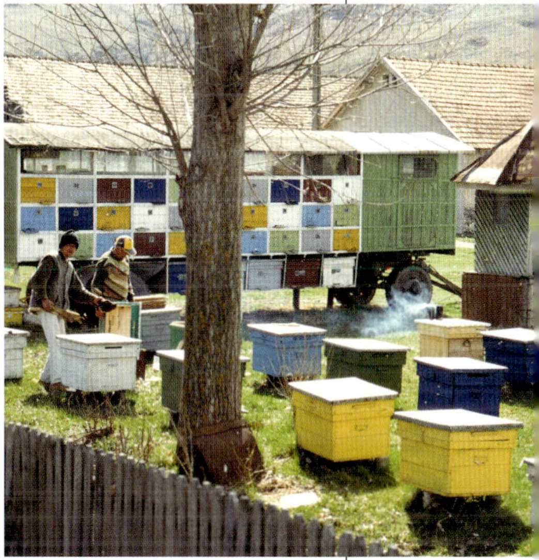

Brightly decorated mobile bee hives

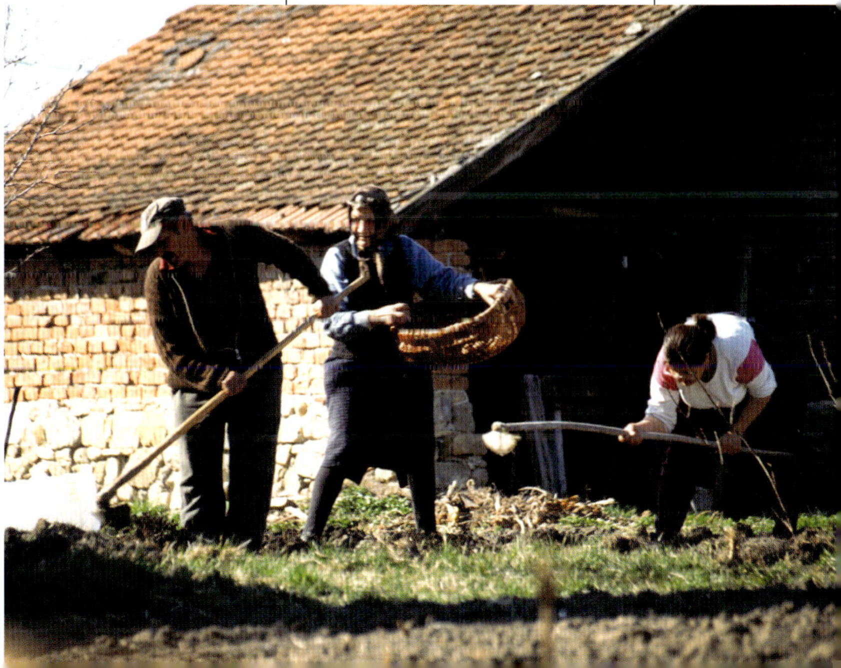

An elderly man is hoeing the soil, while his wife plants potatoes in the furrows he is creating

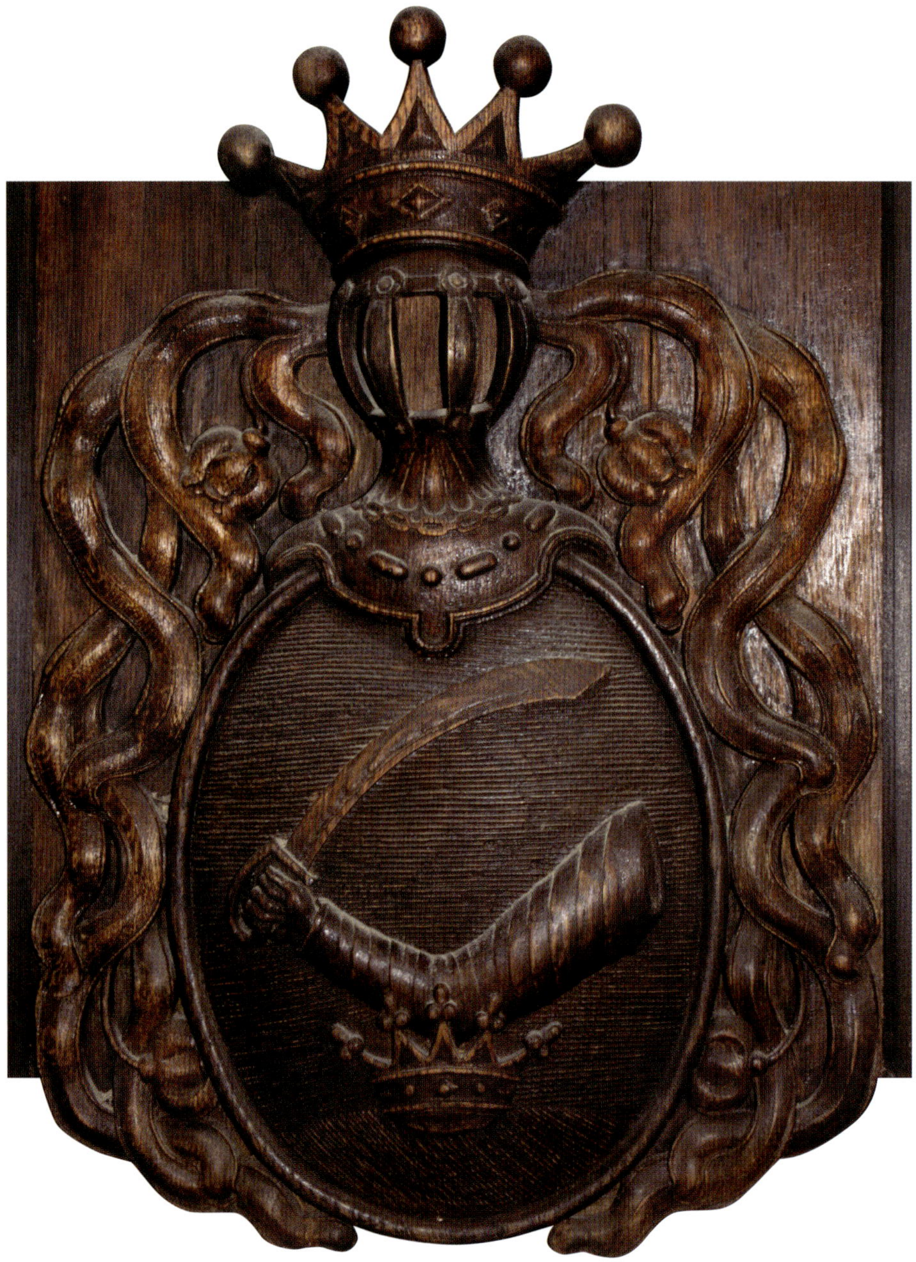

Chapter Nine

'Non est mortale quod opto'
(It is not a mortal in whom I trust)

KALNOKY FAMILY MOTTO

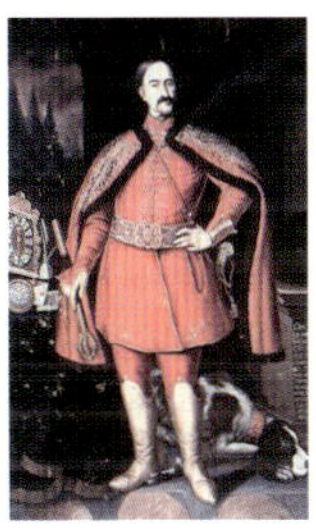

SAMUEL KALNOKY

The Kalnokys are one of the most ancient families in Transylvania, with a history stretching back twenty-five generations. Our absent host, Count Tibor Kalnoky, is of the first generation to have been born outside Transylvania in seven hundred and fifty years.

The Catholic Szekler Kalnoky family became the most prominent landowners in Miklósvár around the middle of the seventeenth century, and the construction of the present-day castle began then. At the end of that century it was further enlarged by Samuel Kalnoky, who was the Vice-Chancellor of Transylvania – he received the hereditary title of Count. But during the communist regime the Kalnokys, like all Romanian aristocracy, lost everything. Some families stayed in their homeland – but were not allowed to study and had to accept menial work as street sweepers or cleaners. Tibor's grandfather, a notorious anti-communist and anti-Nazi, chose to flee. Leaving was far from an easy option however. The family, including Tibor's father, then just eight years old, were lucky to escape with their lives. They fled Romania on the eve of the Second World War – crossing the borders by hitching a ride from Germany in the turrets of General Patten's tank. The Count's family subsequently saw all their property seized by the communists when they came to power at the end of the war.

The family fled to America, where Tibor's grandmother found work as a servant – before eventually going on to run the famous Nuremberg Guesthouse that played host to the victims and the accused during the Nazi war trials. Growing up in

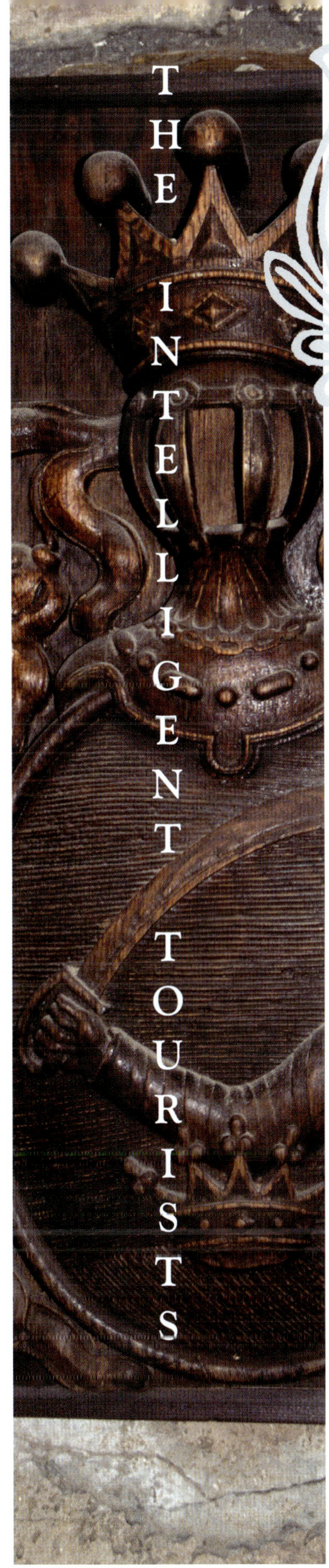

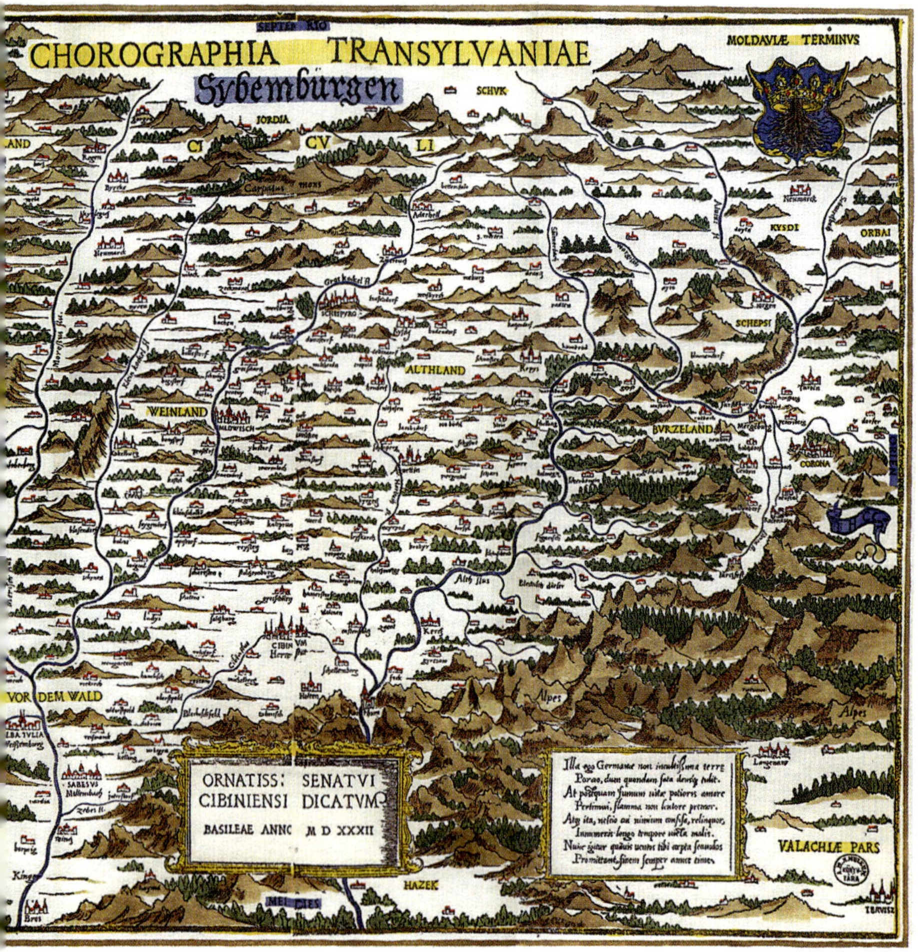

An early map of Transylvania

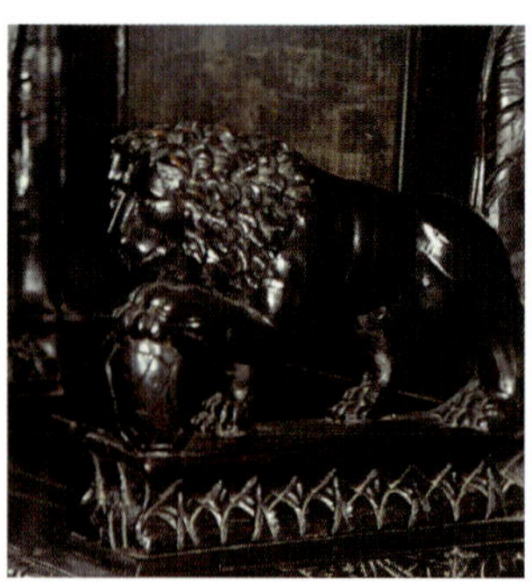

America, Tibor's father found work with IBM, rising to head the eastern European branch of the company. Tibor enjoyed a truly international childhood: born in Germany, he went to kindergarten in Holland, attended an American school in Cannes and a German school in Paris.

In 1987, about two years before the collapse of communism, Count Kalnoky and his father made an illicit visit to their family's former home. As aristocrats travelling in a country where foreign tourism was virtually forbidden, there was a risk that they could be accused of spying or trying to stir up unrest. When they came out of the house they discovered a huge crowd had gathered in the courtyard to greet them and embrace Count Tibor's father. However, the crowd dispersed as quickly as it had gathered as word spread that cars were approaching. As there were no cars in the village at that time, it had to be members of Ceauşescu's much-feared Securitate, the secret police, making their way to the village. The Count and his father were forced to escape by fleeing through the forest. Despite days of questioning, none of the villagers ever betrayed the secret of their visit. Count Tibor has since described this as 'the trigger moment,' when he decided to commit himself heart and soul to taking back his ancestral home.

The revolution in Romania paved the way for people to reclaim property seized by the state and Count Tibor returned to Transylvania ending fifty years of exile for his family. However, it took eight years of suing the government before the castle and hunting lodge were returned – and he discovered that during the communist regime many of the buildings had fallen into ruins while others were in a very poor condition after years of service as

communist party headquarters and community halls. The family set about finding ways to save their heritage and restore the former character of the estate. Since the revolution Count Tibor has been working to fulfil his vision of turning Miklósvár and the surrounding woodland into an environmental retreat and a preserve of the region's architectural heritage.

Day Twenty-seven: evening

We walk up to the main lodge. Once again we have been told that Count Kalnoky will be dining us with this evening – but having failed to meet him throughout our stay we have no real expectations of seeing him. However, this time when we arrive, instead of being taken to the kitchen where we have been eating with Robert and the lodge staff, or the dining-room we shared with our friends from San Francisco, we are shown down a steep flight of uneven stone steps into the wine cellar – the Count has returned and he will be joining us this evening.

The wine cellar is decorated in a modern gothic style – a huge scrubbed wooden table is laid with three place settings and a brass candelabra. Further candles burn in wall sconces and dusty bottles of wine line the walls. A huge wooden panel carved with the Kalnoky crest hangs on one wall. As we sit at the table a fire suddenly blazes spontaneously to life in the huge fireplace behind us.

We sit and wait for the Count. And wait. And wait. After all the build up and delays we have begun to conjure up all sorts of terrifying images in our imagination – so when we hear slow heavy footsteps on the stairs, and the huge wooden door slowly creaks open, it is all we can do not to clutch each other for reassurance.

The Count enters. He exists. He is tall, with dark hair – but not in any way saturnine – in fact he is softly spoken and extremely charming. Urbane and sophisticated he speaks flawless English (as well, it transpires, as Romanian, Hungarian, German and French).

He has just been to inspect the latest work being carried out on the castle, 'Have you

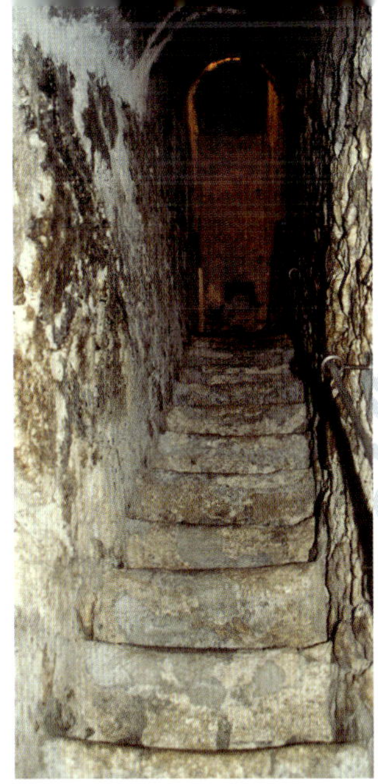

The steep, uneven steps leading to the wine cellar

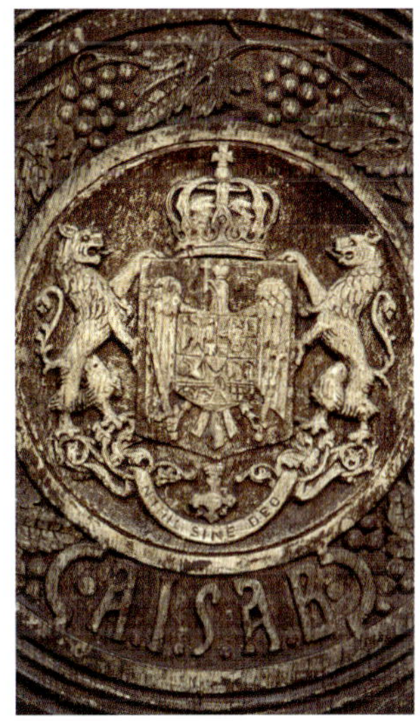

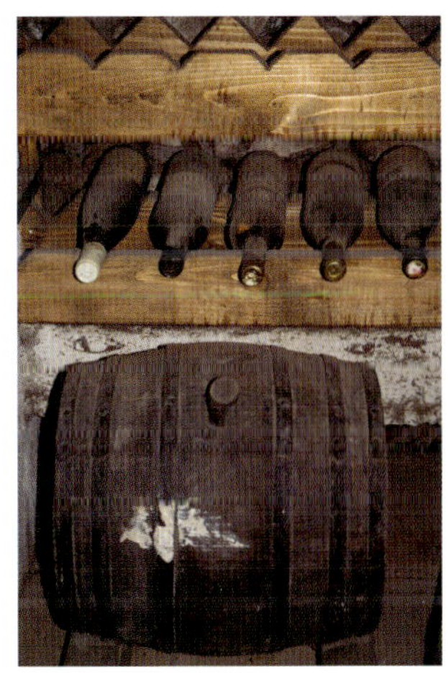

Ancient dusty bottles and barrels in the wine cellar

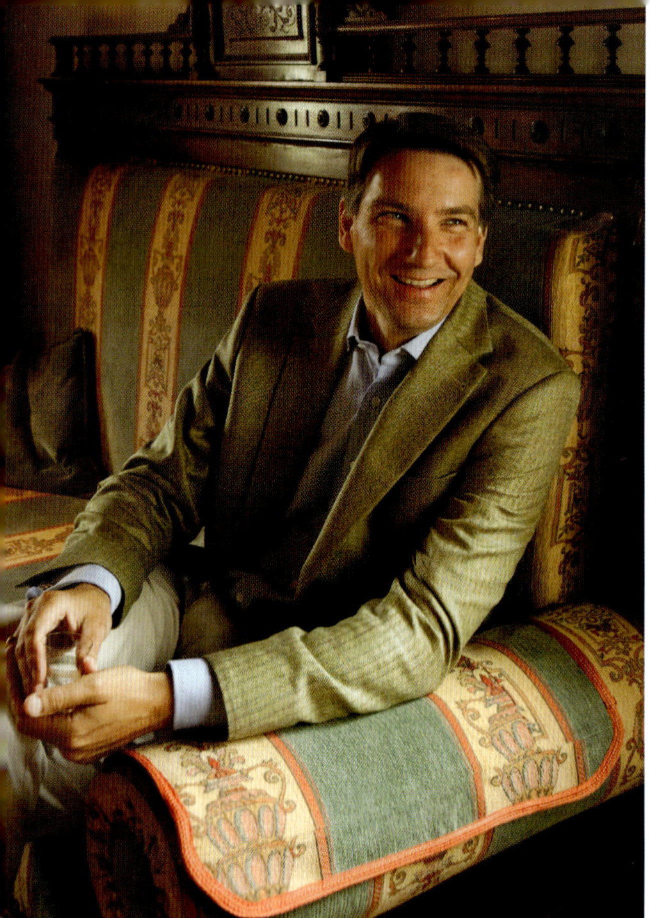

Count Kalnoky

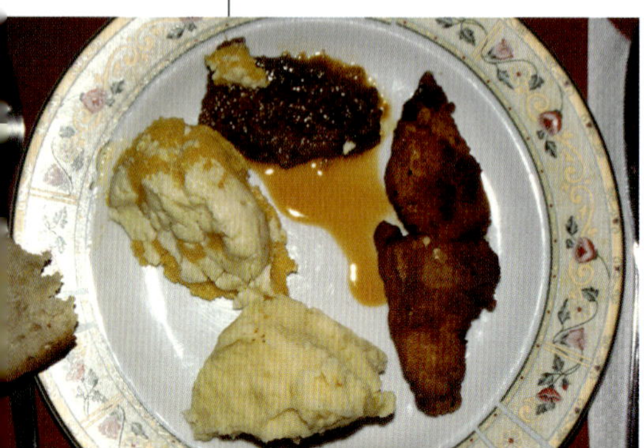

Fried *mămăligă* with mashed turnips and peas in a spicy sauce

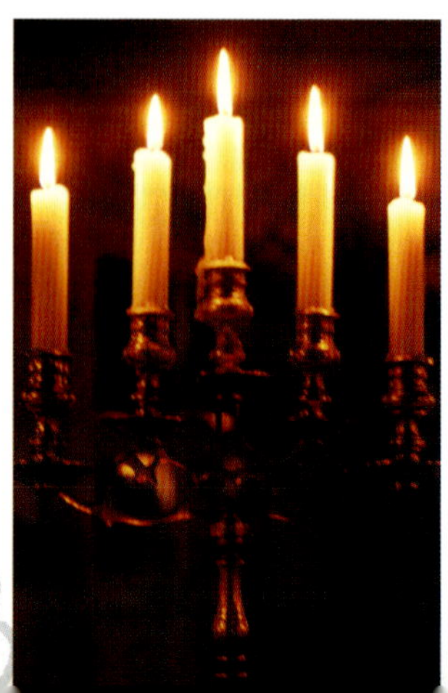

seen the excavations?' And to see if there is any improvement to the amount of water running down the inside of the walls: 'Well it saves on taps'.

The kitchens have provided another delicious meal – hearty vegetable soup, followed by fried *mămăligă* with mashed turnips and peas in a spicy sauce. We finish with fruit salad with crispy sweet pastries dusted with icing sugar – the Count tells us that their Hungarian name translates as 'Things that break into one thousand pieces'. He is a delightful dining companion, charming and funny with a lively laugh which verges on a giggle. He apologises for the fact that his wife will not be joining us – he is married to the great-granddaughter of a Hungarian count (who incidentally was given his title by the Pope in recognition of his work as an artist. Note to self – send some of Pollyanna's sketches to the Vatican).

We chat about our shared interest in wildlife – he is especially keen to hear about the small bird sanctuary we used to run. For fifteen years we rescued, cared for and rehabilitated injured raptors and other wildlife from our home in Derbyshire. Count Tibor had also rescued injured birds, setting up a wild bird hospital while living on the outskirts of Paris when he was just sixteen. He went on to study veterinary science in Germany before moving to Budapest, where he was working for a veterinary pharmaceuticals company when he was summoned back to Transylvania to reclaim the family heritage. He rescued and cared for a wide variety of animals – and even had an area he nicknamed the 'lunatic asylum'. One patient under his care was a cow who had been poisoned and suffered brain damage. She was even unable to recognise food – unless a white rabbit was placed in her enclosure, at which point she would appear to pull herself together and eat and behave quite normally. When the rabbit was removed her brain would re-scramble – as graphically demonstrated by the Count, whose impression of a mad cow has us convulsed with laughter!

We chat about his ecotourism venture – he is very interested to know why we were drawn to this area. The Count is very optimistic for the success of his project – it is

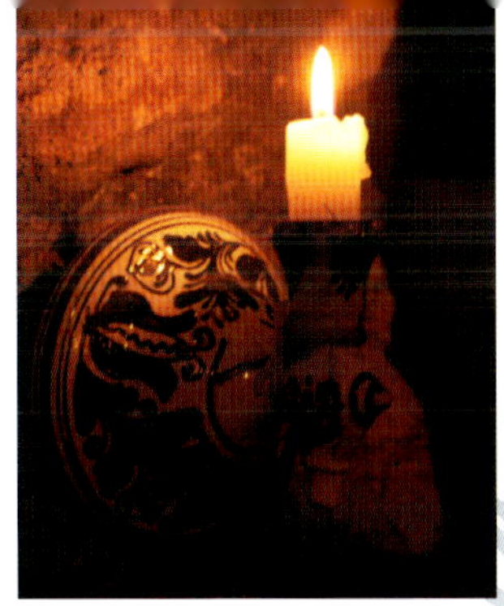

ultimately intended to be sustainable as a provider of jobs for the area rather than a profitable business – and of course to protect and preserve the spectacular natural habitat which surrounds us. He explains that they are aiming to attract visitors who are interested in culture and wildlife, 'We call them intelligent tourists'. (At this point I wonder if I should apologise for any disappointment caused by our arrival.) The project is in its very early stages, and we are part of the 'test phase'. However, a few groups have been here ahead of us and the Count regales us with stories of some of his visitors from America. The first was a gothic rock star – a guy who was about 'Two metres tall and one and a half metres wide', he arrived dressed head to toe in black leather including a floor-sweeping black leather coat and a wide-brimmed black hat. The chain of a pocket watch trailed down to the floor before returning to his jacket – and tattoos were visible creeping out of his cuffs and collar. If we were attracting attention walking through the village I can only imagine the response to this apparition. The Count arranged for one of his colleagues, a guide named Mona, to show the rock star around the cultural attractions of the area. The rock star took a great shine to Mona, which he demonstrated by inviting her to 'See his collection of pickled human embryos'. Initially they were concerned that this was something he was travelling with – but in fact they were back at his home in the USA. (Well that's perfectly normal then.) A tempting offer I am sure, but oddly Mona declined. Apparently the rock star called himself 'Something Von Bismark', the Count paused and with a faintly bemused smile said 'Of course I know the Von Bismark family very well and I don't *think* he was related...'

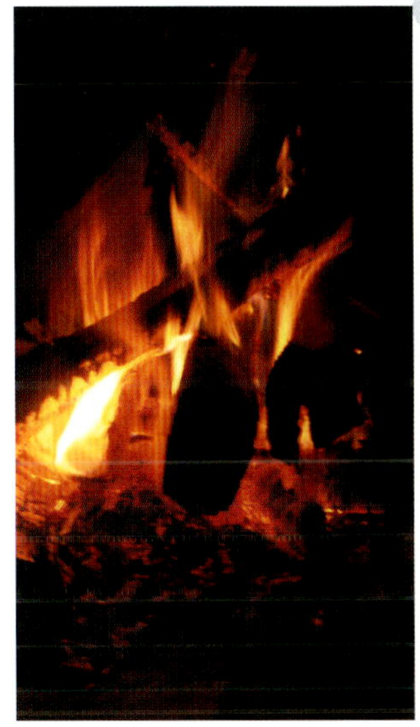

A roaring fire bursts into life

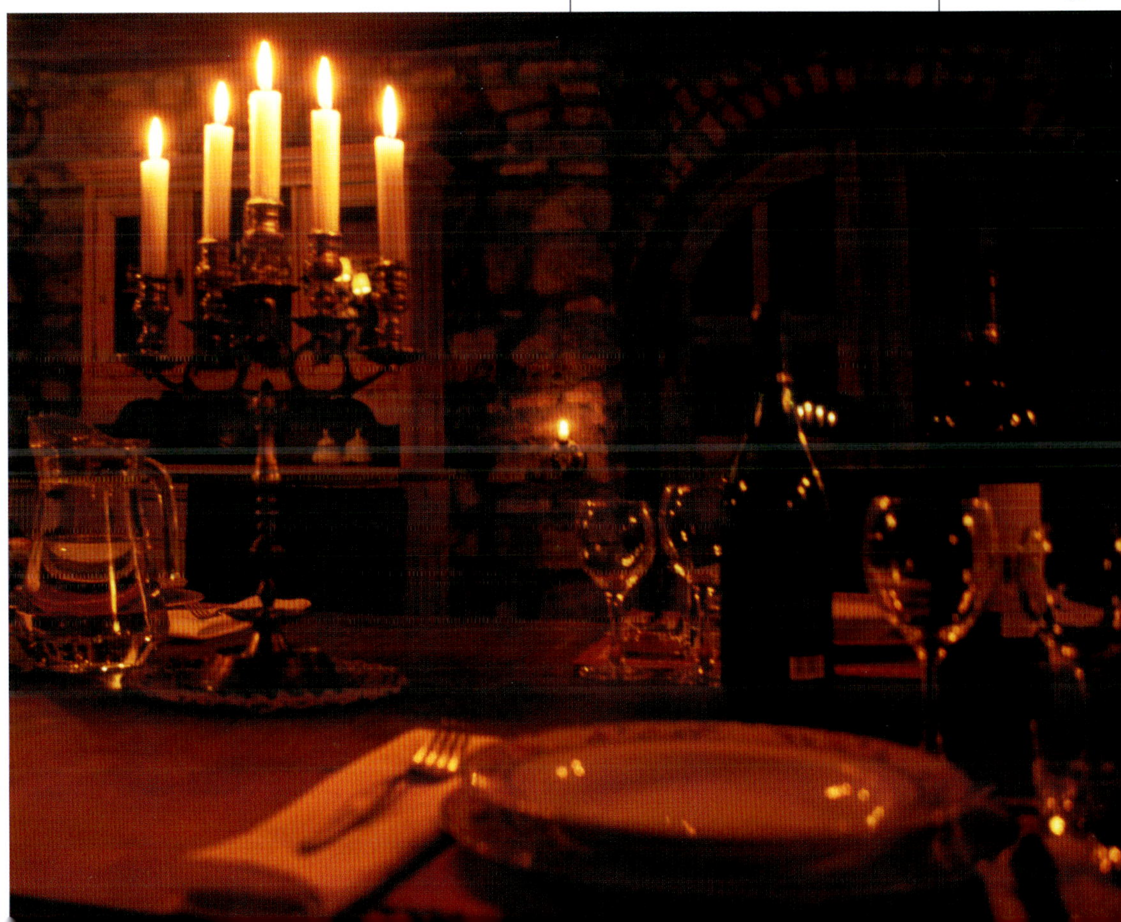

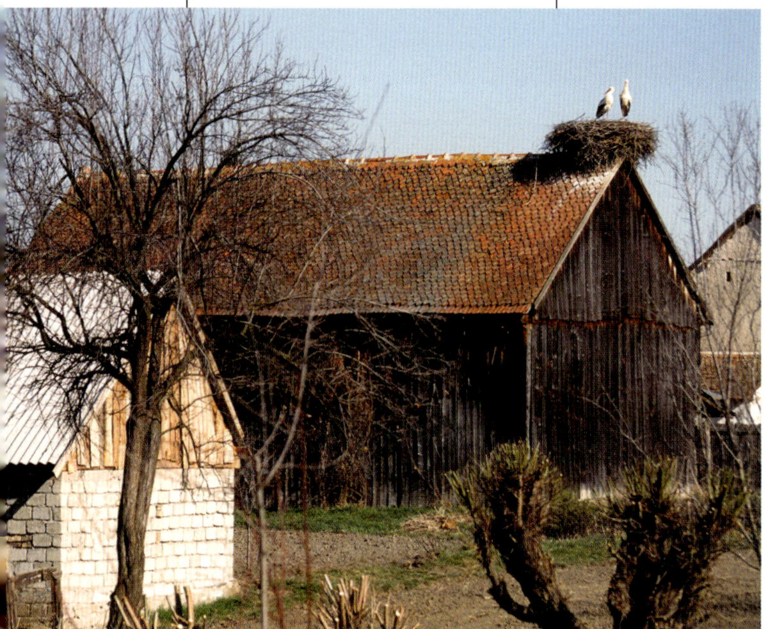

Miklósvár

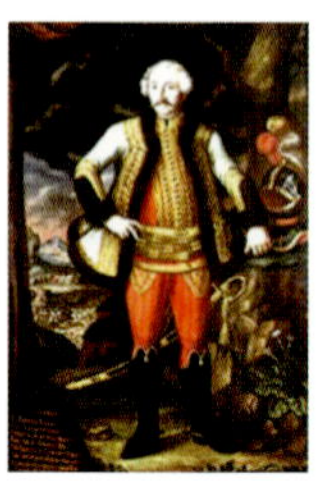

An Ancestor of Count Kalnoky

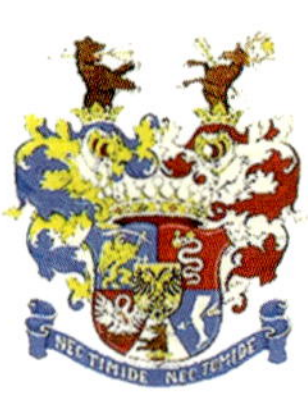

The Kalnoky family crest

The second American to descend on them showed a great deal of interest in the restoration of the castle; he asked if he could meet the Count's architect, and then if he could view the floor plans. Slightly surprised they agreed, only for him to announce that he was planning to rebuild a stone by stone replica of the castle – in Ohio. He went on to ask for the name of a tailor who could make him some traditional royal ceremonial robes, and an artist who could provide him with some freshly painted ancestral portraits to line his new baronial hallway. ('Whose ancestors?' enquired the Count. 'It doesn't really matter', came the reply.) Apparently all of this was sparked by the recent discovery of a Hungarian great-grandmother in his family tree.

The Count also discusses the difficulties of expanding the tourist market after so many years of communist rule. In the scant ten years since the revolution some projects devised by the newly established tourist board have proved more successful than others. He tells us of one of their early initiatives which aimed to lure visitors away from the beaches of Spain and the South of France to their coastline on the Black Sea. According to Count Tibor this blithely overlooked the fact that, 'This area is freezing year round – apart perhaps for an hour one day towards the end of August'. The tourist board's first idea to make the area look like an alluring holiday destination was to purchase twelve twenty-foot-high palm trees to place in rows outside the area's most prestigious hotel – the Palm Beach. All was lovely until the first Romanian winter at which point they realised there was no way to move twelve twenty-foot-high trees undercover – so they all perished in the snows. The hotel is now graced by twelve pots containing blackened and peeling dead palm trunks.

However, the Count talks passionately about his own renovation work and plans for Miklósvár. He is lobbying for the woodland behind the village, home to such rare birds as the lesser-spotted eagle, black storks and white-backed woodpeckers, to be protected under Romanian and European law. He is equally committed to restoring the historical buildings: 'The most important influence on people's soul is nature, but the second is the building they're living in'. By combining wildlife and cultural activities with luxuries such as towelling dressing gowns, feather duvets and hand-picked antiques, he aims to offer something unique in Transylvania – 'A place to experience rural life without living like a peasant'. He has an equally clear vision for his restoration of the buildings: 'I don't like dusty museums – I want this place to look as if the family just popped out for five minutes'. Eventually he hopes to settle his family in one wing of the nearby castle – hopefully after the water has stopped flowing down the inside of the walls.

We leave, wishing the Count the very best of luck with his venture. And indeed the ecotourism project has gone on to thrive since our visit, with six beautiful guest houses now renovated to receive the regular groups of 'intelligent tourists' who travel to this rural wildlife haven.

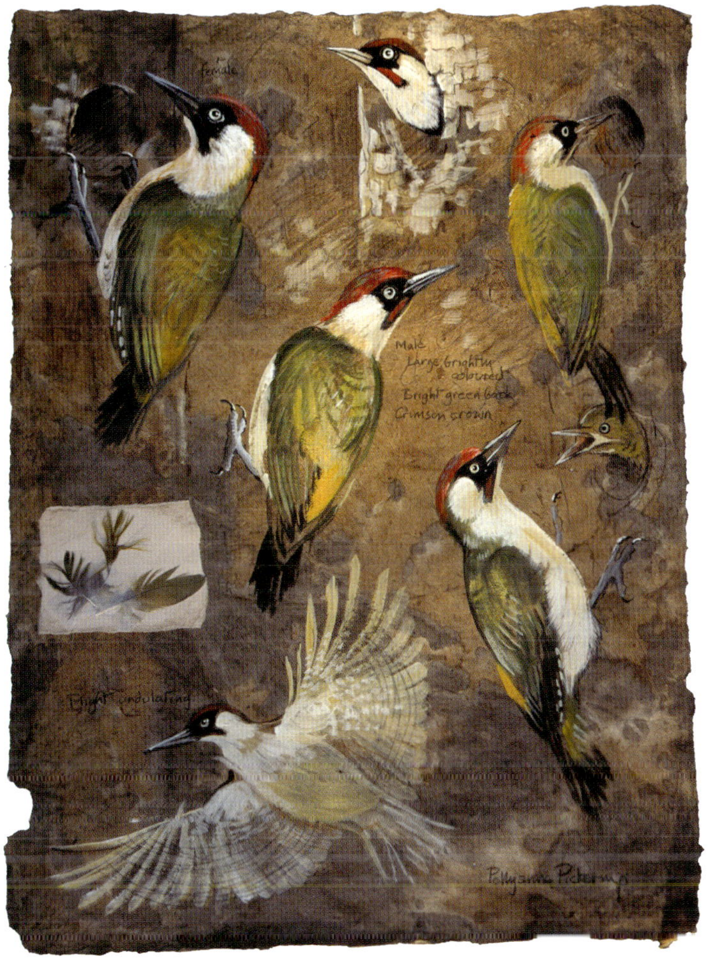

Above and right: Pollyanna's sketches of the green woodpecker

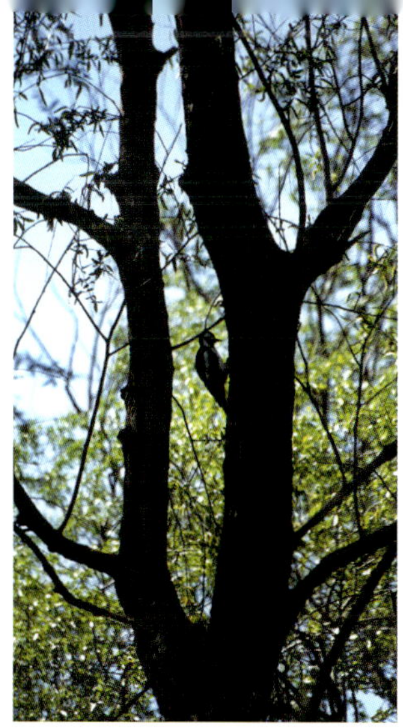

White-backed woodpecker in silhouette

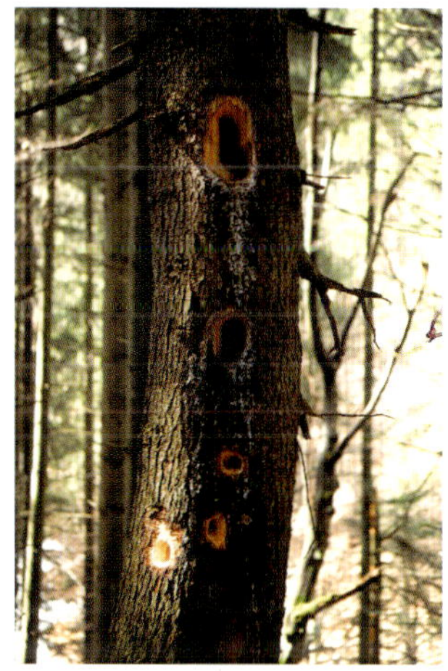

Evidence of woodpecker activity

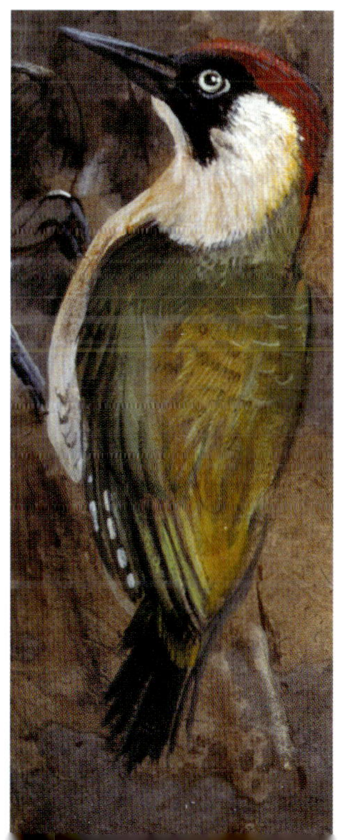

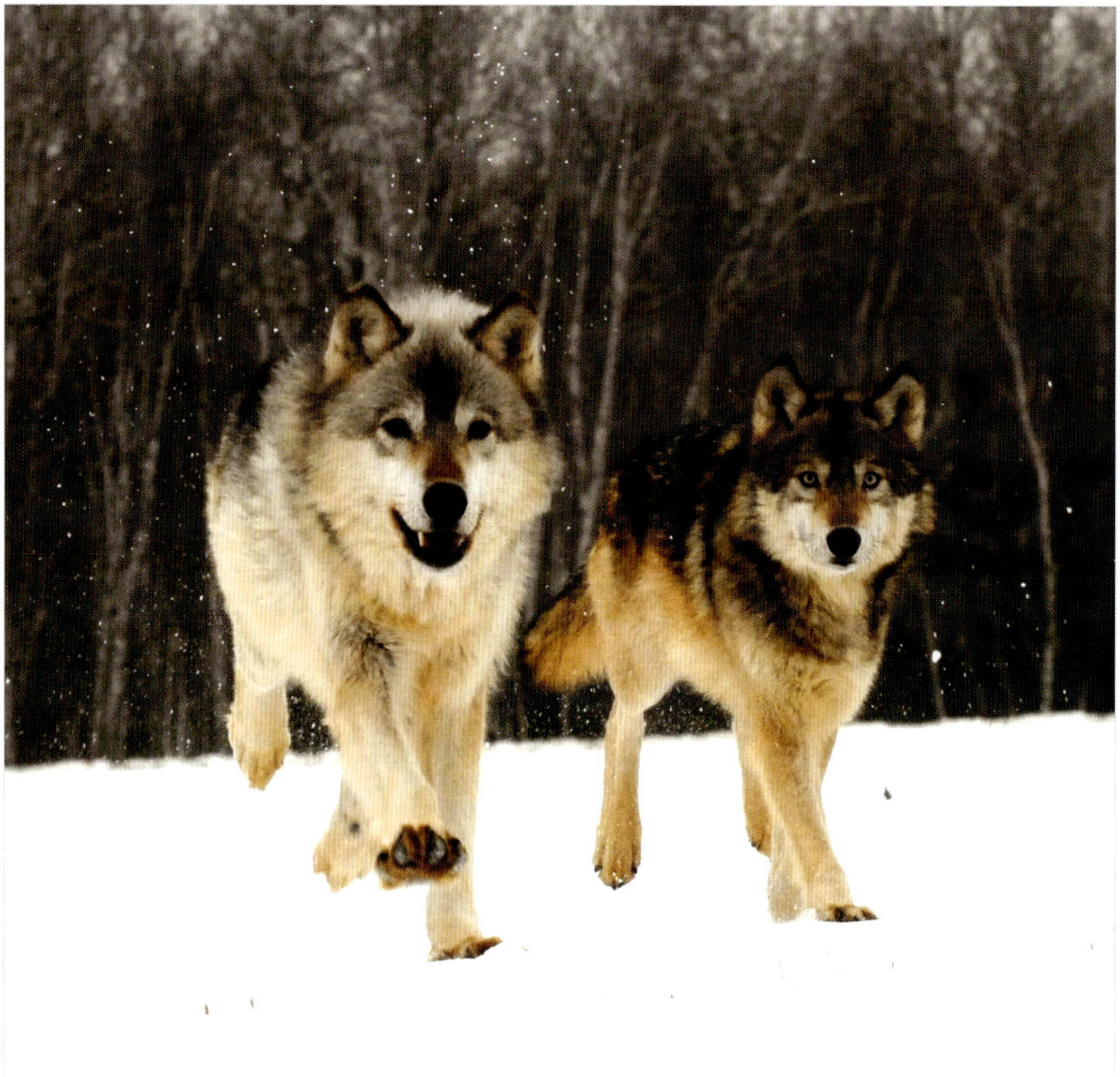

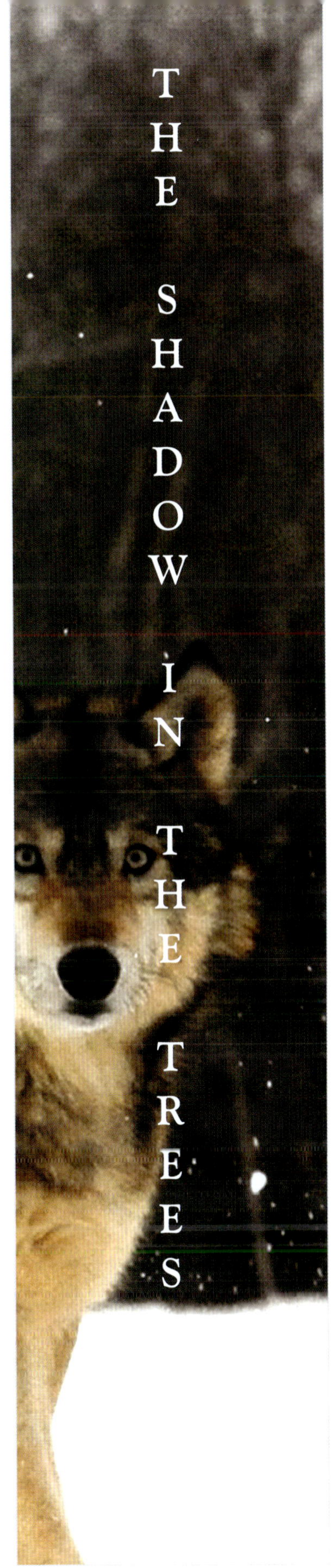

Chapter Ten

'We are in Transylvania, and Transylvania is not England.
Our ways are not your ways, and there shall be to you many strange things.'

Jonathan Harker's Journal, *Dracula*

Almost all of our time in Romania has been spent in rural and wilderness areas. But we are to fly home from the capital city, Bucharest. Located in the south-east of the country it is the industrial and commercial centre. By European standards, Bucharest is not an old city, its existence is first mentioned in a document dating from 1459, as the residence of the ruler of Wallachia – our old friend Vlad Tepes. The founder is traditionally said to be a peasant named Bucur – but there is no record of the settlement until the Middle Ages. In 1862 it became the state capital of Romania. Its eclectic architecture is a mix of elegant historical buildings and modern communist-era tower blocks. During the period between the two world wars the city's beauty and the sophistication of its residents earned Bucharest the nickname the 'Paris of the East'. Although many buildings and districts in the historic centre were damaged or destroyed by war, earthquakes and Ceauşescu's programme of systematization, many survive. The city has grown rapidly, doubling in size since the end of the Second World War.

Economically, the city is the most prosperous in Romania and is one of the main industrial centres and transportation hubs of eastern Europe.

Day Twenty-nine

A couple of days after our dinner with the Count it is time for us to leave Transylvania. Colin returns to collect us in his Land Rover and we make our way back to Bucharest. Having barely passed through the outskirts on our arrival this is really our first chance to see the city – and the suburbs are not immediately impressive. We drive through columns

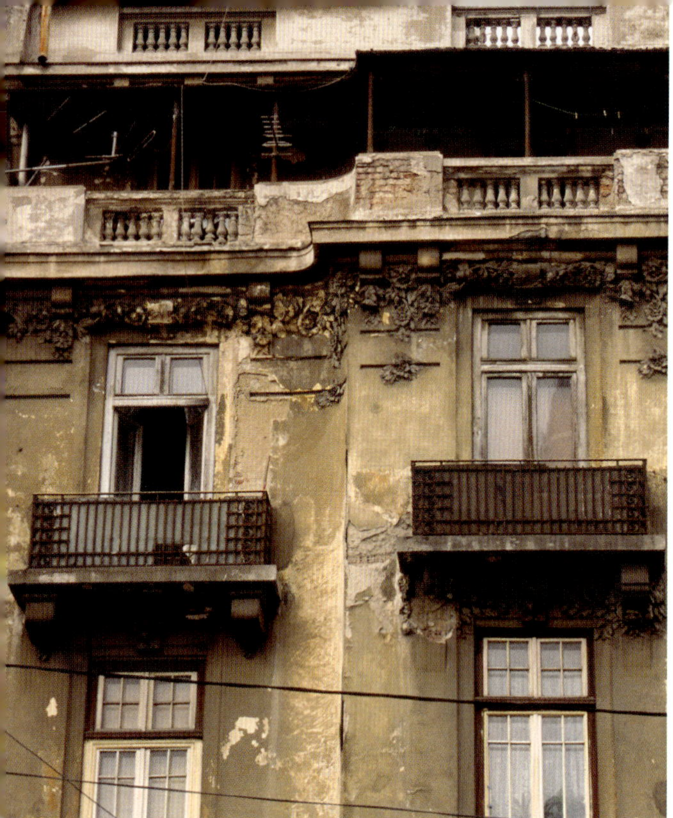

A SCRUFFY EXTERIOR, BUT A SURPRISINGLY BEAUTIFUL INTERIOR

of high communist tower blocks that appear grey, soulless, scruffy and dirty. As Colin views hotels in the capital as 'unreasonably expensive' he has arranged for us to rent an apartment overnight – we fly home the following evening. Our hearts sink when we pull up outside the scruffiest and dirtiest building of all, where Colin collects the keys from the rental agent. We enter the filthy dark entrance hall and cram ourselves and our cases into a minute and incredibly ancient lift. Inside the elevator the buttons to press for the floors have not only lost their numbers, but have melted out of shape – whether from within due to the wiring of the lift or from some external source it is impossible to tell. We narrowly avoid tripping over rusty pipes in the corridor and nervously open the door to the apartment – which is gorgeous. Modern and beautifully decorated, it is immaculately clean with two bedrooms, a beautifully furnished lounge, large bathroom and fitted kitchen. I have rarely been anywhere with such a dramatic contrast between the interior and the exterior.

We walk to a restaurant for our farewell evening meal – it is situated on the boulevard where most of the foreign embassies can be found – and the food is world class. We wash it down with a bottle of 'Vampir' merlot. (*Turistii?* Us?)

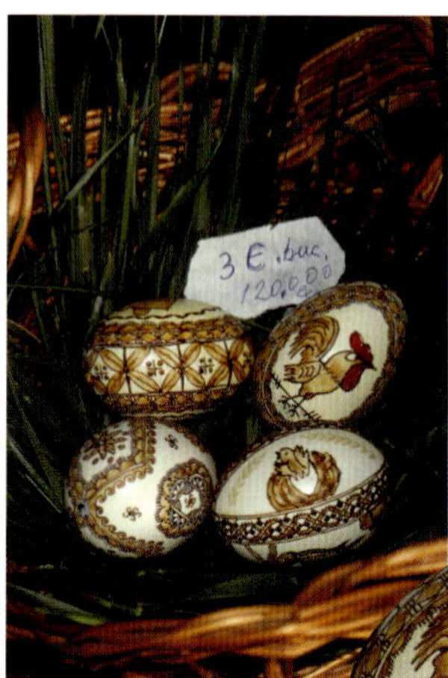

A BASKET OF PAINTED EGGS – A TRADITIONAL ROMANIAN CUSTOM

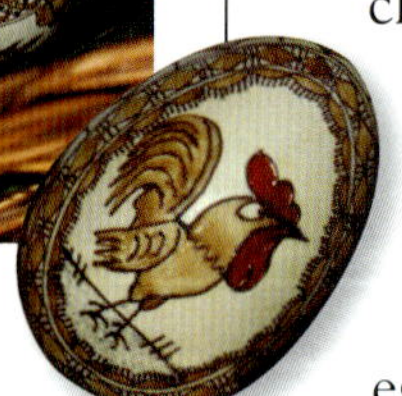

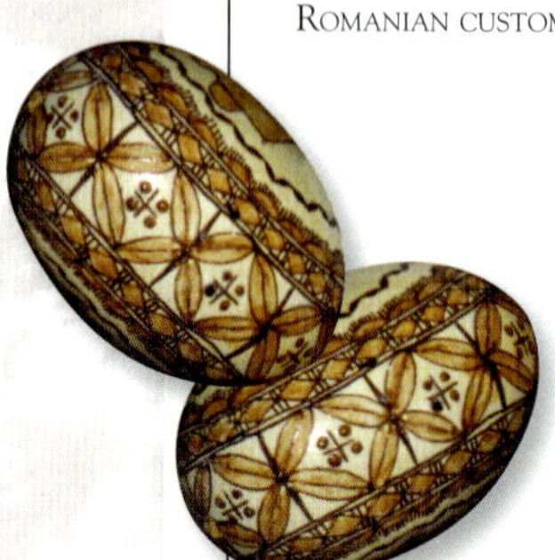

Day Thirty

We have most of the day free before our evening flight, so decide to explore Bucharest a little more. However, this is Easter Sunday and Bucharest is deserted and everything is closed. It is like walking through a ghost town. In the absence of any open cafés, we take the metro to the Ambassador Hotel and eat breakfast in the dining-room. At breakfast we are presented with a basket of painted eggs – a traditional Romanian custom. In churches around the country women will bring baskets of these traditional painted eggs to church, along with bread and sponge cakes which will be blessed by the priest. Apparently some women also conceal other items in the basket, which they will later use to break or cast spells – much to the official disapproval of the church. It is also said you shouldn't take a nap on Easter Day – or you will feel tired all year long!

Our eggs are a startling shade of electric pink – and when we peel them to eat we find the dye has permeated through the shell; the white is alarmingly marbled, giving the egg the look of a heavily bloodshot eyeball.

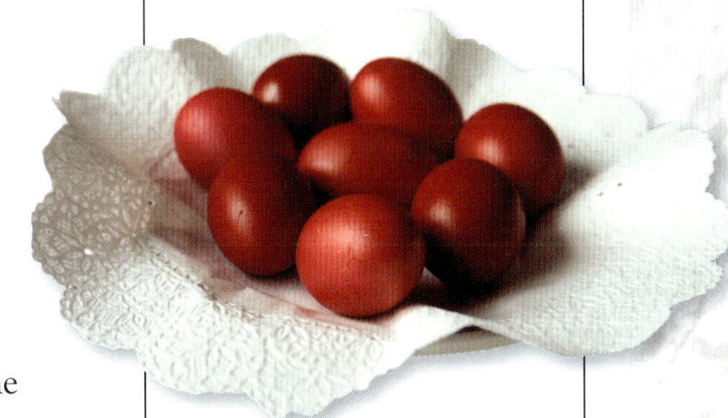

Electric pink painted eggs

We decide to visit the Museum of Peasant Life – but this is also closed. Instead we head for the Museum of Village Life – this is open, but we feel should be named the 'Museum of Village Houses' as there is very little life here either. It is a very accurate recreation of a rural village, but it is odd walking through the immaculate wide paved areas following the journey we have just made. The reproduction houses are all spaced far apart and there are no goats/pigs/sheep/kids/dogs/carts/noise/washing lines/vegetable patches, making the place seem oddly sterile and cold. As the few other visitors to the museum all seem to be Romanian, I am tempted to ask why they don't just travel an hour out of Bucharest to see the real thing. This is not a record of how people would have lived three hundred years ago, it is an empty replica of the vibrant bustling villages which are still very much a reality in the twenty-first century. To be fair the display does include examples of the different types of architecture and buildings from all four provinces – but it does not include the vibrant character of the people. As we drove through a village yesterday on our way to Bucharest, for example, we passed two elderly men driving a donkey cart. They spotted some friends walking towards them down the road. Grinning from ear to ear in delight they called out greetings and one joyously waved the bottle he was carrying, while the other equally joyously waved a dead crow. Not sure if this a traditional Romanian greeting. Here in the museum we see a cart carefully displayed in a roped off area. I can't help but feel if this artefact was to be damaged by human contact, they could source another one with relative ease.

Outside in the real villages we have enjoyed an amazing journey through Transylvania. Staying in Miklósvár we felt we were sharing in a timeless lifestyle in the centre of a fragile environment only just beginning to be discovered by the

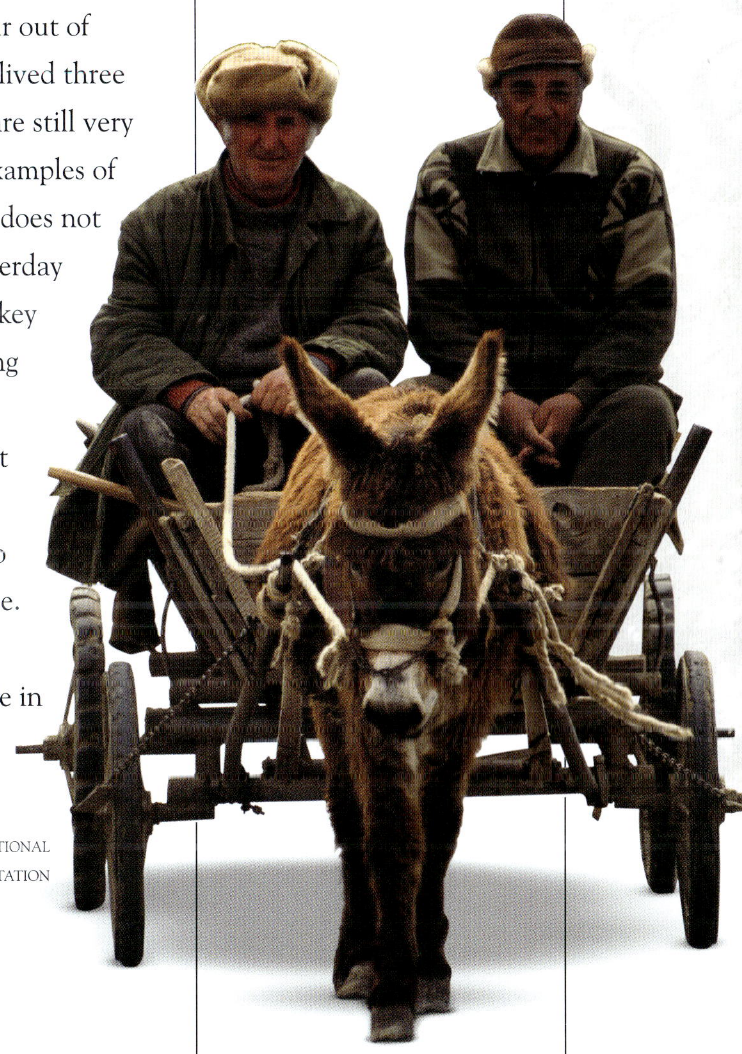

Traditional transportation

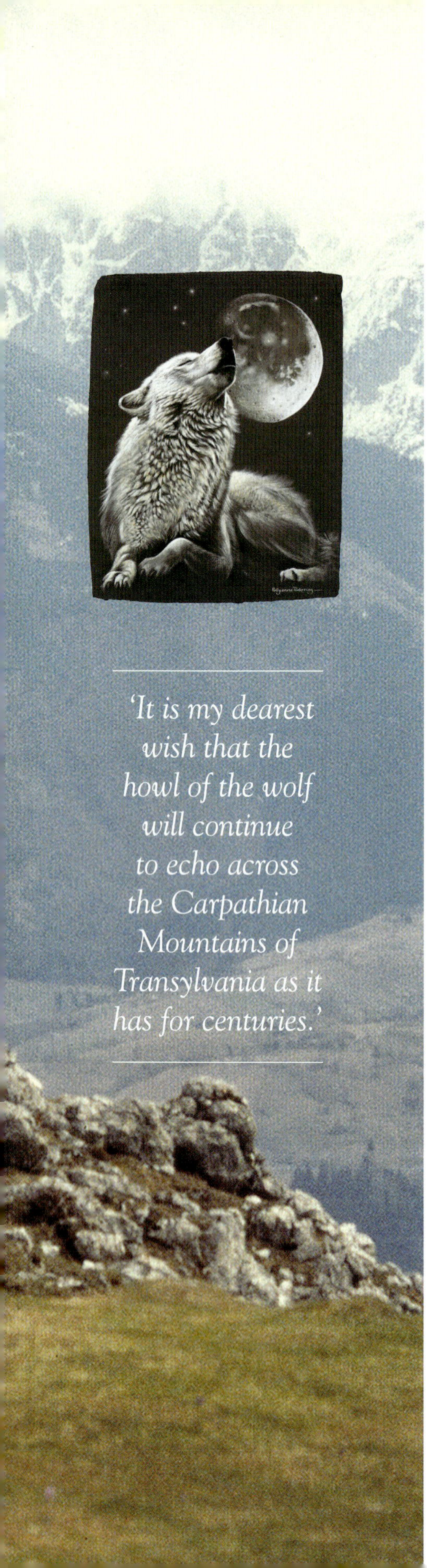

'It is my dearest wish that the howl of the wolf will continue to echo across the Carpathian Mountains of Transylvania as it has for centuries.'

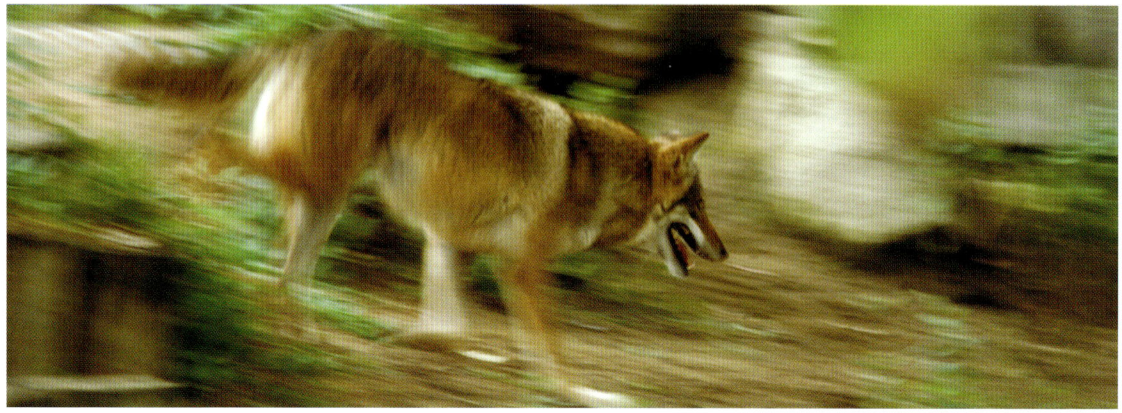

A FLEETING GLIMPSE

outside world. For a few weeks we have stepped into a forgotten world that we thought had ceased to exist anywhere in Europe. We have been moved and inspired by the sheer beauty of the wilderness areas – a haven for wildlife which remains almost untouched by the twenty-first century. We have enjoyed wonderful hospitality and met people dedicated to preserving the wildlife and habitat who have generously shared their time and knowledge with us.

In addition to all this, we are leaving with a huge sense of satisfaction. On one of our hikes with Robert we were rewarded with our own fleeting glimpse of a wild wolf. We had already seen footprints and we had found wolf fur snagged on the wire of a fence, but the wolves had remained stubbornly elusive.

We have been trekking around the shepherd's pastures. The sheep will return within the next couple of weeks to feed on the sweet new grass, wild herbs and crystal clear springwater. There is a warmth in the spring sun which softens the crispness of the air. The first of a sea of wildflowers are appearing in the meadows.

As dusk falls we trek back down the mountainside and re-enter the dark and shadowy forests, the atmosphere cold after the warmth of the open ground. We are walking in single file, following Robert as he leads us through the tangled terrain so familiar to him. Suddenly he stops and points ahead and we see a quick movement – the outline of a wolf, little more than a shadow moving close to the tree line. In a heartbeat he is gone, melting back into the dark cover of the forest. Hearts pounding we move in the direction of his movement, but despite Robert's best efforts we cannot locate him again. The sighting was so brief, it seemed almost dreamlike, but we have seen for ourselves that wolves still roam wild in the forests of Transylvania.

Pollyanna returned from trekking in the mountainsides of Transylvania with a sketchpad full of studies of the forested landscapes, the CLCP's rescued wolves Crai and Poiana and with the indelible memory of a wild wolf glimpsed in the shadows. These combined to create a remarkable portfolio of portraits of wolves in all their guises – from fierce predators to gentle, intelligent protectors of their pack.

Some months after her return she embarked on a lecture tour speaking about her travels in Transylvania and the paintings she completed on her return home. Her audiences travelled with her, sharing every splash of the Danube and jolt of the Land Rover. At the end of the presentations Pollyanna summed up her feelings about the object of her quest with these words: 'It is my dearest wish that the howl of the wolf will continue to echo across the Carpathian Mountains of Transylvania as it has for centuries'.

Wolves are loved and feared, hunted and protected, venerated by some cultures and demonized by others. For as long as humans have walked the earth, wolves have walked alongside us, until in recent times we have pushed them to the very edge of extinction. Surely now it is vital to ensure their habitat is preserved and protected so that these mysterious and charismatic animals can continue on their journey.

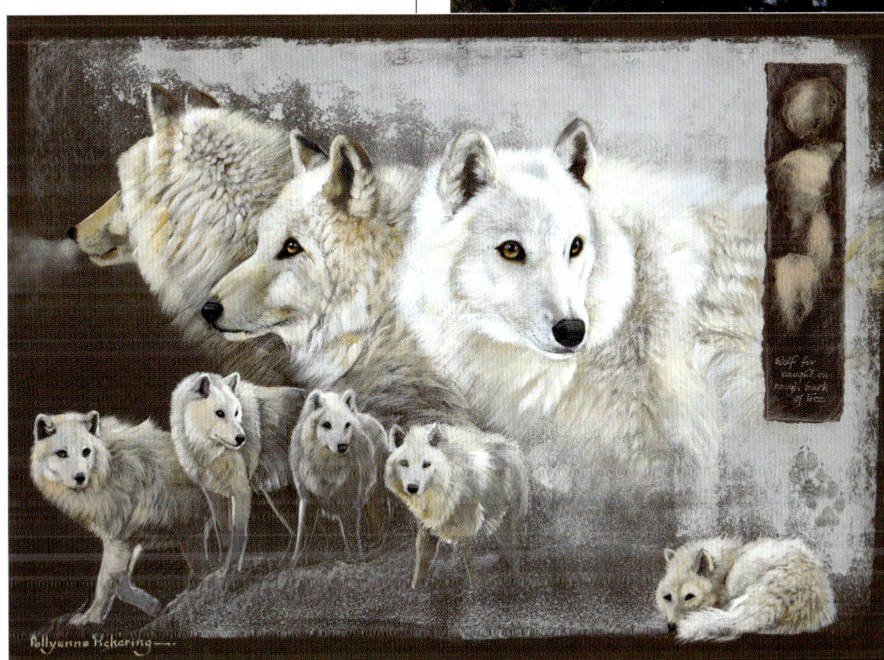

Study of wolves

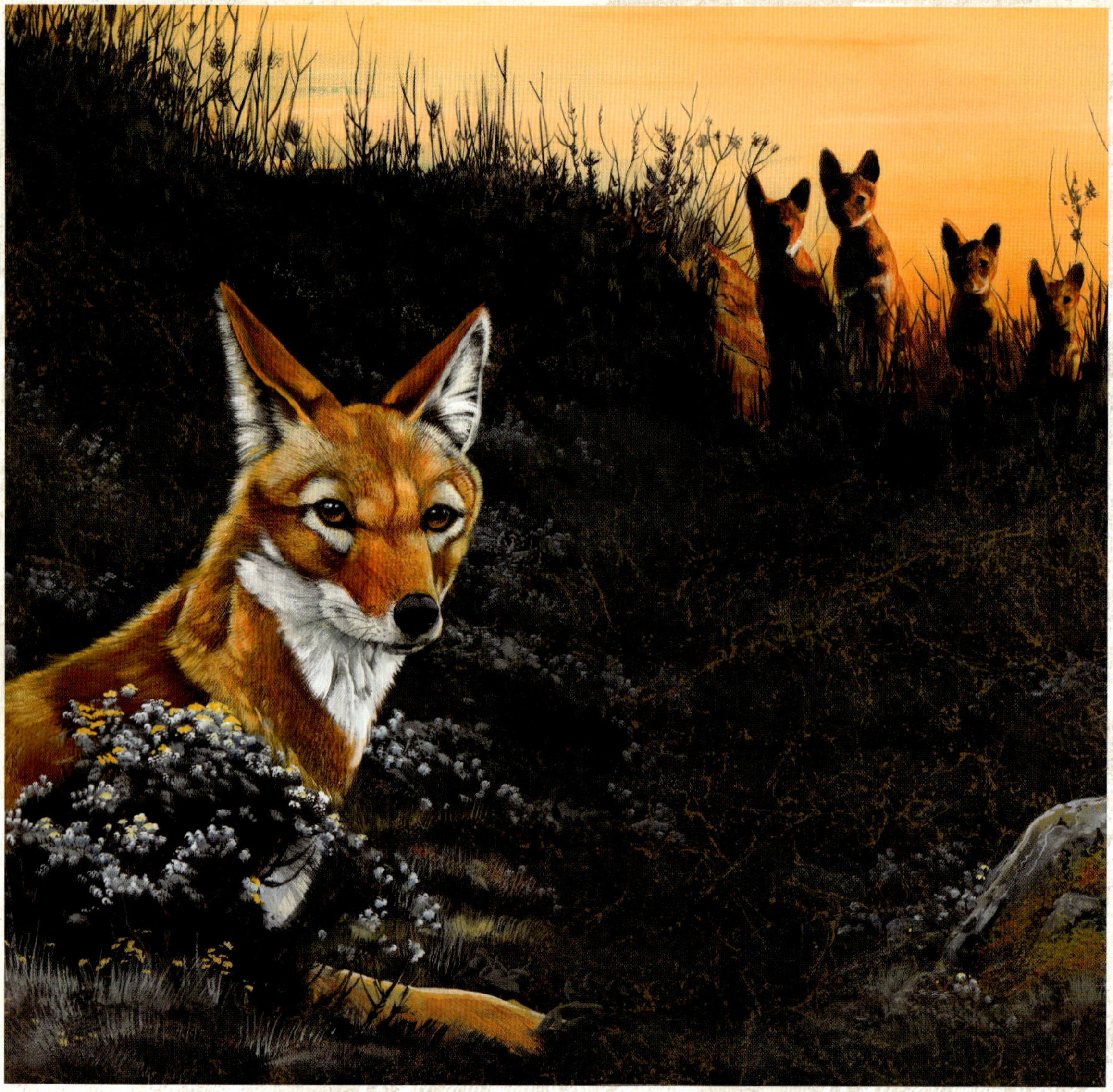

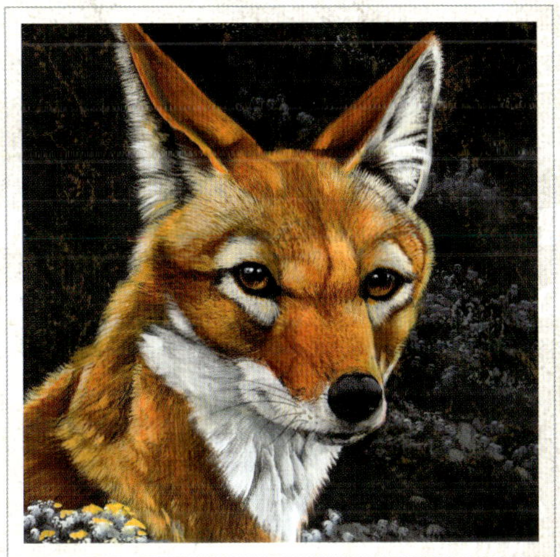

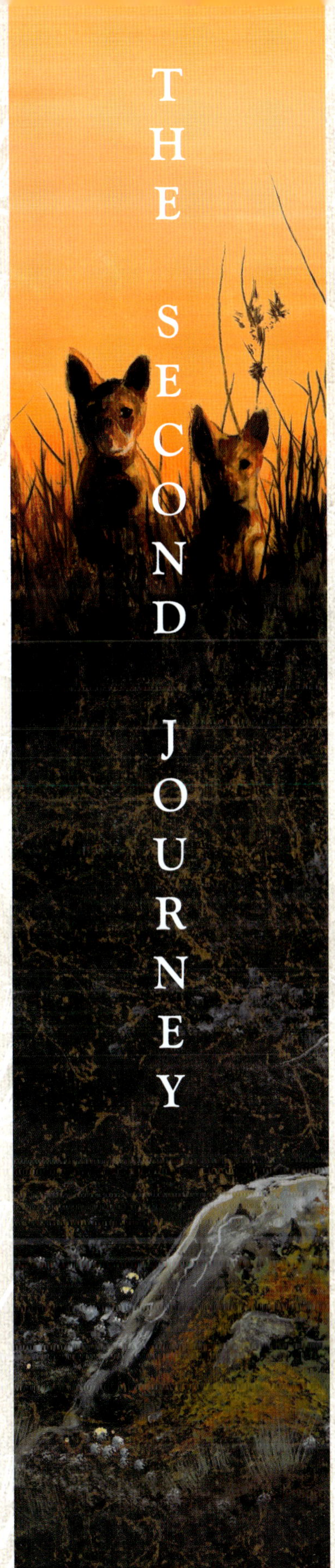

THE SECOND JOURNEY

Foreword

Large carnivores often find themselves at odds with people's interests, and as a result several species are threatened with extinction. While wolves have demonstrated a cunning ability to survive and even thrive in the Carpathian Mountains of Romania and most of Europe, their close relative the Ethiopian wolf could well be the next carnivore to lose its place on earth. It is now the rarest of all canids, with only 450 surviving in half a dozen mountains in Ethiopia. Furthermore, these gorgeous looking mountain dwellers are the most endangered carnivore in Africa.

A wolf in fox's clothing is how I often describe them, due to their fine-tuned adaptations to life in the harsh conditions on the Roof of Africa, where they tap into a seemingly inexhaustible supply of giant molerats and other rodents. Yet, their successful adaptation to the Afroalpine meadows and moorlands that were awaiting the relatively recent arrival of wolves to the Horn of Africa – some 100,000 years ago via land bridges connecting Africa to Eurasia – might be their undoing, since they are now imprisoned in ever-shrinking 'islands in the air'.

What can be done to save this enigmatic icon of the Ethiopian highlands? Conservation actions must be built on sound knowledge, a fact all too often ignored by conservation organizations. For over two decades, Ethiopian and international scientists have devoted great effort to chronicling the life of Ethiopian wolves, helping us to understand the processes that threaten their survival. National Park authorities and conservationists have worked hard to offer protection to the remaining wolf populations, while educating the highland people living side by side with wildlife as to the benefits of protecting Ethiopia's unique flora and fauna. With perseverance and dedication conservationists have sought to protect the wolves and their habitats from encroachment and disease.

The well-crafted story presented in the *Way of the Wolf* by Pollyanna and Anna-Louise Pickering will surely help our conservation efforts by taking the images and thoughts they registered during their visit to Transylvania and the Bale Mountains to a wider audience, and in doing so help us raise awareness and funds for the Ethiopian wolf. Pollyanna is a great wildlife artist, and her magnificent paintings will surely strike the heart of most naturalists the way that they have captured mine. With the seamless integration of

Pollyanna's paintings with Anna-Louise's photographs and her travel-journal styled text this mother and daughter team achieve the right balance of visual messaging and thought-provoking observations. Their passion for wildlife and wild places comes through and invites the reader to join in and enjoy.

The Ethiopian wolf is protected by Ethiopian law, but this by itself cannot guarantee that the Afroalpine habitats where the wolves live are safe. With the main wolf populations found in the Bale Mountains National Park and Simien Mountains National Park, Ethiopian conservationists are well placed to ensure the survival of these magnificent animals. Ultimately, it is the demand for arable land, pasture and firewood that keeps putting pressure on the Afroalpine habitats of Ethiopia, and this process is made more urgent by climate change. More immediately though, it is the domestic dogs that shepherds bring to the highlands, and the diseases they carry, that pose the most urgent threat to Ethiopian wolves.

The struggle to save endangered carnivores like the Ethiopian wolf is a long-term task, requiring constant vigilance, cooperation and dedication. This book and its inspiring pictures and words add to our efforts and the determination of the Ethiopian people to ensure that this iconic species will not vanish from our land.

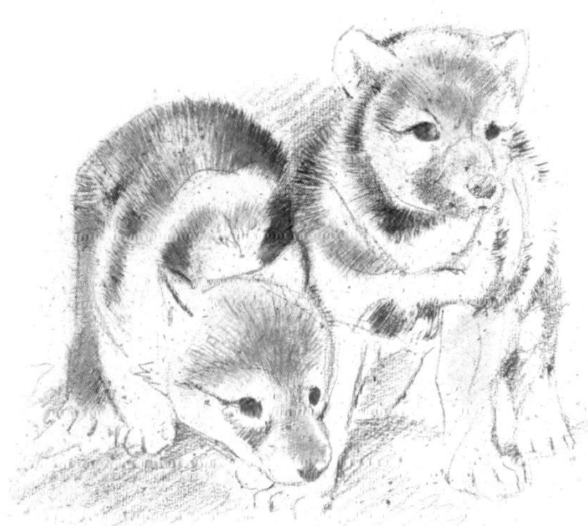

Professor Claudio Sillero

Head of Conservation The Born Free Foundation, and

Director of the Ethiopian Wolf Conservation Programme

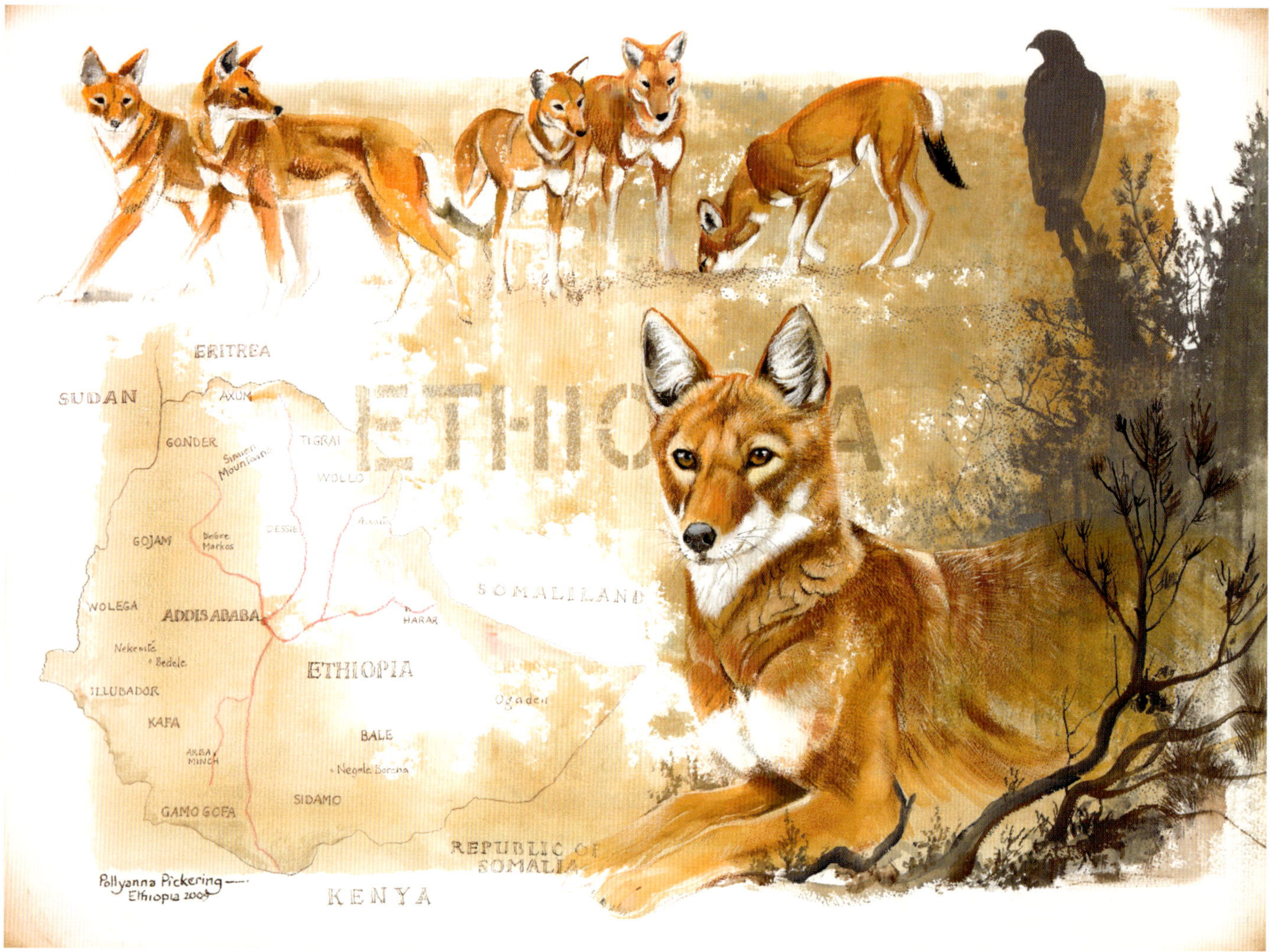
ERITREA
SUDAN
AXUM
GONDER
TIGRAI
Simien Mountains
WOLLO
GOJAM
Debre Markos
DESSIE
SOMALILAND
WOLEGA
ADDIS ABABA
HARAR
Nekemte
Bedele
ETHIOPIA
ILLUBADOR
Ogaden
KAFA
BALE
ARBA MINCH
Negele Borena
SIDAMO
GAMO GOFA
REPUBLIC OF SOMALIA
KENYA
Pollyanna Pickering
Ethiopia

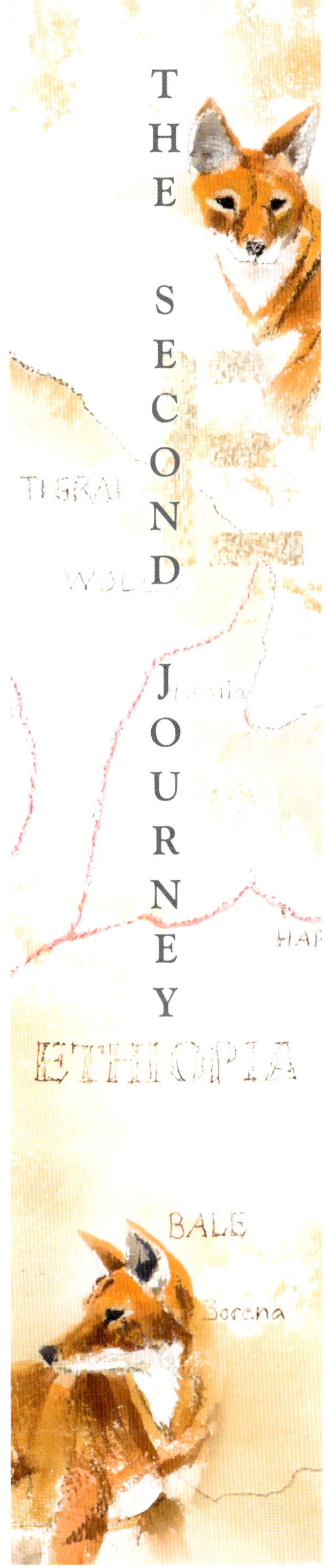

I travelled to Transylvania assuming all my prior expectations and assumptions would be confounded. After years of exposure to the Dracula films and legends I thought I would find a modern country completely unlike the gothic realm of my imagination. In fact I discovered a lost world of medieval villages and looming castles, where ancient superstitions still survive and the horse and cart remains the most common form of transport.

In contrast, any pre-conceived ideas I had about Ethiopia, our most recent destination in Pollyanna's quest to paint the wolves of the world, were completely shattered by the journey we made. Instead of a hot famine-ravaged desert we found cold cloud-covered mountains surrounded by fertile farmland. And rather than searching for the big five 'safari' animals we associate with wildlife travels in Africa we headed out to look for a shy long-legged canine, the only species of wolf in Africa and the rarest in the world.

Usually when we return from our travels with tales of hardship and suffering – camping on the ice in the High Arctic at -40C, removing leeches from our ankles in China, being bitten by monkeys in the rainforests or charged by tigers in India – people often ask Pollyanna why she is so keen to paint her subjects in the wild, rather than visiting a zoo. In fact there are numerous reasons. Firstly, the habitats of the animals. Without experiencing the icy wastelands of the High Arctic or the dripping steamy humidity of the rainforests, Pollyanna's paintings would lack the atmosphere and accuracy which makes

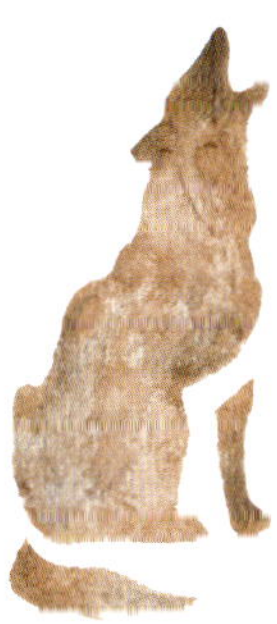

her work so distinctive and recognisable. Secondly, animals in zoos often look very different from their wild cousins. Enclosed in small pens and fed at regular intervals, predators lack the muscle tone of animals which run and hunt in the wild. Captive tigers and other big cats often have huge hanging stomachs unlike their lean wild counterparts. Less immediately obvious is the expression in their eyes – wild animals have an alertness, a wariness, a

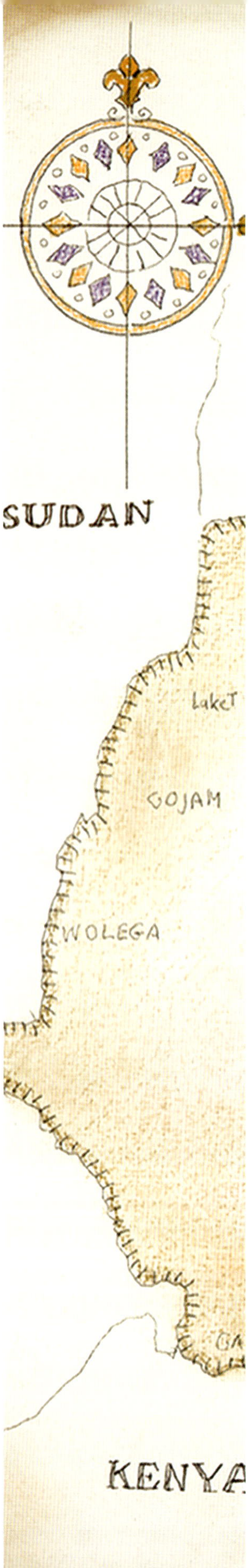

curiosity which they lose to the boredom of captivity. Pollyanna is able to capture this wild spirit in her paintings, giving her subjects a natural vitality.

In the case of the Ethiopian wolf, there is another compelling reason. Only around four hundred and fifty of these beautiful red-coated canids remain – and they all live in the wild in the mountains of Ethiopia. There are no captive specimens in zoos anywhere in the world. Indeed most people are entirely unaware that wild wolves live in Africa. The species was first described in the Simien Mountains in 1835, but even then numbers were undoubtedly already low through most of the country, and few Ethiopians are even aware of their existence. Claudio Sillero of the Ethiopian Wolf Conservation Programme comments, 'The Ethiopian wolf does not play a major role in mankind's culture or economy. So far as we know, there is no mention of it in the culture or folklore of Ethiopia.'

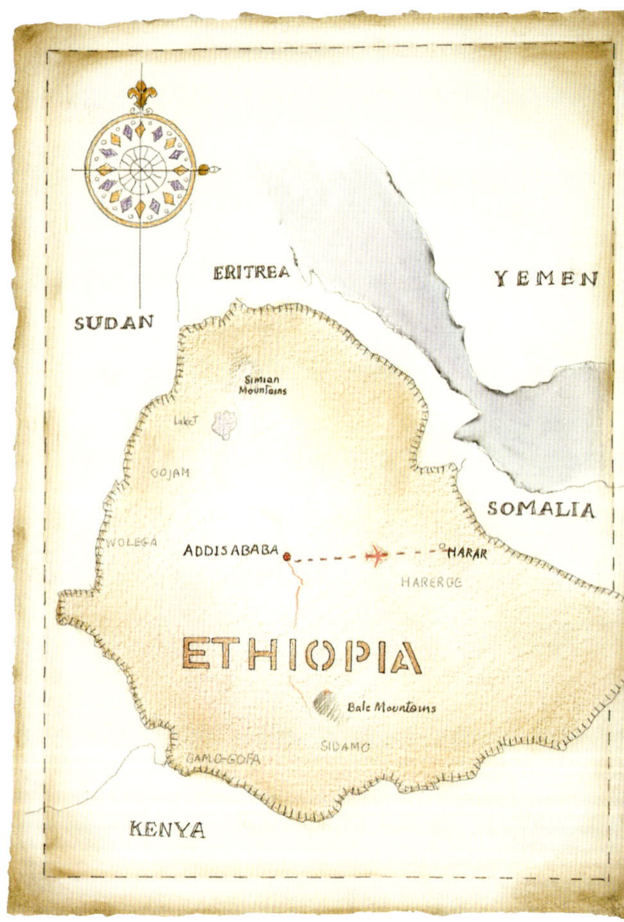

Five years after following in the footsteps of the wolves in Transylvania, we pack our bags once again and set out in search of the rarest wolf in the world.

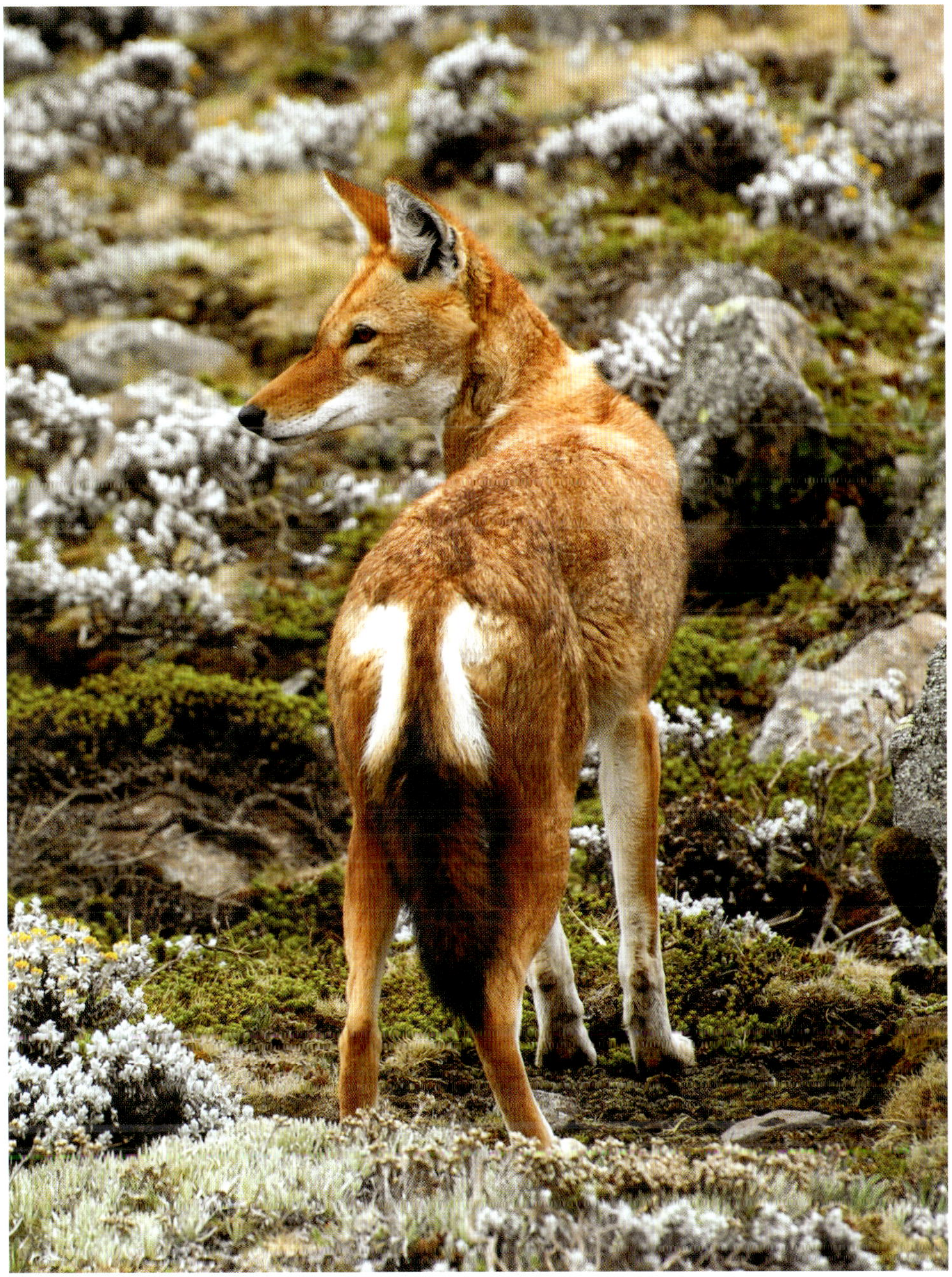

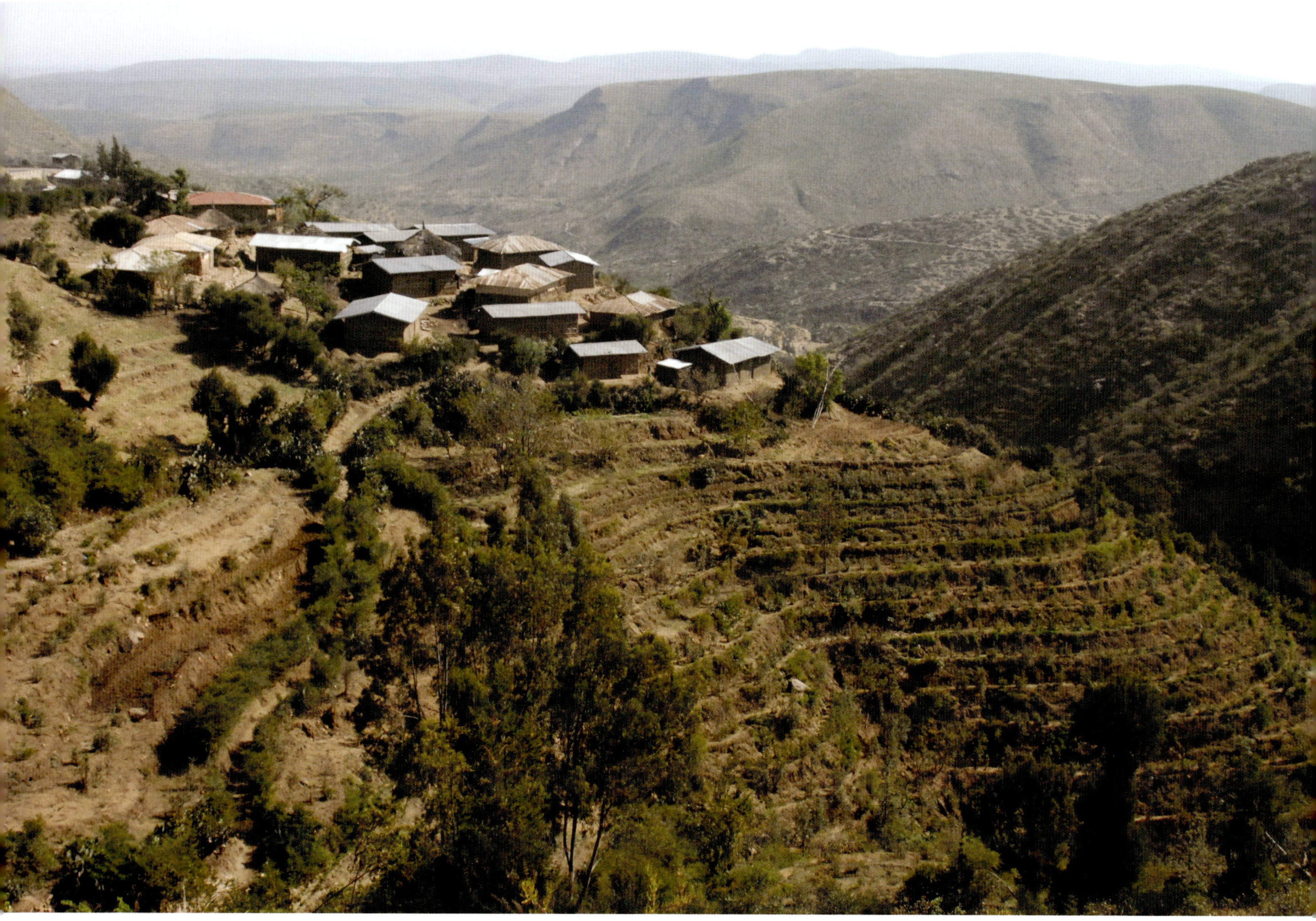

Chapter One

'Where there is coffee, may there be peace and prosperity.'

TRADITIONAL ETHIOPIAN BLESSING

Ethiopia is often referred to as the cradle of humanity – journeying here we are journeying to the very birthplace of man. In 1974, in the Afar region of Ethiopia, archaeologist Dr Donald Johanson discovered part of a female skeleton, believed to be more than three million years old. The Ethiopians named her Dinqenesh, which means 'thou art wonderful' – but we know her better as Lucy. Lucy provided dramatic new evidence for the story of human evolution. Further archaeological finds in 1994 support the theory that our earliest ancestors were born four million years ago along the African Rift Valley, which passes through Ethiopia before continuing southwards through Kenya to Mozambique.

Ethiopia is remarkable in many other ways. Once known as Abyssinia, it is the only country on the continent to have maintained its independence in the face of European colonizers – and even the brief occupation by the Italian army towards the end of the Second World War has left little lasting influence on their unique culture. It is the only African country to possess an alphabet more than two thousand years old and is home to several animals found nowhere else on earth – including, of course, the Ethiopian wolf. It was the location of the most ancient kingdom in Africa, one of the first monarchies in the world, whose kings claimed direct descent from King Solomon and the Queen of Sheba. The most famous ruler of modern times, Emperor Haile Selassie, restored to power after the war, forged good relations with the west and brought Ethiopia into the United Nations. Haile Selassie is also perceived by the Rastafarian movement as a messianic figure who will lead a future golden age of eternal peace, righteousness and prosperity. Many Rastafarians believe that Haile Selassie is still alive and that his 'death' was part of a conspiracy to discredit their religion.

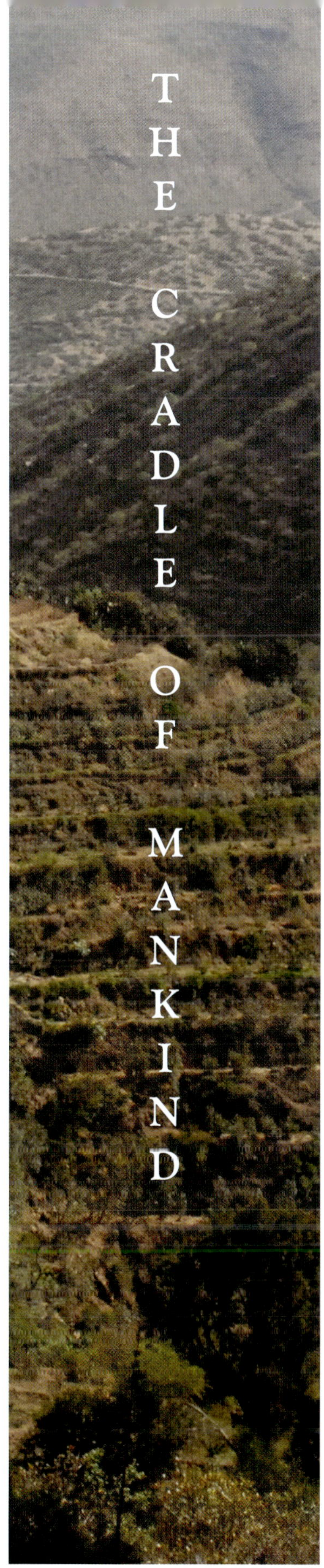

'Testing, awe-inspiring and heartbreaking... you don't visit Ethiopia... because you want a relaxing getaway, you venture here because you want to be moved. And moved you certainly shall be.'

LONELY PLANET

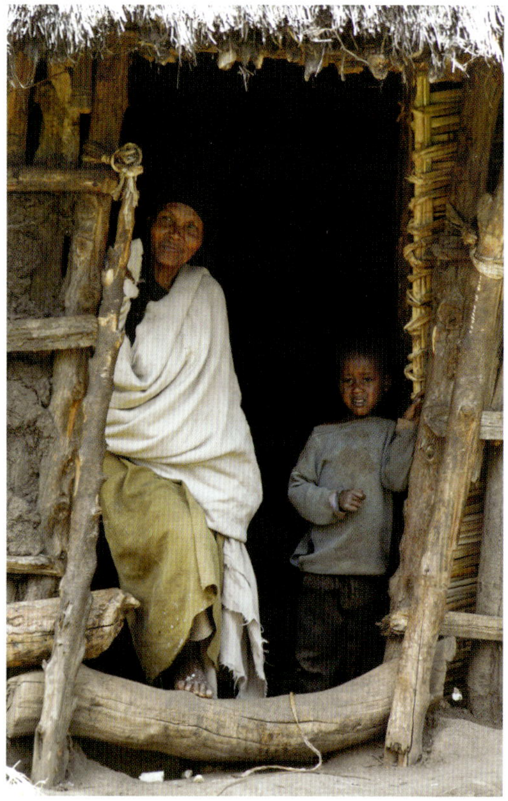

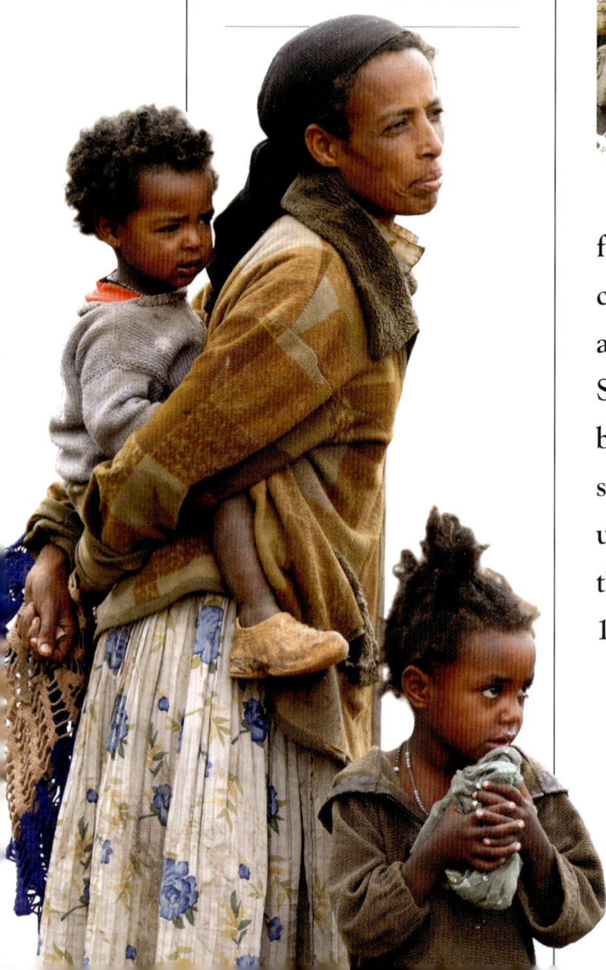

Ethiopia is bordered by Sudan to the west, Somalia and Djibouti to the east, Eritrea to the north and Kenya to the south. The Rift Valley runs through the country from north to south, dividing the country into the Western Highlands, the Eastern Highlands and the Rift Valley Lowlands. This brings huge variations in its climate – from hot tropical valleys to cool rainy fertile highlands. The Ogaden semi-desert plain is dry and bare and the low-lying Danakil Depression is one of the hottest, driest places on earth.

In common with Transylvania, Ethiopia has had little chance to promote itself as a tourist destination. The overriding image of the country internationally is associated with the devastating and widespread famine of 1984 – 85 which is estimated to have affected eight million people and caused the deaths of around one million. The country's unsettled political history and ongoing conflicts have also not helped to tempt casual visitors. In 1974, Haile Selassie's reign ended when the military took control of the country in a coup lead by Lieutenant Colonel Mengistu Haile Mariam who became head of state. During seventeen years of military control, the state of the economy worsened and civil unrest grew until the Ethiopian People's Revolutionary Democratic Front overthrew the regime in 1991. Two years later Eritrea, which had been annexed by Ethiopia in 1962, gained independence following thirty years of conflict.

However, fighting over the disputed border broke out in 1997 and a year later this had become a full-scale war. By the time peace was declared three years later one hundred thousand people had been killed and seven hundred and fifty thousand had become homeless refugees. The war was a disaster for the government's efforts to revitalise the economy and of course for the potential tourism industry, which has also been limited by the lack of suitable hotels and other travel infrastructure.

Recently, however, the Ministry of Culture and Tourism has begun to promote Ethiopia as a land which enjoys 'Thirteen Months of Sunshine' – a claim, which thanks to their unique calendar, is undisputedly true – to try to tempt more international visitors. The most popular destination is Lalibela where magnificent ancient churches are hewn out of massive monolithic rocks. Even so, tourism is very much a fledgling industry here and many parts of the country remain largely untroubled by foreigners. Picking up our *Lonely Planet* guide while planning our journey, we find the first line of the introduction describes Ethiopia as '*testing, awe-inspiring and heartbreaking*' and goes on to warn that '*you don't visit Ethiopia... because you want a relaxing getaway, you venture here because you want to be moved. And moved you certainly shall be.*'

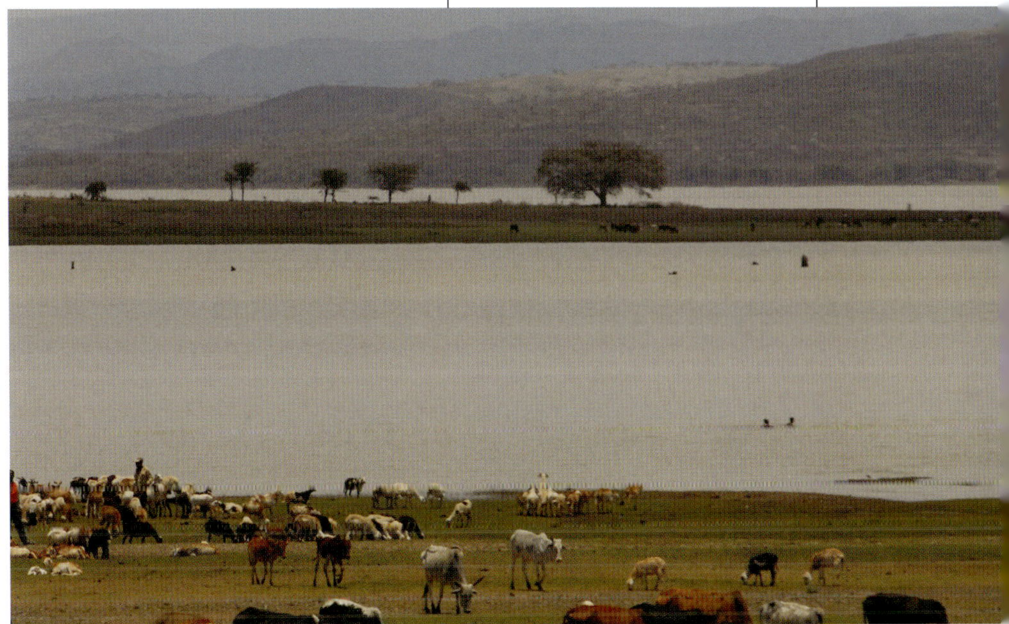

Cattle graze the fertile plains

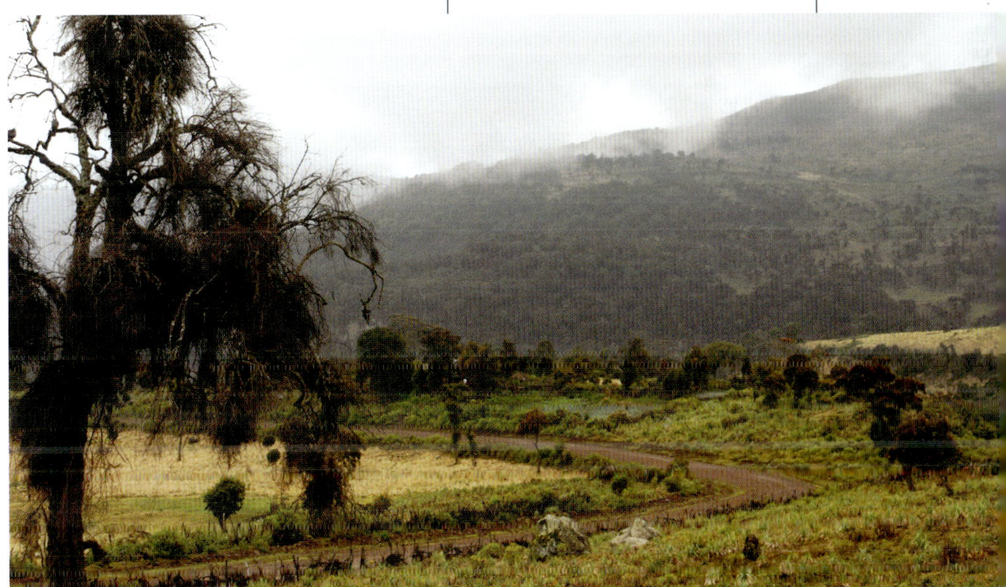

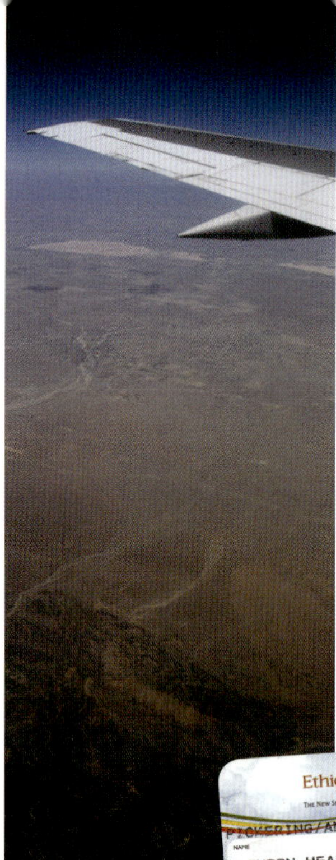

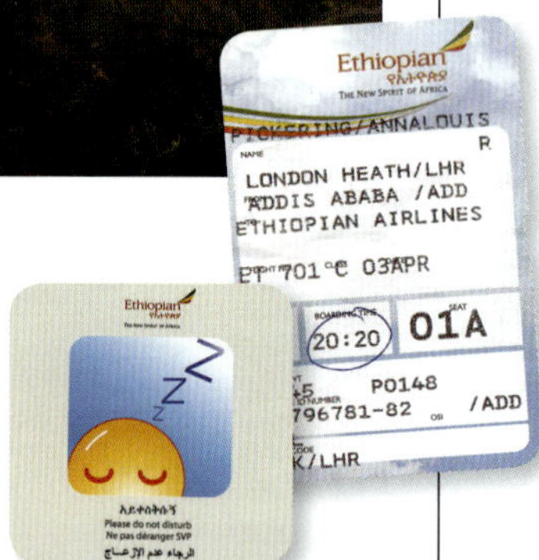

Day One

Night flight to Ethiopia with their national airline. We have arrived at Heathrow in something of a flurry. Just as we were packing the last few bits and pieces prior to leaving home some friends arrived unexpectedly for coffee, with the result that we will arrive in Ethiopia without some fairly essential items including torches and binoculars. However, we feel calmer by the time we board the plane and after poking half-heartedly at the airline food, settle down to try and sleep for the night. We understand we are on a flight which travels via Rome, and at around 2.00 am I am woken by the seat belt sign pinging and an announcement asking us to raise our seatbacks to the upright positions and stow our tray tables. Struggle awake, struggle upright, assume we are landing in Rome as designated. We don't. Just me upright and stowed. Can only assume Rome was a request stop.

Woken again at 4.00 am this time by a woman repeatedly saying 'Scrambled eggs? Frittata?'. In my bleary state I am not sure why this is. I fall asleep. At 4.30 am she abandons the eggs and frittata ploy and instead pushes a warm soggy croissant at me, which for some reason I eat. Try to freshen up. The plane lands. Feel a like zombie.

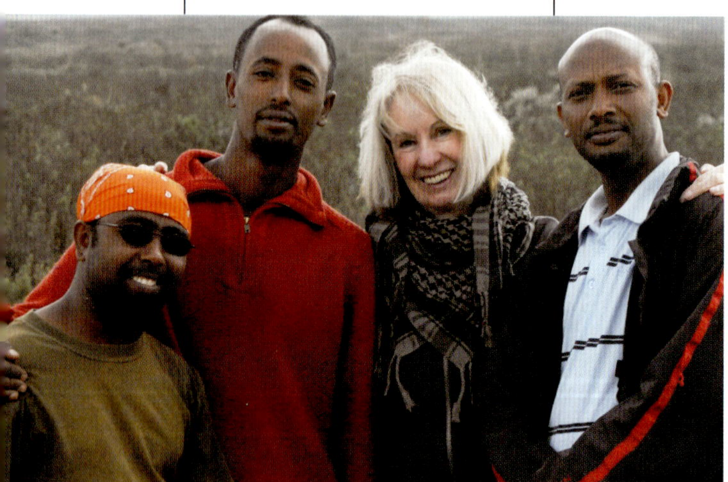

Sami, Abbi, Pollyanna and Tariku – our Ethiopian crew

We queue for visas. We queue for so long that I eventually leave Pollyanna queuing and join separate very long queue for the bank; change some money and rejoin her again – still in the same place in the first queue. We queue even longer to pass through immigration, but then head out to arrivals, where we meet Tariku, who will be travelling with us to act as our interpreter throughout our journey. He introduces us to the rest of our team: our driver Abbi, and Sami who will be cooking for us while we are up in the mountains. We pile into Abbi's four-wheel-drive van, luggage crammed in the back, cameras and sketching equipment on the back seat with us, boxes of pots, pans and supplies already piled high on the roof rack and set out into Ethiopia. Initially we head out through the suburbs of Addis – very busy with the roads crowded with cars, people, cattle and market stalls.

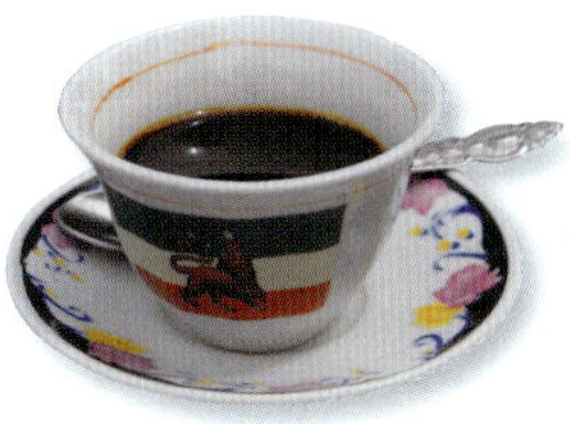

We make a stop for a coffee and bottle of water at a café overlooking a lake in a volcanic crater. A charming setting for our first break – the lake is a haven for bird life. We have only been in Ethiopia a matter of hours and already we spot a fish eagle, on a branch above the water. As we watch he swoops down and plunges his feet below the surface of the water – but returns to his perch without a fish, a failed hunting trip.

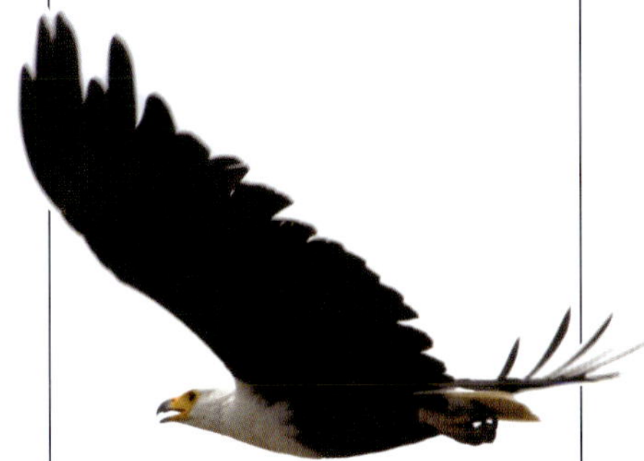

A fish eagle

We also experience our first Ethiopian coffee ceremony. In the foyer, we noticed on our arrival a beautiful young woman roasting coffee beans in a flat pan with a long handle over a tiny charcoal stove, then grinding them with a pestle and mortar, before slowly stirring the grounds into a black clay round-bottomed coffee pot with a narrow spout and straw lid. Tariku tells us this pot is called a *jebena*. When the coffee is served at our table the pot and cups arrive on a bed of freshly cut grass, and our waitress delicately pours a thin golden stream of coffee into each tiny cup from a height of at least two feet. The coffee arrives with a bowl of burning frankincense, which fills the air with a sweet spicy aroma – and a bowl of popcorn, a traditional accompaniment which is roasted on the same fire as the coffee. Our little coffee bowls are proudly embossed with the Lion of Judah, symbol of Ethiopia, and somewhat incongruously perched on floral saucers.

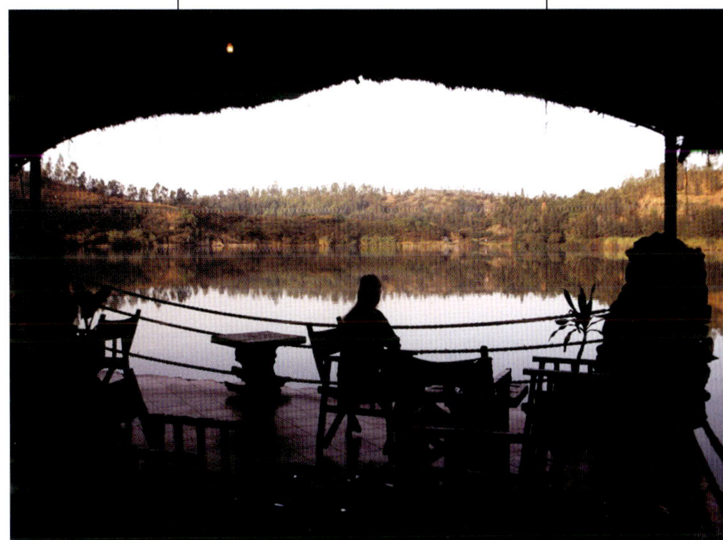

A café overlooking a lake in a volcanic crater

We move on, much revived by the powerfully strong, but smooth-flavoured coffee. As usual when we first arrive in a new country we are stopping our poor driver every few hundred yards to photograph something which is new and exciting to us – donkey carts, birds, crops. In one field Tariku picks us a fresh peppercorn to eat – it has a lengthy after burn, which we relieve with juicy strawberries purchased at a roadside stall. Immediate panic – they might not have been washed – or even worse, they might have been. Despite attracting curious glances from Sami, we prissily wipe the rest with tissues dipped in bottled water – and they are delicious.

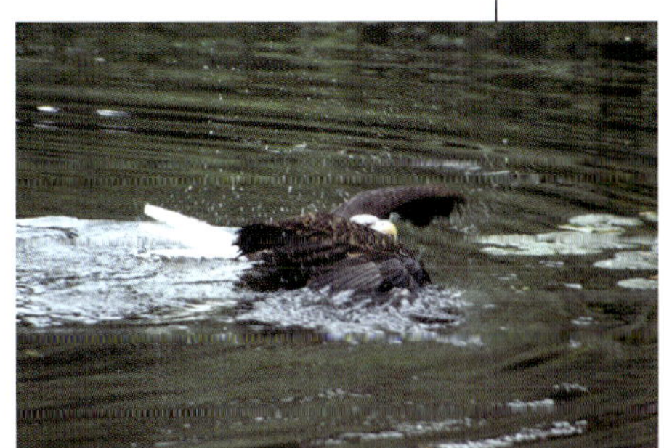

A failed hunting trip

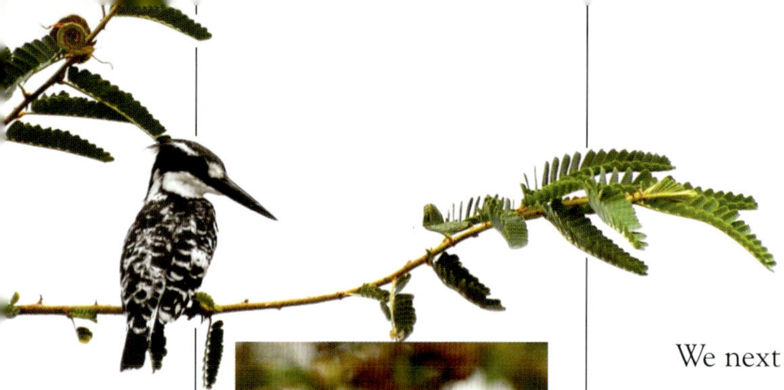

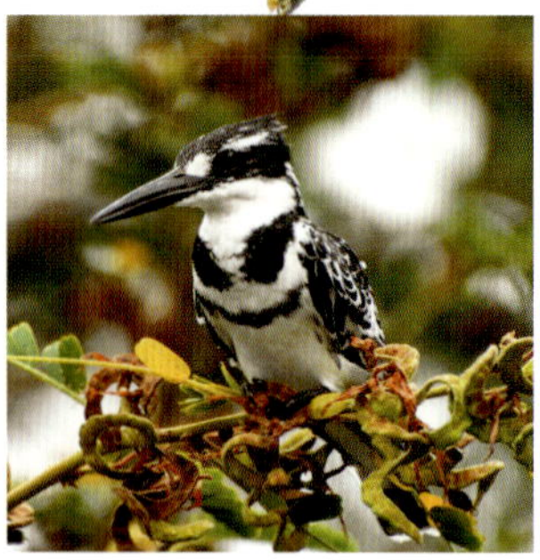

Pied kingfisher

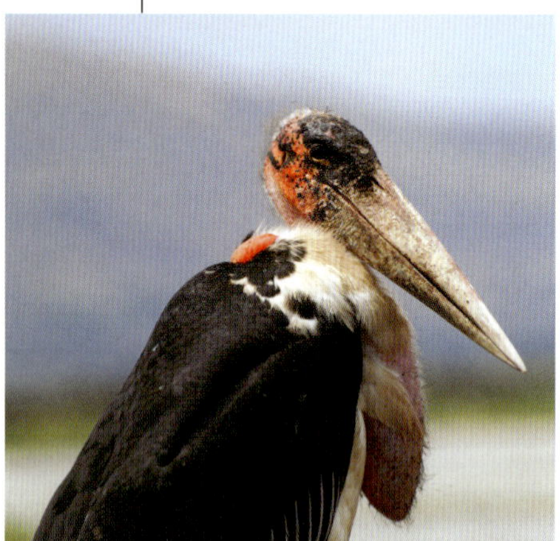

Marabou stork

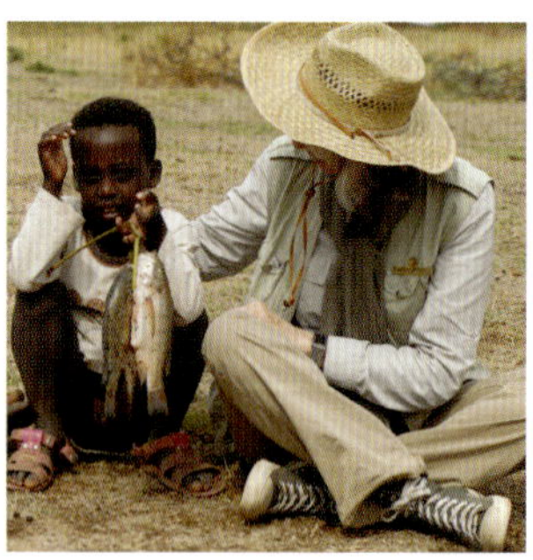

Pollyanna with a young fisherman

We next ask Abbi to stop when we spot an area beside a river where the fishermen are preparing their catch. This is attracting large untidy groups of marabou storks, as well as three species of kingfishers – pied, white-breasted and, Pollyanna's favourite, the malachite. The discarded fish remains are giving off a powerful and distinctive smell under the equatorial sun and my stomach curls into a ball as the aroma hits. It is feeling a little strange anyway, probably because after a disrupted attempt to sleep on the plane so far today I have consumed the following: *1 small soggy croissant at 4.30 am, 1 small handful of popcorn with three cups of ferociously strong black coffee, 1 peppercorn, 2 strawberries (potentially toxic)*. It is now nearly 3.00 pm.

We drive on, pausing to photograph a huge flock of vultures which have congregated over the road kill carcass of a spotted hyaena – more lovely smells – and then arrive at the edge of a huge lake. Here the banks are so full of marabou storks and other bird life it makes the photographs I took at our first stop completely redundant. There are hordes of storks strolling in between the assembled crowds of people; fishermen, people doing their laundry, people bathing in the lake – and both birds and humans appear completely unconcerned by each other's presence. It is possible to walk to within two feet of the storks

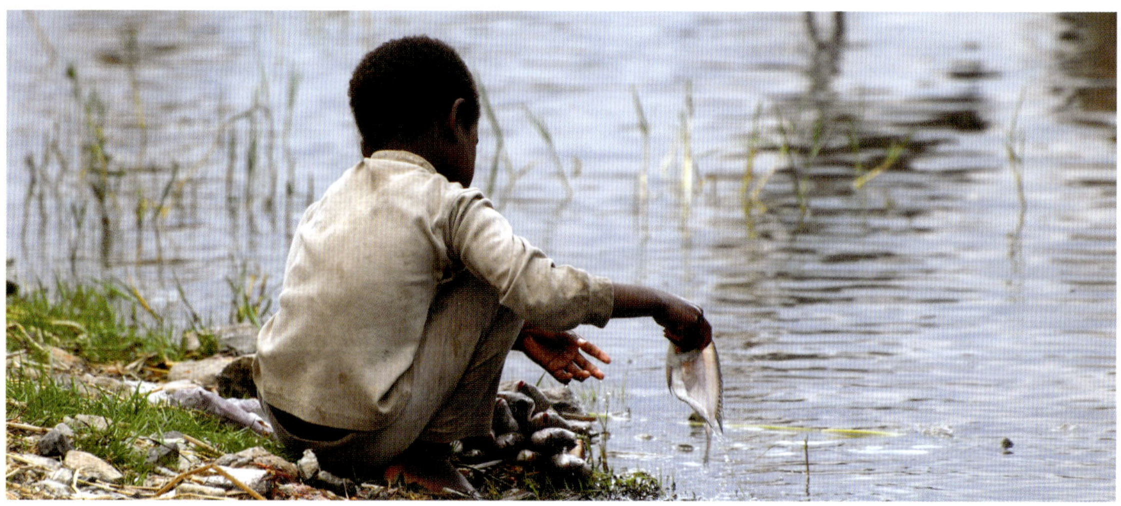

to take photographs – and I also capture close-up shots of pied kingfishers sitting on branches above the water, and many other water birds including egrets, ibis and hammerkopfs. We have only been in Ethiopia a matter of hours and already Pollyanna is in her element sketching. We have been lucky enough to spot pied kingfishers before in our travels, but have never seen them sitting so calmly in our presence.

Eventually we leave the lake and continue our drive. Already Ethiopia is confounding all our expectations. I had imagined a parched desert-like country, but so far we have seen flowing rivers and large lakes, surrounded by fertile farm land. When we stop at a small open-air restaurant I am again surprised by the choice of food on offer. Reading our guidebooks we have learnt that the national staple dish is *injera*. Our faithful *Lonely Planet* guidebook informs us, '*You never forget your first injera experience*', describing it as '*slightly bitter*' with the appearance of '*old grey kitchen flannel*', and suggesting '*with a bit of time and perseverance you may eventually become a connoisseur*'. Oddly, no one has ever suggested I might need time and perseverance to learn to love a melt in the middle chocolate pudding, so this doesn't bode well. Although we are, of course, planning to try this national dish, I am not sure my stomach is ready for it yet, feeling slightly fragile from a combination of forty hours travelling, the already listed selection of snacks and the combined smell of rotting fish and hyaena.

I ask Tariku to recommend something else and he suggests I ask for spaghetti, with a fresh tomato sauce, which sounds perfect and turns out to be delicious. All the other items listed on the menu are very traditional Ethiopian dishes: *kai wat*, a stew of meat in a spicy sauce; *kifto*, lean meat warmed in a pan with butter; and *minchet abesh*, a thicker stew topped with hard-boiled egg. The spaghetti seems an oddly cosmopolitan dish to be listed alongside these national specialities. However, we soon learn we will find spaghetti in every single roadside restaurant and café we stop at, no matter how small, or how remote the area.

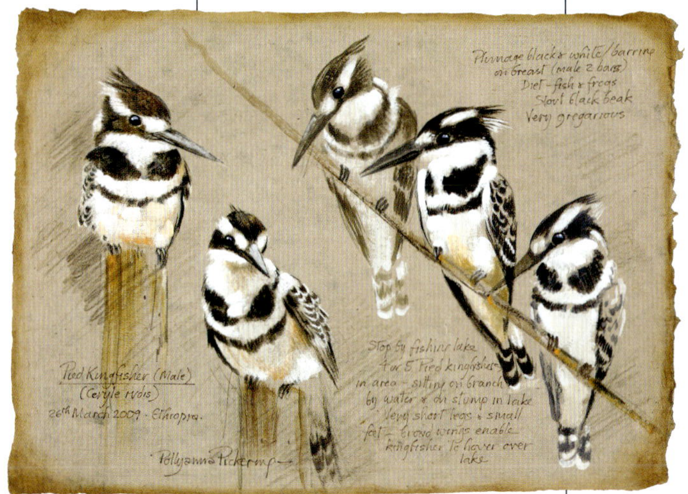

Sketches of pied kingfishers

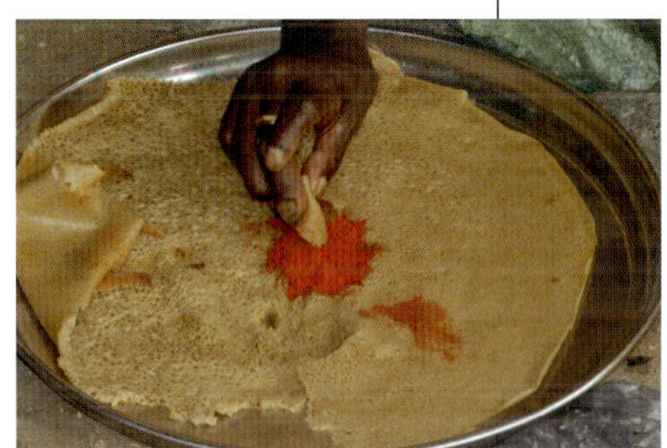

You never forget your first *injera* experience
Below: the alternative

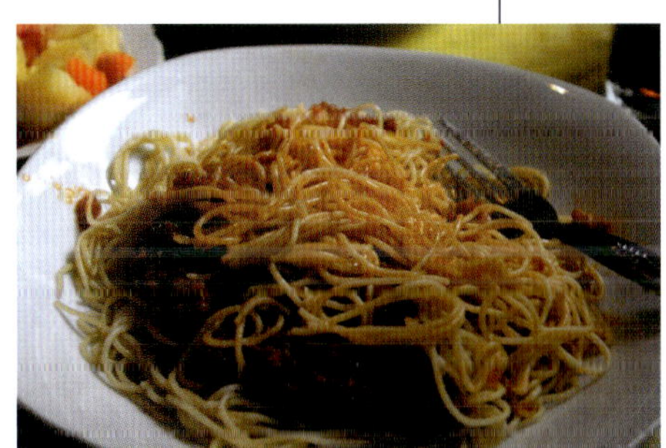

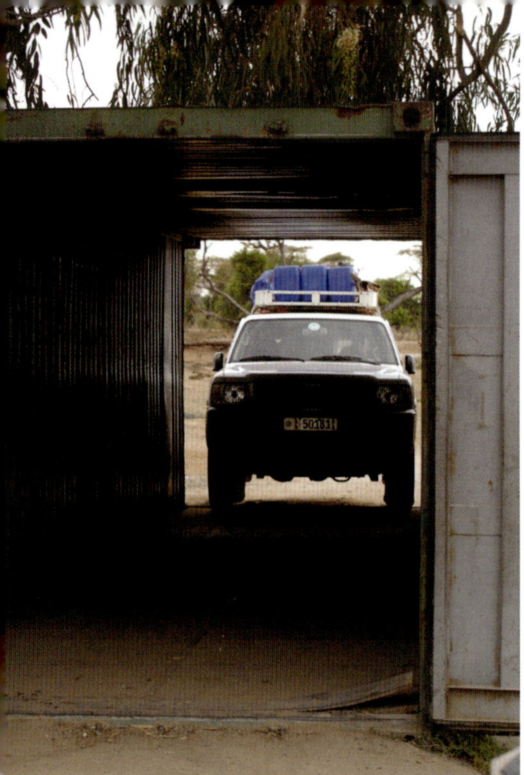

Our car becomes wedged in the millennium bridge

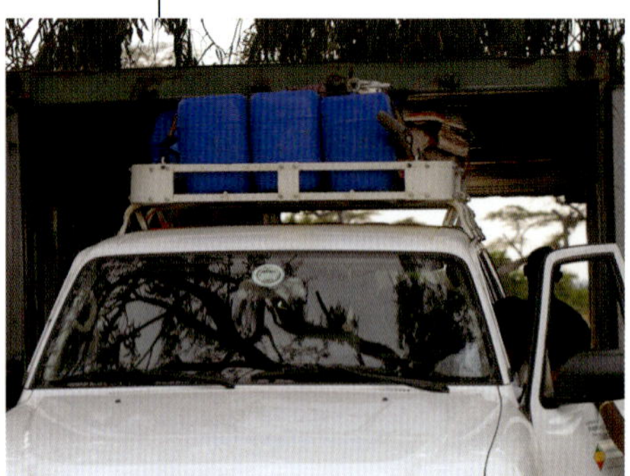

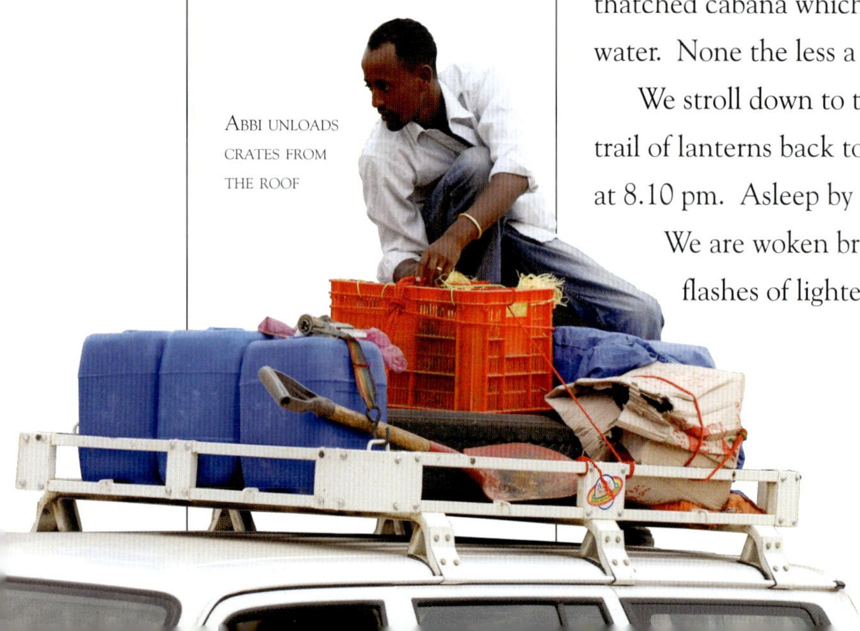

Abbi unloads crates from the roof

Part of the joy of travelling through Ethiopia lies in the fact it has never been colonised; the only African country to remain independent throughout its history, thus developing a unique culture with few outside influences. However, at the end of the Second World War Ethiopia was briefly occupied by the Italians. As we travel we can see almost no lasting effects of this occupation – other than the spaghetti, which is completely ubiquitous, and has become almost a second national dish.

Luckily the spaghetti has the effect of settling my stomach, and I feel a little perkier for having eaten. We set out again, turning on to a dirt road which leads to our destination for the first night – Bishangari Lodge on the shores of Lake Langano in the Great Rift Valley lakes region. We are stopping overnight to break the long drive to the Bale Mountains, home to the Ethiopian wolves.

As we approach the lodge we cross a small river using a bridge which has ingeniously been made out of an old shipping container. With our roof rack fully loaded with provisions our vehicle won't pass underneath. The largest crate is duly unloaded and the car drives in – but will not come out. It is firmly wedged. Even during our first few hours of driving we have learnt that wherever we stop in Ethiopia, no matter how remote and deserted the road may seem, within seconds at least twenty-eight interested people will have arrived out of nowhere to see what is happening. In this instance this is very fortunate, as it provides us with a willing team of volunteers to hang off the back of the car and weigh it down, lowering the roof level enough for the car to squeeze through!

We check in gratefully at our eco lodge and walk to our accommodation – a cute little thatched cabana which even has a private bathroom. Although it is so eco it doesn't have hot water. None the less a very welcome and civilised start to our journey.

We stroll down to the open-air restaurant for a bowl of soup and bottle of beer, then follow the trail of lanterns back to our little house. We wash, slap on insect repellent and collapse into bed at 8.10 pm. Asleep by 8.15 pm.

We are woken briefly around 2.00 am by a torrential storm, with loud thunder and vivid flashes of lightening.

COFFEE

Ethiopia is the birthplace of coffee – the name itself being a corruption of the Kaffa, the region where the beans were first cultivated. The coffee ceremony is a centrepiece of daily life, forming an integral part of the country's social and cultural life. An invitation to attend a coffee ceremony is considered a mark of friendship or respect and is an excellent example of Ethiopian hospitality. According to legend, more than a thousand years ago Kaldi, an Abyssinian goat herder, saw his goats acting strangely. They were eating the red cherries of some wild bushes and were full of energy, dancing on their hind legs.

He gathered several berries, took them to his wife, and believing them to be a gift from God, she sent him to a monastery to show the monks. However, the head monk proclaimed them the work of the devil and threw them on to the fire. The other monks gathered to find the source of the delicious smell of roasting coffee beans. In an effort to stamp out the embers, and save the delicious smelling beans, the monks crushed them underfoot and then to preserve them placed them into a large pitcher of hot water. Later that night they drank the resulting brew and found themselves wide awake and full of energy. They began to drink coffee regularly to help them stay alert during their night prayers. There are many variations of this legend, but at its heart there is some truth; the practise of drinking coffee does seem to have started in Ethiopia and spread gradually throughout the rest of the world.

Ethiopia is the oldest coffee exporter in the world and the crop remains hugely important to Ethiopia's economy, accounting for close to 70% of all export earnings. It is estimated that 15 million Ethiopians are employed in the country's coffee industry. There are few large farms and 90% of the coffee is produced by small growers. Between 200,000 to 250,000 tons of coffee are grown in Ethiopia every year, depending on the climate, prices and the seasonal nature of coffee plants. About half of the total production is consumed in the country.

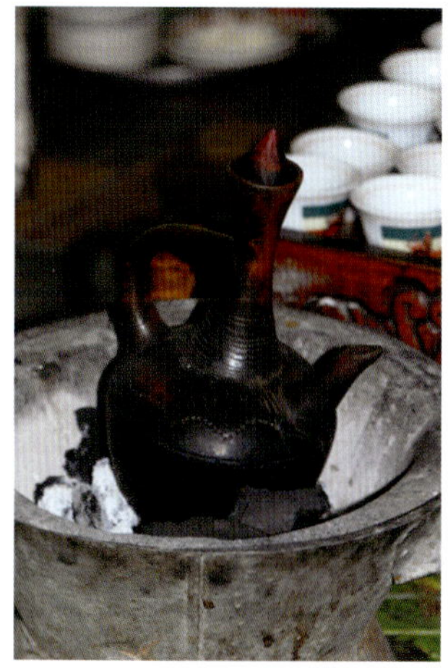

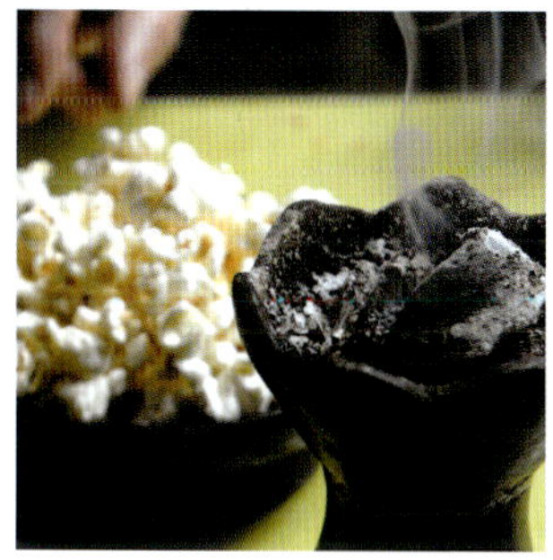

THE COFFEE CEREMONY

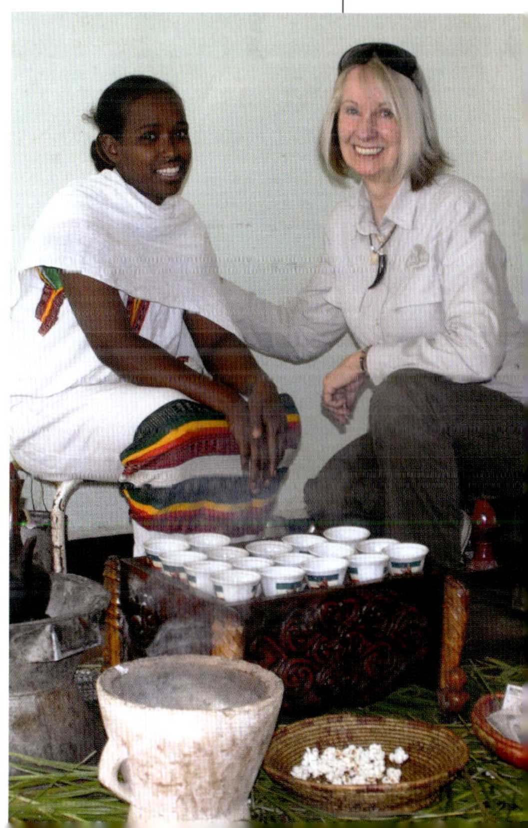

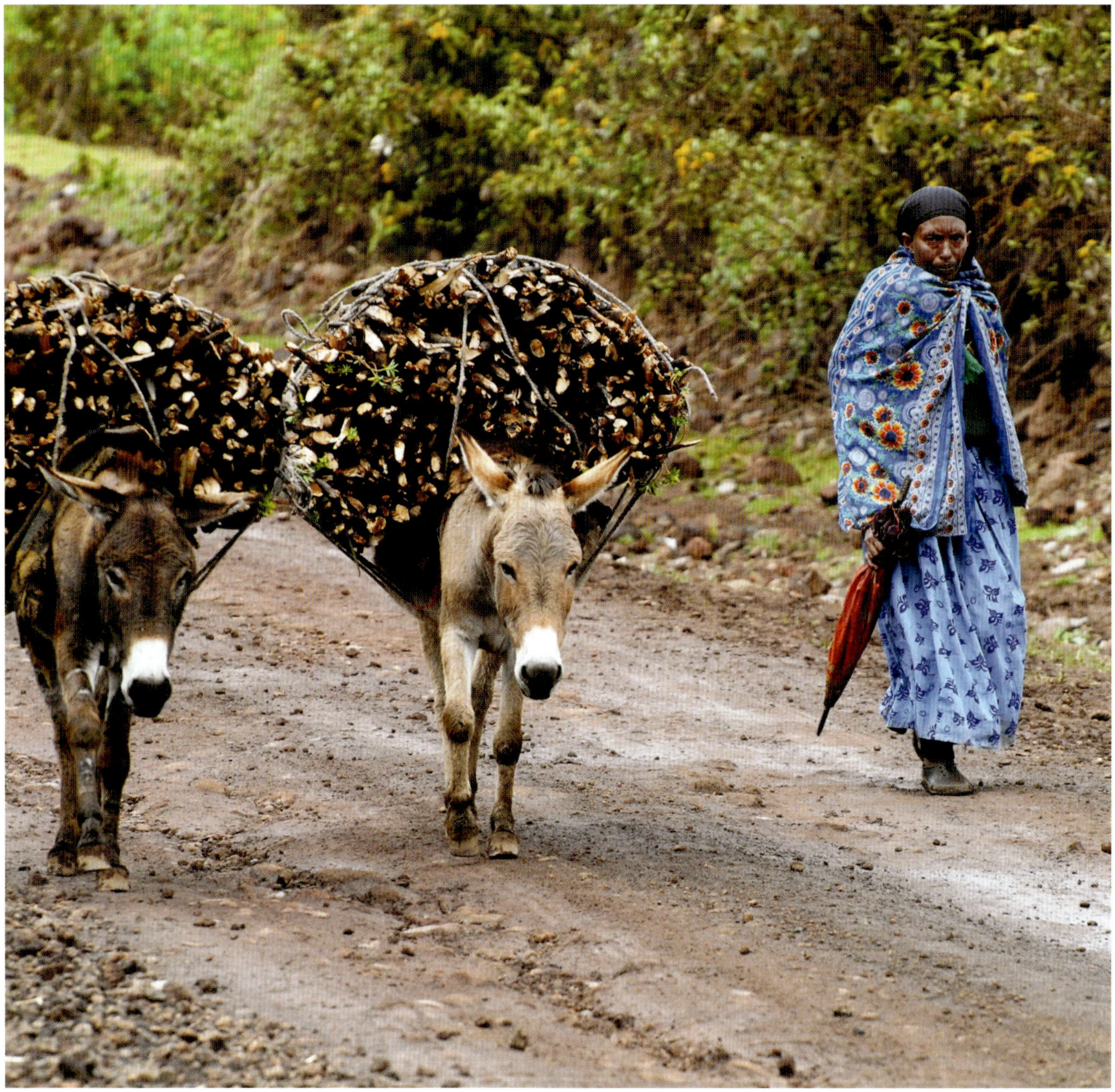

'I travelled throughout Ethiopia, a heaven-blessed land of natural beauty and potential abundance, meeting Ethiopians from all walks of life. I encountered a fascinating culture and a handsome people whose intelligent and courteous ways were beguiling.'

THEODORE M VESTAL

In Transylvania we felt we had travelled back through time. In flying to Ethiopia, we really had. The calendars of the entire world are based on the work of ancient Egyptian astronomers, working as long ago as 4000 BC. They discovered that the solar year lasted slightly less than 365¼ days. However, it was the astronomers of the Alexandrian school who incorporated this knowledge into a calendar around 300 BC, inventing the concept of leap years.

Julius Caesar took his Roman calendar from the Alexandrian science and this Julian calendar spread through the western world. The Copts inherited this and built upon it incorporating a method of computing the date of Easter, which was handed on to their descendant church in Ethiopia.

Ethiopia still uses its own unique ancient calendar which runs approximately seven and a half years behind the Western Gregorian calendar. The Ethiopian year retains the ancient Egyptian system of having thirteen months – twelve of thirty days each – with an additional month at the end of the year lasting five or six days depending whether the year is a leap year or not. The year starts on 11th or 12th September in the Gregorian calendar. Each Ethiopian year is dedicated to one of the four Evangelists – Matthew, Mark, Luke and John. The year of St Luke is a leap year, and therefore always has six days in the thirteenth month.

Time is celebrated on a twenty-four-hour clock that begins at daybreak. The start of the Ethiopian day is 6.00 am, so 7.00 am is 'one', 8.00 am is 'two', and so on.

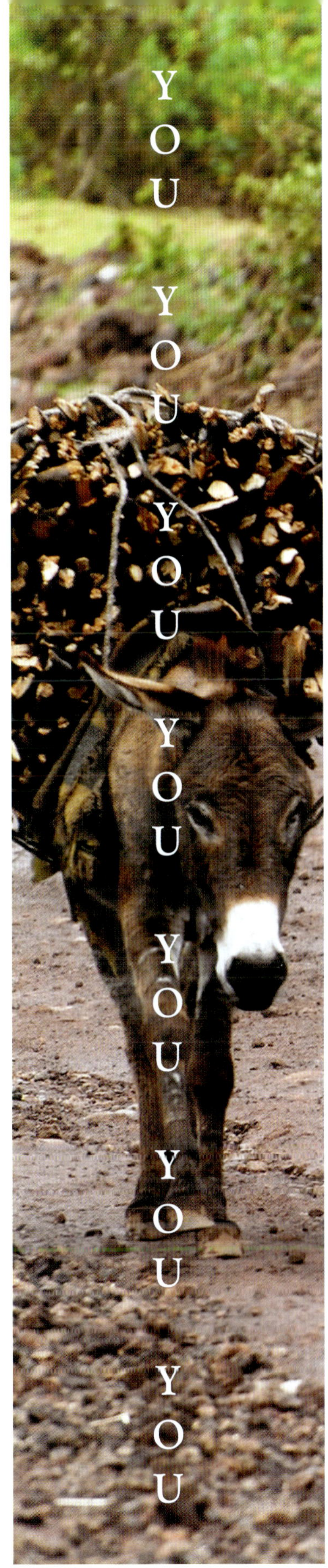

You You You You You You You You

ጠዎሪፍ

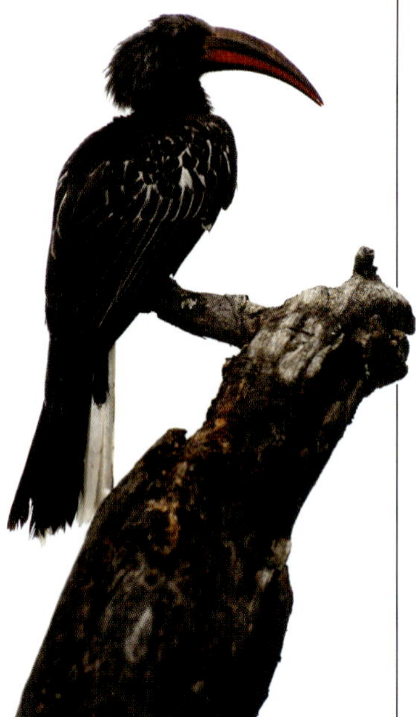

Hornbill

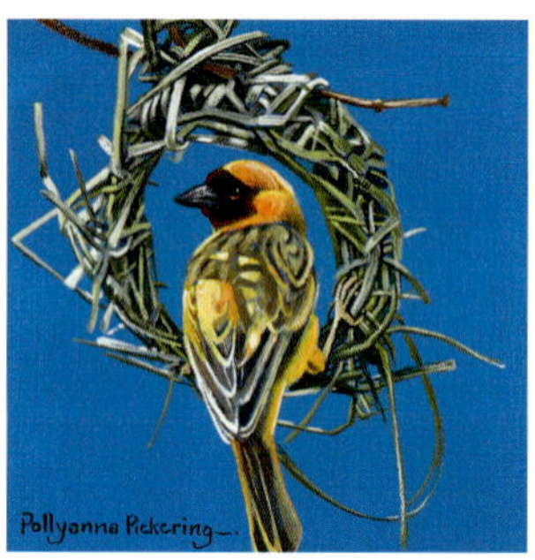

Weaver bird

Cormorants chill out on some rocks

Day Two

With the night flight, yesterday began for us on the 3rd April and sort of blended seamlessly into the 4th April, so I wake up at 7.00 am on the 5th April which is also 1.00 am on the 27th March, as well as being Palm Sunday (or a week before Palm Sunday) in 2009. Or 2002. I am sure I should feel at least seven years younger.

I stagger out of bed – despite having slept for almost ten hours it still feels a bit 4.00 am in the morning – but manage to wash and dress and take the opportunity to have a walk round. We have travelled two hundred and fifty kilometres south of Addis Ababa and chosen to stay at this lodge which has been designed and built on sound environmental principles using solar power for light and heat. The kitchens cook with biogas produced by a large biodigestor filled with food waste. The lodges have been constructed using natural materials and traditional techniques. We are staying in a little wooden cabin called a *godjo* in the middle of a grove of huge fig trees.

The lodge is located in an area of outstanding beauty that combines five ecological zones: wetlands, beach and lake, forest, dry pumice rocks and acacia shrub.

We walk down to the edge of the lake. Langano means 'clear water', though it is the colour of builder's tea. Tariku told us it is one of the few places in Ethiopia where it would be safe to swim, as the lake is free of bilharzia. In most fresh water lakes you can find a species of snail which are carriers of schistosomes – small, white-grey worms between seven millimetres and twenty millimetres long. If you are unlucky these worms will penetrate your skin then move into other organs of the body, such as the liver or bladder, where they lay eggs. Symptoms of bilharzia include tiredness, a fever, aching muscles and diarrhoea – and in severe cases it can be fatal.

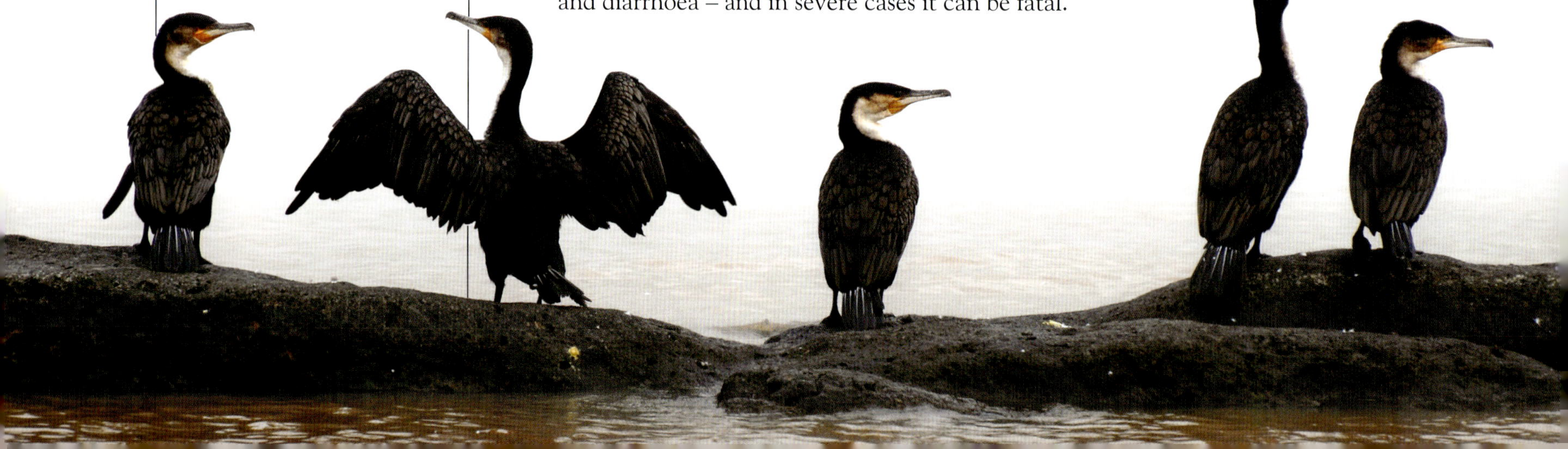

Although we are reassured that the lake is free of these worms, we are still not tempted to grab our bikinis and dive in – because we can see four prehistoric looking hippos wallowing at the edge of the lake. We are delighted to have the opportunity to sit and watch them. Planning our journey we had been focusing on travelling to find the wolves. We were also looking forward very much to seeing the bird life which makes Ethiopia a bird-watchers paradise – but had not been expecting to see too much in the way of other wildlife. Although most of the large mammals we associate with Africa can be found in Ethiopia, numbers are low and generally only found in remnant pockets in the most remote areas. Certainly there are no safari tours to game parks as in countries such as Kenya or South Africa, where wildlife sightings are all but guaranteed.

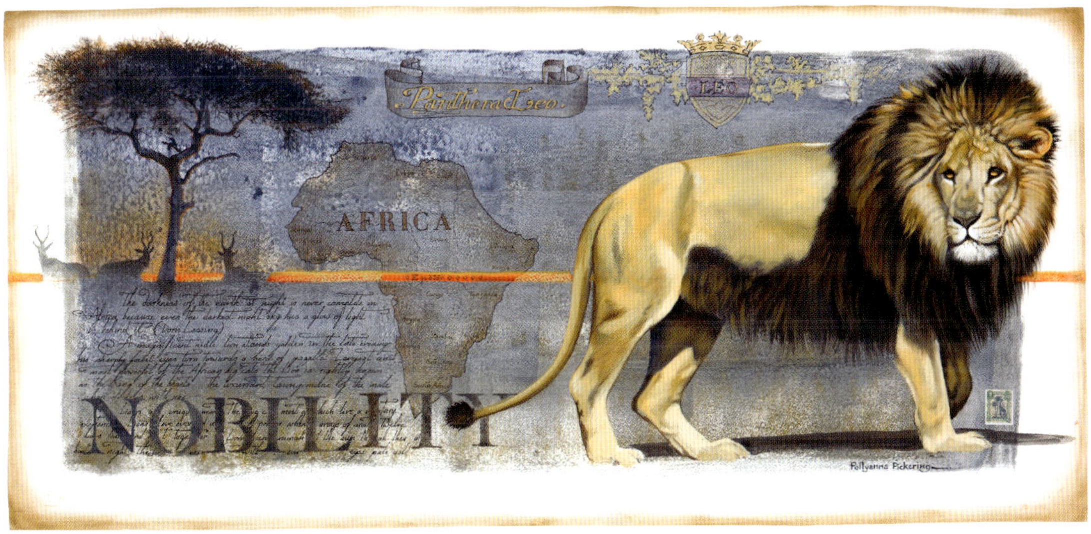

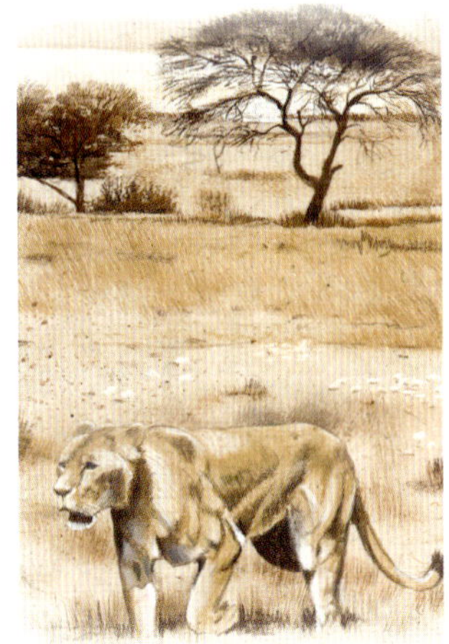

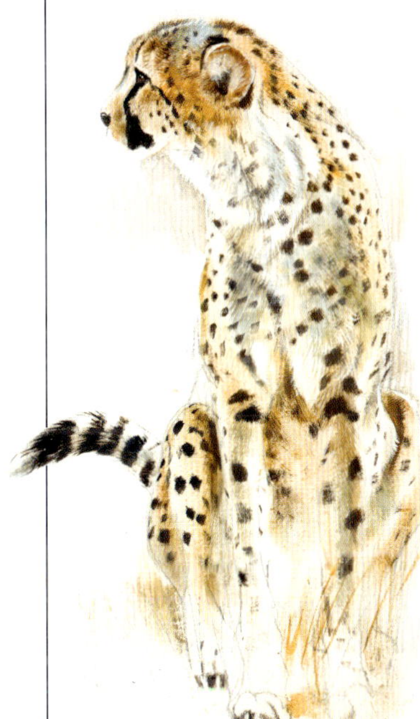

Pollyanna's studies of lion and cheetah

Abyssinian lions are smaller than most lions found on the continent, but the male has an impressive thick black mane. They are still believed to be fairly widespread in the Rift Valley and Somali border regions, but are thinly distributed and very rarely seen. Leopards are probably the most numerous big cat to be found in Ethiopia, but they are highly secretive. Cheetahs have been spotted in some of the national parks, and also probably

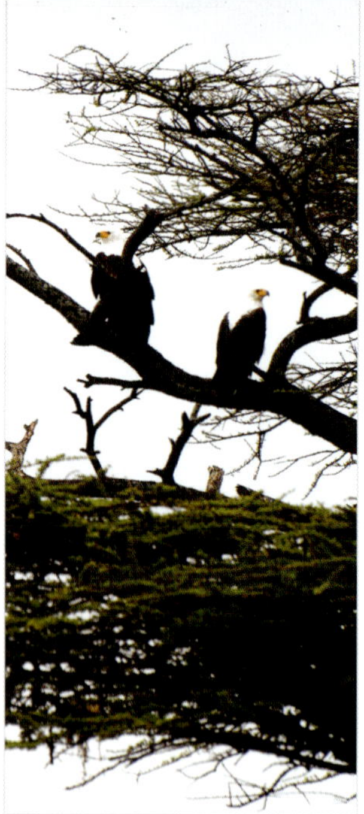

Two fish eagles guard their nest

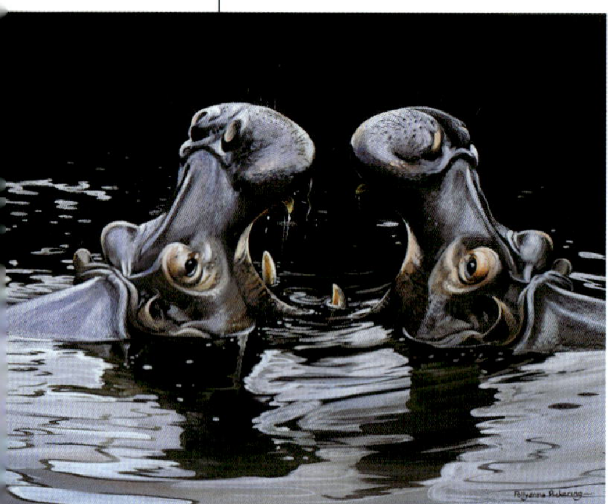

Hippo Hippo Hooray
Private collection (UK)

survive in very low numbers throughout the Rift Valley, but confirmed sightings are rare.

So we are delighted to find ourselves looking at four huge 'traditional' African mammals, as they enjoy the cool waters. Hippos are commonly cited as being the most dangerous animals in Africa, being responsible for more human deaths than any predator – but there is little actual evidence to validate this claim. None the less we keep a sensible distance – despite being herbivores, hippos are armed with long, lethal teeth capable of killing a large crocodile with a single bite, and they can be aggressive if threatened or provoked. Hippos that are not stressed are generally placid creatures, especially if given enough space and sufficiently deep water in which to submerge themselves. Our little group appears perfectly

THE HIPPOPOTAMUS

The name 'hippopotamus' comes from the Greek for 'water or river horse'. Hippos are not related to horses – in fact, their closest living relatives are probably whales. Hippos are the third largest land animal – only elephants and some rhinos are bigger. An adult male hippopotamus typically weighs around 2300 kg – but can weigh as much as 3200 kg. In a single night a hippo may eat as much as 45.5 kg of grass. They can live up to 45 years.

Hippos are found living in slow-moving rivers and lakes in Africa. With their eyes, ears and nostrils positioned on the top of the head, they can hear, see and breathe while most of their body is underwater. They are so well adapted for life in the water they even have a set of built-in goggles: a clear membrane covers their eyes for protection while still allowing them to see when underwater. Their nostrils close and they can hold their breath for 5 minutes or longer when submerged. Surprisingly, adult hippos can't actually swim and are not generally buoyant. When in deep water they propel themselves by leaps, pushing off from the bottom. However, young hippos are buoyant and more often move by swimming – propelling themselves with kicks of their back legs.

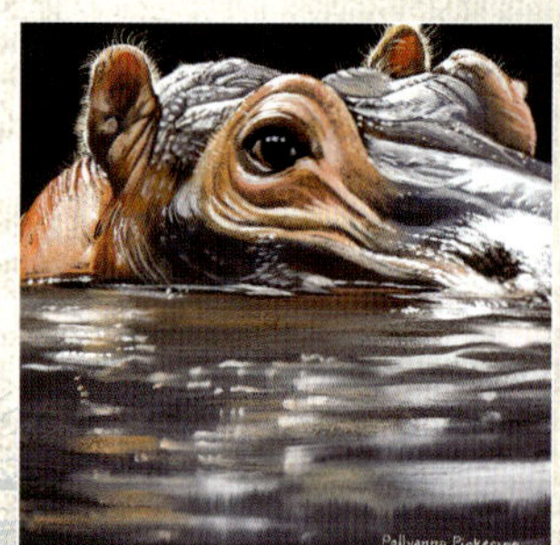

relaxed and content, so Pollyanna takes the opportunity to sit down with her pencils. It is not every day she has the opportunity to sketch four hippos before breakfast.

We walk around the lake for a further hour, watching the abundant bird life, including two fish eagles on a nest, before heading up to the restaurant for a cheese omelette and black tea.

We pile our luggage and ourselves back into the car and set out on the remainder of our drive towards the Bale Mountains, with a few breaks en route. The first of these is to unload the vehicle and ease it back through the container/bridge (which we learn was placed here two years ago, making it the millennium bridge!). We walk on ahead while the roof rack is being reloaded and securely strapped down and soon find ourselves strolling companionably with an elderly goatherd. Our conversation is naturally somewhat limited, but we appear to be getting along famously. By the time our crew catches up with us, we are playing football (actually rolled-up-socks-and-a-shirt ball) with a group of kids. We hop back in, and out again almost immediately when we spot a flock of weaver birds nesting over the river.

Pollyanna sits on the river bank completing a couple of pages of quick sketches, which will form the foundation for the finished paintings she is already planning to complete in her studio back home.

Our next stop is in the largest town we have passed through since leaving Addis. The large church in the central square is in the process of holding a religious service. We climb out of our vehicle – but soon start to find the experience a little overwhelming. There are zillions of people crowded into the streets outside, and we are practically mobbed. As soon as the children in Ethiopia spot a tourist, they run up and immediately shout 'You! You! You! You! You! You! You!' while pointing at you. Occasionally this is interspersed with the word *faranji* – foreigner. Tariku explains that they are just excited to see us and it is more or less the English equivalent of the Amharic shout of *Ante!* or *Anchee!* which is a colloquial way of catching someone's attention. In their native language it does not

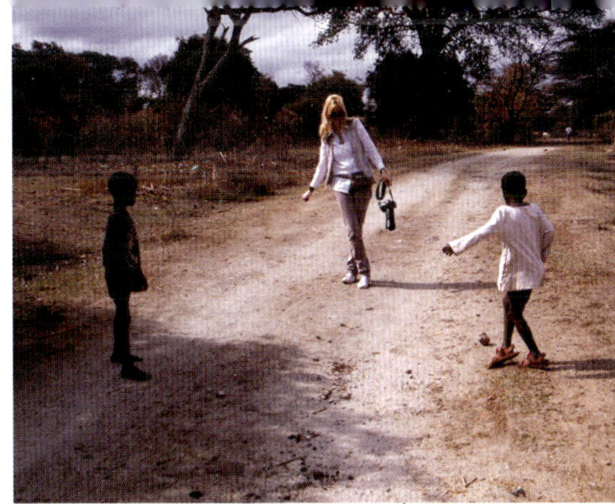

Anna-Louise demonstrates her footballing skills

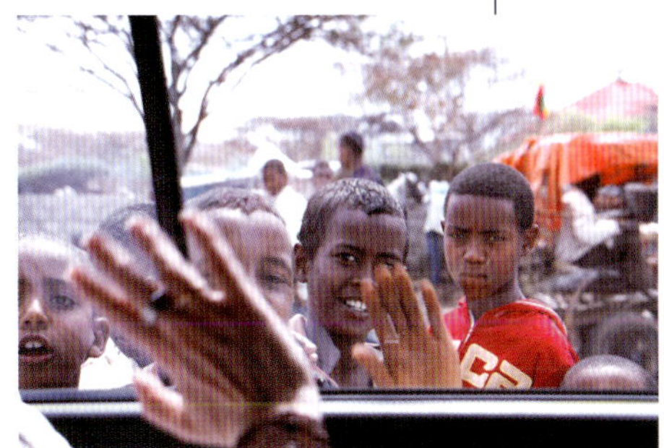

You you you you you!

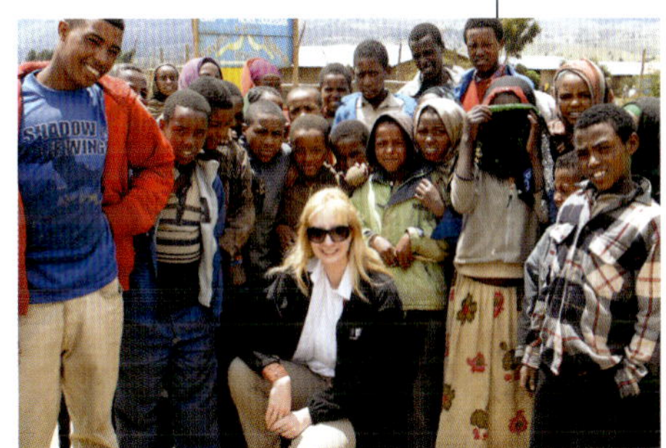

Anna-Louise attracts a large group of curious onlookers

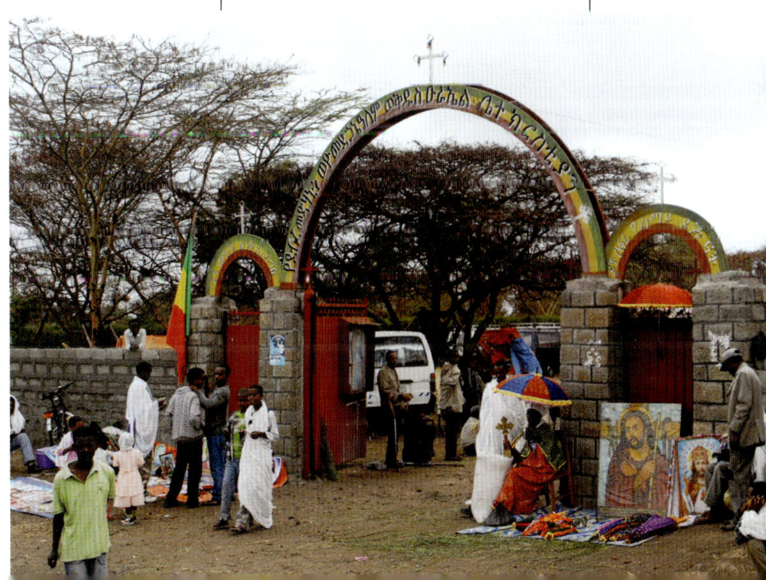

The entrance to the orthodox church

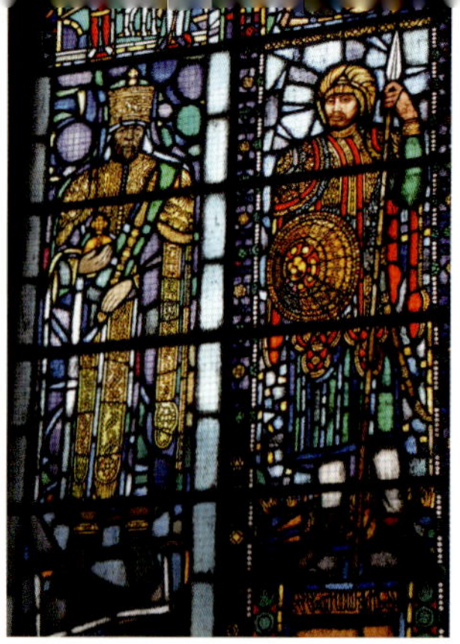

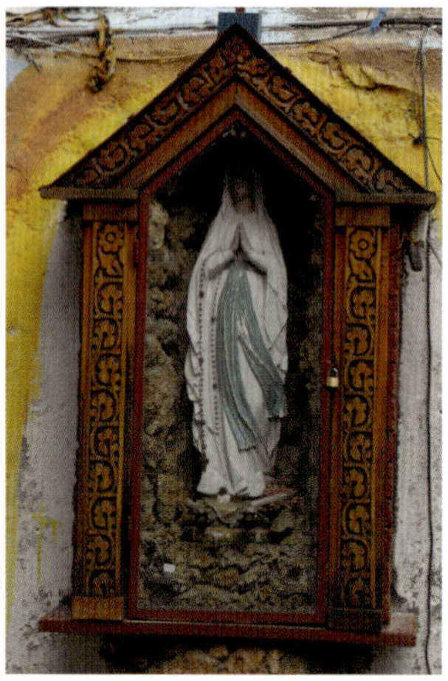

Shrine to the Virgin Mary

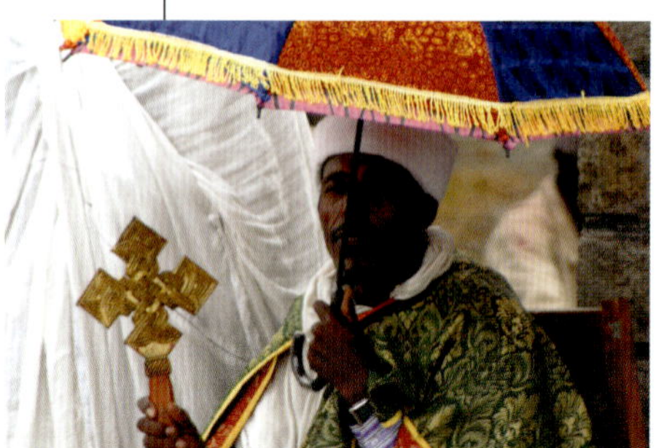

The Ethiopian branch of Christianity developed entirely independently

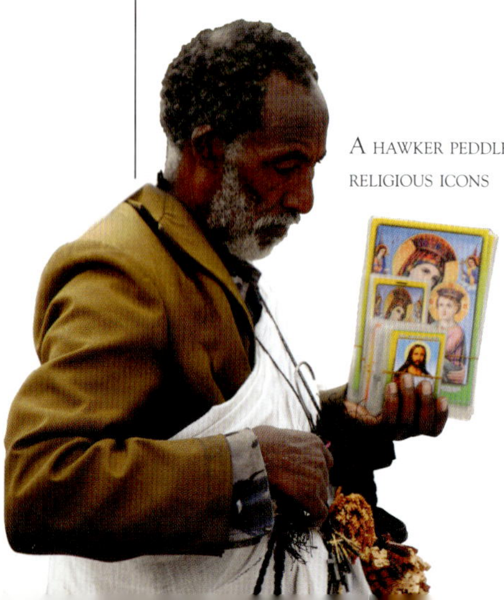

A hawker peddles religious icons

sound as rude or aggressive as it does in English. This chant will become the background noise to any stop we make on our journey. With small groups of children this is not an issue – but here in the town we are surrounded by a large group, and the chanting is quite oppressive. Our already dog-eared *Lonely Planet* calls this phenomenon '*faranji* frenzy'. We struggle our way through to the entrance to the church – beggars line the pathway to the gates – and enter the courtyard which is divided into groups of men on one side and women on the other. Hawkers weave in and out of both groups offering religious icons and incense. This is the Ethiopian orthodox church, so in essence Christian – but it is certainly far removed from the familiar Church of England.

Ethiopians date the coming of Christianity to Ethiopia to the time of the apostles, when a Christian philosopher from Syria was shipwrecked on his way to India. His two students, Frumentius and Aedesius, were washed ashore and taken to the royal palace. Eventually they became the king's private secretary and royal cupbearer. They were also permitted to preach their new religion and contacted the bishop of Alexandria and begged him to send missionaries since the people in Ethiopia had proved ready to receive the gospel. By the fourth century Christianity was recognised as the state religion, and coins were minted bearing the cross – the first in the world to bear this symbol. During the seventh century the Muslim conquests cut the Ethiopians off from the rest of the Christian world; their

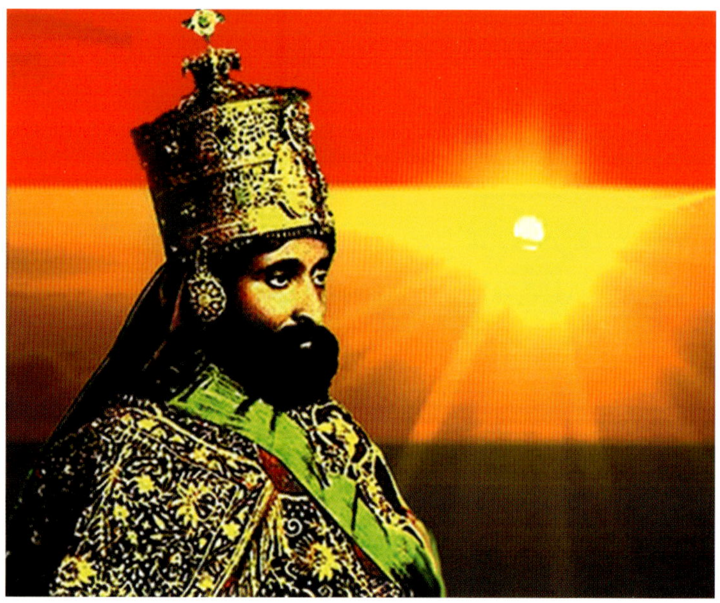

Haile Selassie

branch of Christianity then developed almost entirely independently from any outside influence. It still incorporates deeply ingrained superstitions, rituals and beliefs that include elements of ancient Judaism. Priests carry out animal sacrifices, every Ethiopian Christian male is circumcised, devout Christians keep Sabbath on Saturday (as well as Sunday), and there are restrictions on food. A Tabot, an essential part of every church and its most holy item, is a replica of the Ark of the Covenant which is only removed from the holy of holies for important festivals.

Our next brief stop is in the Rastafarian town of Shashamane. Land around this area has been given to the Rastafarians by the Ethiopian government as a homeland, and many Rastafarians have emigrated here from around the world. Everywhere we turn in Ethiopia we are confronted by another remarkable piece of living history. The Rastafarian movement began in Jamaica during the 1930s when Marcus Garvey, a black political leader, prophesized: 'Look to Africa where a black king shall be crowned, he shall be your Redeemer'. The prophecy was rapidly followed by the crowning of Emperor Haile Selassie I in Ethiopia, an event hailed as the fulfilment of Garvey's prophecy. The religion takes its name from Haile Selassie's original name, Jah Rastafari, and he is regarded by Rastafarians as the Black Messiah. Just a few weeks before our arrival in Ethiopia Barack Obama had been sworn in as president of the United States, and many Ethiopians – not just the Rastafarian community – saw him as the next link in the chain, another fulfilment of the prophecy. His image was everywhere, and many people wore T-shirts bearing his name, and his optimistic slogan of change: 'Yes We Can'.

When Tariku mentioned that we were to stop in a Rastafarian town, I had lazily stereotyped it in my imagination as a very laid back place,

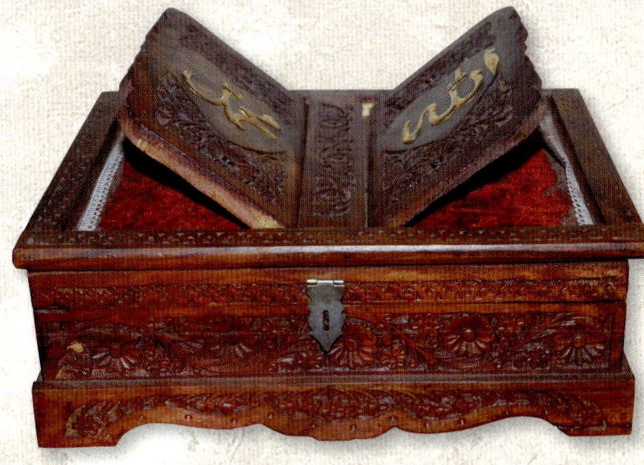

ETHIOPIA AND THE ARK

The Ark of the Covenant is the name traditionally given to the coffer into which the tablets of the Ten Commandments were placed by Moses. In the Bible the Ark is described as measuring almost 1.15 m in length and made of acacia wood overlaid by gold. It was carried by means of two long gold-plated bars. The Ark is believed by the devout to hold great powers – Biblical sources describe it blazing with fire and light, and levelling mountains. Its power helped the Israelites to defeat their enemies and they carried it with them to the Promised Land.

According to the Ethiopian legend the first King of Ethiopia was Menelik I, son of the Ethiopian Queen of Sheba and King Solomon of Israel. As a young man Menelik went to Jerusalem to visit his aging father. King Solomon gave his son the Ark of the Covenant; he carried it back to the town of Axum in his home country, where he placed it in the safe keeping of the local priests. It has remained there until the present day in the Church of Our Lady Mary of Zion, and today is guarded by an elderly half-blind priest, the only human permitted to look upon it.

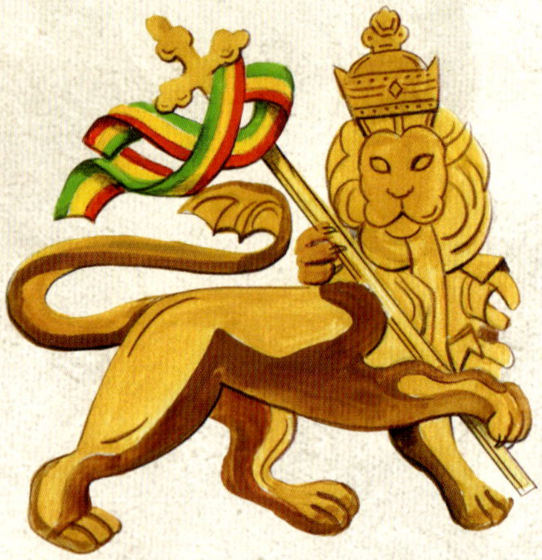

LION OF JUDAH

Wildlife experts estimate that approximately 1,000 Ethiopian lions remain in the wild. Smaller than other African lions they are famous for their black manes. Lions, and specifically 'The Lion of Judah', are the symbol of Ethiopia's monarchy. The Lion of Judah is originally described in Genesis. In Judaism, the lion is seen as symbolizing the tribe of Judah while in Christianity it is assumed to represent Jesus Christ.

In Rastafarian belief, Haile Selassie is considered to be the reincarnation of Christ – therefore literally 'The Lion of Judah'. Images of lions can be found everywhere in Ethiopia, on statues and the local currency, adorning tea cups and T-shirts. Meanwhile Pollyanna's artwork has also been reproduced in many forms – from postage stamps to cross-stitch kits – and now even on the cover of a CD! New York based DJ Brandon Striker selected one of Pollyanna's majestic lion portraits as the artwork for his 'Ultimate Culture Mix' album which features a wealth of reggae tracks.

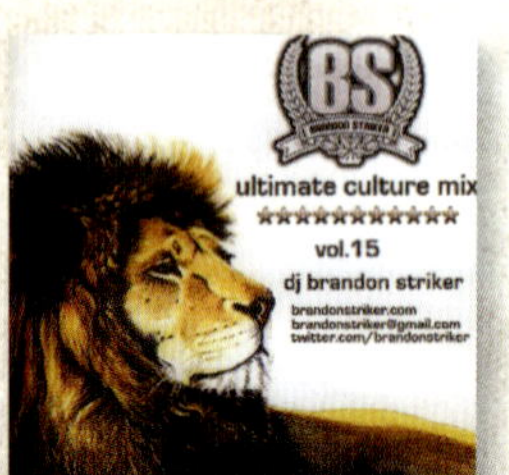

Reggae music developed within the Rastafarian culture, so her painting of a noble lion makes the perfect cover image.

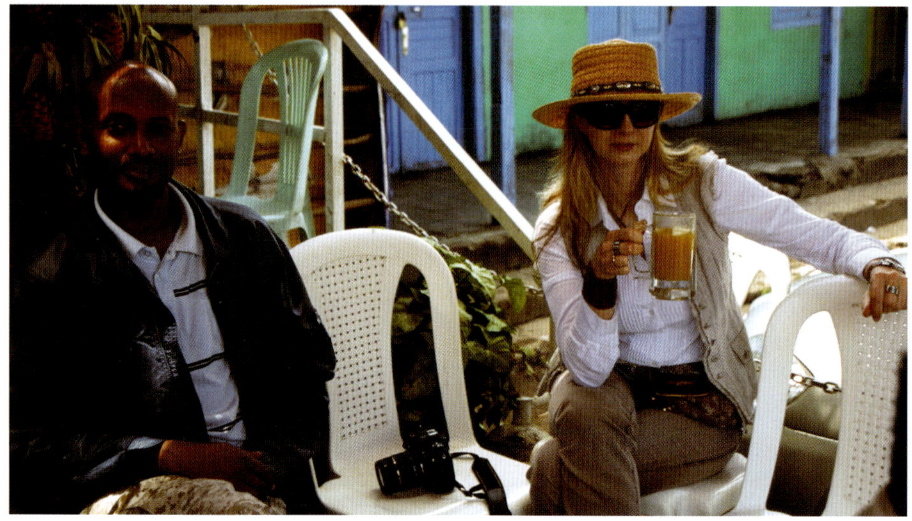

Tariku and Anna-Louise enjoy a roadside break

with relaxed citizens chilling out in the streets. In fact it is clearly a very prosperous and industrious town, with a great many new buildings under construction. We stop in a café and order glasses of freshly squeezed orange juice – but the owner tells us through Tariku that today's oranges are 'not that good' and suggests we have the fresh pineapple juice instead. It is a good recommendation – the juice is divine, the naturally sun-ripened pineapples providing a mellow sweet rich flavour quite unlike that of the imported fruit we buy in supermarkets back home. We call into the bank to change the money we obtained at the airport into some smaller notes, more suitable for day to day purchases. There are lots of people milling about in front of the counter, but no obvious queue. I approach a window that appears to be free, but the teller shoos me away pointing vaguely towards a desk where people are busily filling in forms in triplicate. I don't know if I need a form and as they are all written in Amharic I am none the wiser when I have picked one up. I try a different window, where this time my one hundred birr notes are split down into a huge wodge of very freshly minted five's. Unusually for a bank, on the customer's side of the dividing glass there are huge brick-like stacks of low denomination notes piled on the counter, bundled together with rubber bands.

We drive on through the countryside – now we are truly out into the rural regions where villages of mud huts sit in lush green pastures. Donkeys and cattle graze along the roadside. Here, as in Transylvania, the majority of the people are subsistence farmers, living off the land.

We get out to stretch our legs in one village ('YouYouYouYouYouYou') and Tariku falls into conversation with a lady who has arrived to see the *faranji* – and she motions that we are welcome to step inside her hut to see the interior. The dirt floor inside is partitioned into two areas, the second of which has an open fire in the centre of the floor. There is no chimney, the smoke simply escapes through the thatched roof. The smoke will naturally fumigate the hut, protecting against insect infestations. There is a third smaller area which is home to the goats and cattle and an upstairs storage area. Through Tariku, the lady tells us that her entire family of seven live in this dwelling, along with their animals. A lot more than seven people have gathered to see us, however, and are crowding around the entrance. The women are fascinated by our hair, and want to take turns touching it – although it is not so much the blonde colour which attracts their attention, but the smooth fine texture. Outside a nearby hut a lady is sitting plaiting her children's hair into a complicated pattern. She also attempts to plait Pollyanna's – although initially she cannot get a grip on the smooth surface, and we can see she is struggling to know where to begin. This is all much to the amusement of our crew who have stepped out of the vehicle to watch and call out comments and encouragement. Pollyanna meanwhile is wincing with pain as her hair is firmly twisted and plaited; but this is a woman with experience of plaiting the hair of wriggling three year olds, so a blonde *faranji* won't defeat her.

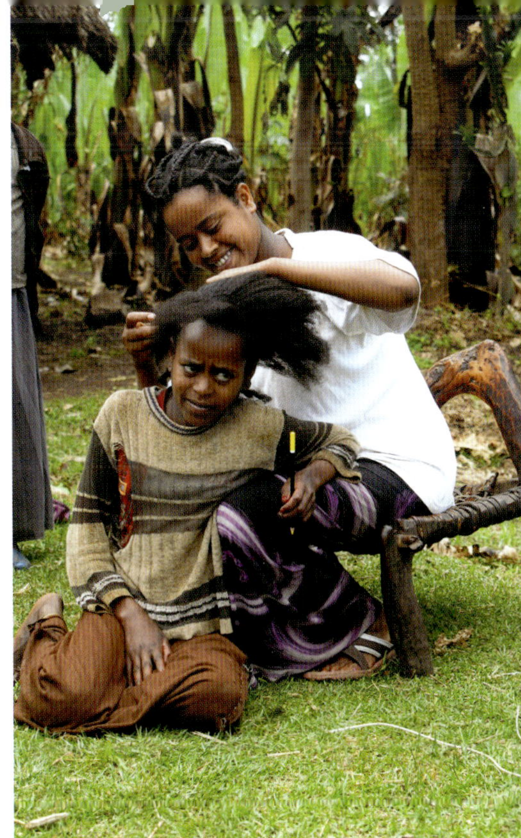

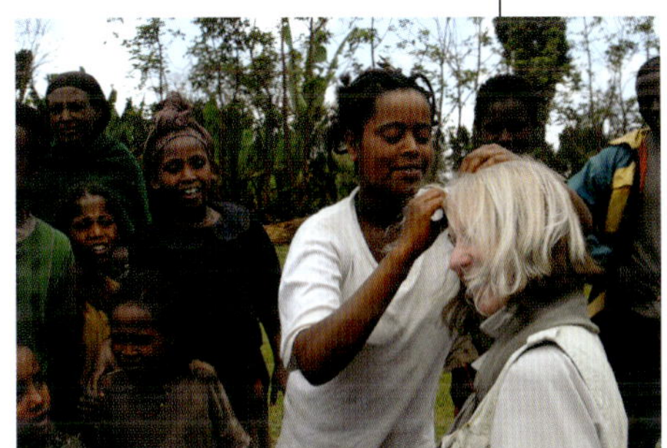

A WOMAN PLAITS HER DAUGHTER'S HAIR – THEN TRIES THE SAME TECHNIQUE ON POLLYANNA

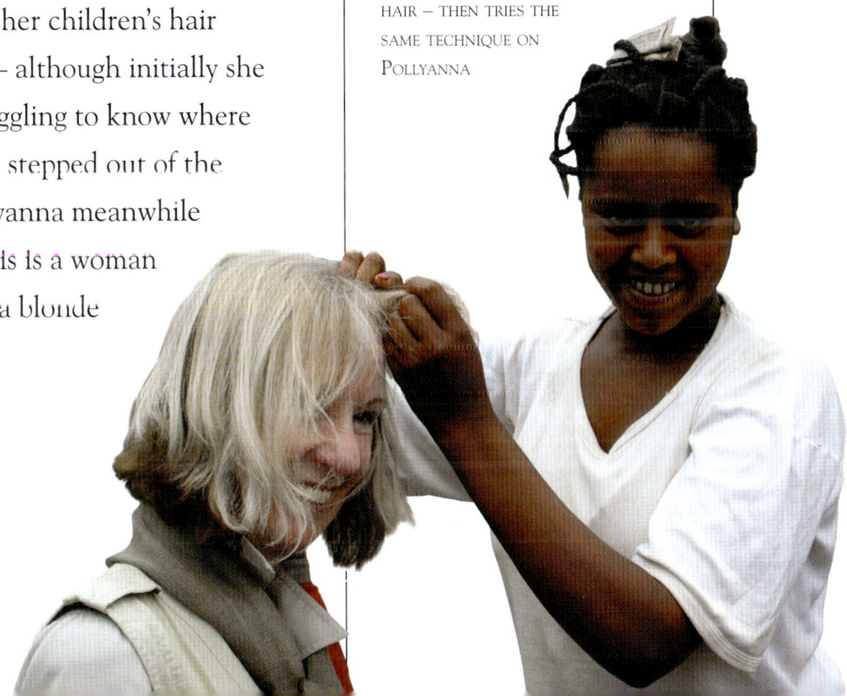

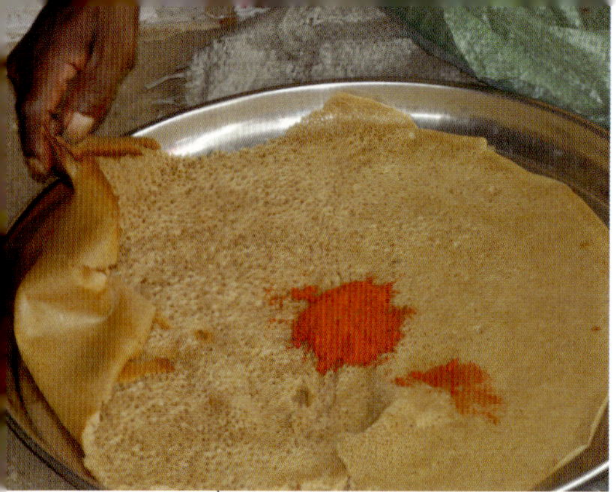

Injera

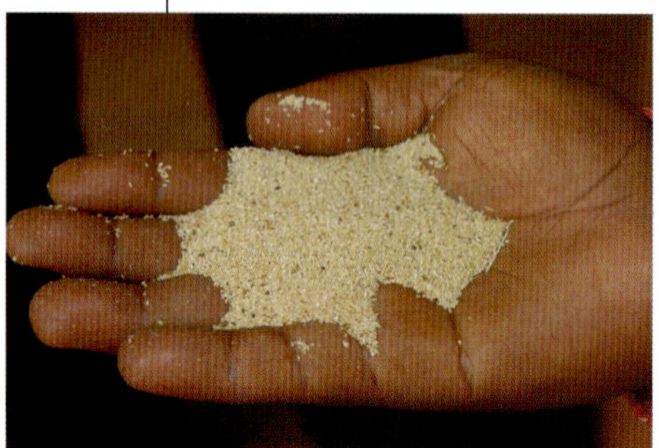

Teff

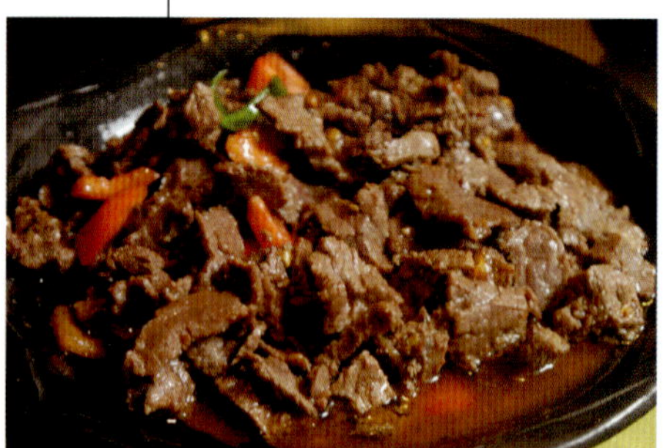

Kai wat

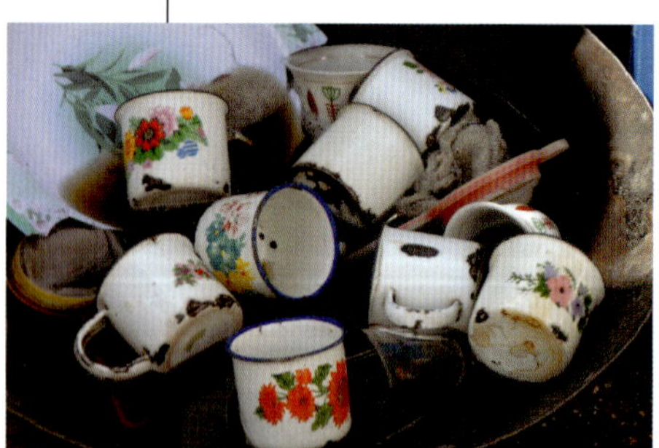

The washing up

We stop for lunch at a small lodge and decide we are ready to try the infamous *injera*, staple of all Ethiopian meals. *Injera* is used as a plate and a utensil, mopping up and carrying the food to your mouth. The concept of eating from individual plates strikes most Ethiopians as deeply eccentric – food here is shared from a single plate without the use of cutlery. Much of the Ethiopian diet is heavily meat based, but as we are here in the run up to Easter, which is also a time of fasting, there are more vegetarian dishes available than usual. We opt to top our *injera* with a vegetarian version of *kai wat*, the classic stew made with beans. *Injera* is unique to Ethiopia and is made from a cereal called teff. Teff is one of the smallest grains in the world, measuring about 1/32 of an inch in diameter. Approximately one hundred and fifty grains together would equal the size of a kernel of wheat. Ethiopia is the only country to use it as a food (as the grains are so small I imagine that no one else noticed it) – and they have been preparing *injera* from it for at least four thousand years. Presumably this has given the inhabitants time to develop a taste for it. I suspect it may take me around the same length of time. When the *injera* arrives it is in the form of a thickish grey soggy pancake, rolled up like a warm flannel. The method of eating *injera* is to spread it out like a plate, dollop on the spicy stew, then fold bits in your right hand only and eat it with your fingers. (The left hand is viewed as unclean and it is frowned upon to eat with it. However, we are so used to using both hands when eating that, for much of my time in Ethiopia, I end up sitting on my left hand while in a restaurant to avoid causing offence to my fellow diners.)

'Like eating sour foam rubber that's been stored in a farmer's armpit.'

The *injera* looks like it should be very bland and tasteless. It isn't. It has a very bitter taste, almost sour, which adds a sharpness to the already hot and spicy bean stew. Travel journalist Stephen Bleach described his first taste of *injera* in an article in *The Times*: *'Like eating sour foam rubber that's been stored in a farmer's armpit'*. I struggle bravely through the best part of one – and can see Pollyanna also fighting her way doggedly through her meal. Along with the bean stew we have ordered another Ethiopian favourite – *enkulal tibs*. This is a version of scrambled eggs, usually made with peppers, although we request ours with onions and tomatoes – and this is delicious. As we never quite get a taste for *injera*, despite several more attempts, this, along with the spaghetti, will be the mainstay of our meals while we travel.

Onwards into the countryside. First we drive into Dinsho, a small town around an hour and a half's drive from our eventual destination, the Sanetti Plateau in the Bale Mountains. Here we have to report to the Bale Mountains National Park office to let them know of our arrival in the area, pay our fees to enter the national park, and indicate the number of days we intend to stay.

The road we have been travelling on throughout the day is in a very poor state of repair. Mostly it appears to have been paused mid-construction with long detours on to parallel single dirt tracks. The surface is pot holed and ridged and, where I imagine it would usually be a dirt road, after last night's storm it is now sticky mud – almost swampy in some places, and flooded in others. We see several buses and lorries struggling as their wheels stick or

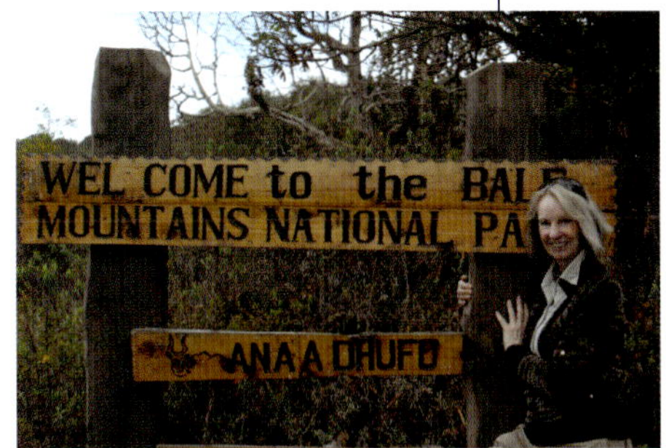

We pay our fees at the Bale National Park office and inform them of our length of stay

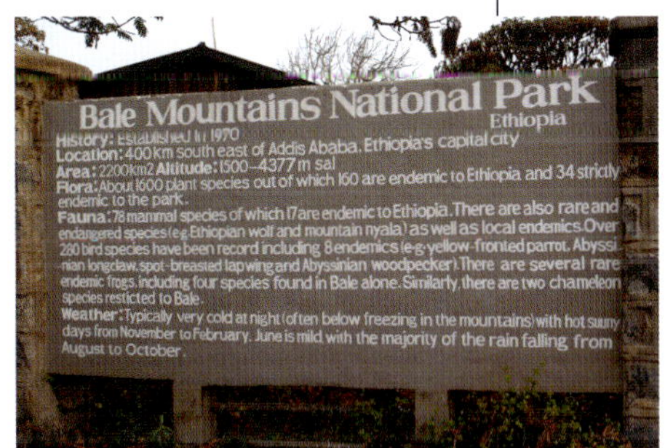

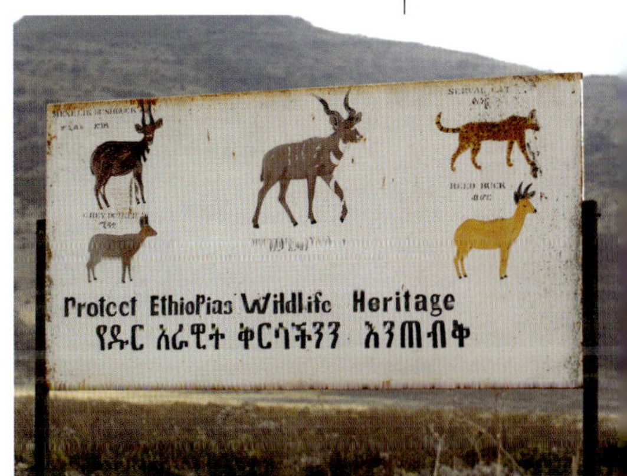

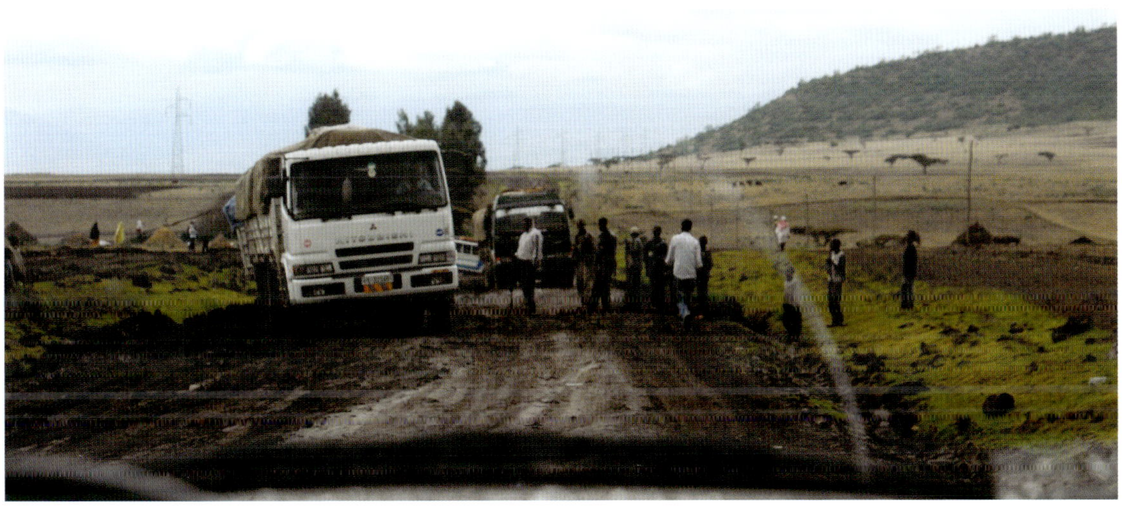

An overnight storm leaves many of the roads difficult to negotiate

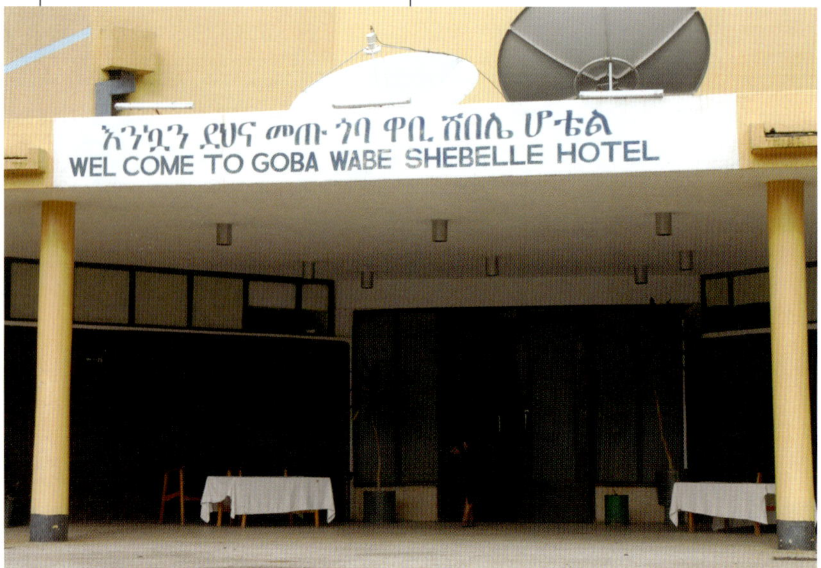

The Goba Wabi Shebele Hotel

slither – but Abbi is an excellent driver and he guides us skilfully around all the obstacles.

Eventually we arrive at our destination – the Goba Wabi Shebele – a government run concrete block hotel on the edge of the Bale Mountains National Park. We are given the key to Room 29, but when we follow the directions to the door, we find there is no Number 29, just a brass sign saying 'Saloon'. However, our key fits, so we step inside – and find a large room with a coffee table, sofas and a television. We back out, assuming we are in a public lounge area – but when we head back to reception we are sharply directed back to the saloon. On further exploration there is a small adjoining bedroom, around a fifth of the area of the huge sitting-room. There is hot water so we are able to freshen up and unpack before heading up to the dining-room. As previously mentioned an Ethiopian menu does tend to be heavily meat based; the hotel is very happy to cater for vegetarians, but not necessarily used to doing so. This results in some slightly odd combinations, like tonight's menu consisting of a bowl of delicious onion soup followed by rice cooked in garlic with chips and boiled carrots, followed by an orange chopped into quarters accompanied by a pitcher of the local beer, fermented from wheat. This is good, but unusually chewy by our standards – not all the husks have been strained out during production. We eat to the soundtrack of random cheers from the adjoining bar – poking my head round the corner to find the cause of the celebration, I discover Manchester United are beating Aston Villa on a snowy television screen balanced precariously on a battered chest of drawers.

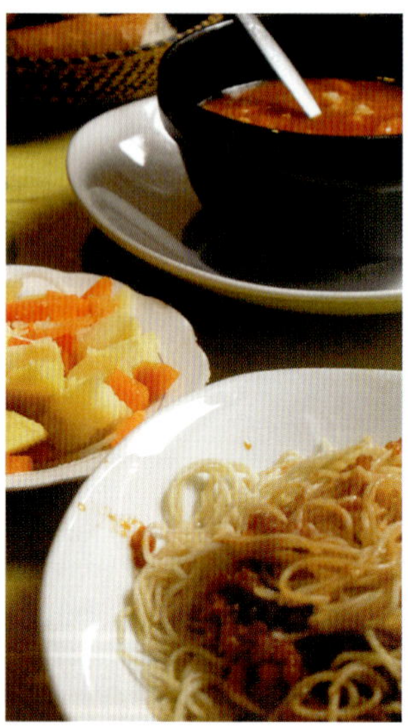

Our vegetarian diet results in some odd combinations

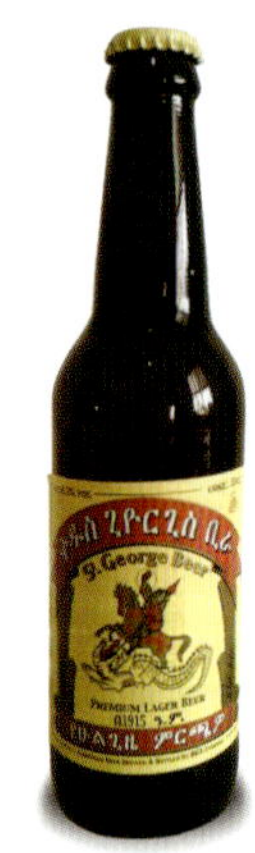

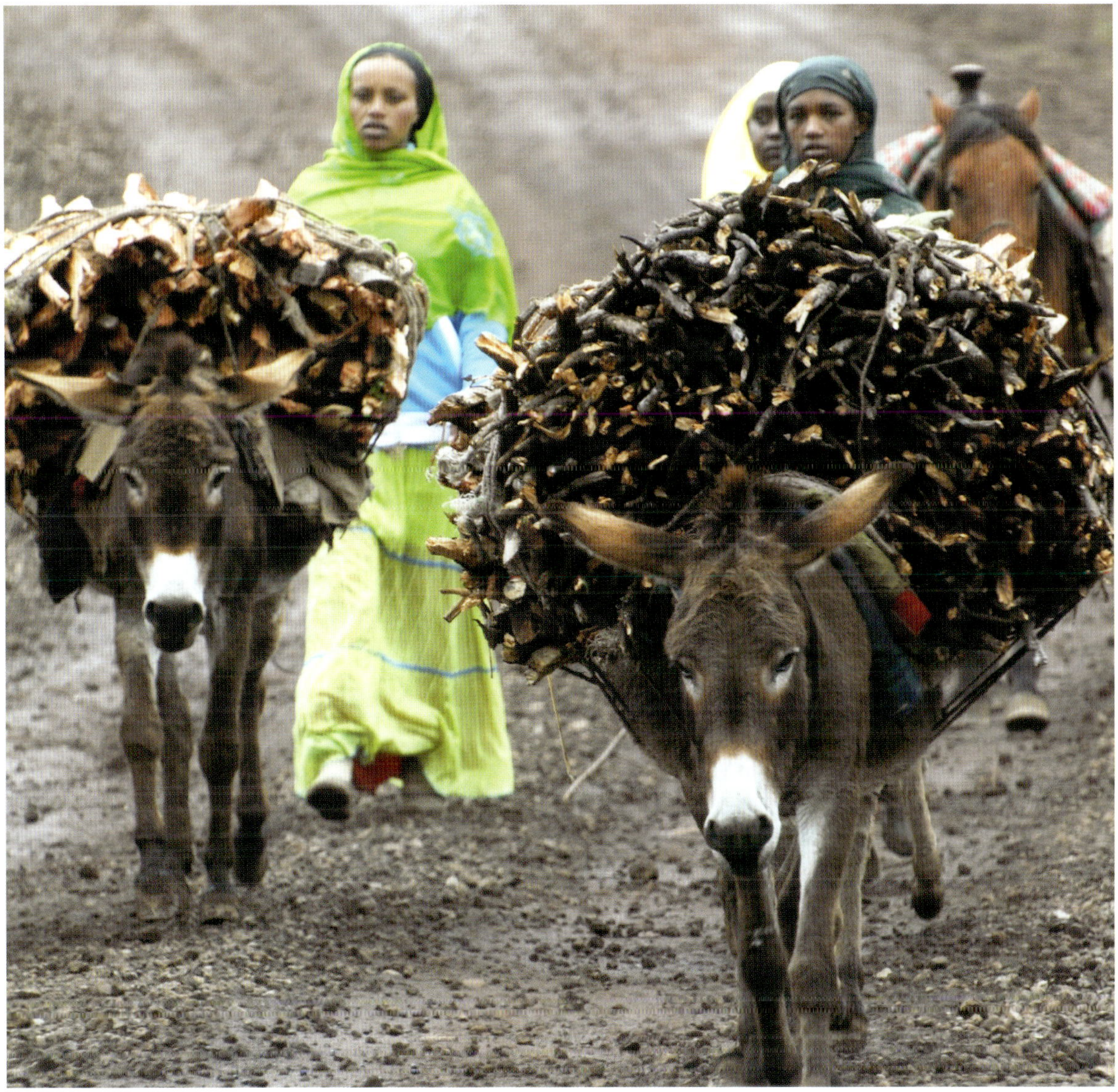

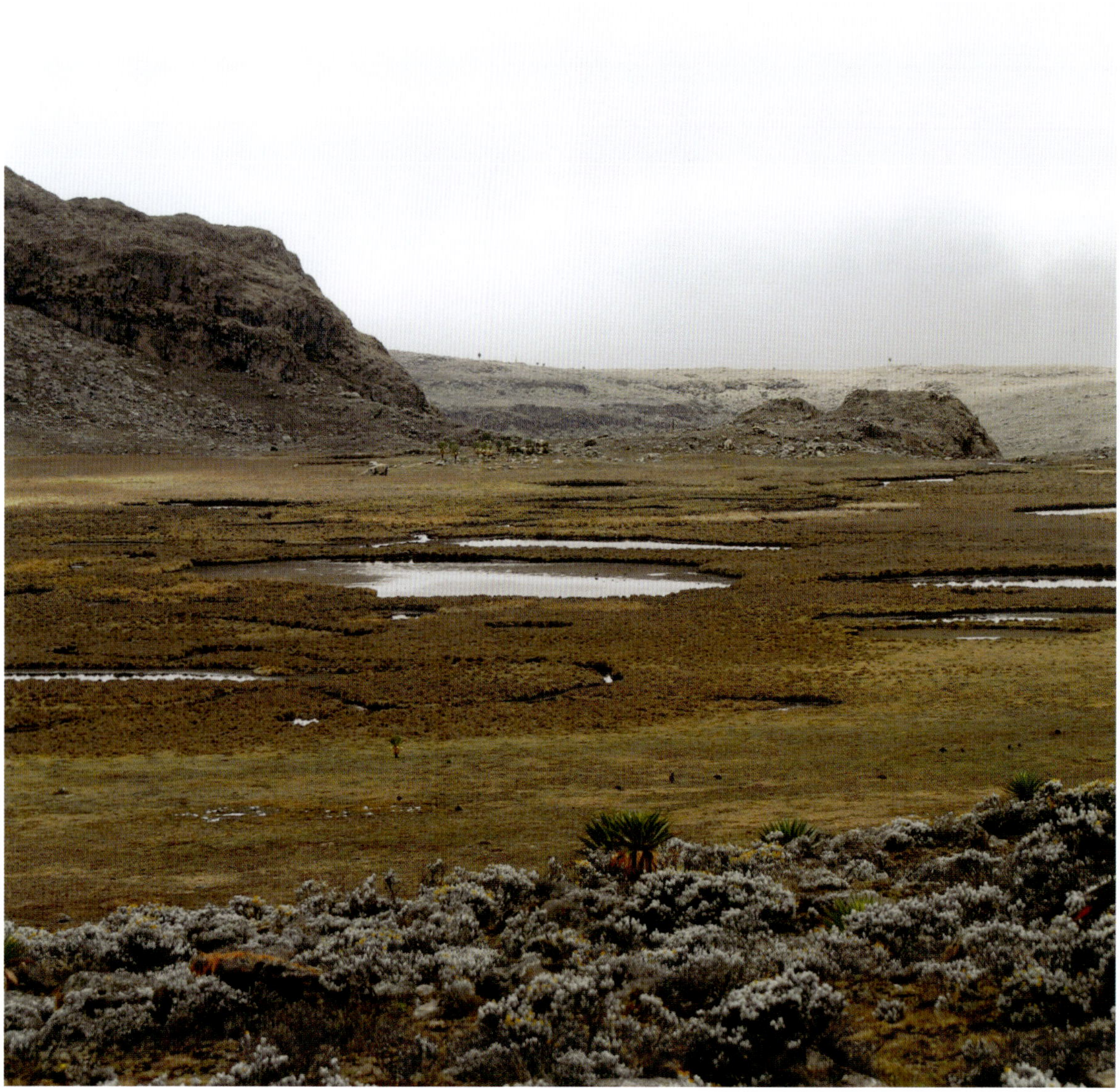

Chapter Three

RED WOLVES ON THE ROOF OF AFRICA

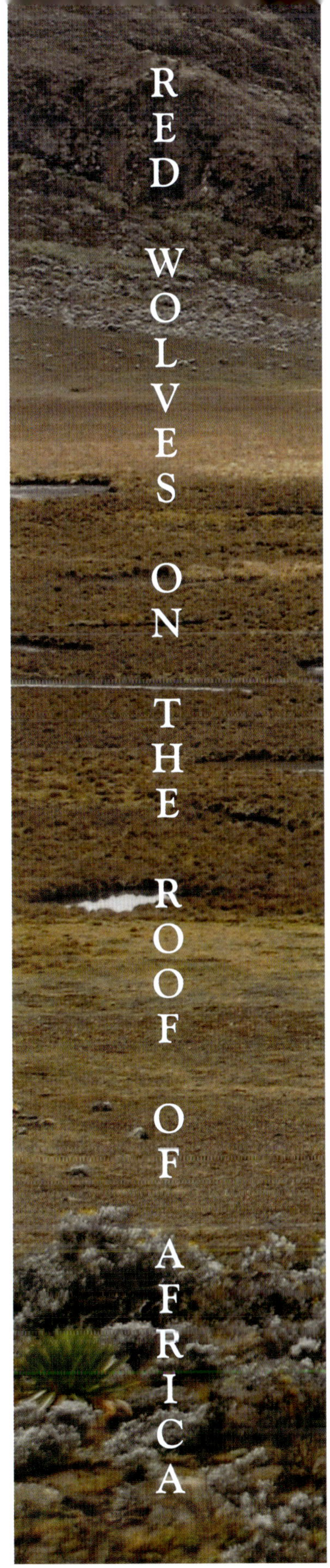

'All through the valley the people are whispering
The wolves are returning, returning
To the narrow edge of our fields, our dreams'

ANITA ENDREEZZE, *RETURN OF THE WOLVES*

Ethiopia is the most mountainous country in Africa. We are heading for the Bale Mountains National Park, an area of high-altitude plateau and spectacular volcanic peaks. The largest area of Afroalpine habitat in the whole of the continent it is home to many of Ethiopia's endemic mammals, including the mountain nyala and, of course, the Ethiopian wolf.

The national park area is split in two by the spectacular Harenna escarpment that runs from east to west. North of this escarpment is the bleak Sanetti plateau area at four thousand metres above sea level. The plateau is one thousand square kilometres of ancient volcanic rocks through which rivers and streams have cut dramatic gorges over the centuries. In the north west an enormous single crag of rock stretching for over sixty kilometres forms the boundary – in the south the land slopes gently down into a thick heather belt, which in turn gives way to the moist tropical Harenna forest.

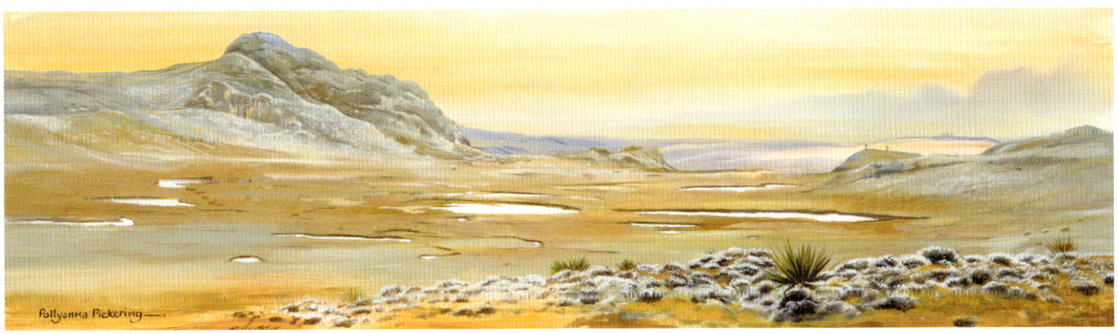

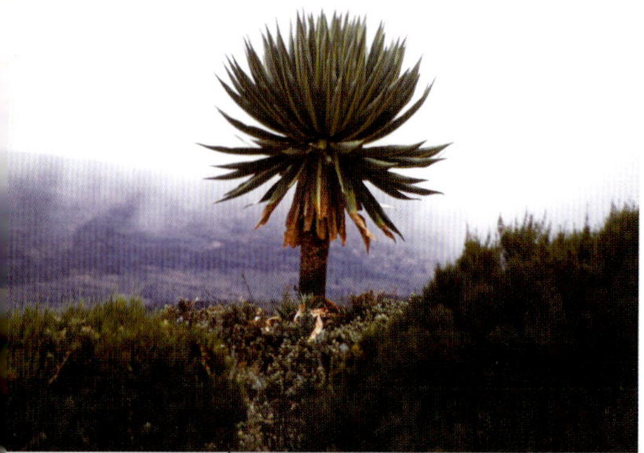

Giant lobelia

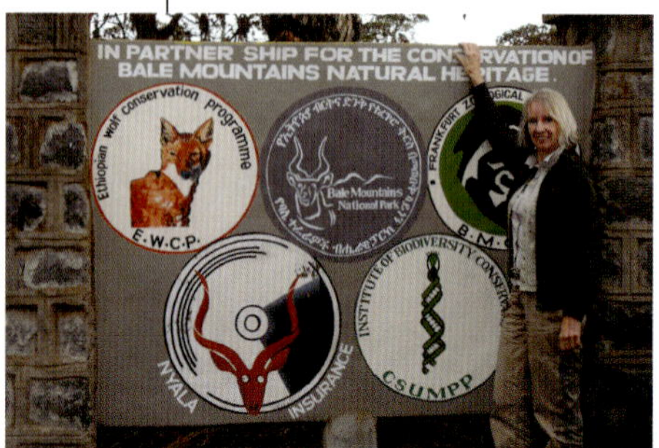

Pollyanna at the entrance to the national park

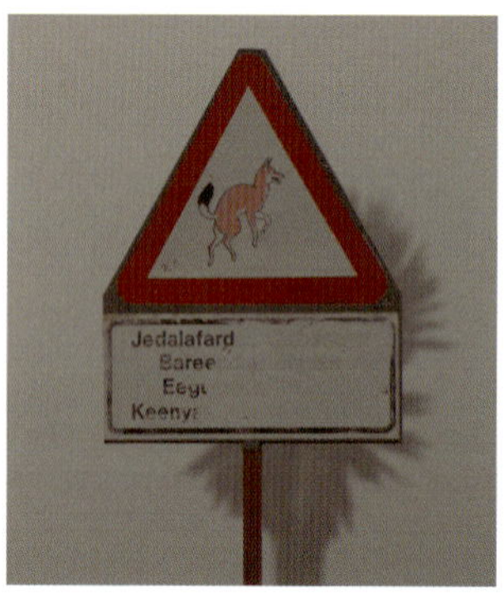

Wolves crossing

In the rainy season shallow depressions on the plateau are filled with water, forming small lakes which provide habitat for water birds, especially ducks migrating from Europe during our northern winter. Before our visit, doubtlessly influenced by images of famine, we had thought of Ethiopia as a hot, dry desert country. But this mountainous region enjoys high levels of annual rainfall and lengthy periods of damp cloudy weather, interspersed with days of sparkling sunshine with brilliant blue skies. We are visiting at the beginning of the wet season when the nights are still star-filled, clear and cold, usually with heavy ground frosts, and temperatures may fall between minus six degrees and minus fifteen degrees centigrade in the main peak area of the park.

This region is home to one of the largest remaining populations of Ethiopian wolves. Only around four hundred and fifty of these elegant, long-legged creatures survive today, living in seven scattered mountain ranges. The populations are generally small and isolated. Around three hundred remain in the Bale Mountains, threatened by loss of their highland habitat, disease and persecution. For thousands of years the wolves lived side by side with the Oromo Highlanders, the native people who graze their livestock on the hillside. In the space of one thousand square kilometres of grassy plateau, heaths, moorlands and steep rocky slopes the paths of humans and wolves seldom crossed. But in the early 1990s, following the overthrow of the government, mounting unrest in the countryside caused many of the highlanders to move into the boundaries of the national park, where they cut down trees and burned and cleared areas of heather and grassland to provide grazing for their cattle. They began to shoot and kill wolves and other wildlife. Claudio Sillero and the Ethiopian Wolf Conservation Programme (EWCP) stepped in to initiate discussions with the village elders and succeeded in putting an end to the shootings by demonstrating that the wolves only preyed on small rodents and did not hunt livestock. They began their educational programme to raise conservation awareness among the people of Ethiopia.

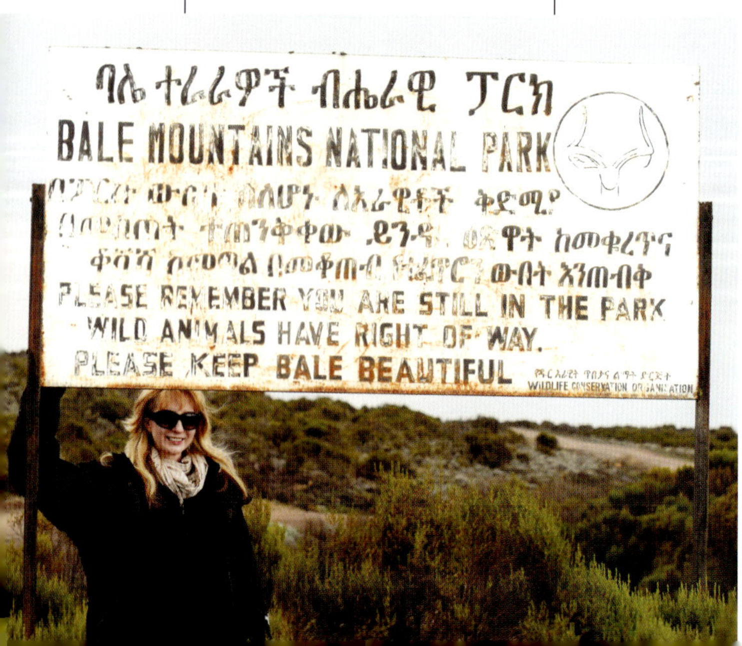

Anna-Louise takes note of the rules

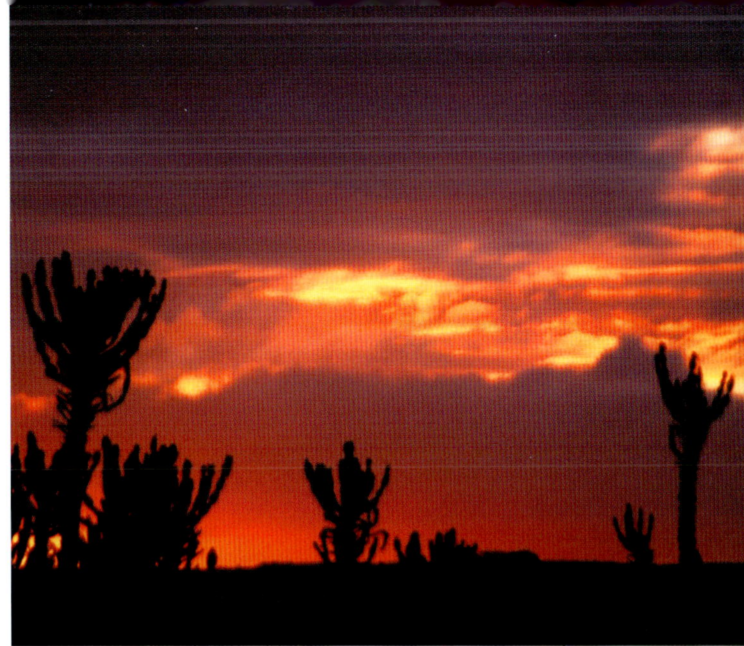

SUNSET ON THE PLATEAU

Day Four

Up at 6.00 am to take advantage of the hot water and brave the bath and wash my hair. However, almost as soon as I step out of the bathroom complete with dripping hair, there is a power cut – so I have to gently drip dry through the morning.

Across to the dining-room for toast and black tea, then we pile into the vehicle with our crew to begin our journey towards the Sanetti plateau to spend our first day looking for wolves. The plan is to spend a week here searching for the wolves, and to allow Pollyanna plenty of time to study and sketch their mountain habitat, before we move to the EWCP headquarters.

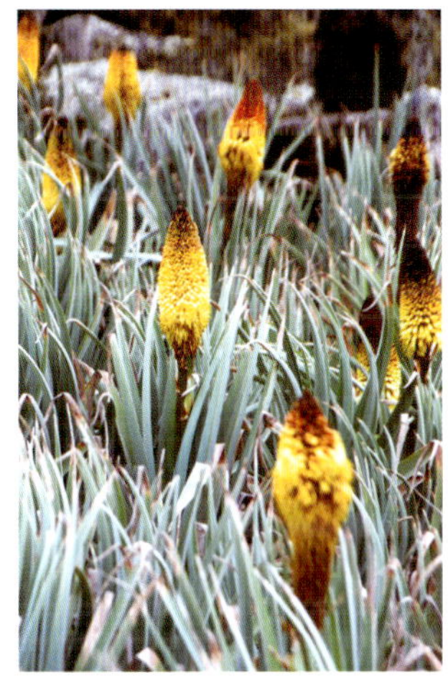

WILD RED HOT POKER

The road from Goba to the Sanetti plateau is the highest all-weather road in Africa and we start to climb rapidly as soon as we leave the outskirts of the town. We drive through plantations of exotic trees, passing small settlements of villagers who leave their huts to watch us as we pass. In just forty-five minutes we reach the edge of the plateau.

Sanetti has an ethereal beauty. Clumps of everlasting flowers cover the ground, their reflective silvery leaves giving a pale softness to the landscape. These plants have dry, papery flowers which allow them to survive the desiccating winds up to 4,300 metres high in the mountains. Other species of Afroalpine plants cope with the extreme temperatures by remaining very small or becoming large. The best example of the latter is the curious looking giant lobelia. We see their stems standing high against the skyline. Young lobelias protect themselves from the extremes in temperature by forming a 'nightbud' – closing their leaves tightly as the temperatures fall. In older plants the old leaves hang down to protect the thick stem which is full of water. Once the plants have flowered they die, and we see many blackened 'skeletons' of dead plants.

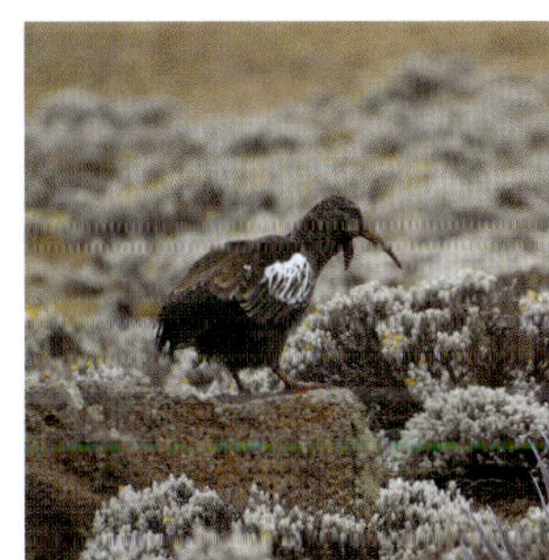

As we drive on to the plateau a road sign looms out of the mist – an Ethiopian wolf enclosed in a red triangle, warning us to look out for the wolves crossing the road. We can only hope!

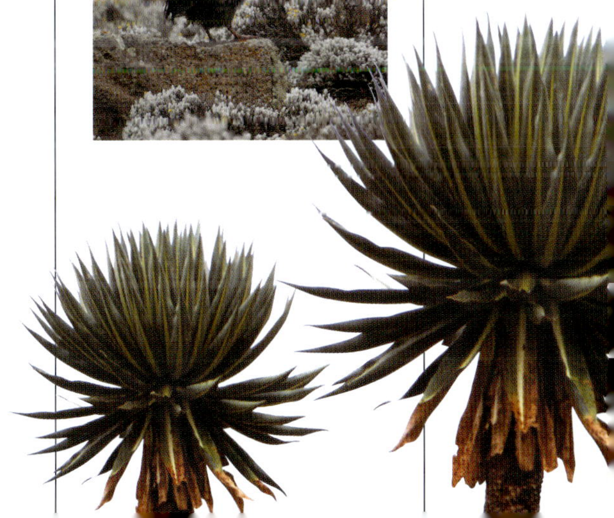

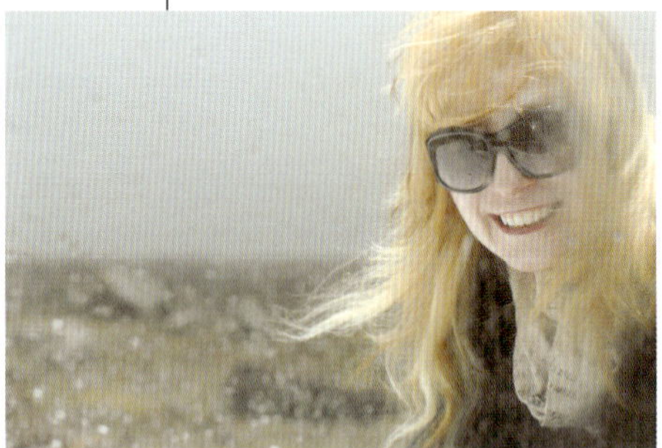

Anna-Louise emerges from the mist

The telecommunications tower is shrouded in fog

We will spend some time hiking in this region, but this morning we drive directly to the highest point, the peak of Tullu Deemtu – 4,377 metres above sea level. Although we knew from our research that the wolves lived on a mountainous plateau, for some reason we hadn't imagined the mountains being quite *this* high. We are only a few hundred metres lower than Everest base camp, and when we get out of the vehicle and start walking we are immediately aware of the effects of the altitude. When walking on a level path, or slightly downhill, we feel fine. As soon as we start walking uphill we are quickly breathless.

We are so high we are in the clouds. Mist is rolling in over the slopes and the peak of the mountain is in dense fog. The peak is home to a telecommunications tower – and we make our way up to it. Tariku informs us that there are magnificent views from here – but we have to take his word for this. However, on the rocks around the tower we see splintered bone fragments – signs of lammergeiers, also called the bearded vulture or bone-breaker. These enormous birds, which have a wingspan of over two metres, drop bones from a great height on to the rocks below to break them open so they can eat the bone marrow within. In this thick fog we would have no chance of spotting a lammergeier far above us, and I hope that a huge chunk of bone will not suddenly drop without warning from overhead.

Not only has the visibility dropped to zero, it is also extremely cold. Tariku tells us that the name Sanetti means 'where the strong winds blow', and I am starting to think that checking this in an English – Amharic dictionary before we left would have been a valuable exercise. In fact it is freezing – I am already starting to wish I had brought more in the way of padded clothing.

We knew that it might prove very difficult to find the wolves. A survey in 2002 suggested the plateau is home to around two hundred and seventy-five individuals, around half the world population, but they range over a large area. In my communications with the researchers at the EWCP they have told us that they can sometimes spend weeks in this area without seeing any wolves at all. However, I hadn't imagined that they might be hard to see because of thick fog – when we walk a few metres away from the vehicle it vanishes

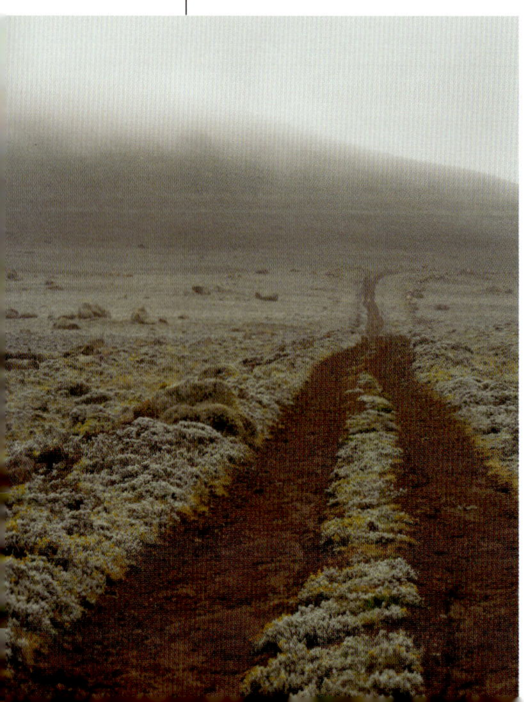

The Sanetti plateau is only slightly lower than Everest base camp

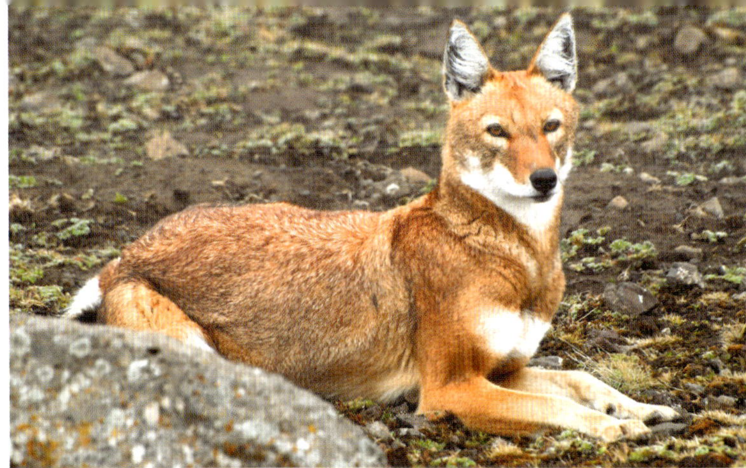

Our first sighting of an Ethiopian wolf

from sight – so the odds of spotting a distant wolf seem very slim. Having spent many months of my life in remote locations not seeing the wildlife we have gone to study, I mentally settle myself in for disappointment, or at best another long wait.

We drive slowly down the track from the peak. We are scanning the misty horizon through our binoculars, avidly watching for any hint of movement of red fur in the distance.

And almost immediately, I see a red shape. Feeling almost foolish I enquire: 'Erm… is that a wolf?' We have only been on the plateau for a couple of hours – and this is one of the rarest animals on the planet. Surely it can't be possible simply to drive up here and find one straight away. And this is not a distant speck – this wolf is just sitting quietly by the side of the road looking at us. Had he been any further away he would have been lost in the mist, but there he is, apparently completely unconcerned as I fumble for my cameras and Pollyanna produces her sketch pad. His rich red fur is muted in the soft grey light, but we have plenty of time to observe his delicate features – the white fur under his chin and on his chest, the fine pointed muzzle, and the huge ears which turn to listen to sounds inaudible to us.

'Feeling almost foolish I enquire: 'Erm… is that a wolf?''

'Ethiopian wolf' has been selected as the official English name for the *Canis simensis* we are currently observing, but we can clearly see why at different times they have been referred to as the red jackal or red fox. Genetic research has shown the animal we are watching to be a wolf and only distantly related to the jackal – but with the colour of the coat, their slender legs and bushy tails, they have the look of overgrown foxes.

When Pollyanna's completed painting of this wolf hangs on a gallery wall, she hopes very much that the observer will be able to feel the same sense of wonder which she experiences when sketching wildlife on location. Although Pollyanna's work is fundamentally traditional, she is not aiming for a hyper-realistic photographic reproduction of an animal,

ETHIOPIAN WOLF

The Ethiopian wolf has a bright chestnut-coloured coat of short, soft fur with dense white underfur. The throat, chest, lower part of the neck, the underside and inside of their legs are white. They have a thick black bushy tail, pointed ears, slender snout and long legs. In many ways they resemble overgrown foxes – but are genetically wolves. An adult weighs 11 to 19 kilos. Ethiopian wolves live in packs of between 3 and 13 adults that share and defend an exclusive territory. Unlike most wolves, which hunt together for large prey, generally they forage and feed alone on small rodents.

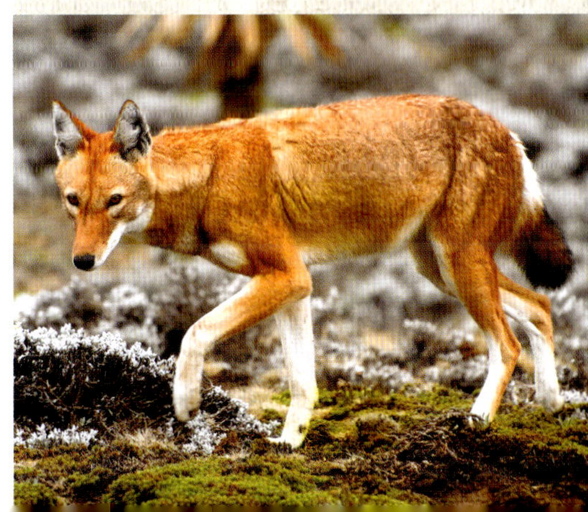

Ethiopian wolves resemble large foxes

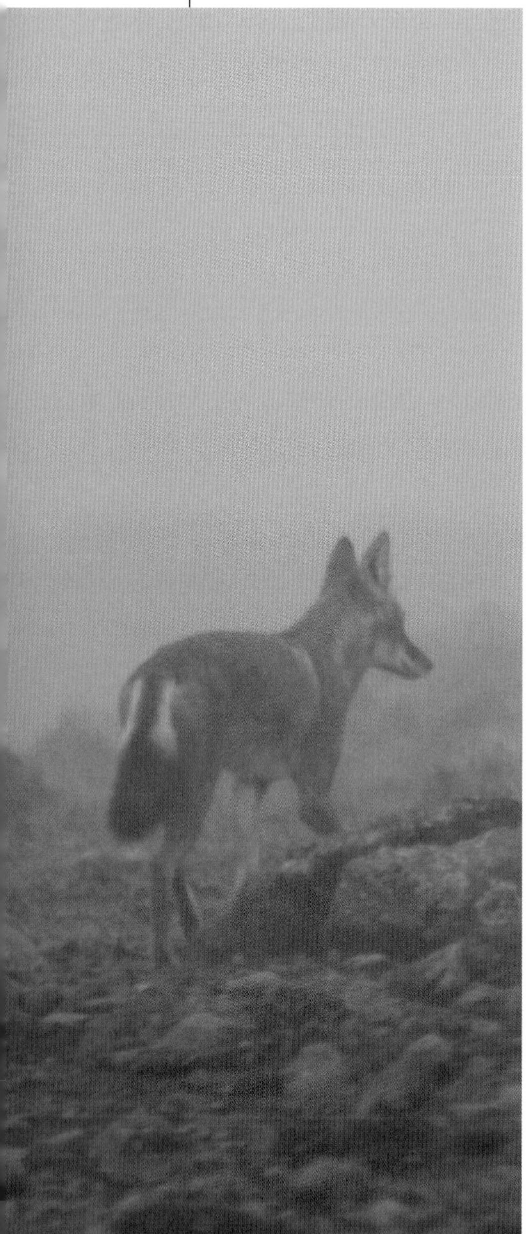

The wolf strolls away into the mist

but in some way to go beyond the purely factual image a camera can produce. Although viewers of her work often comment on the fine detail in her paintings, this is most often concentrated in one focal point. Other areas of her canvases remain loosely worked, almost impressionistic – and this combination of styles creates a richness of emotion and beauty that is somehow never captured through a camera lens.

Eventually, the wolf stands up, stretches, yawns, and strolls languidly away into the mist. Having been silent while watching the wolf we immediately start chatting ten to the dozen about our amazing good fortune – at which moment a second wolf crosses the road in front us. Abbi starts the engine and drives slowly towards a curve in the road – and the wolf emerges from the gloom walking incredibly close to the road and our vehicle. This flies in the face of the experience of pretty much every endangered wildlife watching expedition I have ever been on. The wolves appear to have been misinformed about their role in this joint venture – don't they know that they are supposed to vanish from their regular habitat as soon as we arrive, leaving behind just a few tantalising footprints? Usually we will spend long hours watching and waiting for an absent animal, to be told: 'Oh you should have been here yesterday/last week/last month… There were five of them playing together with their young/carrying out their annual mating display/ tapdancing'. In fact on our first day we will have sightings of no less than fourteen wolves, at varying distances – including one which will again venture very close to our vehicle just before we turn to head back to the hotel for the night.

We also see clearly the reason for the abundance of wolves on the plateau. It is teeming with little lemming-like rodents. They are everywhere – all across the grasslands we can see them popping nervously (unsurprisingly) out of holes, to dash across the ground before diving back underground. Mostly we are spotting the Blick's grass mouse along with a few black-clawed brush-furred rats. I attempt to photograph them, but they are very difficult to get close to – diving for cover at the slightest noise or movement. The word 'abundant' does not even begin to describe this rodent population. We learn that there are staggeringly large numbers here in these Afroalpine meadows. In some areas the combined density is estimated at three thousand kilos in a square kilometre. To put this into perspective, the weight of the rodents in the meadows of the Sanetti plateau is comparable to that of all the large ungulates – including the wildebeest and zebra – of a typical national park in East Africa!

We also spot the largest of the rodents, the giant molerat. Like the wolves they are endemic to these mountains, so are found nowhere else in the world. These are the favourite prey of the wolves as they can weigh up to six hundred grams, thus making a slightly more substantial meal than the other rats and mice we have seen scurrying across the plains. The molerat has to come above ground to feed on the vegetation, so during the evolution process its eyes have migrated to the top of its head. This allows it to scan for predators while exposing as little of its body as possible.

The availability of these prey animals means that, although the wolves live in close-knit territorial packs, they usually forage and feed alone on small prey, in contrast to most wolf and wild dog species which hunt in packs for larger animals. Rather than using their speed and strength to bring down large prey they use patient stealth. The fact that the rodents are active in the day also means the wolves are diurnal – unlike the European wolves in Transylvania for example which have become increasingly nocturnal. This does demonstrate the remarkable adaptability of the wolf species. Wolves can be found from the freezing Arctic tundra to the deserts of Mongolia and the United Arab Emirates and are able to adapt their behaviour to these incredibly diverse environments.

Eventually, slightly dazed by our successful wildlife sightings so soon after our arrival, we wend our way back down the mountain to Goba and stop for a welcome cup of coffee in a café on the main road. We perch at a small table on rickety white plastic chairs as the coffee is prepared and presented to us. From our European perspective this is actually a bit more like a coffee hovel than a coffee shop – there is an uneven dirt floor, the walls are rough concrete with patches of peeling paint, and the window a grid of thick

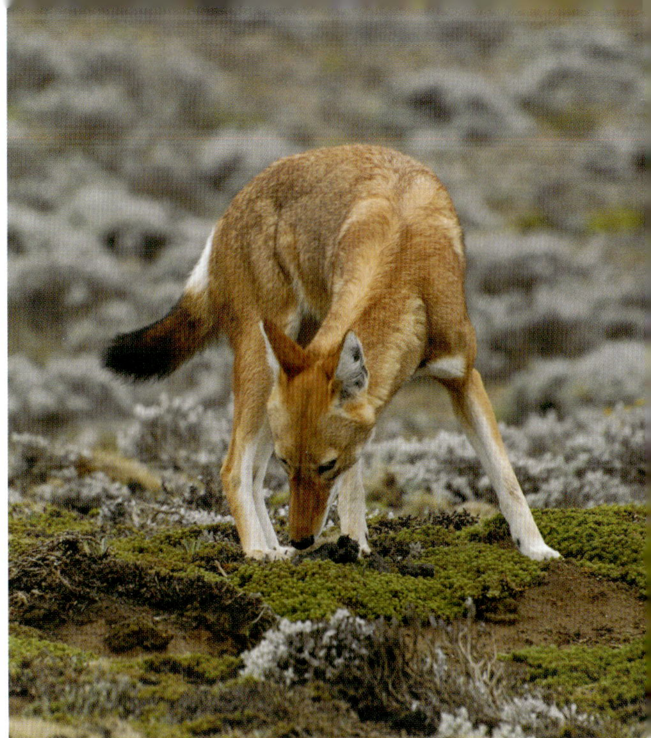

Wolves feed on abundant rodent populations on the plateau

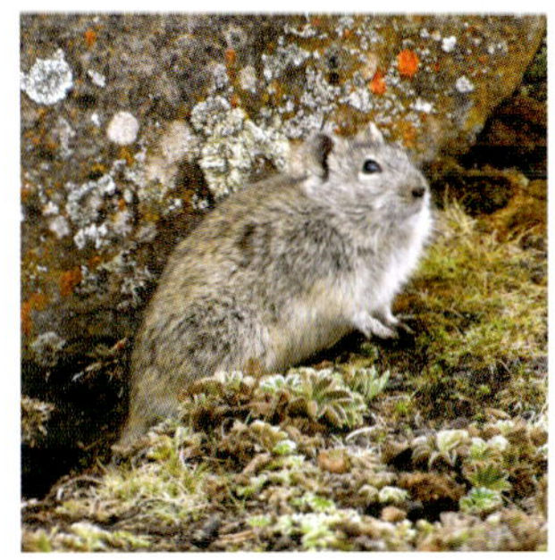

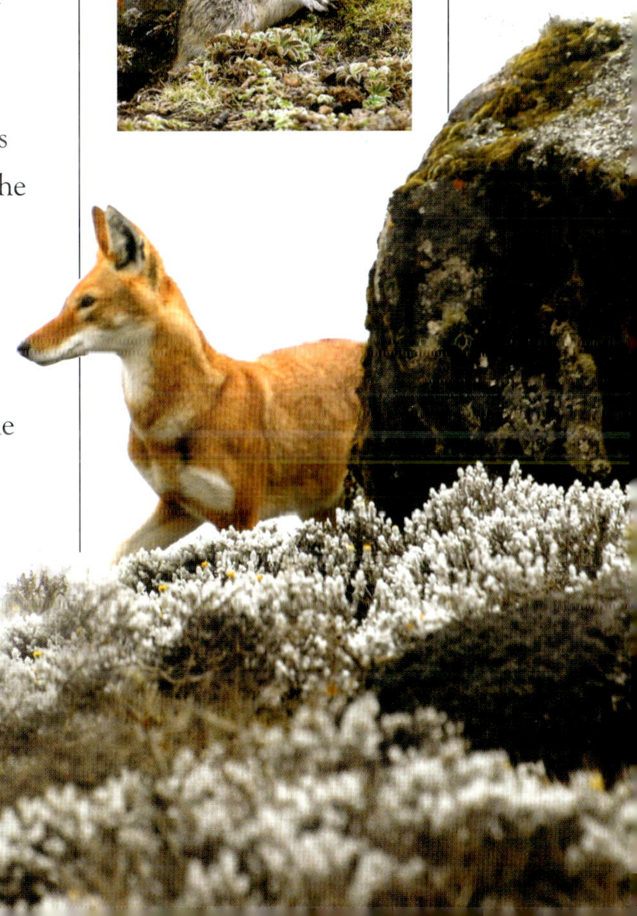

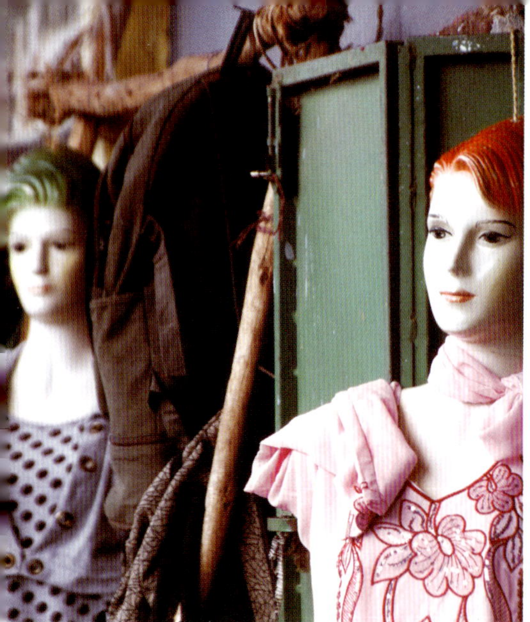

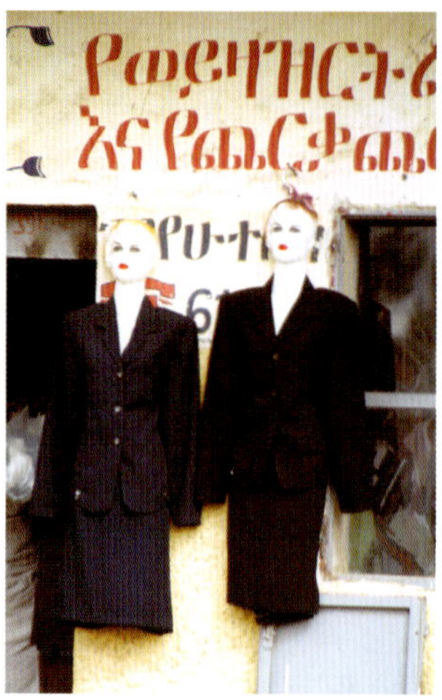

Outfits are displayed on unearthly mannequins

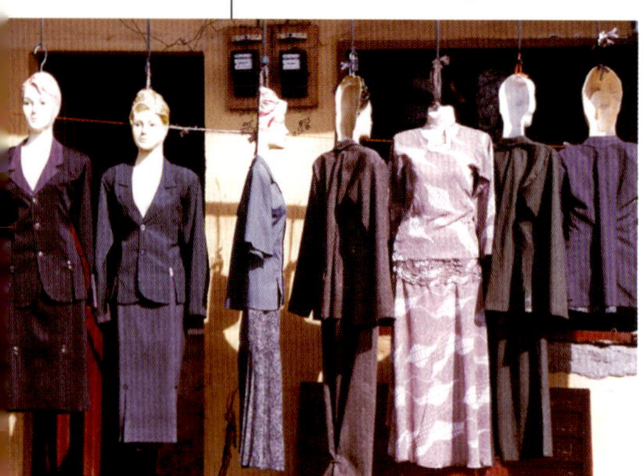

metal girders. Nonetheless we thoroughly enjoy our very welcome hot drink and, duly refreshed, decide to stroll back through the streets to our hotel, taking in the local atmosphere. As Tariku helpfully points out, we are the only white people in the entire town, so we are not surprised to attract a great deal of attention, and before we have gone too far a little retinue of children start to follow. Here the chant of 'YouYouYouYouYou' is interspersed with fits of giggling, and there is much excitement as we give them pens and pencils. However, very few tourists reach this part of Ethiopia, so we are able to pass relatively unmolested. In the short time we have been in the country we have been left with the impression that everyone in Ethiopia is constantly walking somewhere, so in that respect at least we fit in. Like the rest of the pedestrians we have to hop into the edge of the road to dodge the occasional donkey cart or even more occasional bus or truck which rattles down the dirt of the main street. The shops on either side of the road are fascinating to us. We are particularly taken by the dress shops, which display their outfits on unearthly mannequins. Moulded in white plastic they are as pale as I am; the shops hang them above the storefront, so they blow in the breeze looking for all the world like fashionable ghosts.

There is still a power cut back at the hotel, so I write my journal by candlelight, then walk across to the very dark dining-room for oat soup (like a very thin porridge or gruel) followed by the same meal as last night, with the addition of cooked spinach. We then head back to our sleeping saloon and go straight to bed.

Days Five – Ten

We fall into a routine of driving towards the plateau and spending our days partly hiking and partly driving while looking out for the wolves and other wildlife. When the clouds lift we are able to appreciate the spectacular views in this region. The flattened mountain peaks and volcanic outcrops rise from the mists of the Harenna forest into the clear blue skies, empty save for a lone buzzard wheeling overhead. The alpine lakes and deep, rocky gorges create a landscape resembling the *The Lost World* as depicted by Sir Arthur Conan Doyle in his novel describing an expedition to a plateau where prehistoric animals still survive. It has been suggested that Unesco designate The Bale Mountains National Park as a world heritage site of Outstanding Universal Value, and we are beginning to fully realize why it deserves this protection.

The area is universally valuable for a number of reasons, not least its spectacular scenery which creates a diversity of vistas unique to the Ethiopian highlands. The Bale Mountains ecosystem is an internationally recognized centre for biodiversity and no fewer than a quarter of mammals and six per cent of birds found here are endemic to Ethiopia. The area

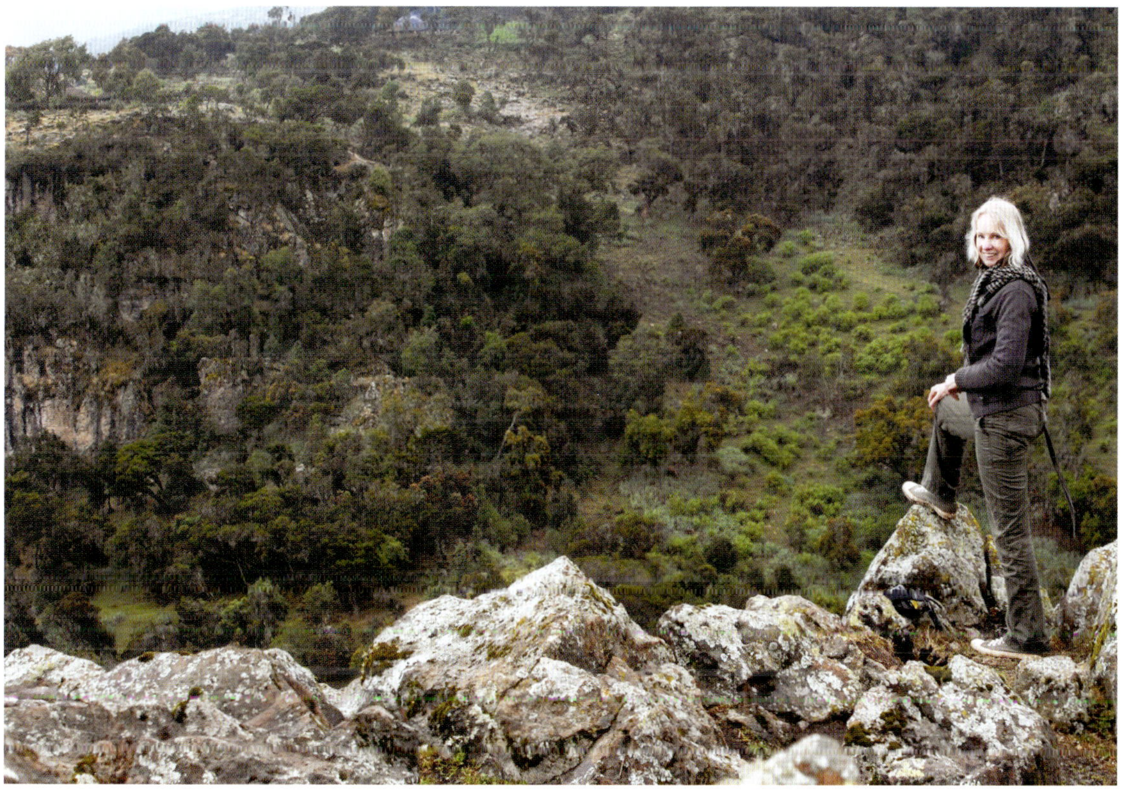

The spectacular deep rocky gorges of the Bale National Park

'Over the plains of Ethiopia the sun rose as I had not seen it in seven years. A big, cool, empty sky flushed a little above a rim of dark mountains. The landscape 20,000 feet below gathered itself from the dark and showed a pale gleam of grass, a sheen of water... I knew how far below in the swelling heat the birds were an orchestra in the trees about the villages of mud huts; how the long grass was straightening while dangling locks of dewdrops dwindled and dried; how the people were moving out into the fields about the business of herding and hoeing.'

Doris Lessing

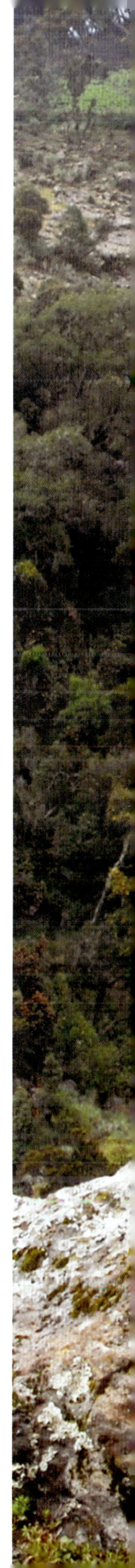

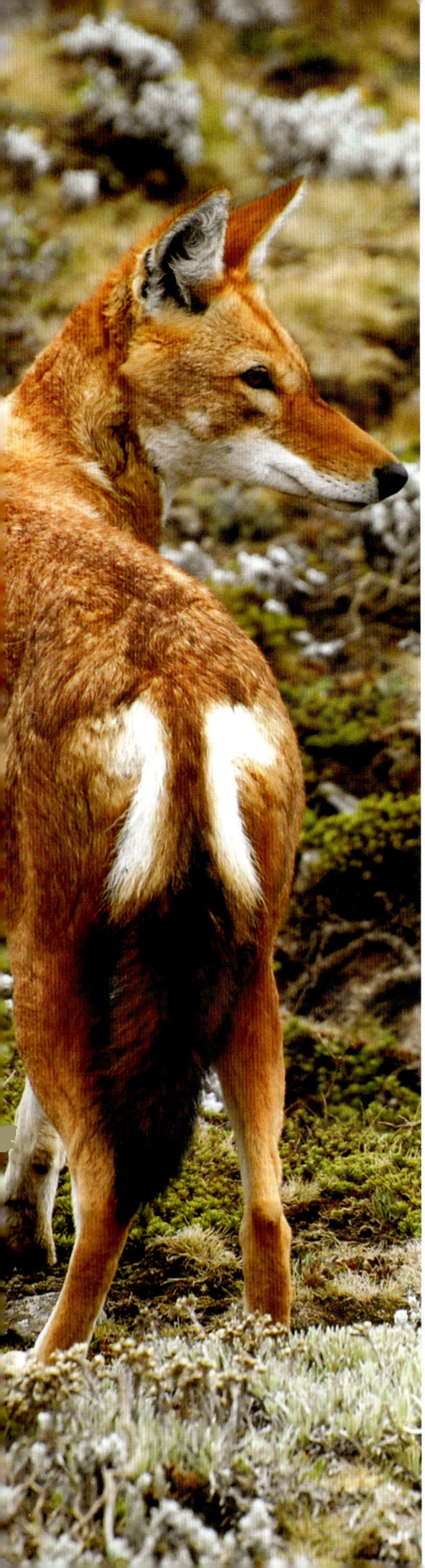

contains the entire global population of the giant molerat and the largest populations of mountain nyala and the highly endangered Ethiopian wolves. It has been estimated that more mammal species would become extinct if the habitats of the Bale Mountains were to vanish than if any other area of equivalent size on the globe disappeared. The mountains harbour forty per cent of the one thousand known species of medicinal plants in Ethiopia, and the largest remaining natural plantations of wild coffee.

Amazingly, while in this area we see at least one wolf every day – although we never quite equal the record of fourteen sightings on our very first day. However, Pollyanna's sketchbook is filling with studies which will help her represent the true essence of these beautiful wolves. Maybe it is just because we know how incredibly endangered they are, but they seem fragile and vulnerable and we are beginning to feel very protective towards these unusual red animals. She is also outlining the landscapes of the plateau – the ever-changing shadows and colours, swirling clouds and fleeting mists, which suddenly separate to reveal red-tinged rocks and silvery grasses. Pollyanna finds that the colour, lighting and atmosphere are all altered by the humidity and she wants to capture the plateau in its many different moods. Each part of the world has its own soul and spirit – these fragile open plains are in complete contrast to the dense forests of Transylvania. But both habitats are vulnerable and valuable in their own way and Pollyanna will recreate the strength and fragility of the contrasting environments in her finished canvases.

The abundance of mice, rats and other rodents on the plateau also attracts large numbers of raptors, and we enjoy some excellent sightings, including golden eagles, Verreaux's eagles and lanner falcons. Our wildlife expert and interpreter Tariku has, of course, been accompanying us up into the mountains every day, with Abbi driving the vehicle. At this point Sami, who will be cooking for us over the coming week, would have been free to stay in Goba. However, on the first day he decided to come with us as he wanted to go to the peak of Tullu Deemtu, the highest point in Ethiopia – the first time he had been given this opportunity. Sami is a city boy through and through – he grew up in an orphanage in the sprawling incoherent metropolis of Addis Ababa, and still lives in the capital. As such he has never had a great interest in the wildlife of his country – but on the first day he is so thrilled to see the wolves, he chooses to join us almost every day on our trips to the plateau and gets just as excited as us with every new sighting. One afternoon we have a lovely view of an eagle on a dramatic rocky outcrop. Just as I am lining up a moody photograph of said eagle it raises its tail and ejects a perfectly horizontal stream of white faeces before taking off. All our crew collapse into giggles – especially Sami who is practically squeaking with hysteria in the back. Throughout the whole day, just when we think he has composed himself, he will once again collapse into laughter at the memory, taking the rest of us with him.

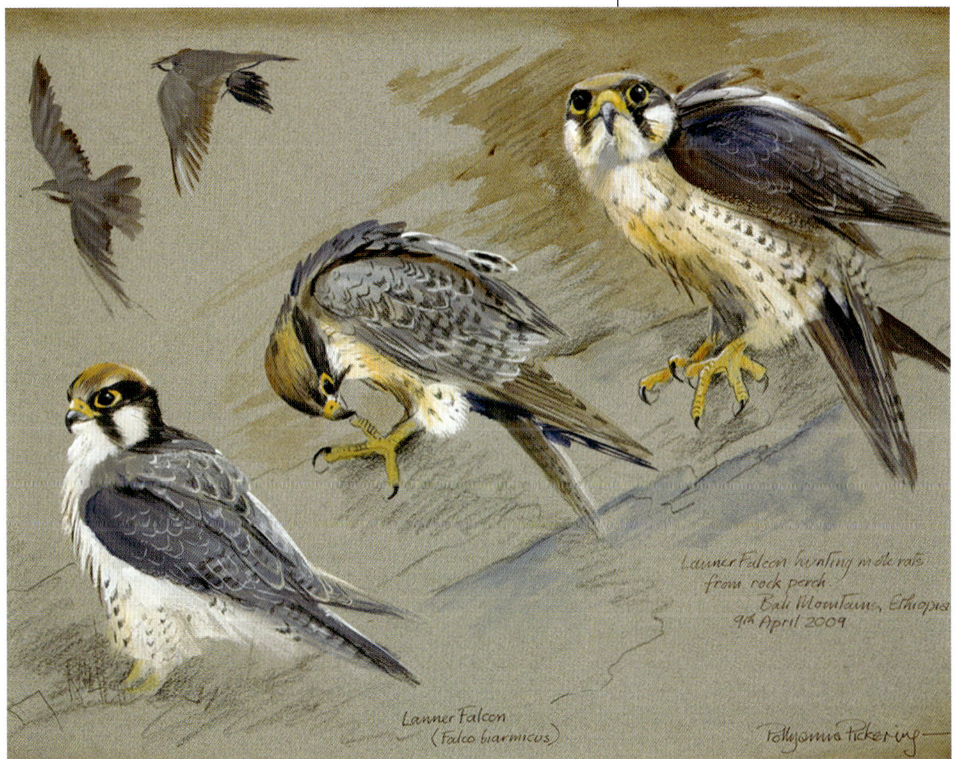

Sketches of the lanner falcon

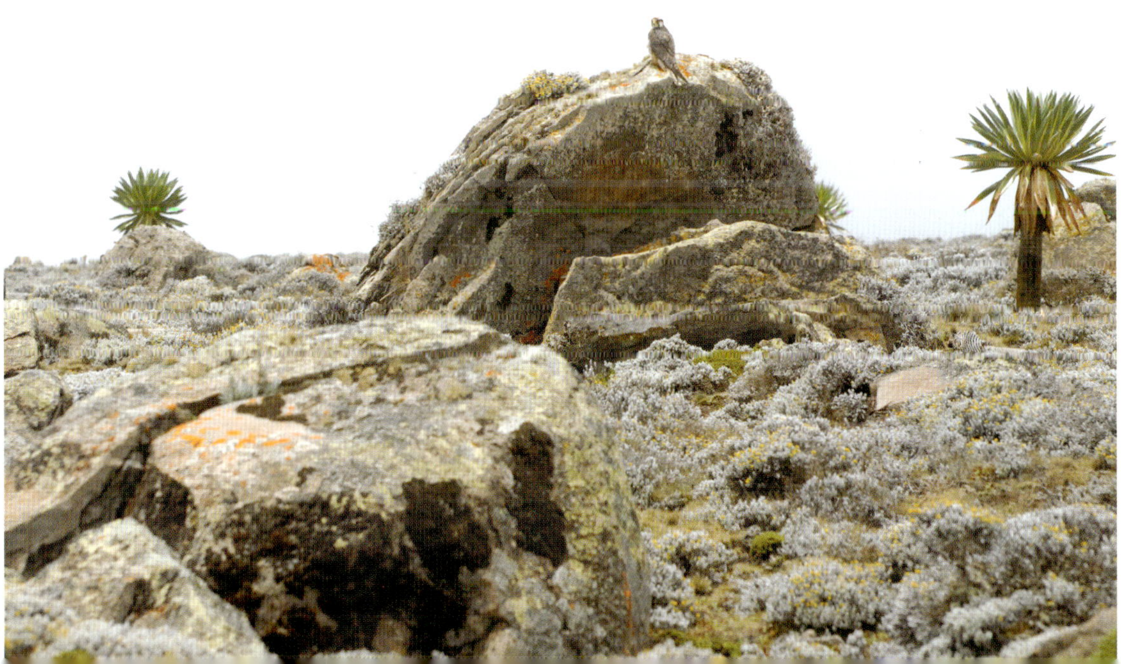

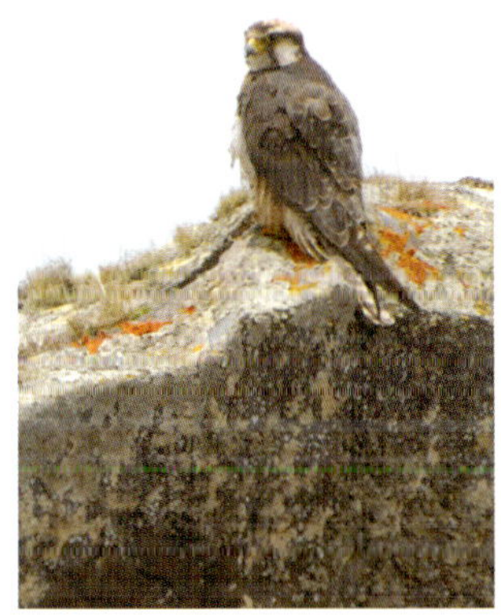

Lanner falcon

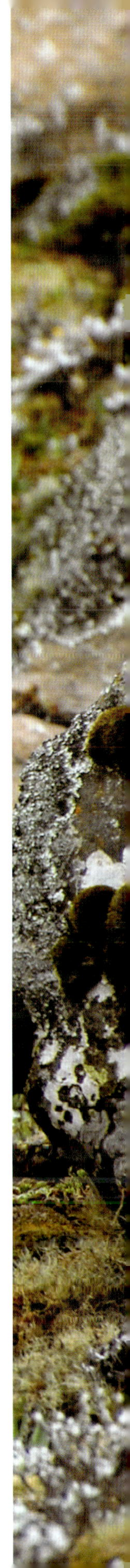

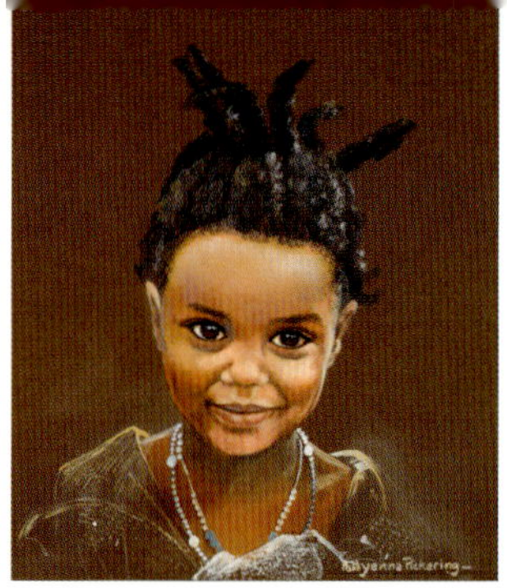

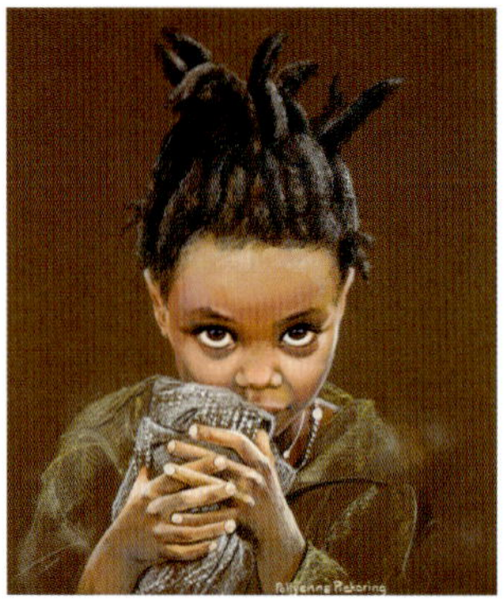

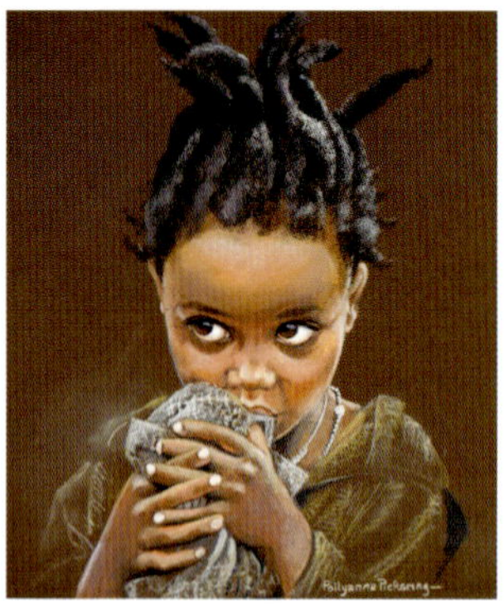

Pollyanna's studies of a local child

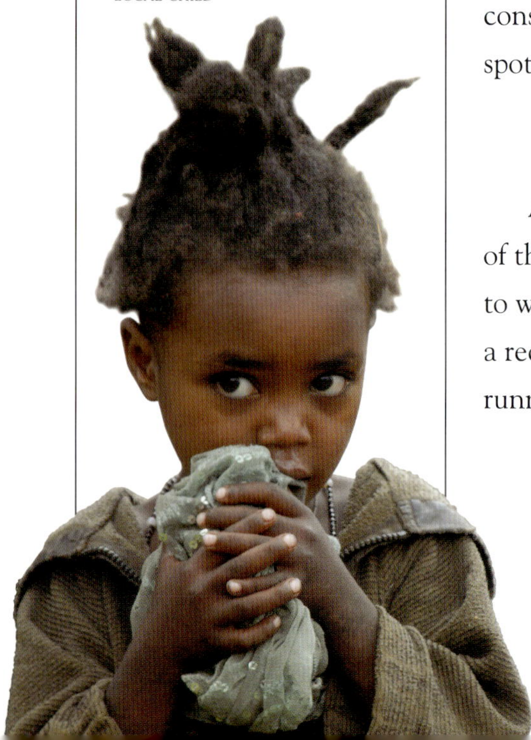

Pollyanna is sketching the birds as eagerly as she has been recording the wolves. It may seem a lot more time consuming to sit and sketch, rather than just capturing a quick snapshot with a camera, but one advantage of this is that we have a greater opportunity to observe interesting behaviour – birds interacting with each other, preening, displaying, hunting and gathering nesting materials.

We hardly see another vehicle on this high remote road – a bus passes through occasionally, but little else. On our third day, however, we have stopped to watch a buzzard when a four wheel drive draws up alongside us and stops. The driver is Claudio Sillero the founder of the EWCP. With him are Graham Hemson, who has been the field coordinator since 2006, and the two newest members of the team, Anne-Marie Stewart and Chris Gordon. It feels incredibly bizarre to meet someone I recognize unexpectedly on the top of a mountain in Ethiopia! And of all the people you could choose to meet here, these would be top of the list. Claudio is undoubtedly one of the leading canid experts in the world. Founder and director of the Ethiopian Wolf Conservation Programme and Head of Conservation at the Born Free Foundation he holds the post of Bill Travers Fellow for Wildlife Conservation at Oxford University, and is the chair of the International Union for the Conservation of Nature (IUCN) Canid Specialist Group. Claudio supervises several conservation projects worldwide and in 1998 he won the prestigious Whitley Award for conservation for his work with the EWCP. We chat for a while before we head off to try to spot more wolves, and they head off to try to save them.

Day Eleven

A slow drive up towards the plateau today – we can see that it is very foggy on the peaks of the mountains up ahead so, as we have on many of the days, we take the opportunity to walk part of the way, through a small village of mud huts. These are constructed in a rectangular style, which is unique to this region. On our first walks the children were running to hide inside the doorways when we passed – scared by the paleness of our skin –

but now they will hang around outside to watch as we pass. We have been invited to look inside one of the huts – we step inside to see the fire burning, the room full of juniper wood smoke which is oozing slowly out through the thatched roof. Tariku translates for us, and we learn that the family we are visiting farm their own fields and have cattle and goats. There is a large store of barley in the hut – at least eight or nine bulging sacks.

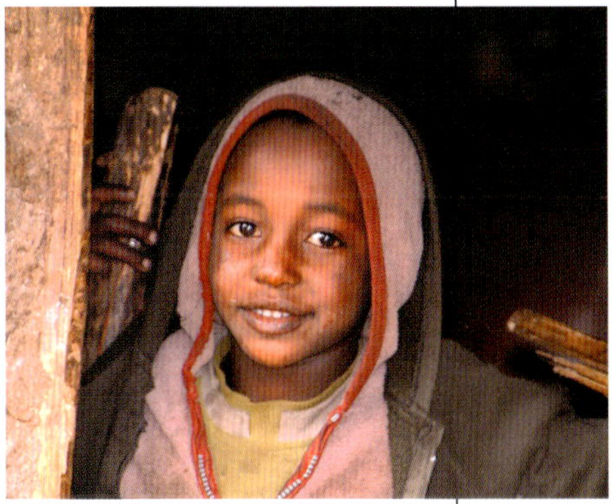

In Europe our strongest association with Ethiopia is probably the haunting images of the 1984-5 famine which affected the inhabitants of four provinces which received record low rainfalls in those years. Close to eight million people became famine victims during the drought and it is believed around one million died. A BBC news crew was the first to document the disaster, with Michael Buerk describing '...a biblical famine in the twentieth century' as '...the closest thing to hell on Earth'. The report shocked Britain, motivating its citizens to bring world attention to the crisis in Ethiopia. We have struggled to reconcile the remembered film of pitiful starving children with the lush fertile land we have been travelling through. Tariku tells us that not only was this area of the country unaffected by the drought, with plentiful food throughout this period, but the lack of communication infrastructure within the country combined with the low literacy levels in these remote villages means that most of the people here still remain completely unaware of the famine which ravaged another area of their country.

We walk further up the mountainside towards a hut set back from the roadside. We have waved to the lady who is standing outside her home most days as we passed either on foot or in our vehicle. Today we introduce ourselves. Through Tariku we learn she is called Ayesha and that she considers herself blessed to live in this beautiful place which provides all she needs. Looking at the beauty stretching out below her garden, we cannot disagree.

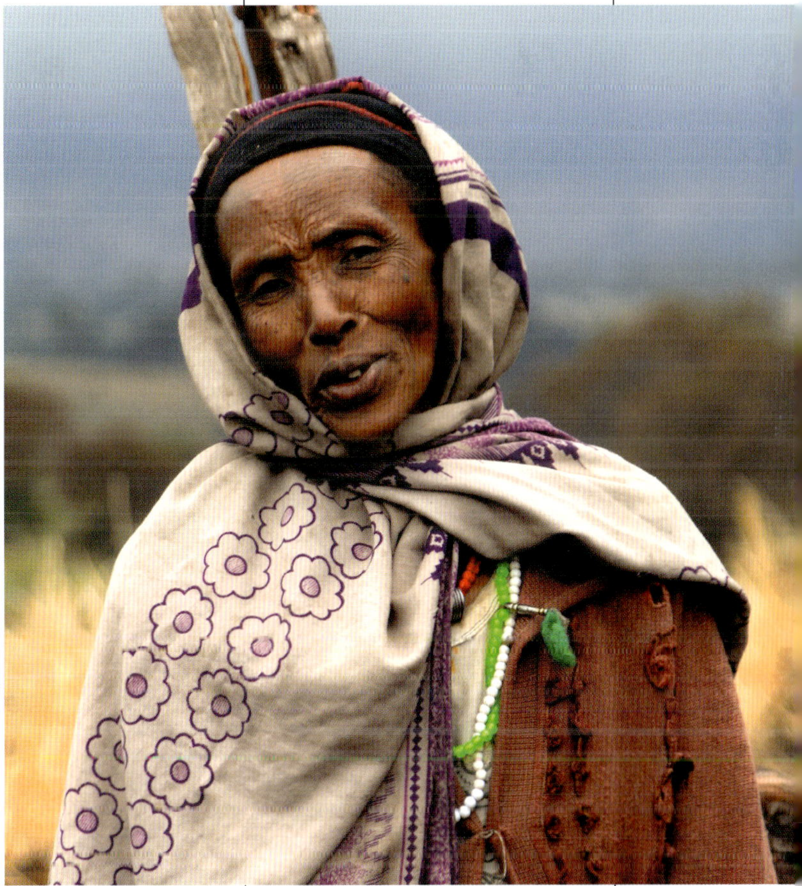

Ayesha

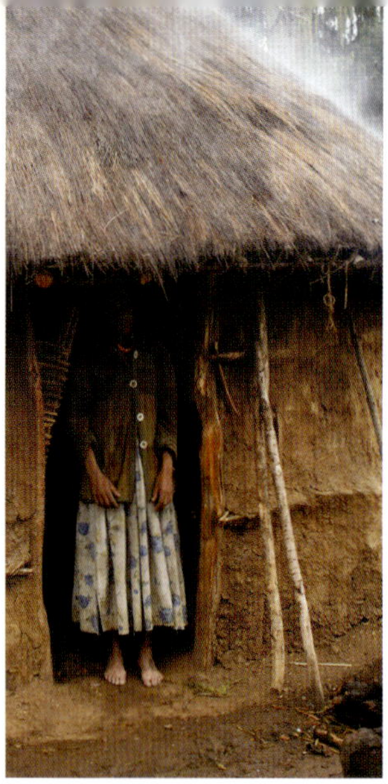

Ayesha welcomes us to her home

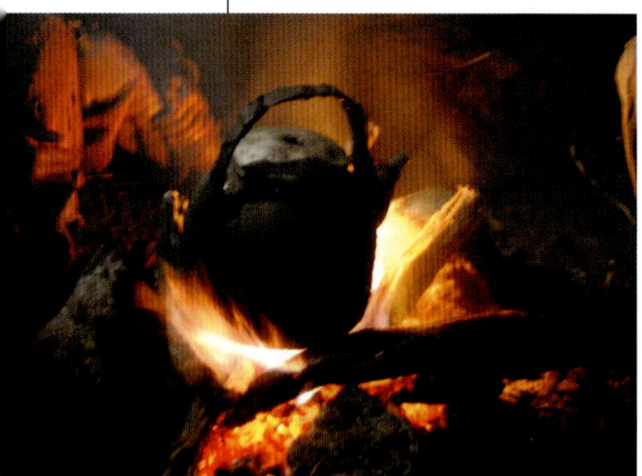

Water is boiled over an open fire

Ayesha's son Hassan

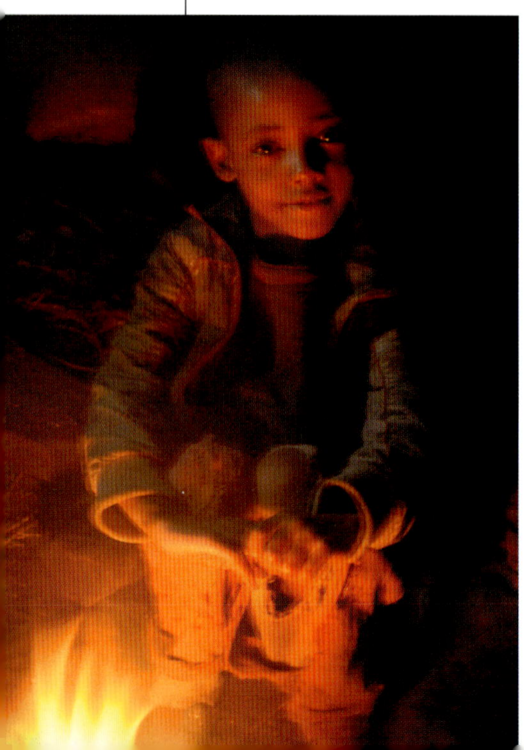

She invites us into her home, although she is concerned it may not be tidy – which shows that whether you live in a stone farmhouse in Derbyshire or a thatched hut in Ethiopia the immediate reaction to unexpected visitors is exactly the same. We squeeze through the tiny door and then through into the second room where she has her fire, and a small bed/seat. In a partitioned corner there is a small white goat bleating away. Ayesha invites us to join her for coffee – we are delighted to accept this generous hospitality. She boils water in a blackened kettle over the open flames, while grinding coffee beans with a mortar and pestle, which she then drops into the kettle. She reaches into a bag by the fire and breaks off a lump of a substance we can't initially identify, which she also adds to the pot. Tariku explains this is salt – a common addition to coffee in this area. (Interestingly this isn't a practise unique to the mountains of Ethiopia. There is a tradition for adding a pinch of salt to coffee in northern Scandinavia, Turkey and Hungary. Pollyanna's father, a tea and coffee dealer, would often add a small sprinkle of salt to his coffee, saying that it took away the bitterness.) While the coffee is brewing Ayesha turns her attention to preparing the cups. We watch her nervously.

Obviously as sensitive stomached Europeans we are completely unable to drink the tap water in Ethiopia. Our much consulted *Lonely Planet* guide also advises against drinking from streams, rivers and lakes, and even wells or pumps as the water source can also be contaminated by animal waste or human sewage. Boiling water for one full minute (or for at least three minutes if you are more than one mile above sea level) kills most types of disease-causing organisms and we have been using boiled water for drinking and cleaning our teeth with no problems. (If you are reading this before travelling – please make sure it's a full, rolling boil!) Therefore we should be able to enjoy the coffee without worrying about any adverse consequences. However, Ayesha is rinsing the cups in cold water poured out of what appears to be an old container for engine oil. I nervously enquire where her drinking water comes from and she explains that it is from the river close by. This is excellent news. This means the stream is fed by spring water from the mountain tops and as she is in

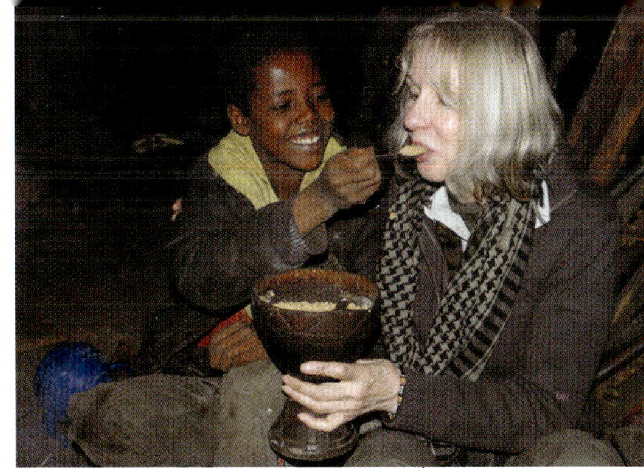
Hassan feeds barley to Pollyanna

practically the highest dwelling in the village there is every chance it should be relatively uncontaminated at that point – in fact about as pure as water ever gets in Ethiopia. This doesn't stop me trying to discreetly wipe the rim when she passes my coffee.

My coffee is served in a battered enamelled mug. Ayesha sends a child scurrying into the back room to bring out a yellowing plastic bag, inside which is carefully wrapped a single bone china cup patterned with roses with matching saucer – obviously only for use on special occasions, it has been brought out for her foreign guests, and she pours Pollyanna's coffee into this precious cup. I have never seen Pollyanna treat anything with more care – she is absolutely terrified of accidentally breaking it. At one point when Pollyanna moves across to the log opposite her seat to take a photograph of Hassan, one of Ayesha's sons, she places the cup on the ground for a moment and I am scared that someone might accidentally kick it over in the darkness of the interior. Fortunately, we finish our coffee without incident and a very relieved Pollyanna passes the precious cup back to its owner.

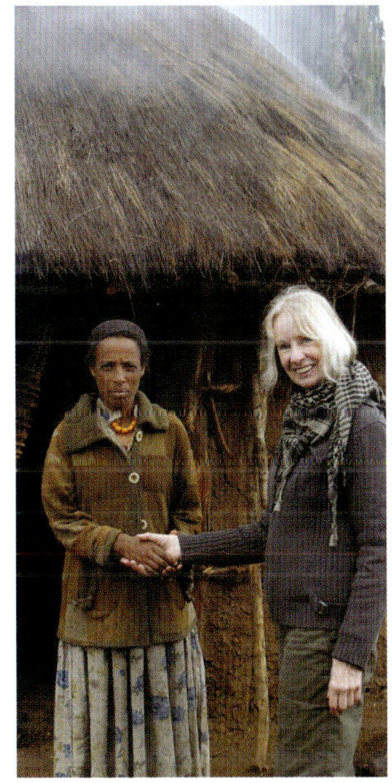
Pollyanna bids farewell to Ayesha

Ayesha then prepares a bowl of ground barley for us to try. Although I have never tried eating dried stuffing mix, I would imagine that it is the closest substance in taste and texture we might have as a basis for comparison. The barley is deliciously flavoured with herbs and spices, but is very dry – it sucks all the moisture from my mouth immediately. I pass the bowl and communal spoon to Pollyanna and she gingerly tries a small mouthful – her mouth visibly puckers, but she chews and swallows bravely, smiling and nodding, then passes the bowl to Ayesha's son. Hassan takes the spoon and begins to feed the barley to Pollyanna. Tariku explains that this practice of hand feeding a guest is called *Gursha,* and is an expression of friendship and acceptance. We are very touched – and very moved by the whole experience of being made so welcome by Ayesha. Although this area remains untouched by famine, by our standards the people here have almost nothing. Yet Ayesha invited two strangers into her home and made us feel like welcome guests.

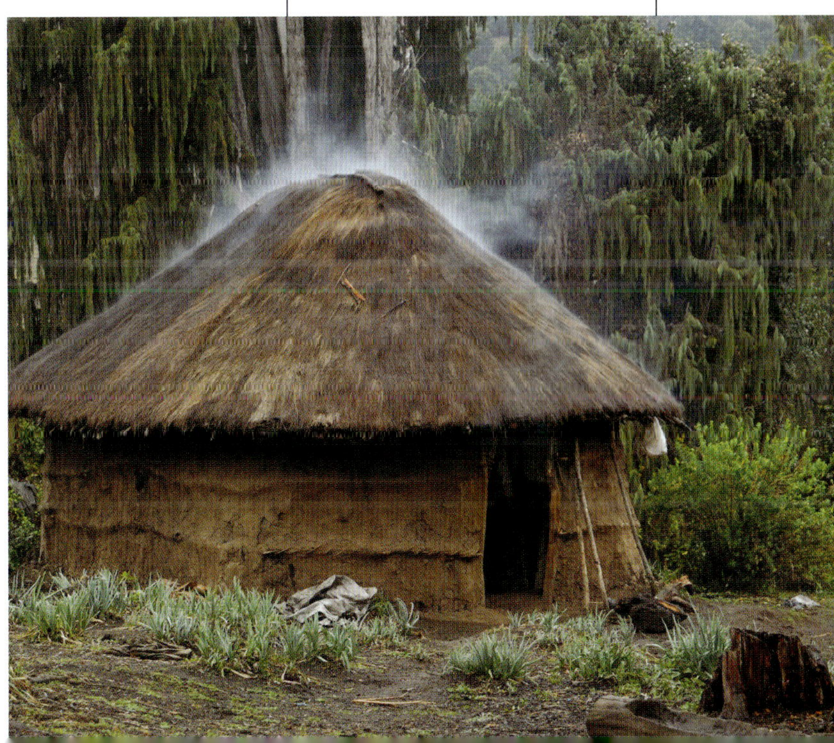
Ayesha's hut – smoke rises from the open fire

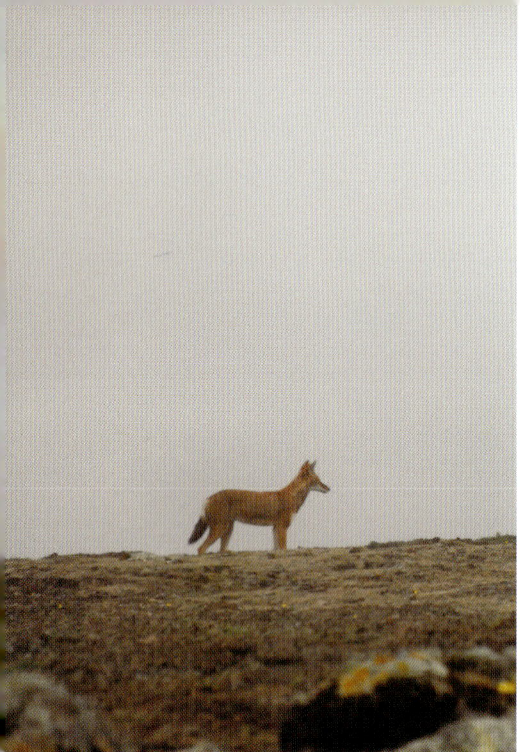

A LONE WOLF APPEARS OUT OF THE MIST

BELOW: A YOUNG WOLF HUNTS AMID THE PLATEAU'S ROCKY OUTCROPS

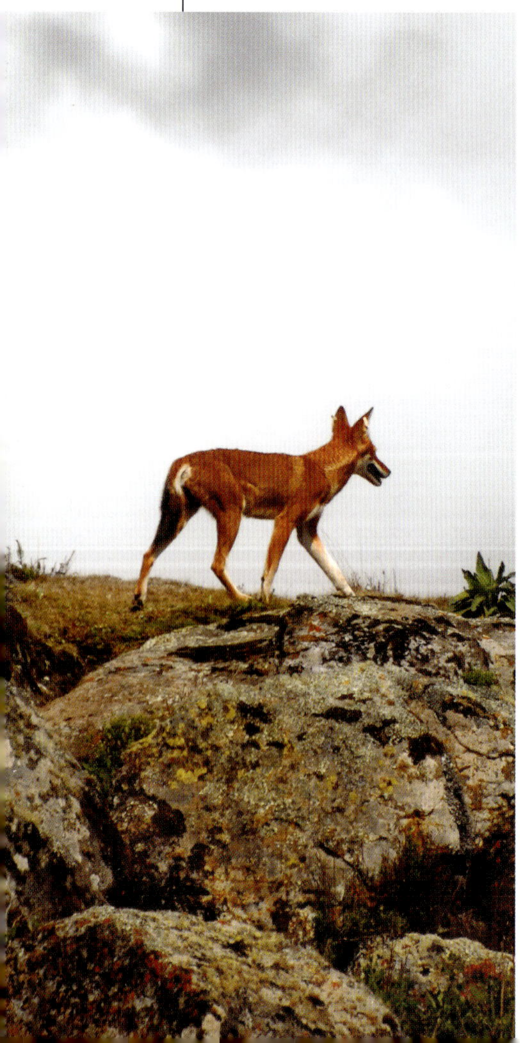

We emerge from the smoky interior to find that a thick soup of a fog has descended. We say our farewells to Ayesha and drive slowly up the road – where we again meet Claudio and Graham making their way slowly down! We stop to make arrangements for our arrival at the EWCP headquarters the following day. They have been walking the plateau all morning trying to carry out a wolf count, but have not seen any in the fog. We head up on to the plateau and, as we can barely see a foot in front of our faces, stop immediately for our picnic lunch of cold omelette sandwich with a slab of cheese.

I then walk into the wilderness for a 'bush break' as they have been euphemistically christened. I realise that today the fog is so thick I barely need to find a rock to hide modestly behind – just a few feet away from the road I am completely invisible. I crouch delicately behind a small plant – at which moment the thought occurs to me that on occasion this cloud cover can lift incredibly rapidly and at any moment I could be left entirely exposed in bright sunshine.

In fact just moments after I have returned to the vehicle the fog does begin to clear and before we have driven more than a quarter of a mile we have spotted two wolves. Our wolf-spotting skills are actually improving. You would imagine that their bright red coats would stand out clearly in the silvery green of the plateau; but the area is also full of large rocks which are covered with a rusty coloured lichen, which mirrors the colour of the wolves' coats perfectly. On many occasions one of us has brought the vehicle juddering to a halt because we have spotted a rock loping across the plains.

We drive on into the bright sunshine – for once we can see the peak standing clear of any clouds – towards the 'edge' of the mountain, the Harenna escarpment where the Sanetti plateau drops away. Below the escarpment is four thousand kilometres of moist tropical forest – the second largest area of forest in Ethiopia. As we travel away from the plateau, the main habitat of the wolves, I spot a young wolf drinking from a large pool of water close to the road – he is quite some distance from the wolves' usual habitat. Clearly a youngster, he seems to be quite curious about these visitors to his home and he edges closer to the

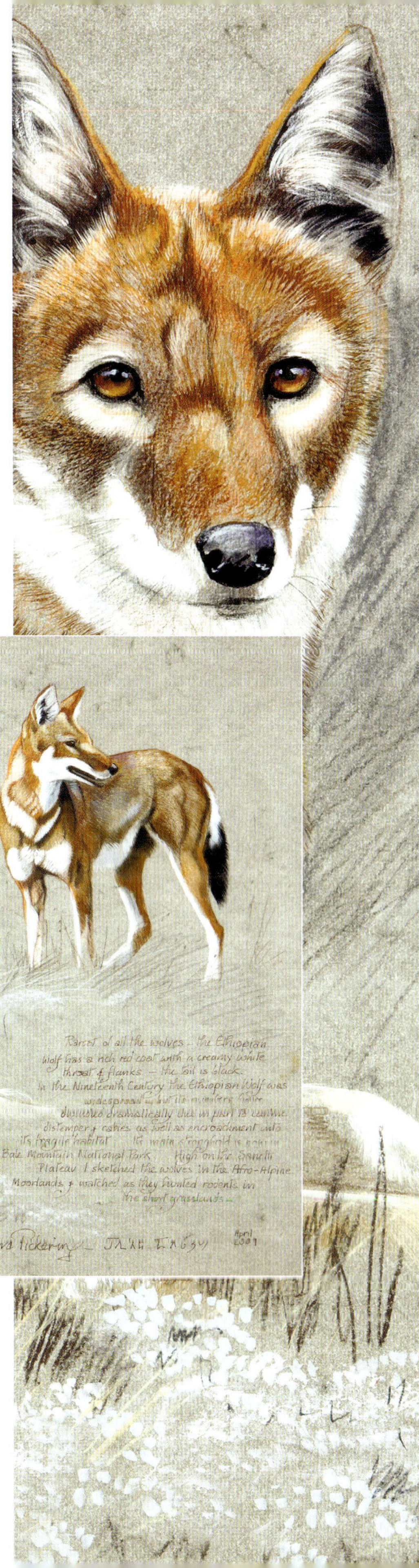

vehicle for a better look at us, head cocked on one side. Unfortunately two big lorries trundle past and spook him – he runs off and is soon out of sight.

In the rocky outcrops of the escarpment we spot klipspringers – small antelopes whose name translates from the Afrikaans as 'rock jumper'. They are remarkably agile among the steep rocks. The klipspringer stands on the very tips of its almost circular hooves. Strong back legs can project the klipspringer up a smooth wall, and they can land and stand on the tiniest of outcrops.

As we head back the fog has lifted completely and for the first time we can see the promised spectacular views from the Roof of Africa. In many of Pollyanna's paintings the subject is actually the light – the way the sun catches the fur of the wolf, highlights one rock and shades another. She is fascinated by the way the sunlight defines the form of her subjects; the richness of colour created when the light passes through a leaf. She cannot wait to capture the natural beauty around her and bring it to life on canvas. Her goal is always to try to recreate the energy and mood of one transient moment in the ever changing natural environment.

We drive slowly back to the hotel, spotting one more wolf en route. To our joy the power is on and there is hot water to wash the powerful smell of wood smoke out of my hair.

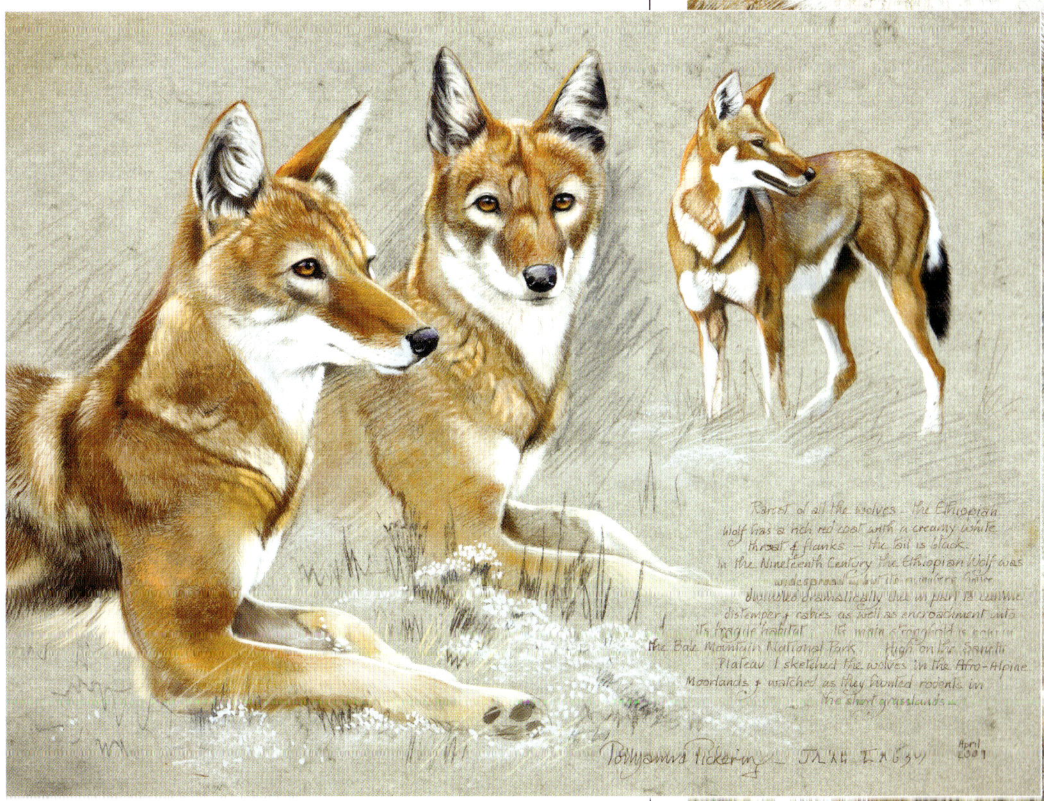

A PAGE FROM POLLYANNA'S SKETCHPAD

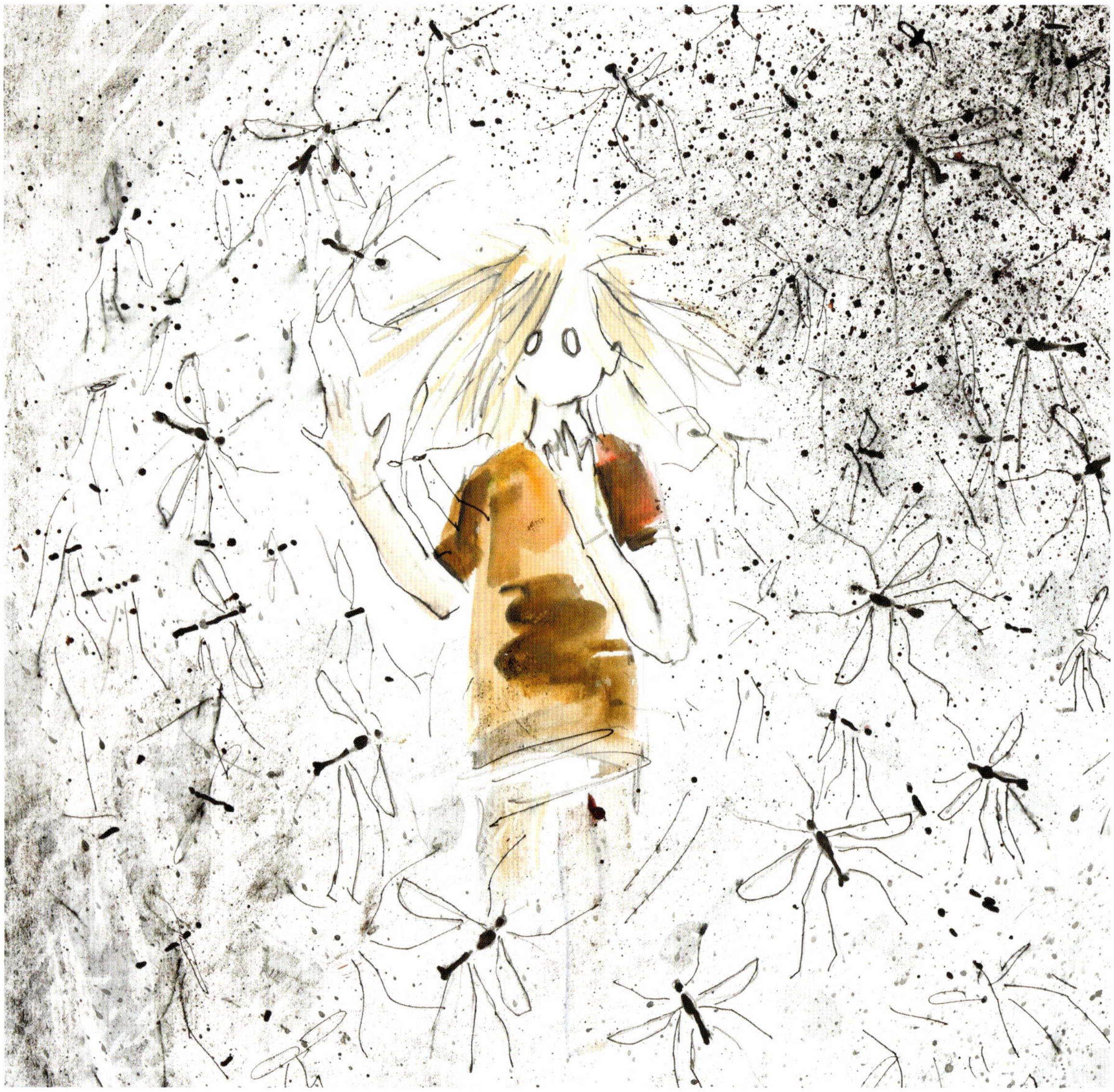

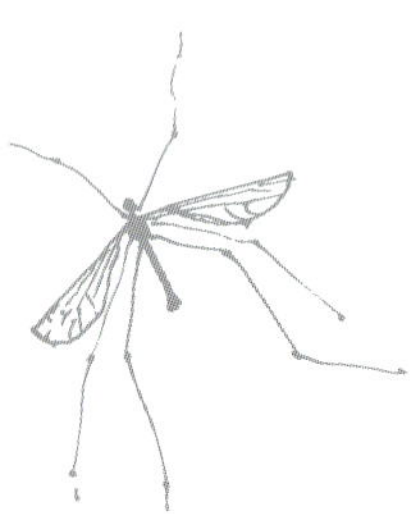

Chapter Four

MY OWN PERSONAL ROOM 101

'You asked me once, what was in Room 101.
I told you that you knew the answer already. Everyone knows it.
The thing that is in Room 101 is the worst thing in the world.'

GEORGE ORWELL, *1984*

I have a confession to make. It is something I am somewhat ashamed of, something I don't talk about in public. I have a phobia. Quite an extreme phobia, a shaking, screaming, unreasonably overreacting phobia. And the worst thing is that I am afraid of a living thing. My entire life has revolved around the natural world; I have always lived with animals, worked with animals. I like to think I love and respect every living creature... but there is just one that fills me with terror. When people hear about the journeys I make with Pollyanna to study wildlife, they often comment on how brave we are, especially when we get close to wild animals – we have been charged by tigers, chased by a king cobra, I have found scorpions in my camp bed and tarantulas in my bath towel. But the thing is I am not especially brave, I am just not afraid of these things. Of course my heart was pounding when a tiger exploded out of the undergrowth a few feet away, and I have nothing but the utmost respect for the elephant which without warning turned to charge my jeep in Namibia – but ultimately the momentary terror was overcome by the wonder of seeing these beautiful powerful animals defending their wild kingdoms.

Naturally I am sensibly wary and respectful of venomous snakes, but they are not my phobia. I think snakes are exquisitely beautiful with their smooth cool bodies and unreadable eyes. I love the frogs and toads that many people think are slimy and revolting – and have picked up countless toads as part of a road patrol close to my home to prevent them from being squashed during their annual migration. I have lain on my stomach in the middle of the Namib desert to photograph a scorpion at close

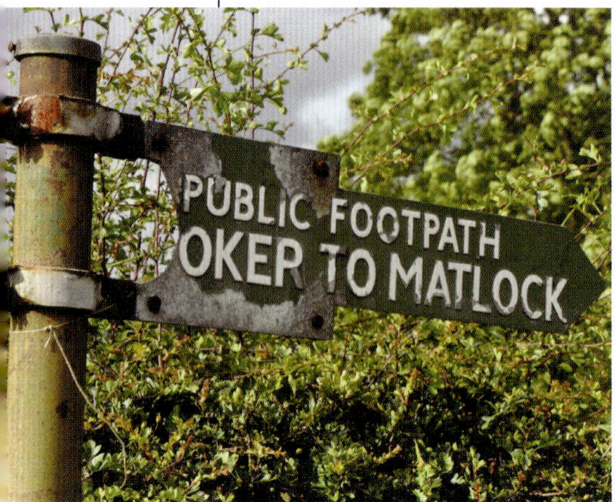

Walking in the Derbyshire hills

quarters, and while I might not pick spiders up with my bare hands I have no problem removing them with an upturned glass and piece of paper (or in the case of some of the larger rainforest species, an upturned waste bin and a sketchpad...). Crocodiles are dangerous, but incredible to observe from a safe distance. Mice are cute, rats are intelligent. Piranhas are exquisitely beautiful with their glittering sparkling scales, and sharks have an awe-inspiring streamlined efficiency. A healthy instinct for self preservation might mean that I am not going to rush up and cuddle many of these creatures, but I am not irrationally terrified of them.

No, the one creature which strikes fear and terror into the very heart of my being is... The Daddy Long Legs. More accurately called crane flies, they are poor fliers with a tendency to 'wobble' in unpredictable patterns during flight. They are also completely harmless. They don't sting, they don't bite. I know this. They do, however, completely FREAK ME OUT. If one comes near me in an enclosed space I have to leave until some wonderful volunteer has removed it for me. I have been unable to enter the stock room at our gallery because a daddy long legs sitting on the door has meant I can't reach out for the handle. It's so bad I even find it quite hard to talk about them. Yes I know, pathetic, but that's phobias for you. Irrational, but still real. (I once met someone with a genuine terror of mushrooms, so it could be worse.)

Over the years I have managed to control the phobia to the point that I can mostly cope in open spaces, even if they fly near me – I may freeze for a moment, but as long as there is space for us both to move away, I can usually prevent a full-scale panic attack. There was an exception to this – once when walking in the Derbyshire hills I crossed a field which for some reason was absolutely crawling with daddy long legs. It is common to walk through a few in the grass, but secure in my thick hiking boots I can pass through them as long as I don't think about it too much. But this one field was swarming. A few yards in I realised every blade of grass was alive with them. I have no idea why this one particular area had become such a haven – I hadn't seen any unusual numbers up until this point. Their larvae, called leatherjackets, consume the roots of grasses and other vegetation before entering a pupal stage and then hatching out. I can only imagine an enormous simultaneous hatching had just occurred. I was so far into the field that it seemed more sensible to head for the stile on the far side and then continue the planned route of my walk – and so I tried to calm myself and continue steadily and rationally to the other side of the field.

Obviously I failed miserably. After taking three or four calm, purposeful steps I completely lost it and eventually reached my destination via a series of leaping panic stricken pirouettes, springing on tiptoe while uttering a series of high pitched shrieks and wails. I eventually collapsed in the daddy long legs free sanctuary of the far field gasping for breath, heart hammering wildly. This wouldn't have been quite so bad if my uncontrolled panic at something so entirely benign had gone unobserved – and thankfully there were no other walkers in sight. However, as I crossed the field a hot air balloon drifted serenely overhead, riding the gentle summer breeze over the tranquil landscape. The passengers in the basket must have been treated to the sight of a fully grown woman suddenly leaping frenziedly through an innocuous field as though being attacked by an invisible deadly foe. In the still air my screams would have reached them easily, but from their elevation above the fields they would have been unable to see the tiny insects which were causing my terror.

So why, in a book which is primarily about wolves, am I suddenly talking about my deepest, darkest and most embarrassing fear? These little introductory passages which lead into each chapter are intended to provide helpful background information to the extract of journal which follows. Usually this is about the place we are visiting, or the animals we are searching for. But without this insight into the darkest of my fears the following diaries wouldn't make much sense.

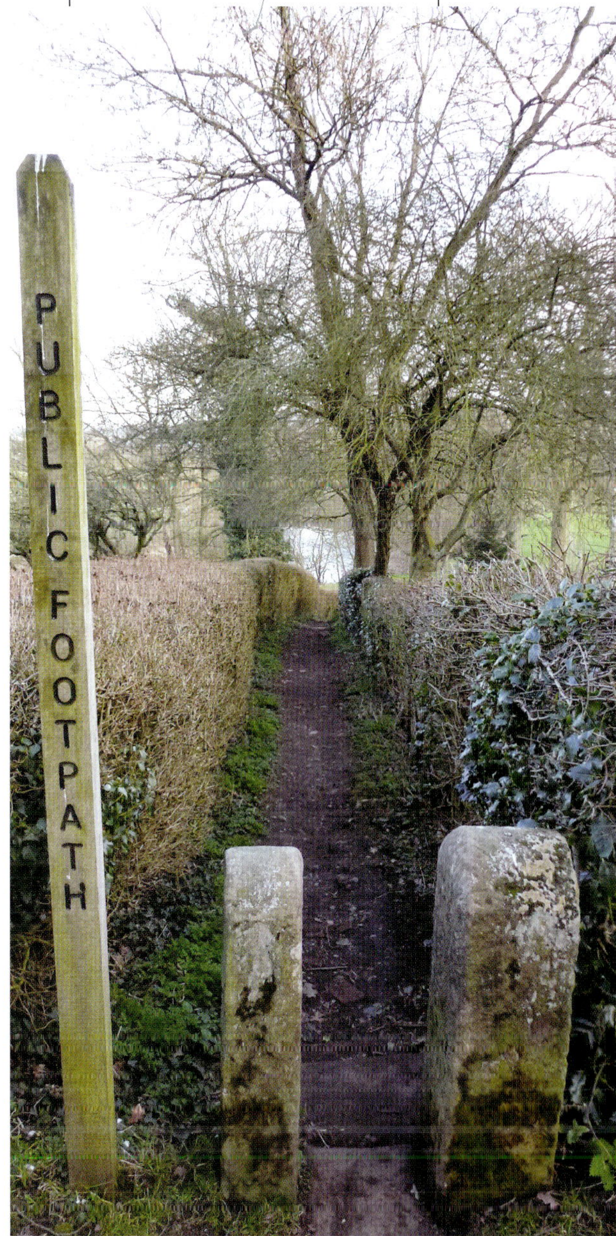

Glorious Derbyshire countryside

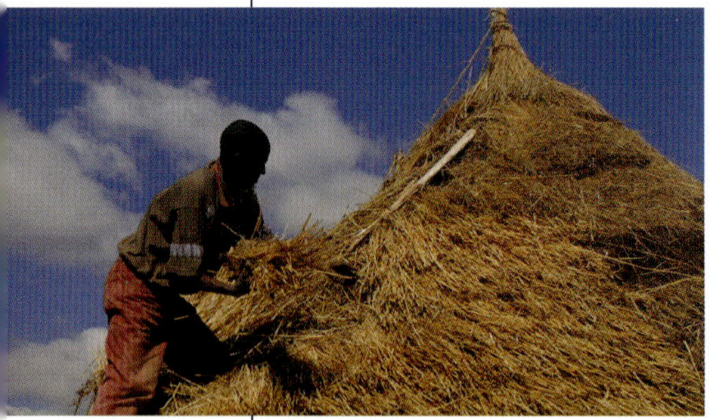

Thatching the roof of a traditional *tukul* hut

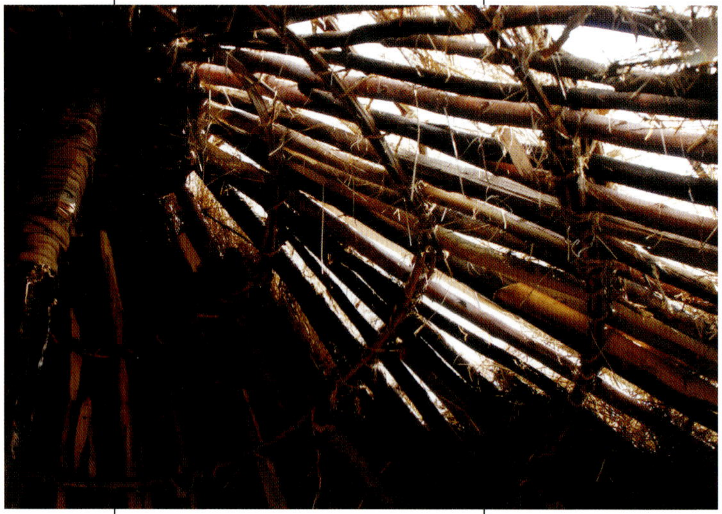

The view from inside

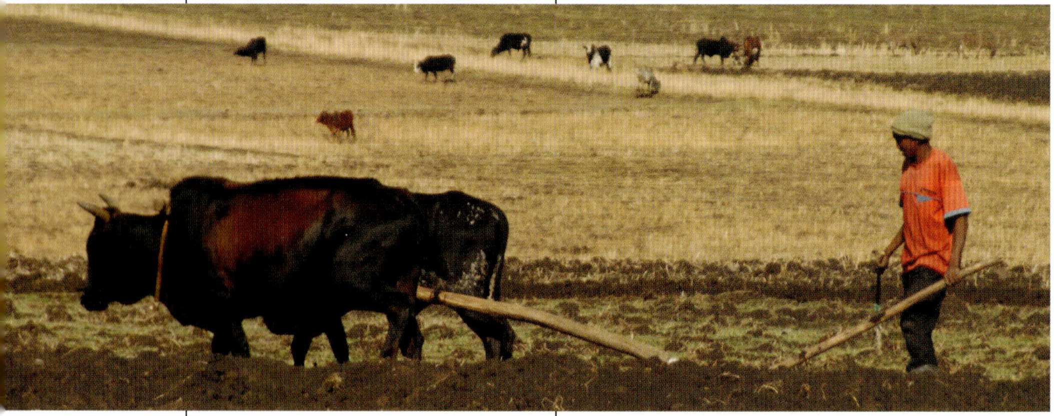

Ploughing the fields using oxen

Day Twelve

Toast and coffee for breakfast, then we drive back down towards Dinsho. The research buildings based here function as the office for the EWCP and various other conservation projects.

As usual the drive is fascinating, and we make lots of stops en route. We watch a man thatching the roof of a round *tukul* – a hut built in the traditional style, deftly weaving the new material into the wooden construction. Tariku tells us that an anthropologist who visited this area announced that the shape of these huts represents a woman's breast. (In my experience anthropologists say this about almost anything circular they are confronted with. I am unconvinced, given that a circular structure with a conical roof logically provides the best protection against the elements – if the roof was flat the rain would pour through. I suspect it tells us more about what is going on in the heads of the anthropologists than the more practically minded indigenous people they are studying.)

We also pause to watch a man threshing grain in the traditional way, tossing it high in the air on rough wooden paddles to separate the chaff. In fact we have little choice but to stop, as he is standing in the middle of the road to carry out this work. He indicates that Pollyanna should have a go – which she does, much to the amusement of the assembled villagers. They all seem more than happy to let us interrupt their hard work – probably on the grounds that our efforts are the funniest thing they have seen in weeks. One of the ladies indicates she should have a go using another method – tossing the grain on a flat basket. This seems to go quite well, so if the painting should fail, Pollyanna can probably find employment here. The same lady also invites us

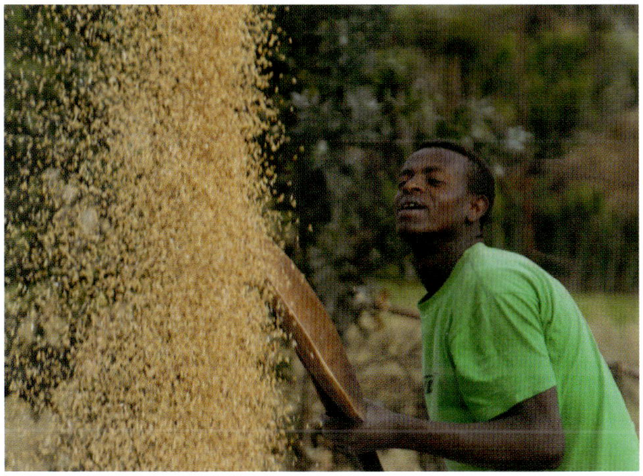

Tossing grain to separate the chaff

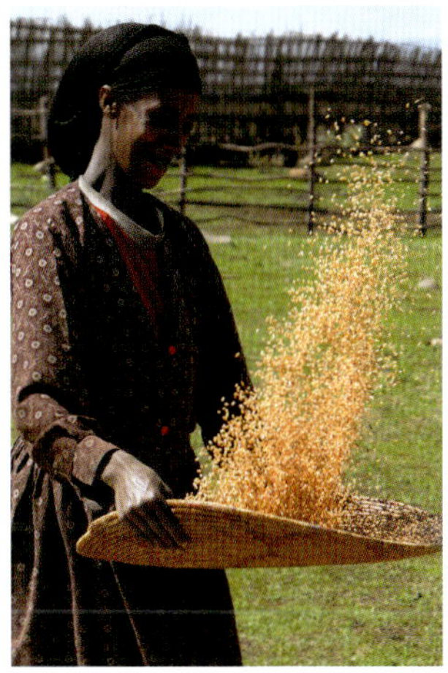

A local lady shows how it's done

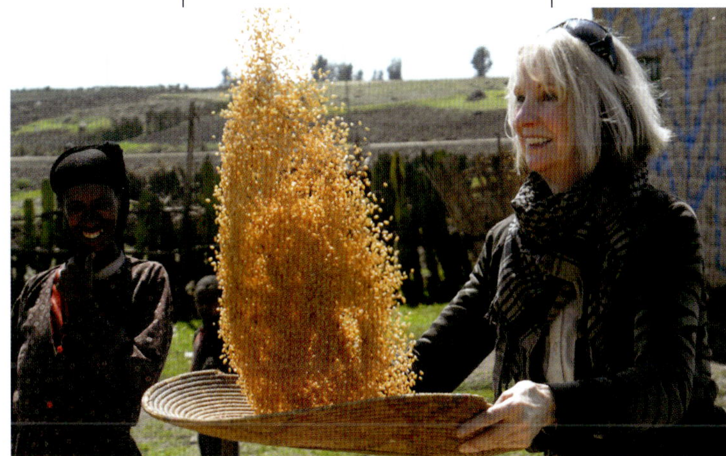

Pollyanna follows suit

to look inside her home – which is much wealthier than any we have visited before. She has a separate cookhouse and storeroom away from the main building.

Eventually we arrive at the park headquarters at Dinsho. We have been invited to stay in the field workers' accommodation at the EWCP, up in the mountains, but there is also a small lodge here and Tariku suggests we look at it as a possible alternative. The available bedroom looks fine – spartan but perfectly serviceable – but the communal bathrooms are another matter entirely. Not only are they a very healthy walk from our room, they are absolutely filthy. In the one I visit the loo seat is sopping wet, and there is dirt banked up against the walls. Worse still it is a small space and there are at least thirty daddy long legs in it. In fact soon after arriving at the headquarters, I come to realise I have been dropped off in my own personal Room 101 – there is a plague of them. It is a living nightmare. I am trying to shoo one way from the open door of the car – I don't fancy sharing the rest of the day's drive with it – when Tariku sees me, and Pollyanna explains that I am afraid of them. He gives me an odd look.

Meanwhile Sami has set up home in the communal kitchen and cooks some delicious veggies and rice. The dining-room at the lodge has an unusual ambience; the tables

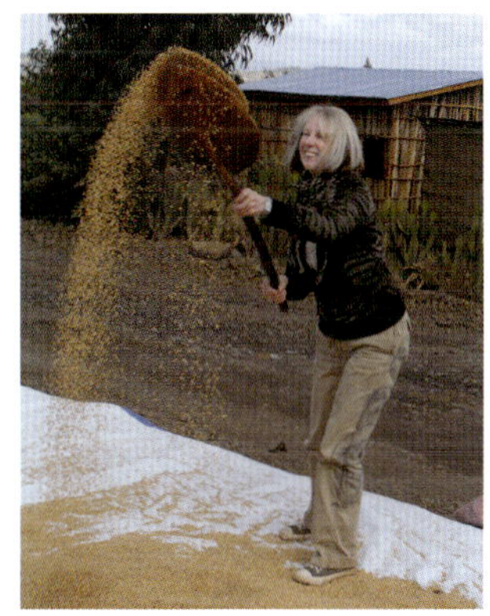

'I come to realise I have been dropped off in my own personal Room 101'

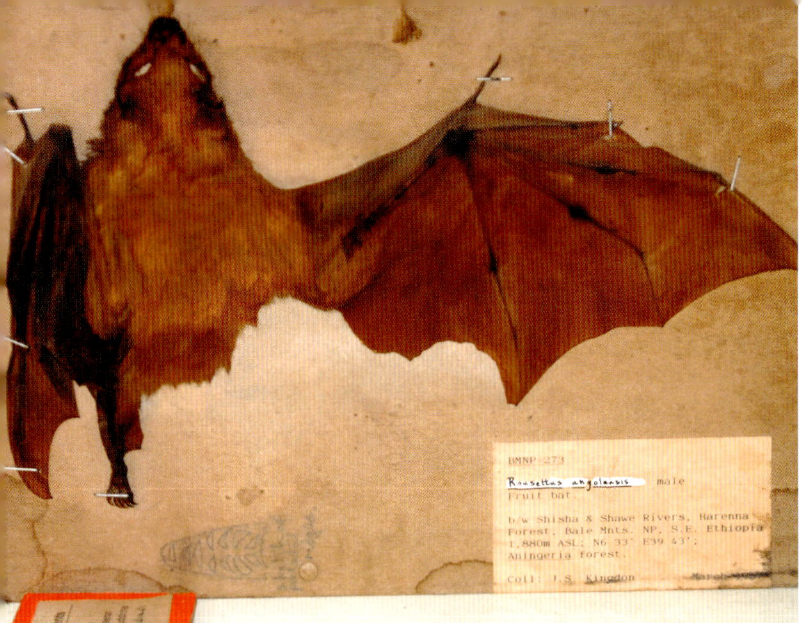

An 'empty' bat pinned to cardboard

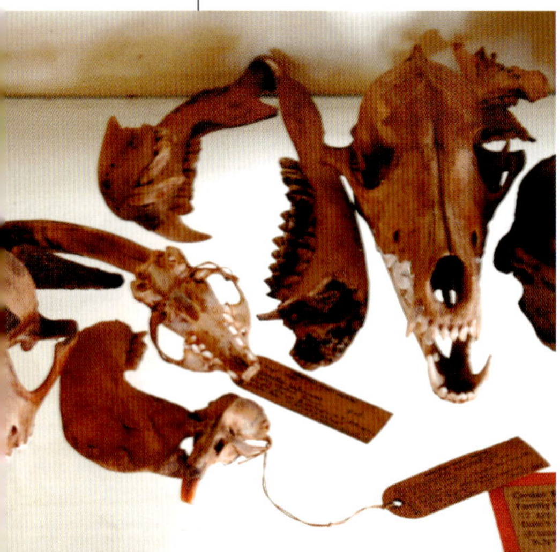

A selection of dusty skulls

Specimen jars full of pickled creatures

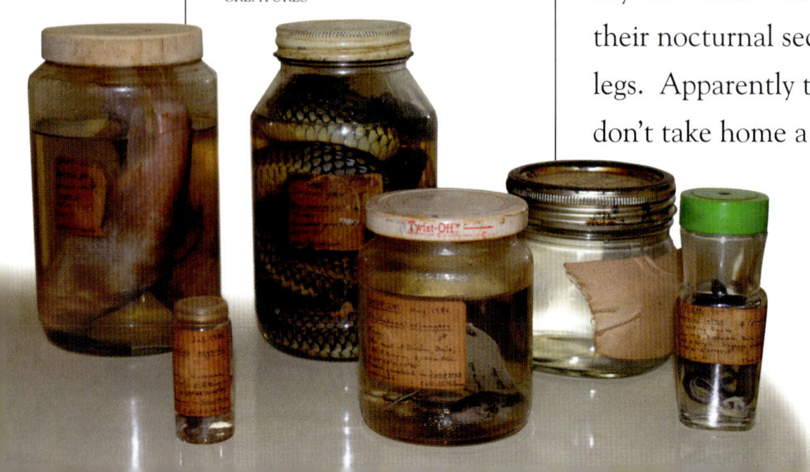

are surrounded by shelves of grimy specimen jars displayed behind yellowing glass doors. The jars contain a diverse collection of pickled embryos, internal organs and folded snakes. Also on display are arrays of dusty skulls, empty bats and mice stapled on to sheets of cardboard. At the far end there is a sun-faded collection of stuffed animals, which the taxidermist has decided to mount in such a way as to disguise as much as possible the original shape of the creature. So we have a serval in the shape of a mongoose, a crow in the shape of roadkill and a couple of eagles in the shape of bouquets of flowers.

After lunch we set out on a walk to explore the area around the park headquarters. Although we are considerably lower than we have been on the Sanetti plateau (hence the plague proportions of the daddy long legs) the air is still very thin and we are gasping as we walk uphill. However, our struggles are well rewarded as the wildlife in this area is amazingly tame. A family of warthogs runs past us on the trail, tails pointing comically into the air. We get very close to the mountain nyala, another endemic species we were very much hoping to see. They used to be common throughout the southeast of the country, but habitat loss and poaching have seen the numbers fall to around three thousand individuals, the majority of which are found in the Bale Mountains. They are impressive animals – the males stand around one and a half metres high at the shoulder and sport an imposing pair of twisting horns. Male mountain nyala are a sepia brown colour that slowly darkens with age, while females are paler with a scattering of spots and stripes. They live in small family groups of females with calves. We also get our best sighting yet of another endemic animal – Menelik's bushbuck. This subspecies, which can only be found in Ethiopia, is normally shy and elusive and no accurate estimate has been made of their total population because of their nocturnal secretive habits. We also get even closer to around eight million daddy long legs. Apparently the dominant hagenia tree in these forests attracts them. (Note to self – don't take home a cutting for the garden.)

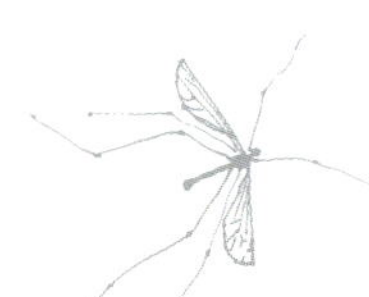

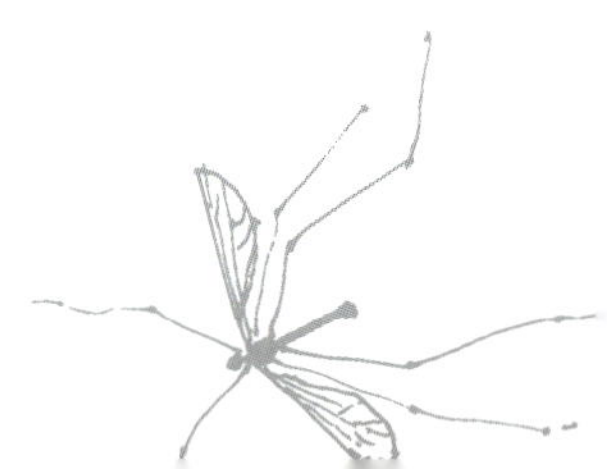

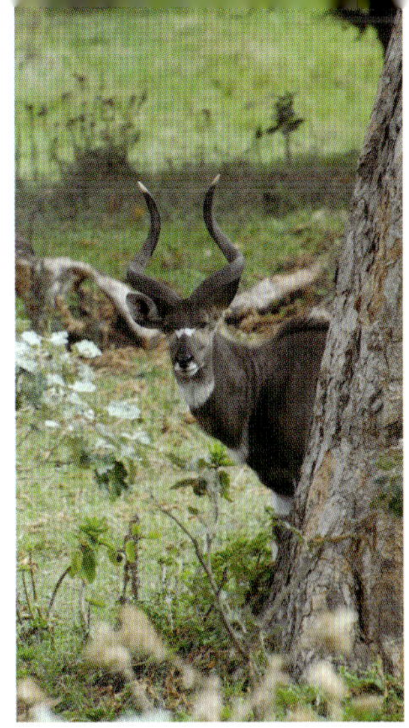

Mountain nyala

At one point as I am ploughing manfully (and twitchily) through a cloud of the spindly insects, Tariku turns to me, 'You know those things you are scared of?' he asks, 'Yes' I reply, hoping for a bit of Ethiopian wisdom which will tell me how to repel them using the sap from a nearby tree. 'This would be a really good time to get over it' comes the reply. I suspect the next few days will either act as aversion therapy or I will be carried away in a straight jacket by the end.

We head up to the EWCP field base, which is just a couple of miles away. We say hello to Claudio, Graham, Chris and Anne-Marie, and they show us the available accommodation. The bedroom is way more basic than at the lodge at Dinsho, but we decide to take them up on their offer of lodging – swayed mainly by the bathroom facilities. Rather than a sit-down loo in a filthy bathroom there is a little squat loo – a hole in the ground in a small cubicle – a little way up the hillside behind our room. This is a much cleaner and more hygienic option. We also have a little shower-room next door to our bedroom. There is another bedroom next door which our crew can share, and a communal kitchen which Sami can use to prepare food, so we move in.

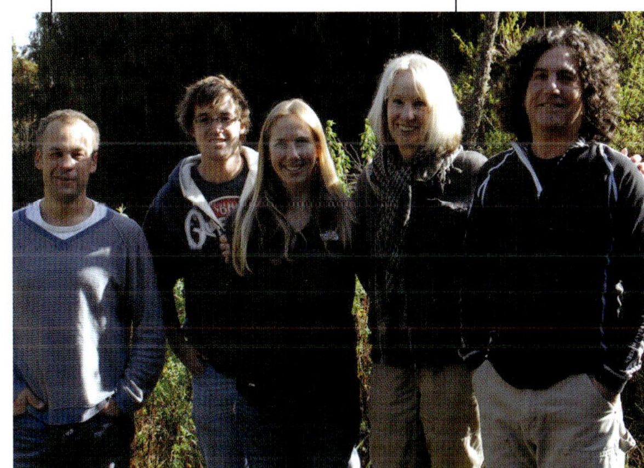

Pollyanna at the EWCP field base with (from left): Graham, Chris, Anne-Marie and Claudio

We are very grateful to have been invited to stay here at the heart of the conservation project and have this wonderful opportunity to learn more about their work, so it seems

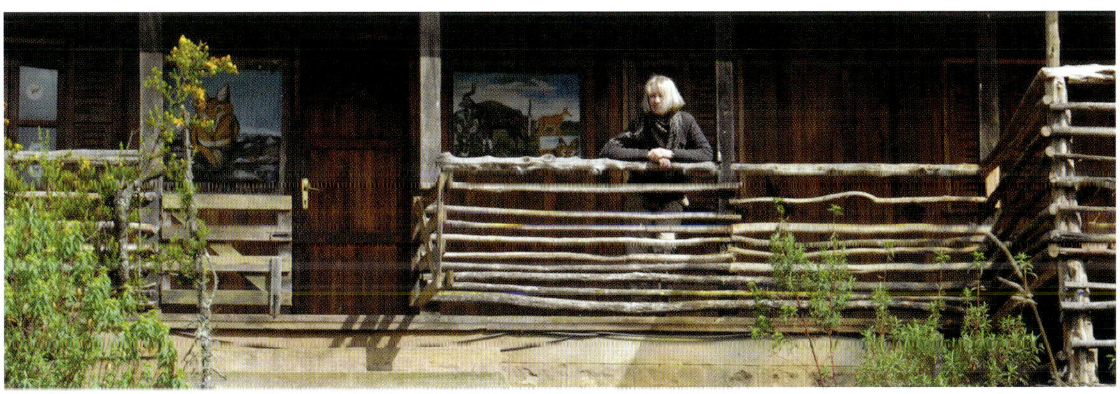

Pollyanna at the EWCP headquarters

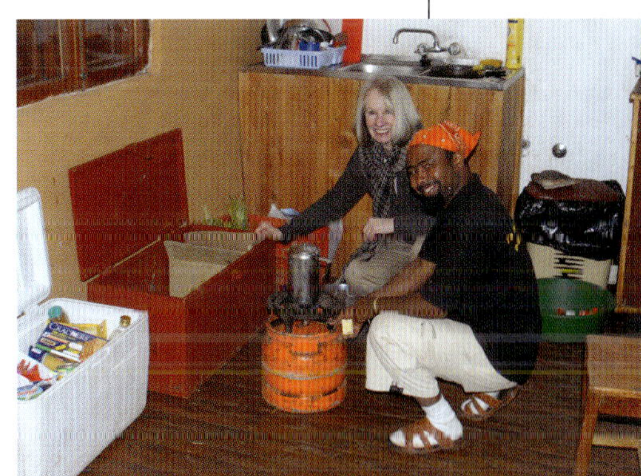

Pollyanna helps Sami in the kitchen

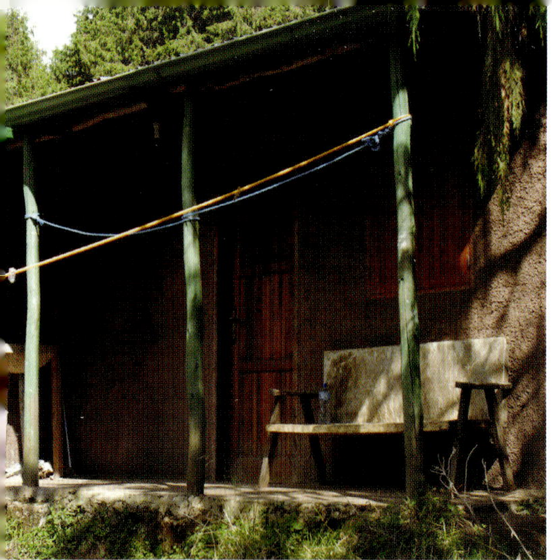

Our accommodation

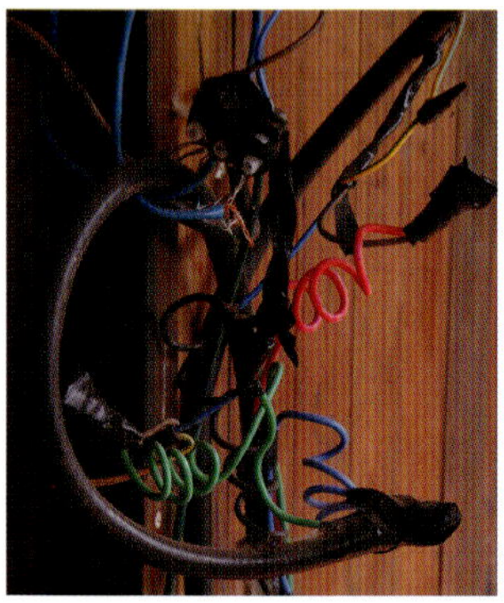

Electrical wiring outside our room

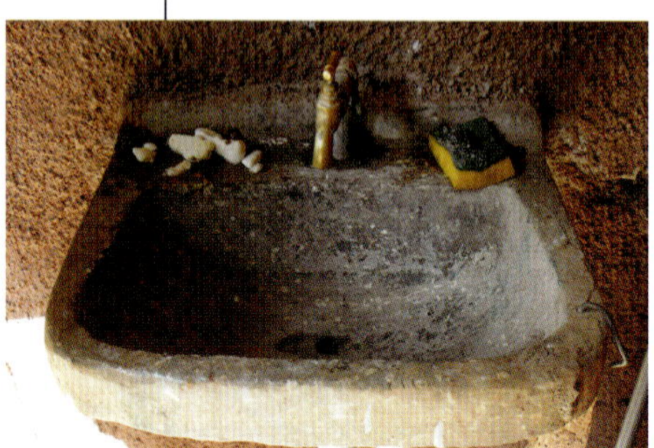

The sink has vanished under layers of rust and disuse

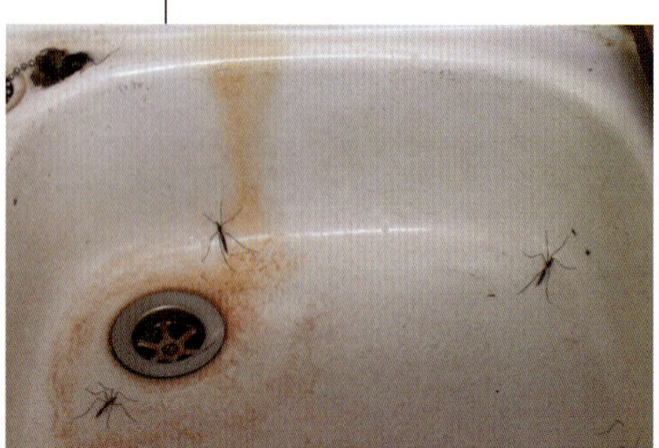

There are daddy long legs everywhere...

ungracious to whine about our sleeping quarters, which are usually used by field researchers or students. However, the word 'room' is actually a slightly flattering description. We enter through a door plastered in daddy long legs, into a small room with two metal-framed beds pushed against the wall, topped with a thin mattress, sheet and blanket. One bed has a small inflatable cushion as a pillow, the other has what appears to be a plank in a pillowcase. Most of the floorspace is taken up with a collapsed metal-framed bunk bed – the top bed has fallen on to the one below, so we use this to store our bags. The frame of what would have been the top bunk encircles the structure at chest height making it extremely difficult to reach anything in or out. The ceiling is sagging disturbingly, and where I lie in bed there is a large hole directly over my head with wires emerging out from filthy darkness, surrounded by a huge cobweb. It is quite difficult to close my eyes and sleep with this portal to the unknown directly above me. On the upside there is also a big wooden table and we appear to have electricity provided by a car battery clipped to a junction box – and the interior is blessedly free of daddy long legs (DLL).

I devise a method to continue to keep the room DLL free. The door sticks, so has to be forcibly kicked open or yanked shut. In order to prevent the DLL being attracted into the room by the light, when we want to leave we count to three, then one person turns out the light while the other jerks the door open and we both leap out as quickly as possible slamming the door shut behind us. Problems arise when people knock at the door and expect us to stand and chat with the door open. Are they insane??? Instead we go through the same leaping out and slamming shut procedure, so probably leave everyone at the EWCP with the distinct impression that we are hiding something extremely suspicious inside our sleeping quarters.

Right outside our bedroom is something which used to be a sink but has vanished under layers of rust and disuse – but luckily next to that is the shower-room which also boasts a much newer basin. This basin is still recognisable, but is decorated with insect legs and a very dead and somewhat mashed and sodden large spider along with blobs of unidentifiable black slimy grunge. However, there is hot water so we can wash. At which point we discover another slight problem – we have managed to leave our towels in the bathroom at the last hotel. I improvise by ripping one of my nightshirts in two – which provides us each with a towel stylishly complete with a single sleeve.

Another issue soon arises – attentive readers will have remembered that due to unexpected visitors at the start of our journey we managed to leave home without packing a torch. The toilet facilities are behind our room, reached by crossing a slippery wooden plank across a drainage ditch, then walking up a steep muddily treacherous slope through the forest. This is difficult to negotiate in the daylight – it is completely impossible in the pitch black of an Ethiopian night. Luckily Tariku comes to the rescue, lending us head torches so that we can navigate our way through this obstacle course. This is fantastic as it means we can reach to the loo without breaking an ankle, but also my worst nightmare as DLL are attracted to the light. And the light is on my head.

'I am a mere whisker away from a complete nervous breakdown'

We have already discovered that we need to visit the toilet in pairs, as the door doesn't stay closed without being locked, the only lock is on the outside (!) and of course it is a shared facility. So in order to enjoy any privacy we have to take it in turns to lock each other in, then release each other. This means that while Pollyanna is using the facilities, for several minutes I am standing alone on a mountainside in the pitch dark with a DADDY LONG LEGS ATTRACTOR STUCK ON MY FOREHEAD. I am a mere whisker away from a complete nervous breakdown.

Things can only get worse. Our cell is cold. Freezing cold. So cold it is impossible to sleep, despite piling all our coats, fleeces and clothing on top of our beds. I spend the entire night huddled into a shivering ball dreaming about electric blankets and hot-water bottles.

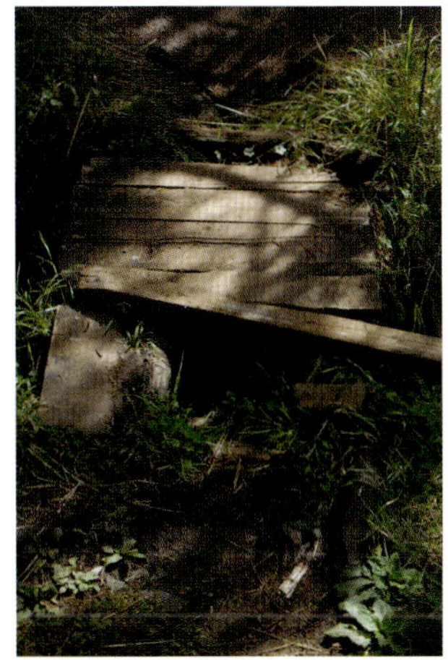

Slippery wooden plank across a drainage ditch

Toilet facilities

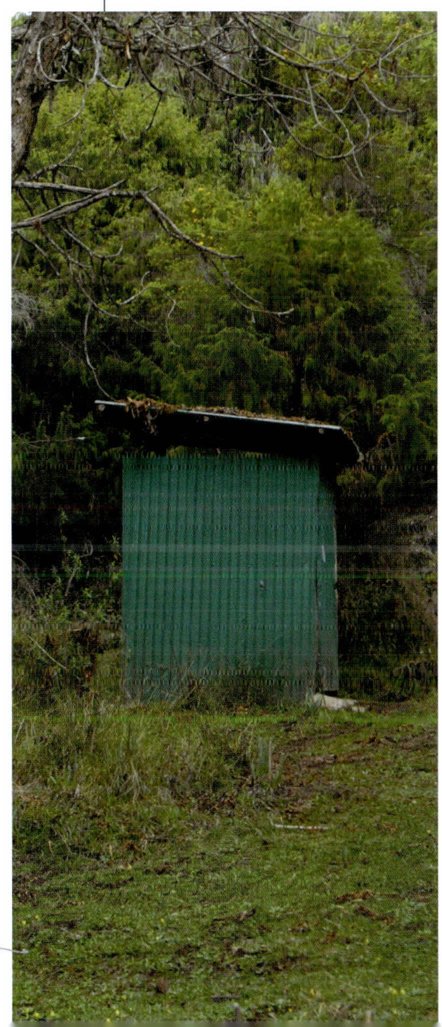

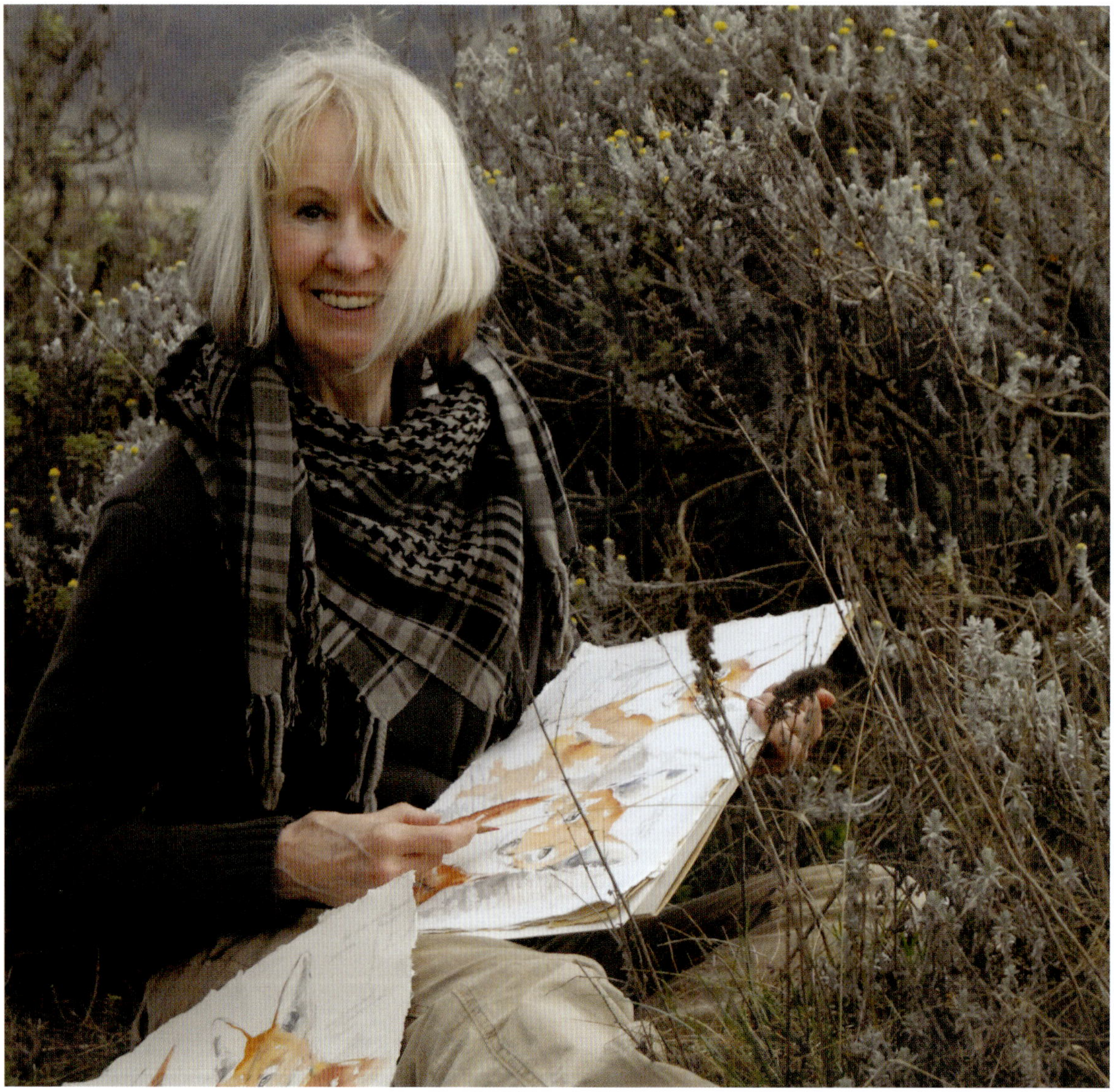

Chapter Five

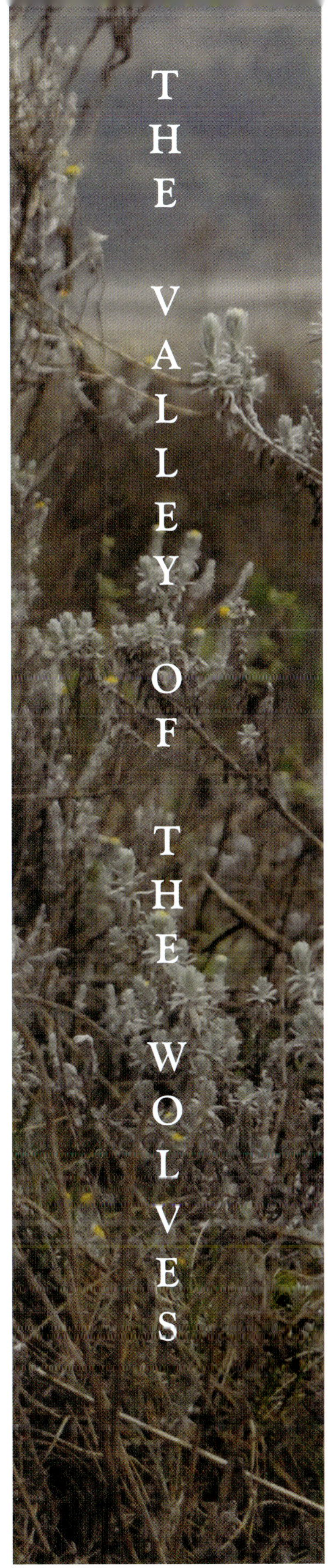

'It is much easier to show compassion to animals. They are never wicked.'

HAILE SELASSIE, EMPEROR OF ETHIOPIA

Until recently the Ethiopian wolf was notable simply for its precarious status on the verge of extinction. But in 1988, Claudio Sillero and Dada Gottelli of the University of Oxford's Wildlife Conservation Research Unit (WILDCRU) set up a field study in the Bale Mountains – the stronghold for the species. The researchers initially focussed on the behavioural ecology of the wolf and the measures needed to protect them effectively from extinction, focussing especially on the ability of wolf packs to coexist with mountain farmers and their livestock. The study unveiled a fascinating story, with wolf packs dominated by an alpha pair but all pack members actively helping to rear the young despite hunting individually – this was considered to be unique behaviour.

After a hiatus imposed by warfare and civil unrest, in 1995 they returned to Ethiopia and set up the Ethiopian Wolf Conservation Programme (EWCP) in collaboration with Ethiopia's Wildlife Conservation Department. The EWCP was endorsed by the Ethiopian President, Dr Nagassu Gidada and, using data compiled in extensive field studies, works closely with the Ethiopian authorities and the local community to actively conserve and protect the wolves and the unique wildlife of the highlands.

Ten years before our visit rabies swept like wildfire through the Bale wolves reducing the population in just three years from an estimated four hundred and forty to under one hundred and fifty individuals. The disease was spread by cattle dogs belonging to the highland villagers, now living in closer proximity to the wolf packs. The EWCP carried out an emergency intervention vaccinating seventy-two wolves to contain the rabies outbreak in the Web Valley subpopulation – one of three areas

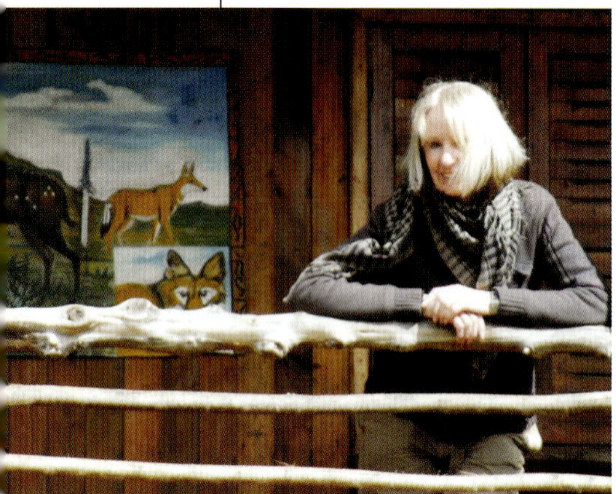

Pollyanna takes time out from sketching

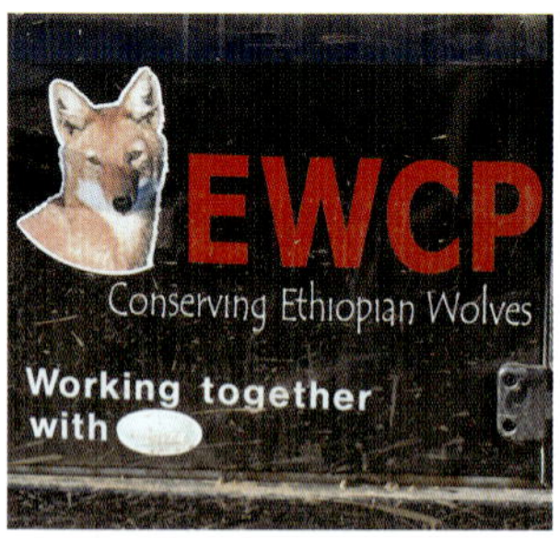

The EWCP Land Rover

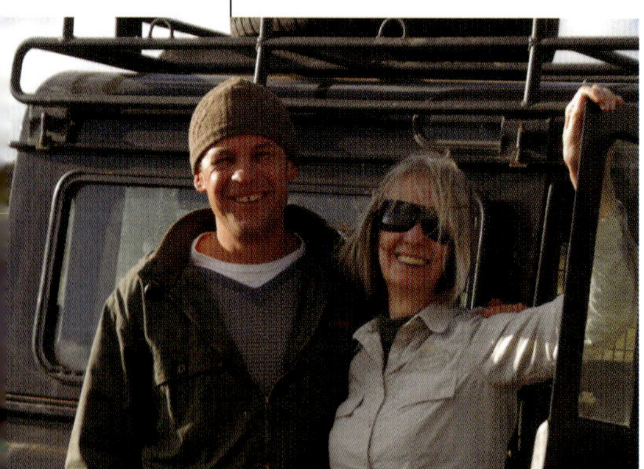

Pollyanna with Graham

with high densities of wolves in these mountains. Since then more than sixty thousand domestic dogs surrounding the wolf population in the Bale Mountains have been vaccinated in a campaign against rabies and distemper – a programme which has also been of great benefit to the human population in the region.

In parallel with the vaccination campaign the population of wolves within the mountains has been closely monitored – and the researchers have recorded a recovery to pre-epidemic levels in two key areas. This leaves little doubt that the EWCP has contributed to the conservation of the Ethiopian wolf.

The EWCP also runs an education campaign which targets both children and adults, helping them to learn about and understand their native wildlife and encourage them to live in harmony together. The education team visits schools and makes presentations at meetings in the local communities. 'Wolf Day' is an annual sports day organized for primary-school children. In the long term the EWCP is hoping to increase ecotourism in the region, working with local guide associations, creating trekking routes, establishing campsites and constructing mountain huts. It hopes that by using the Ethiopian wolf as a flagship species wildlife conservation throughout the Afroalpine ecosystem, home to the vast majority of Ethiopia's endemic plant and animal species, will be boosted.

Day Thirteen

After a difficult night, we have an absolutely wonderful day. When Chris and Anne-Marie learn how cold we were the previous night they immediately dig out spare sleeping bags. They are now officially my favourite people in the world. And Graham very generously offers to drive us into the Web Valley area, where he hopes to be able to show us some Ethiopian wolf pups! This is incredible – a fantastic surprise – we thought we were far too late in the year to see any pups. Graham explains that due to unusual weather conditions experienced in the region this year, this one particular alpha pair had their pups unusually late.

Most matings in the Bale Mountains occur between August and November. After a short courtship, which primarily involves the dominant male accompanying the female constantly, the wolves copulate over a period of three to five days. The female discourages attempts from all but the pack's dominant male but she is receptive to any visiting male from neighbouring packs. Graham tells us that the studies carried out by the EWCP indicate that up to seventy per cent of matings involve males from outside the pack.

Here the alpha female of each pack gives birth once a year between October and December in a den which she digs under a boulder or inside a rocky crevice. There are between two and six pups in a litter. Pups are born with their eyes closed, without teeth and with dark fur. They emerge from the den after three weeks, at which time their fur also begins to lighten to the chestnut brown of the adults. For the first four weeks they are entirely dependant on milk, after which they begin to eat some solid foods. All the pack members guard the den, chase potential predators and bring food to feed the pups. Adults have been observed providing food to juveniles up to one year old.

After a fairly gentle start, passing through arable land we cross a natural bridge over the Danka river. After this point the few huts and trees thin out, and the rest of the drive into the valley is a bone rattling roller coaster ride down a rough track, fording the river at two points and jolting over ditches. As we begin to career wildly down one specially steep rocky slope Graham shares with us the fact that the brakes on the EWCP Land Rover have been a bit dodgy recently, but he thinks they should be OK. Should??

In the event we reach the part of the valley Graham is aiming for without incident – unless you count Pollyanna making Graham reverse back though a ford he has just successfully navigated so that she can photograph the crossing as an incident for her talk. Graham sweetly obliges, although I think his smile may be a little gritted.

The Web Valley cuts a swathe through the central area of the Bale Mountains. The Kotera plain area in the valley is the other key habitat for the Ethiopian wolves in this region. Graham and the field workers have been closely observing the pups at the den site

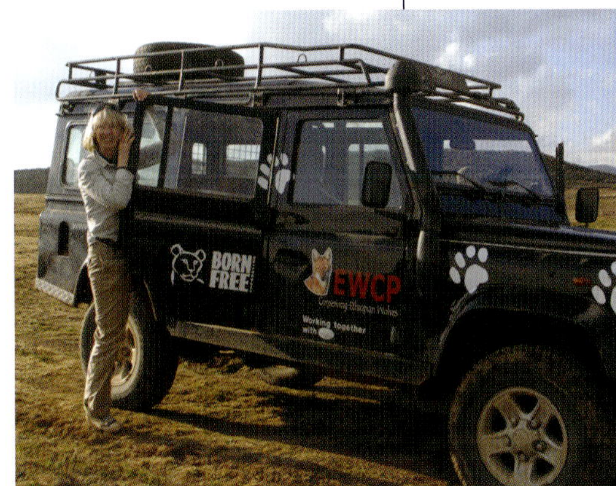

Our trusty Land Rover

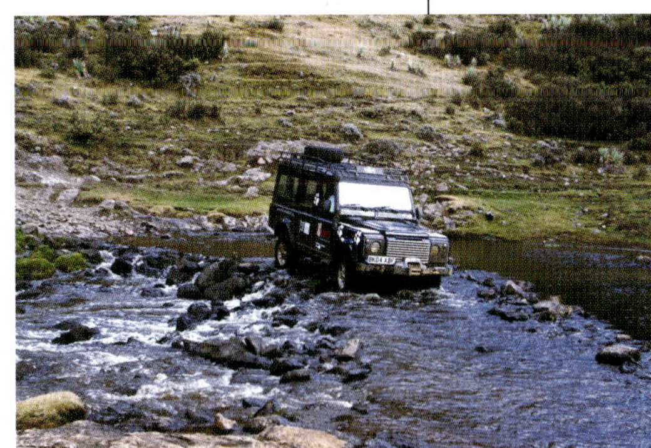

Fording the river

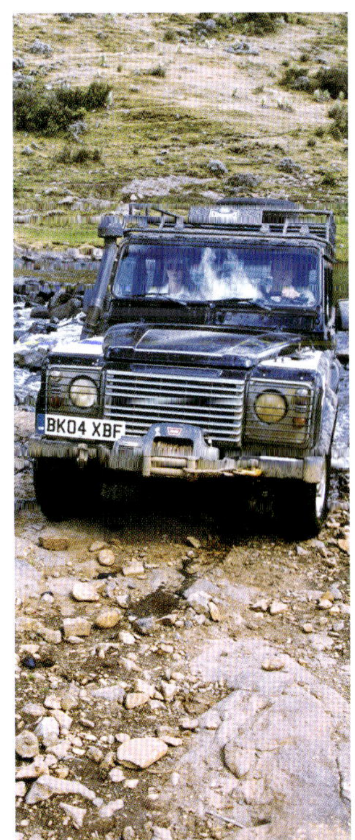

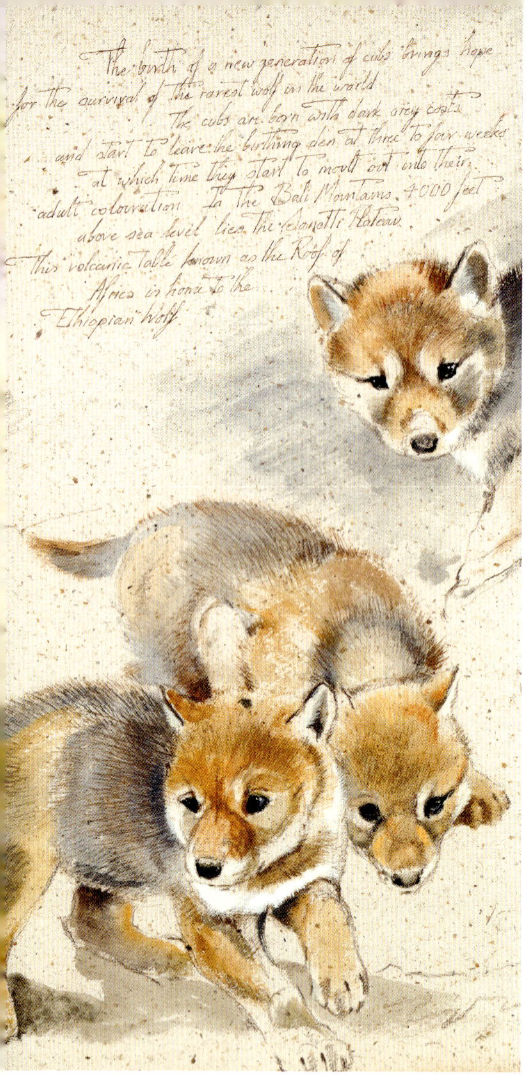

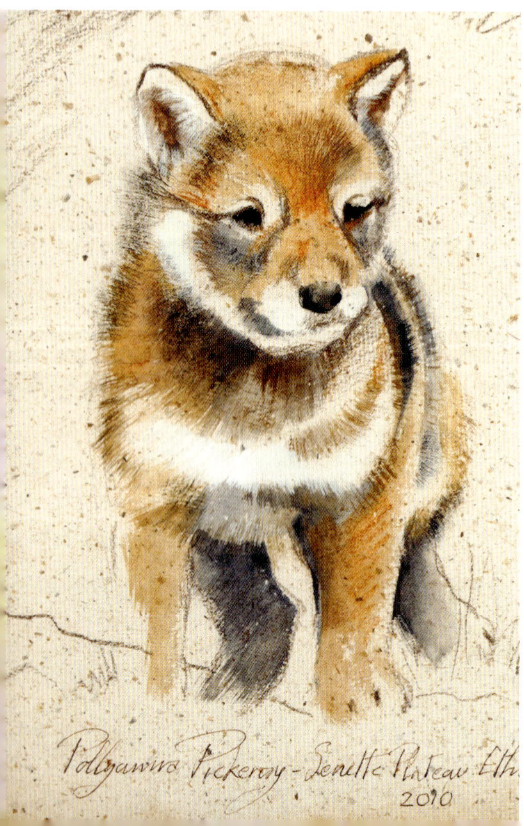

he is taking us to – and I think he is pleased to drive us into the valley today, as tomorrow will be his last day in Ethiopia with the EWCP. After three years as the field coordinator he is leaving and this is his chance to say goodbye to the wolves. Although Graham tells us he has thoroughly enjoyed his time working here, he has a wife and a young child in Australia, who he rarely gets to see, so he has made the difficult decision to leave his job and join them there. Anne-Marie will be taking over as field director, with Chris working alongside her as the technical coordinator.

As soon as we near the den site we see two little pairs of ears pop up out of the foliage on the ridge above the den – the pups are out and about! The alpha male and female are both close by, but they quickly wander off to hunt. The pups remain behind. Graham parks the vehicle and we climb up on to a plateau where we can observe the pups for a while – they are adorable. Wolf pups are very playful – they mock wrestle, nipping each other, running, jumping and rolling in the long grasses. Zoologists will always tell us that this play is essential for their development, strengthening their muscles and fine tuning their senses in preparation for their adult lives – but it is hard not to believe it is also plain and simple fun.

This is also a wonderful place to sit and enjoy the sunshine (and fairly brisk breeze). It

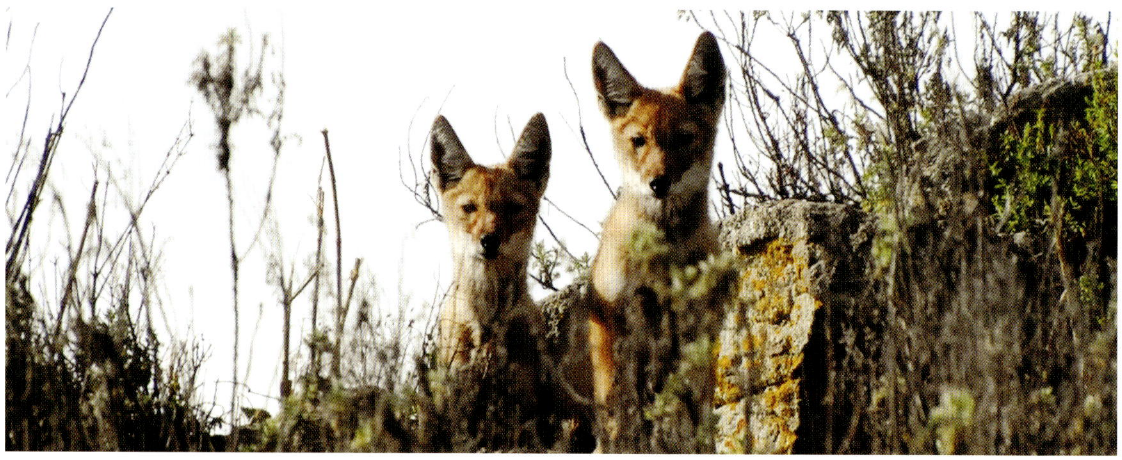

Two wolf pups appear on the ridge above their den

is an incredibly quiet and peaceful spot, overlooking the Web Valley with the mountains encircling us, and we enjoy sitting in silence, while I try to frame photographs of the cubs as they pop in and out of the vegetation and Pollyanna sits busily sketching.

Pollyanna finds that painting 'en plein air' forces her to simplify her work – the limited amount of materials she can carry, the attempt to catch the split-second movement of a bird or animal before the moment has passed for ever. It is in these brief instants that she says she truly experiences nature – and this can never be reproduced by looking at a photograph taken by somebody else. Back in her studio she will then use all the wide range of materials available to her to try to recreate her essential impression of the bird or animal.

Eventually we walk back down, and Graham drives round to the other side of the ridge to a different vantage point – and from here we have the opportunity to see all five of the pups before Graham leads us across to the birth den. The entrance is a surprisingly small hole, with a neat little pile of wolf poo outside. Graham tells us that the female will regularly move her pups between dens, in part to keep the pests down. Wolves will often begin digging a birth den months ahead of time – although sometimes they will enlarge burrows previously made by other animals, or use existing rocky caves.

We return for a final look at the pups. Three of them venture quite close, and we can't believe how incredibly fortunate we have been to have the opportunity to observe this late litter. Finally we return to the vehicle – and just as we start to drive away, the alpha female (Graham can identify her by the two white tags in her ear) returns to give us a perfect display of the Ethiopian wolf's hunting techniques directly in front of our vehicle. She has located a grass rat under the ground and she is sticking her nose into a series of holes trying to find it. Every now and then she stands upright with both ears cocked forward, listening for movement under the earth, then starts to scrabble at one or another of the holes. Eventually she starts to dig more purposefully, and even grabs chunks of earth in her mouth and throws them aside. Finally she exposes the grass rat, which makes a run for it, but she pounces and catches it – we can see two little legs hanging out of her mouth and a

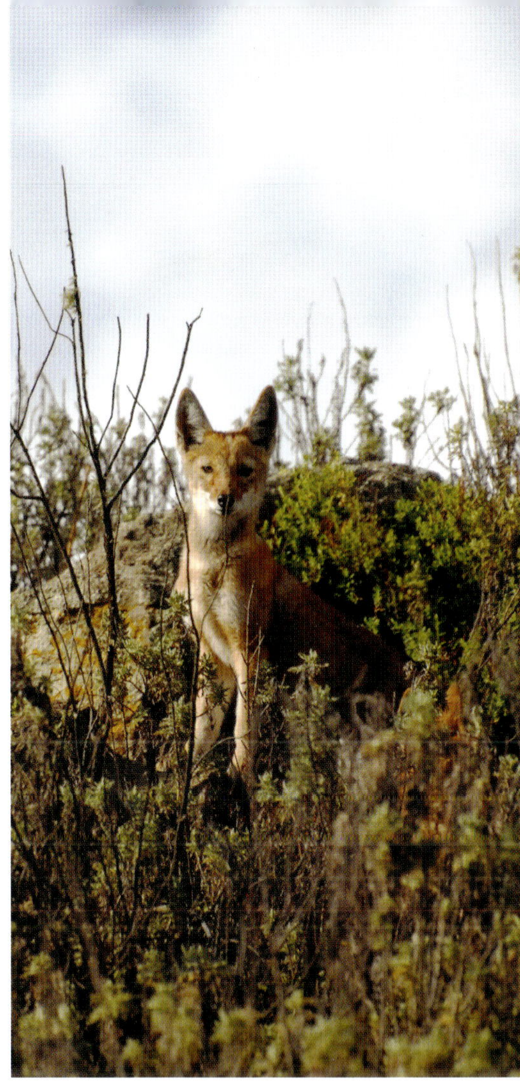

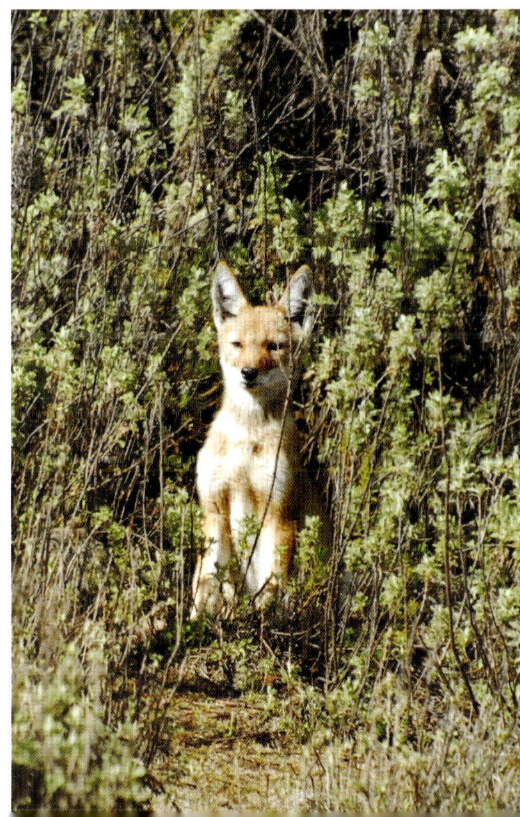

THE VEGETATION PROVIDES GOOD COVER

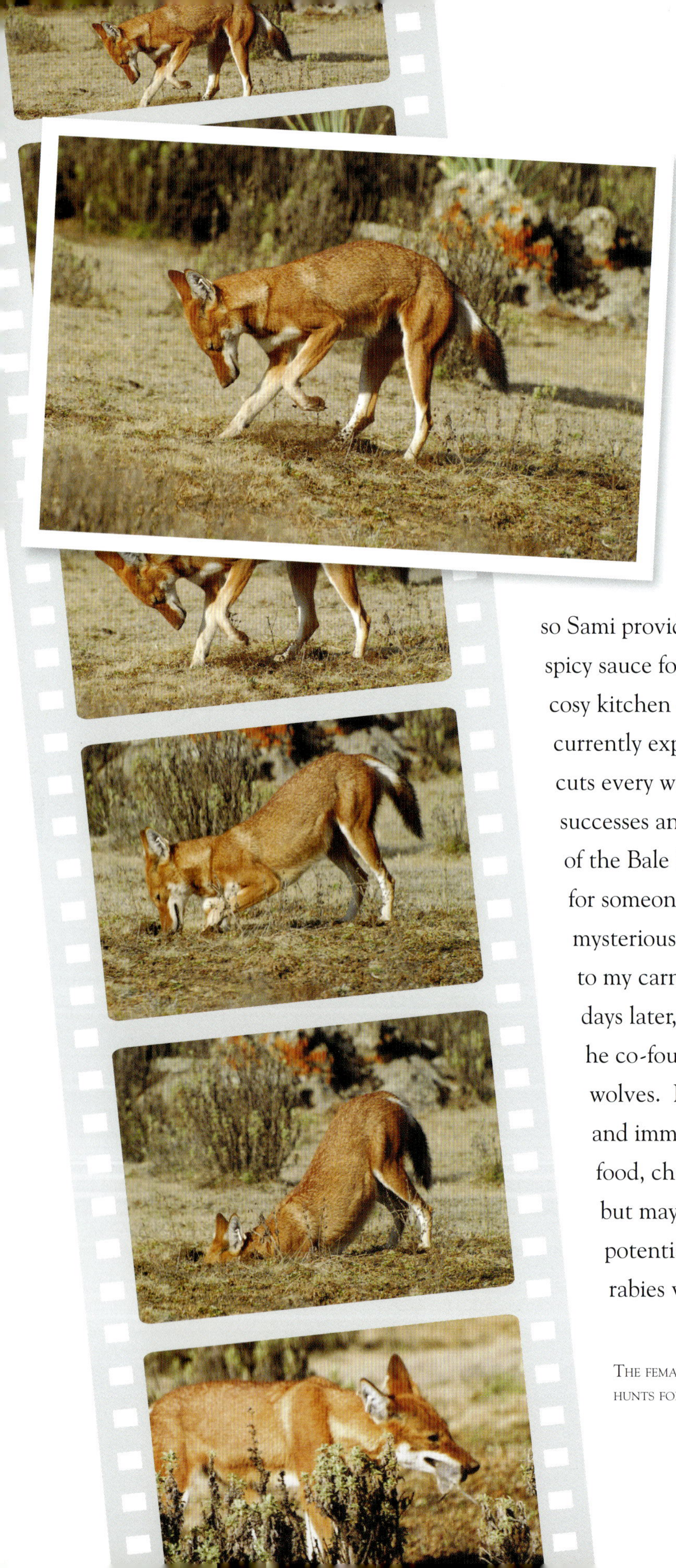

split second later it has gone entirely. The female trots purposefully back towards the den. At this stage of their development she will be regurgitating food to give to the pups to supplement their milk.

We drive back to the EWCP, stopping at one of the river crossings to admire the Finch'Abera waterfall which has carved a cave under its ledge – Graham tells us that these deep narrow caves lead some distance under the ground.

We join Claudio, Graham, Anne-Marie and Chris for dinner in their lodge. In fact we bring our own as they are having beef stew – so Sami provides vegetable soup for the whole group, followed by spaghetti in a very spicy sauce for us. We enjoy a really fabulous evening sitting around the table in their cosy kitchen – there is another power cut, so the table is lit by candles. (Ethiopia is currently experiencing severe power shortages and there is a series of rolling power cuts every week.) We learn all about the work of the EWCP – the challenges, the successes and their hopes for the future. Claudio tells us that he had not even heard of the Bale Mountains until 1987. The New York Zoological Society was looking for someone to study the Abyssinian wolf or Simien fox (as it was known then), a mysterious carnivore found only in a few Ethiopian mountains. 'That night I rushed to my carnivore books to learn more about this enigmatic animal and, less than ten days later, I was in a plane bound for Bale.' Since then the work of the programme he co-founded has gone a great way towards ensuring the survival of the Ethiopian wolves. However, Claudio confirms that domestic dogs still pose the most 'Real and immediate threat to wolves,' saying, 'Domestic dogs not only compete for food, chase them, and transmit rabies and canine distemper to their wild cousins, but may even hybridize with them'. The EWCP has worked hard to eliminate potentially dangerous contact between dogs and wolves, including their ambitious rabies vaccination project. Conservationists are also working to ban dogs within

The female wolf successfully hunts for grass rats

the national park. We discuss other conservation projects, our work with wildlife and much more. The evening ends when Claudio has to go to take a conference call with some major American donors from San Francisco – and we bid everyone good night and head back to our cell to sleep.

First, of course, we have to make the dreaded trip to the loo. Having had such an appalling experience the evening before, I spend thirty minutes converting my hat and scarf into a form of DLL protection somewhat in the style of a beekeeper's headgear – much to the amusement of Pollyanna. Eventually I settle on the combination of my waterproof jacket zipped up as high as it will go, with my hood pulled tight around my face, and the scarf filling the remaining gap, with the whole ensemble then topped off by the head torch. This is fairly effective in keeping the DLL away from my skin – but the downside is that I can barely see anything. So I make my way across the plank/bridge and up the slope clutching on to Pollyanna's arm in the manner of an aged crone, then stand on the mountain slope presumably still attracting just as many insects as the evening before but mercifully unable to see them. Back to the cell, and into our toasty warm sleeping bags. Chris and Anne-Marie, we love you.

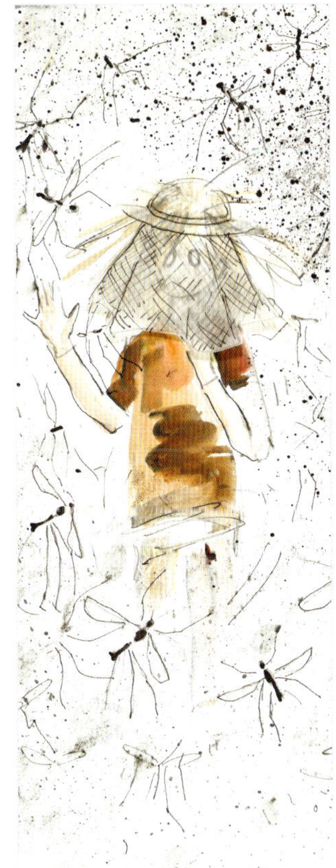

Protective headgear in preparation for the trip to the loo!

As we are now warm we generally sleep much better. However, just as we are settling down to sleep, our crew move into the room next door with much noise and merriment. We manage to doze off until... At midnight the electricity unexpectedly comes back on, suddenly plunging the room into intense brightness. I switch the blazing strip light out and go back to sleep.

1.30 am: My phone, which has been unable to pick up a signal since we arrived at the EWCP, suddenly and for no readily apparent reason locks into the Ethiopian Mountain Network, and helpfully texts me to announce this exciting news with a series of loud pings. I switch off the phone and fall asleep.

3.00 am: Something awakes Tariku and the crew

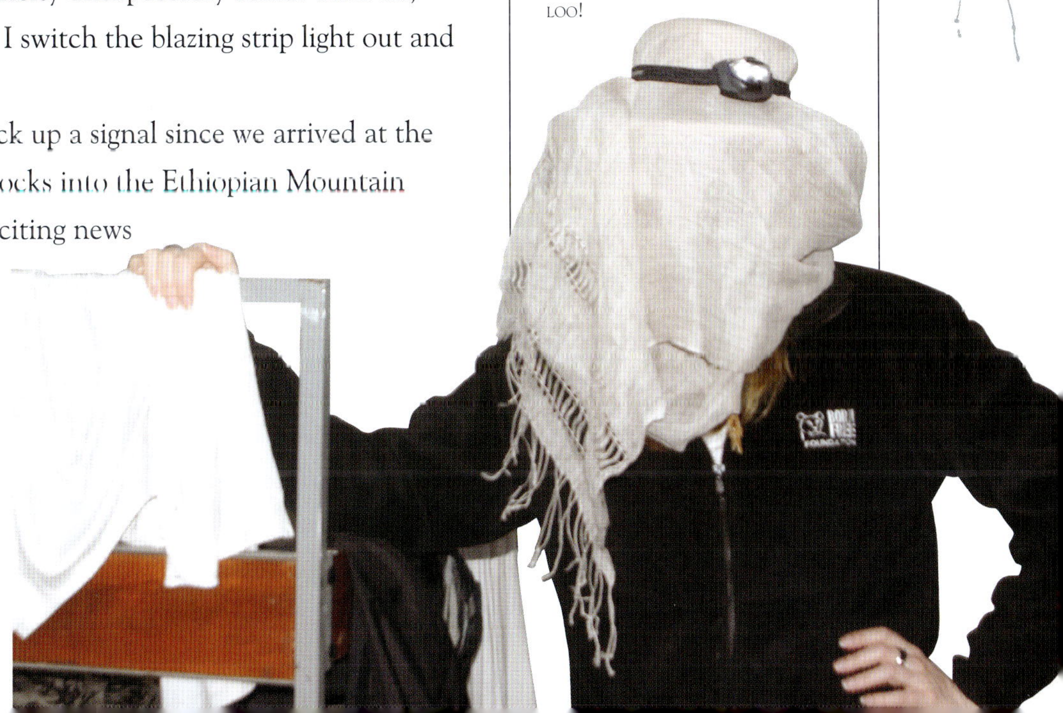

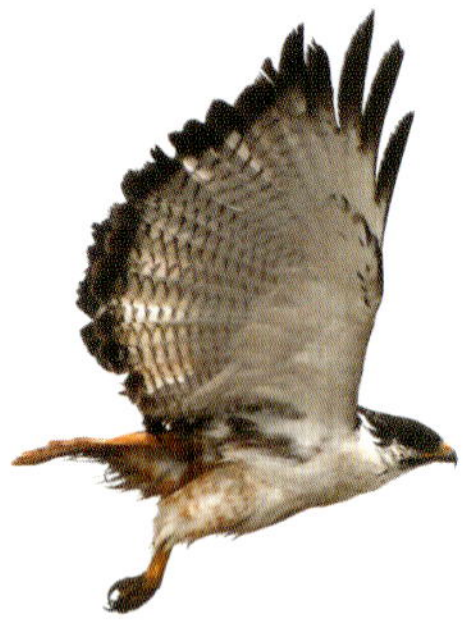

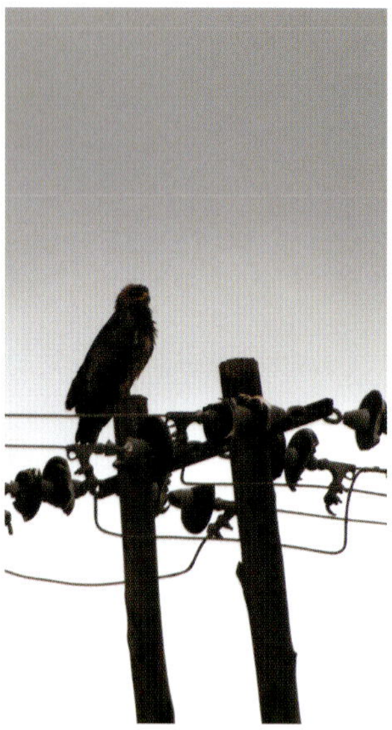

Augur buzzard

Ethiopia is suffering frequent power cuts

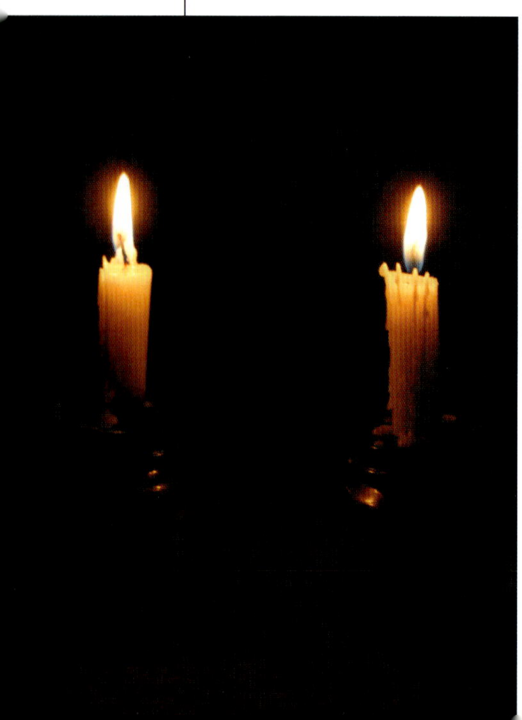

in the cell next door, and they start chatting and giggling loudly, playing music like teenage girlies on a sleepover. Assume I must have fallen asleep at some point because:

5.30 am: I am woken by the sodding Dawn Chorus.

Day Fourteen

At breakfast, Tariku asks if we slept well, and we reply that we had a much better night because we were warm in our borrowed sleeping bags. He informs us the crew slept well, except for the fact that they woke up at 3.00 am and couldn't stop talking. 'Oh Really?' we enquire, as though amazed.

Sami cooks us the most delicious breakfast of toast with local honey, followed by eggs scrambled with tomatoes, onions and peppers. (We will take a short pause here to note that I originally wrote that sentence in my journal with a DLL SITTING ON MY SHOULDER.) We pop up to the lodge to say goodbye to Claudio, Graham, Chris and Anne-Marie, who are leaving for Addis Ababa today. Graham is starting his journey back to Australia, Chris and Anne-Marie are sorting out some paperwork in the capital before returning to Dinsho to take up their new posts with the EWCP and Claudio is returning to Oxford. We feel incredibly fortunate that our visit coincided with one of his now fairly rare visits to Ethiopia.

They have left us in the capable hands of Zegeye Kibret, one of their co-workers. Zegeye tells us that he grew up in Dinsho, where he became actively involved in the nature club in his school. A visit from a young researcher called Claudio Sillero, who spoke to the children about the Ethiopian wolves and about protecting their natural heritage made a huge impression. Some years later Claudio founded EWCP, and Zegeye, who qualified as a primary school teacher, has worked for them since 1998. He runs the EWCP education campaign in and around the Bale Mountains National Park, talking to the local community and school children about the importance of protecting the afroalpine environment and making them proud of their endemic Ethiopian wolf. He tells us 'I became even more enthusiastic about wildlife conservation as my career progressed because I saw that the benefit of wildlife conservation is also a community benefit, creating a better, healthier and more sustainable environment for all.'

For Pollyanna, a keen interest in natural history has always run parallel to her work as an artist, so she especially values the opportunity to learn more about the wolves from someone who is so committed to their protection.

Today he leads us into the Gaysay area, the northern most region of the national park. Nestled in a basin between the Gaysay Mountains and the other hills, it is an area of mixed habitat – grassland, scrub, swamp and river. Although small, this area is very rich in wildlife, and on our trek we are rewarded with excellent sightings of warthogs, mountain nyala, Menelik's bushbuck and a lovely grey mongoose. Zegeye tells us that this is the only area in the mountains where reptiles, including snakes, have been found. All the snakes are harmless and include the three-lined grass snake, a diurnal snake which lives in open grasslands, and the Abyssinian slug-eating snake, which sounds like a very handy souvenir to take home for the garden. A pack of wolves lives in this area and are apparently quite commonly spotted – but they remain elusive today – although after all our wonderful wolf sightings to date we can hardly complain.

Zegeye also spends a lot of time pointing out birds – augur buzzards, Rouget's rail and Abyssinian longclaws among others – but I do not seem to have my bird-spotting eye tuned in today, so consequently spend a lot of time staring vaguely at trees.

Back into the vehicle to drive to an area which is very good for spotting rock hyraxes – and indeed we see several. When they were first discovered hyraxes were grouped with rodents, largely due to typically large incisors. However, unlike rodents they use these teeth for defence and grooming, while using their molars for biting off and chewing food. They have for many years famously been referred to as the closest living relatives to elephants, due to fossil and morphological evidence which shows that the two seemingly different creatures share many common features – but

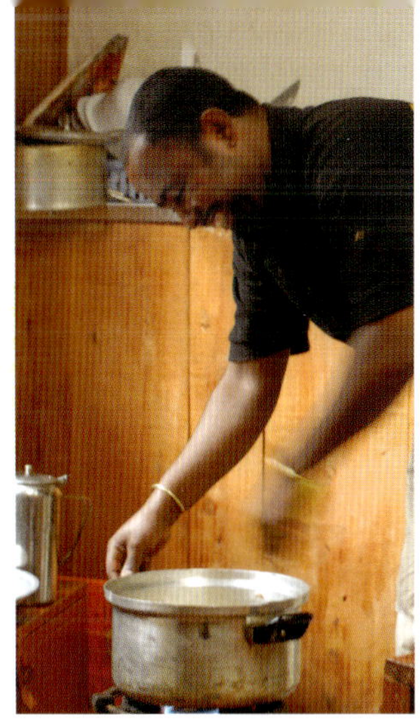

ABOVE: SAMI IN ACTION IN THE KITCHEN. BELOW: THE END RESULT – DELICIOUS SCAMBLED EGGS

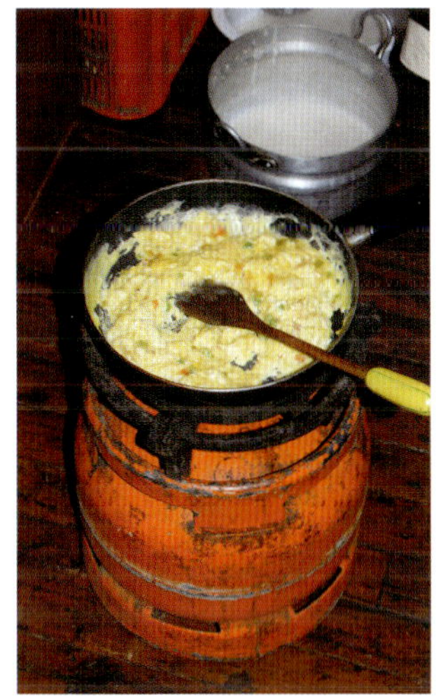

MALE WARTHOG

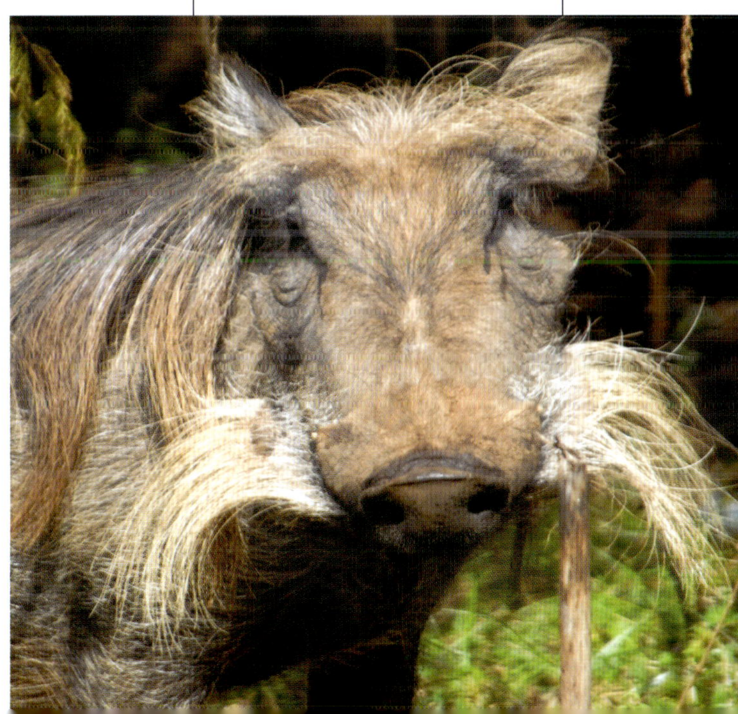

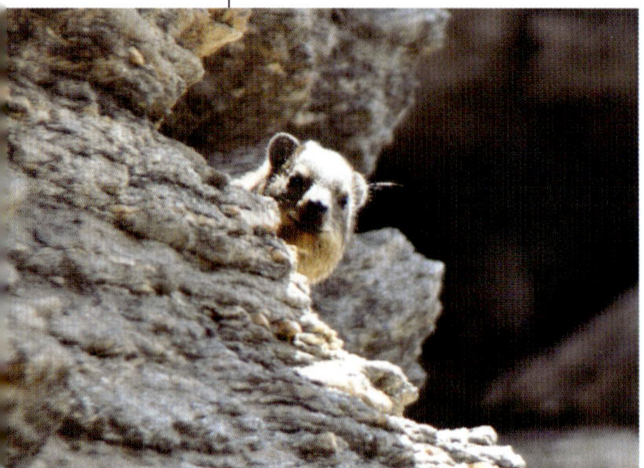

Rock hyrax

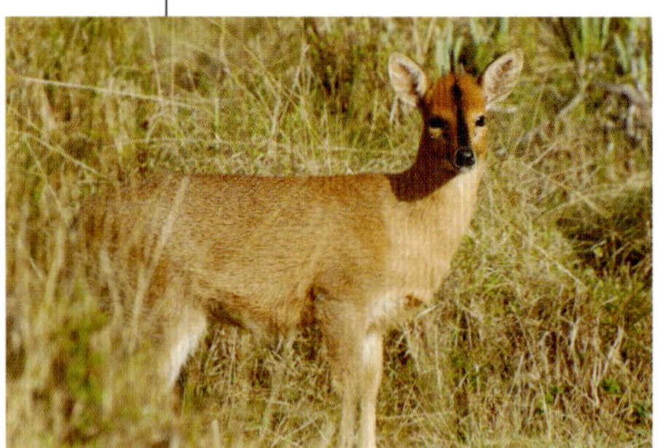

Menelik's bushbuck

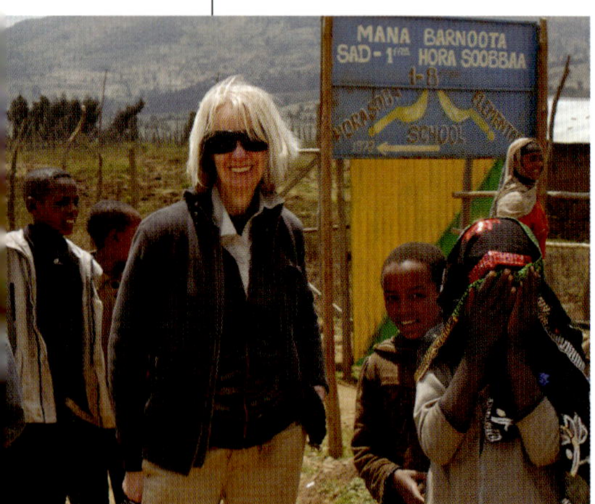

The local children are initially shy...

biologists have recently classified them in an order of their own. Something we can all agree on is how cute they look basking in the sun and grazing between the rocks.

A little further on we stop at a school. The kids all come running out to greet us (no 'YouYouYouYouYou' in this remote area, just shy smiles and giggles) – and gather into a huge semicircle to watch us – until Pollyanna tries to stand with them for a photo at which point they all scatter. We regroup them and try again. Once one or two have been brave enough to shake hands they all crowd round to touch our hands and arms – Tariku says they want to see if our white skin feels the same as theirs!

On the drive back to the lodge we have a wonderful sighting of a mother and baby nyala grooming each other. Pollyanna is able to sit and sketch – another addition to the folder of work she is rapidly accumulating. Although a few of her working sketches are framed and included in her exhibitions, the majority are stored in plan chests in her studio, where she can refer back to them over the years and use them as the inspiration for new paintings. They form a treasure trove packed with chance moments like this one, acutely observed and described on paper in a few deft lines. These pages represent Pollyanna's personal experience of the wildlife encounter.

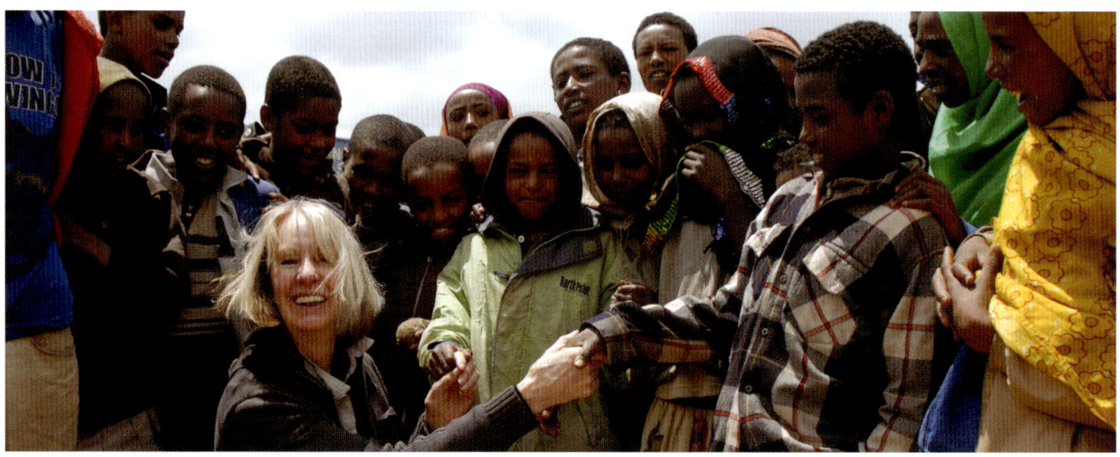

... but they are keen to feel white skin

Sami does us proud this evening with a meal of pasta and veggies, and we then sit on the veranda outside the EWCP lodge where Pollyanna adds the finishing touches to the sketches she has been working on through the day, and I write my journal. Pollyanna is sitting calmly and peacefully on a bench enjoying the last of the day's sunshine. Meanwhile I am sitting opposite twitching every eight seconds like someone with a severe case of Tourettes and asking, 'Did one just land on my head?'

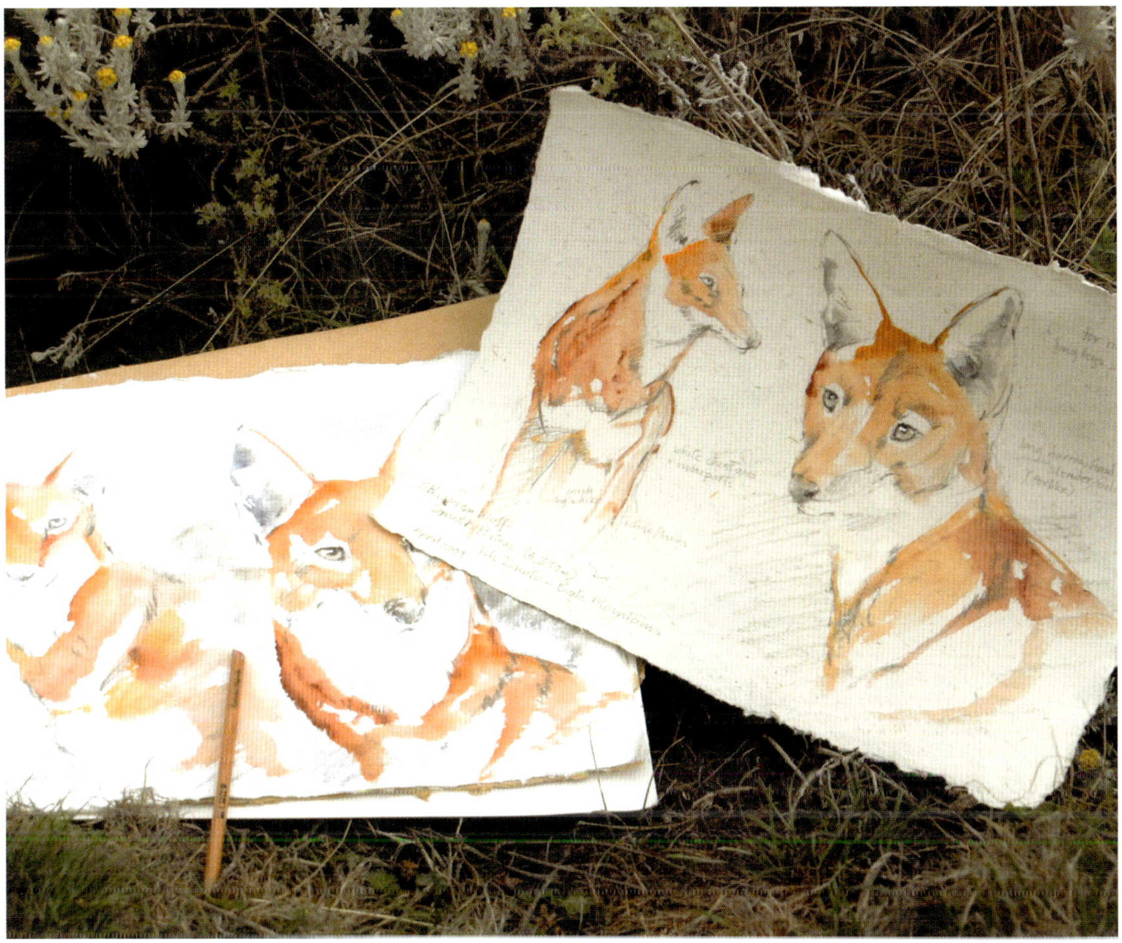

Pollyanna is amassing a wonderful collection of field sketches

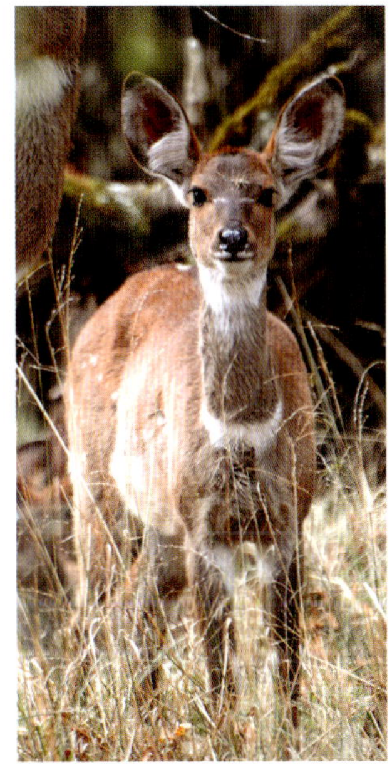

Nyala

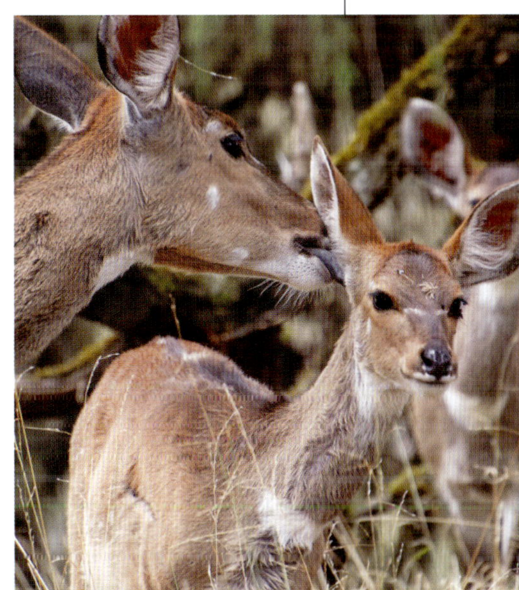

Nyala grooming her calf

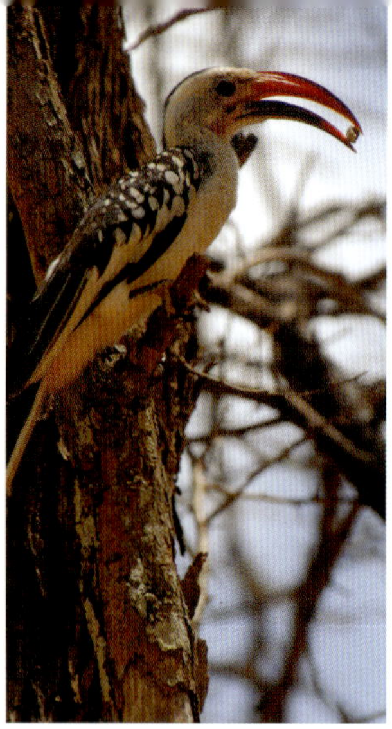

Hornbill

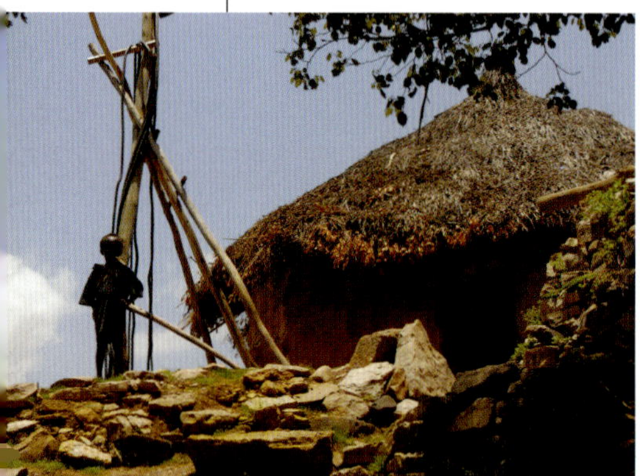

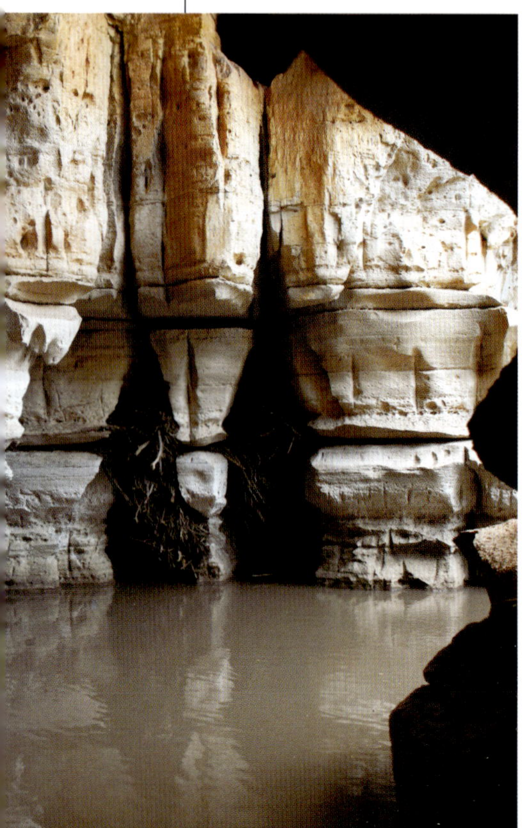

Day Fifteen

This morning we are taking a break from wolf watching to drive one hundred and twenty-five kilometres to visit the Sof Omar caves, a famous local attraction. As ever we make a few stops en route, to photograph a hornbill, a vulture, and to take a bush toilet break. I realise that the entire journal is in danger of turning into an account of bush breaks, but this one merits mention, because having left the mountains we are no longer in the bush. This means that when I get out of the car and attempt to slink discreetly behind a hedgerow, around forty people immediately gather out of nowhere to watch. Tariku, bless him, attempts valiantly to distract them, practically performing magic tricks at the side of the road to try to appear more interesting than a white woman answering the call of nature. I try to squidge further back into the undergrowth, only to find I am hovering over what appears to be (and smells like) the ankle of a dead cow.

Moving on, we have been joined for the day by a young guy from Dinsho, Abera Yilma Megose who has been volunteering with the EWCP. He has offered to act as our cave guide, as he has walked all fifteen kilometres of the cave system twice – but we think he has really just come along for the ride. He is a big supporter of the EWCP and has played for their football team ('Number nine, like Wayne Rooney'). Unfortunately he also has a streaming cold, and is sneezing incessantly in the confined space of the car. Fortunately my nose is so incredibly blocked by Ethiopian road dust that I can barely breathe through it, so I am hoping this will also act as a natural filter for any germs.

The caves are located in a spectacular river gorge, lined with limestone cliffs. I had somehow got the impression that they were quite a mainstream tourist attraction, as well as being a centre for Muslim pilgrimage, but judging from the condition of the almost non-existent muddy track which leads towards the entrance I am wrong about this. As we approach, our vehicle gets stuck in the mud, wheels spinning – and Abbi suggests we walk the rest of way to the cave entrance while he frees the vehicle. By the time we arrive, we have attracted a couple of extra people who join us to head down the rickety uneven stone steps which lead underground.

Inside the Sof Omar caves

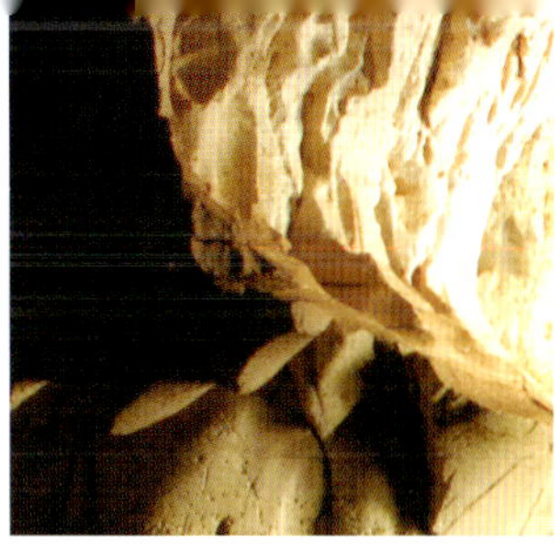

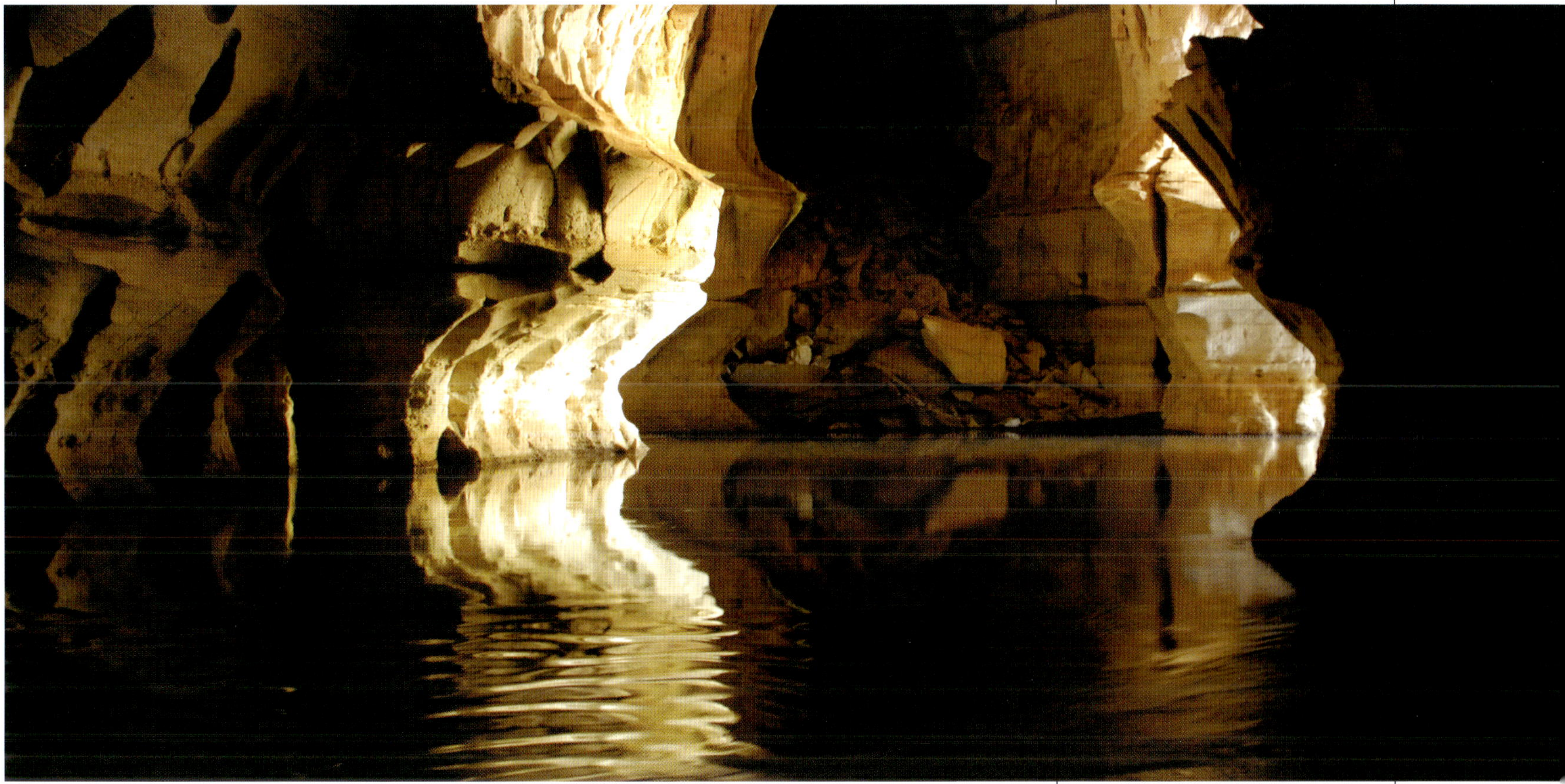

The spectacular caves are named after Sheikh Sof Omar, a holy man who sought refuge in these impressive formations

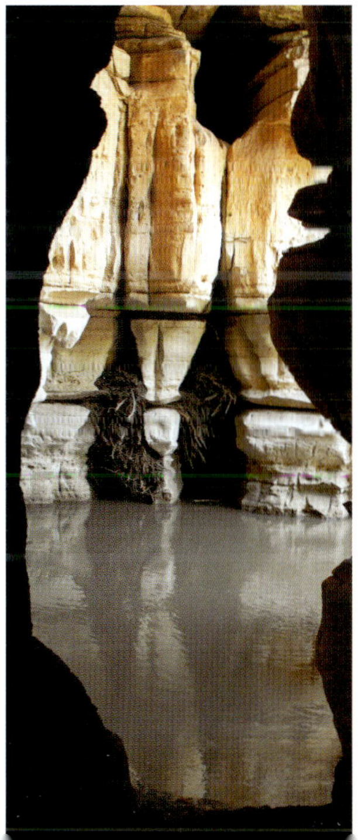

The caves are quite spectacular. Rough stony paths connect huge high cathedral-like caves with curving pillars and huge underground lakes. They are named after Sheikh Sof Omar, a holy man who sought refuge here many centuries ago, and apparently spent many months praying and meditating in the complete darkness.

On the upside these caves are high enough to stand up in, at least as far as we venture into the system – and we see a lovely little horseshoe bat, which flits away before I can photograph it. On the downside caves are not my natural environment. It is very dark

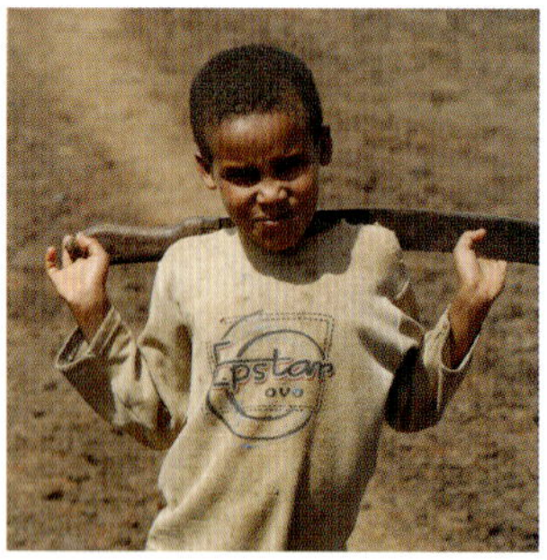

One of our young followers

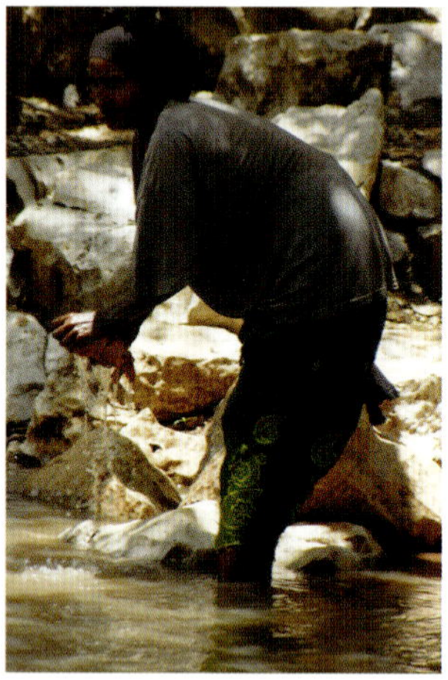

A woman washes clothes in the river

(obviously) – despite my borrowed headlight – the paths are stony, slippery and very treacherous, and this would not be a great place to break an ankle – and deeper in the walls turn out to be home to weird and unnecessary cave-dwelling insects with long twitchy antennae. When we come to a point where the River Web courses through the caves fifteen metres below the edge of the cave we are standing in, I am not too sorry to discover that the bridge has collapsed and we will have to turn back.

Once outside in the lovely fresh air and sunshine we sit on a shaded rock and watch some of the village children playing and swimming in the river, jumping off the rocks and splashing in the strong current. Close by is a group of women washing their laundry on the rocks and collecting the brownish river water in containers to take home.

We wander through the nearby village which is also home to troupes of vervet monkeys, and one of our followers (since emerging from the caves we have gathered an additional eighteen adults and twelve children) helpfully fetches some grain to entice them closer. Pollyanna is about to hand feed them, but to her disappointment I pull her back hissing 'Rabies!!' – it has been impressed on me that while travelling through Ethiopia we should assume that all animals whether wild or domestic are carrying rabies unless we have been one hundred per cent assured that they are not. We find a lovely grassy spot and settle down on sun-warmed rocks to enjoy the picnic Sami has prepared for us. I photograph a lovely tortoise in the grasses – and then photograph Sami photographing the tortoise on his mobile. Our city boy has become completely fascinated by the wildlife during our journey and will be returning to Addis with a phone full of images of wolves, eagles and monkeys. And a tortoise.

Long dusty jolting sneezing drive back to our cell at the EWCP, where of course the electricity is off once again, so I write my journal by the light of head torch.

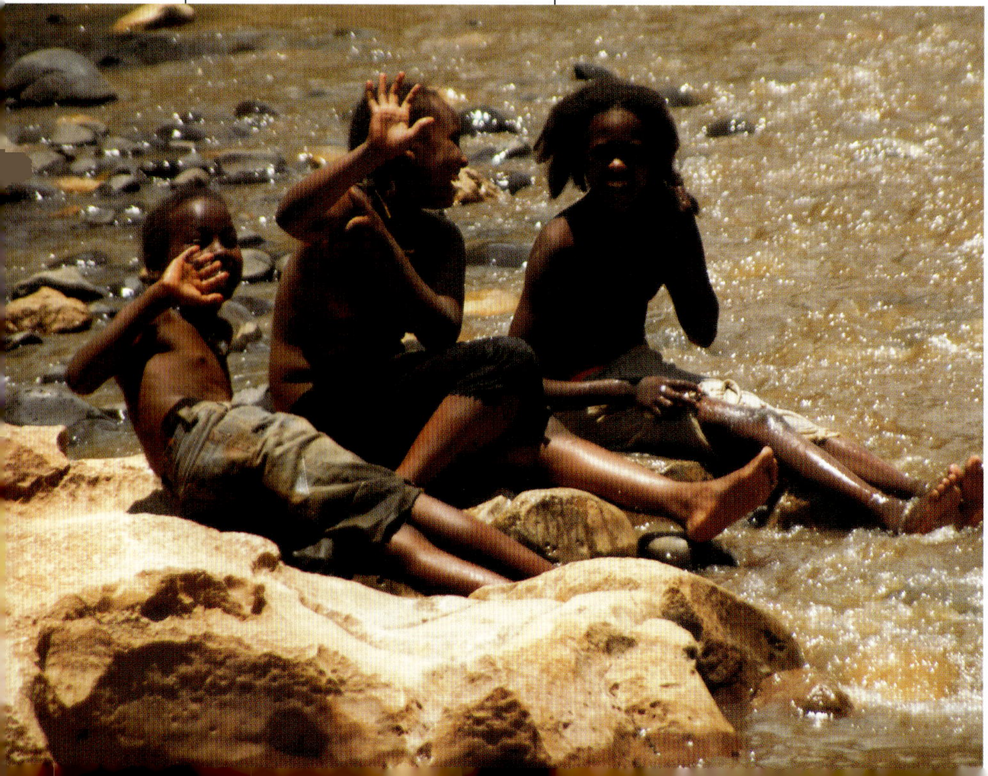

Village children playing in the River Web

Days Sixteen – Twenty-one

We again settle into a routine in the mountains, travelling out with our crew and Zegeye into various areas to observe the wolves and other wildlife. Our good fortune in seeing the wildlife persists – as well as more wolf sightings we see olive baboons, jackals and, driving back through the Gaysay grasslands one evening, a wonderful sighting of a serval, sitting a little way back from the track, eyes gleaming in our headlights.

During our time spent with Tariku he has told us about the work he does as a country representative for a group called Citizens for Africa, a non-governmental, non-profit organization established in 2008. Its broad aim is to help those fellow human beings who are suffering in troubled areas of the continent of Africa. Tariku has been particularly involved in trying to help a small child-care centre in a town called Harar. We had planned to spend a few days in Addis at the end of our journey to see some of the cultural attractions – but decide to alter our plans. After returning to Addis Tariku is flying down to Harar for a few days, so we decide to fly down with him, visit the centre, and spend our few days in what sounds like an incredibly fascinating Muslim walled city. (One of the attractions we hoped to see in Addis was 'Lucy', the oldest and most complete hominid ever found, who now resides in the Ethiopian National Museum. It emerged that Lucy was away 'on tour' in America, so we would have missed her anyway.)

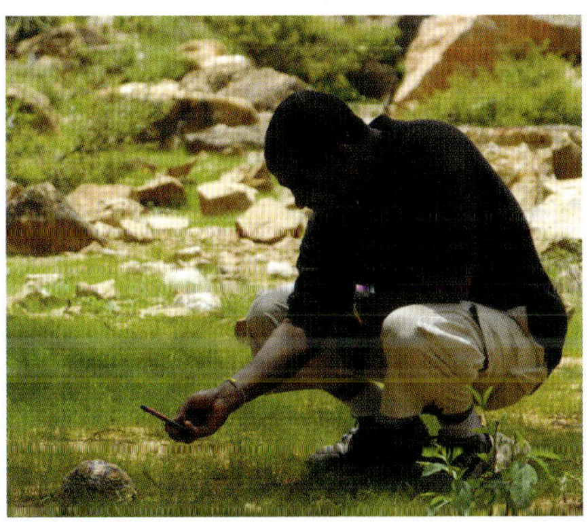

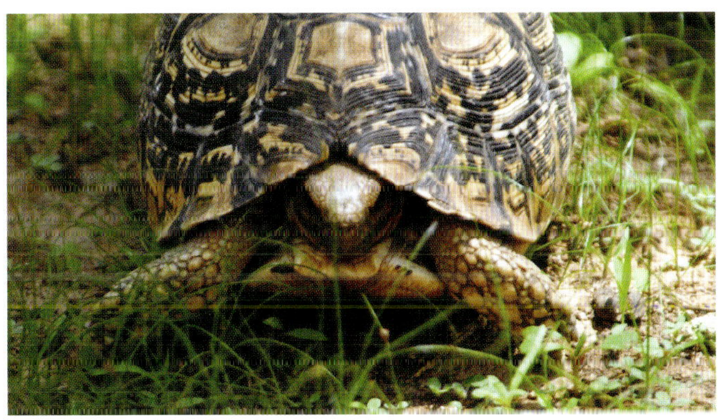

Sami snaps the tortoise on his mobile

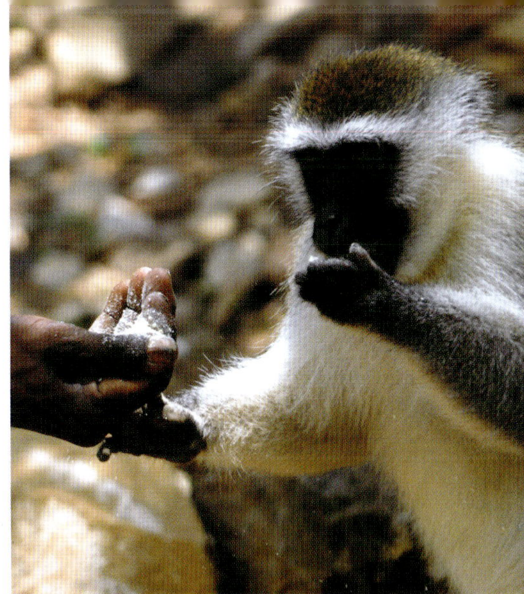

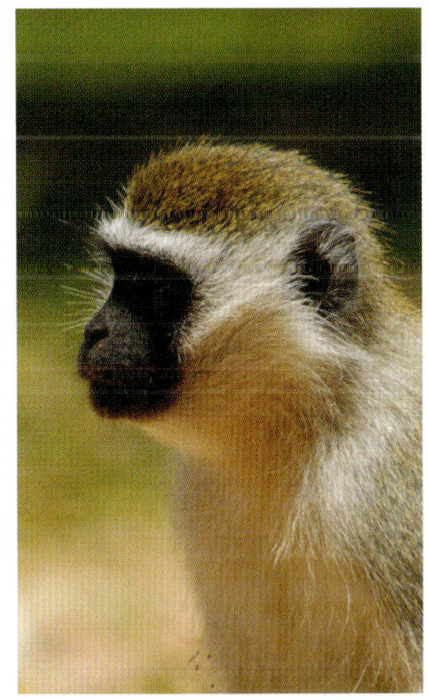

Vervet monkeys

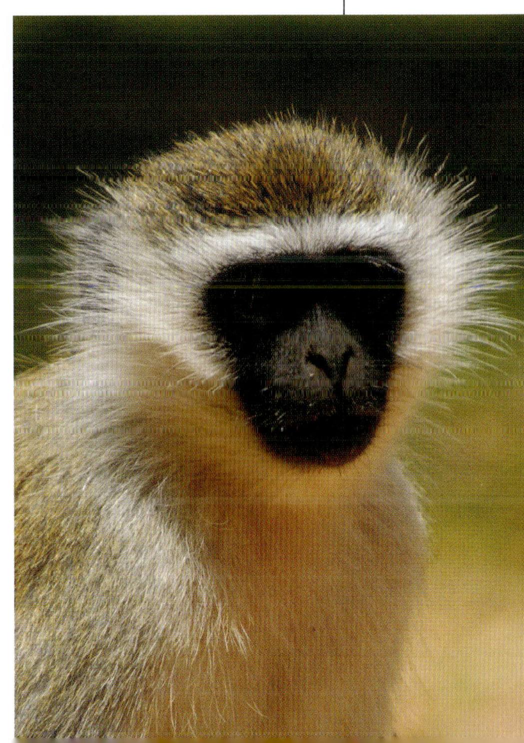

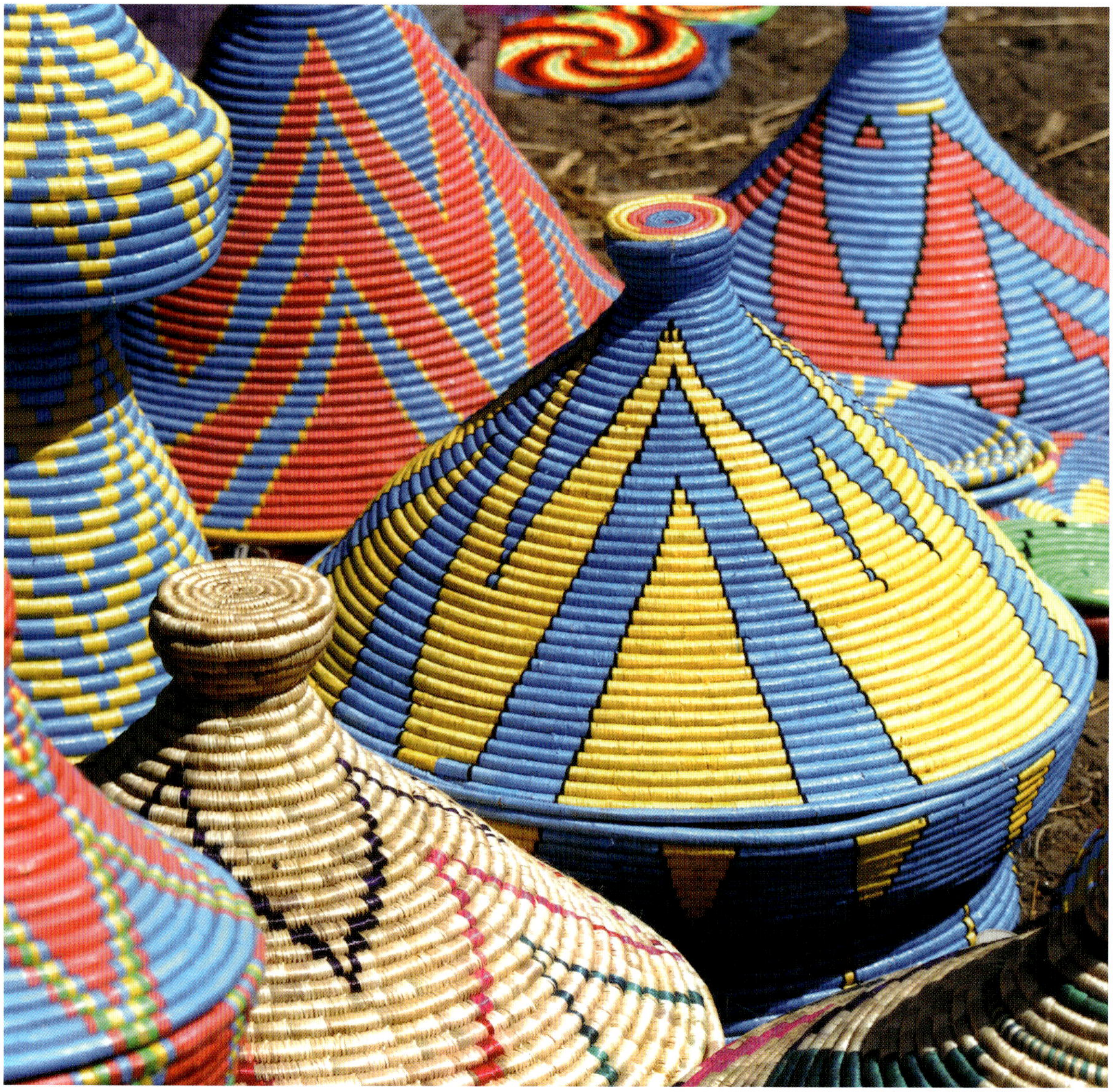

Chapter Six

'I'm not the type who wants to go back to nature,
I am the type who wants to go back to the hotel.'

FRAN LEBOWITZ

The name Addis Ababa poetically translates as 'New Flower'. Situated in the foothills of the Entoto Mountains and standing two thousand four hundred metres above sea level it is the highest capital city in Africa – and the third highest in the world. The city was founded in 1887 by Menelik II after he had established temporary capitals at six different locations. Each of these was deserted when supplies of wood for fuel were exhausted. Addis itself was in danger of being abandoned until the introduction of fast-growing eucalyptus trees from Australia provided the city with a regular source of fuel. The city currently has a population of about two million.

Addis Ababa is an important administrative centre not only for Ethiopia but also for the whole of Africa. The headquarters of the United Nations Economic Commission for Africa was established here in 1958.

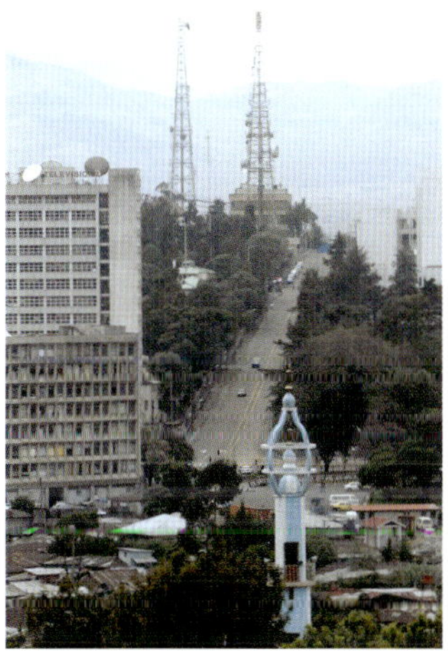

DOWNTOWN ADDIS ABABA

We are returning to fly from Addis to the medieval walled city of Harar, which stands among green mountains on the east wall of the great Rift Valley. This area is not a popular tourist destination, largely due to its relative proximity to the Somali border. Animosity between Ethiopia and Somalia dates back centuries, with wars and conflicts over borders and territories. The last major cross-border warfare occurred between 1998 and 2000 during the chaotic warlord-led era, but as recently as 2006 Ethiopian troops moved into Somalian territory and tensions continue.

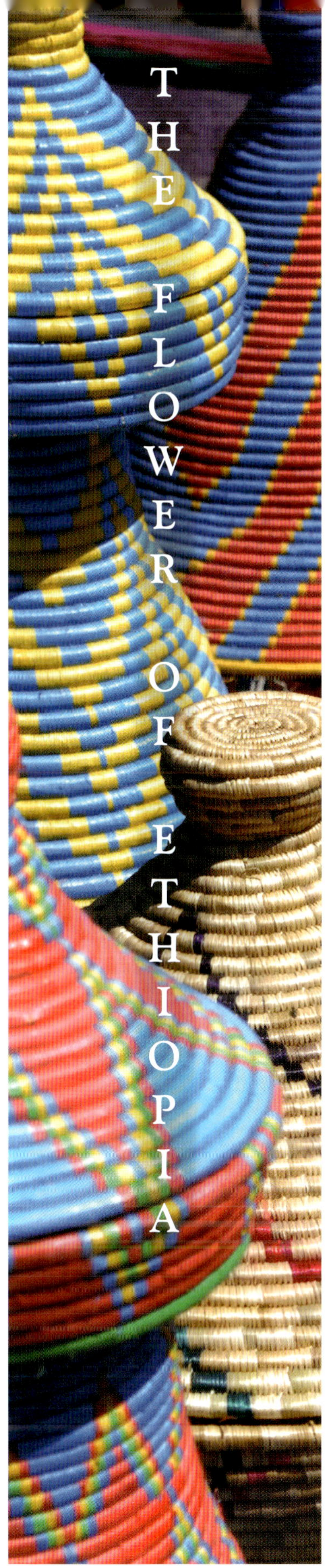

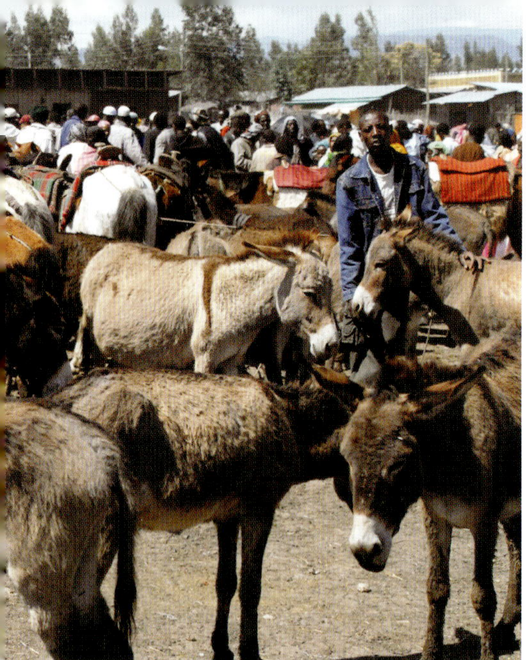

Donkey parking area

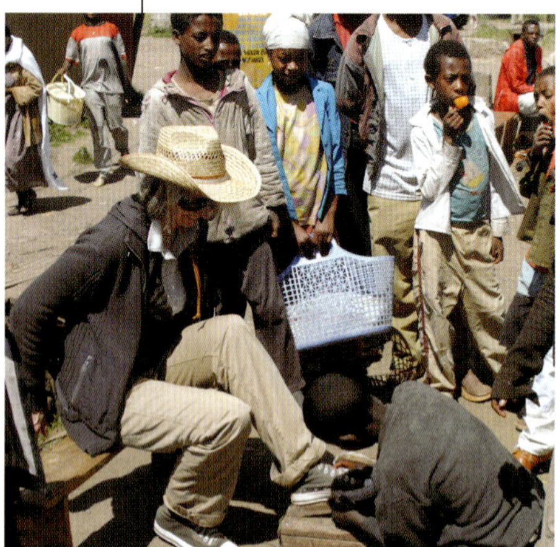

Pollyanna gets her shoes cleaned

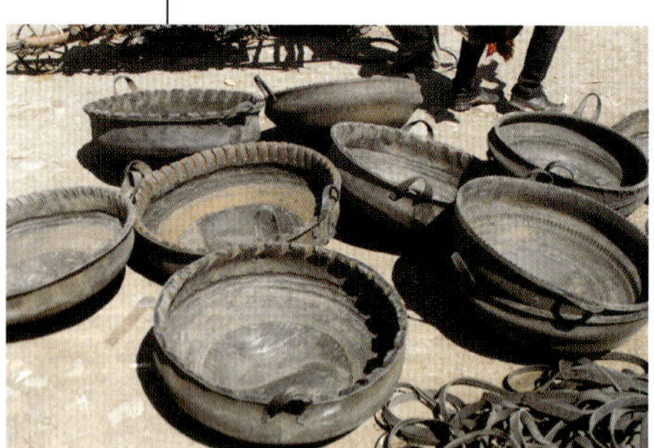

Bowls made from discarded car tyres

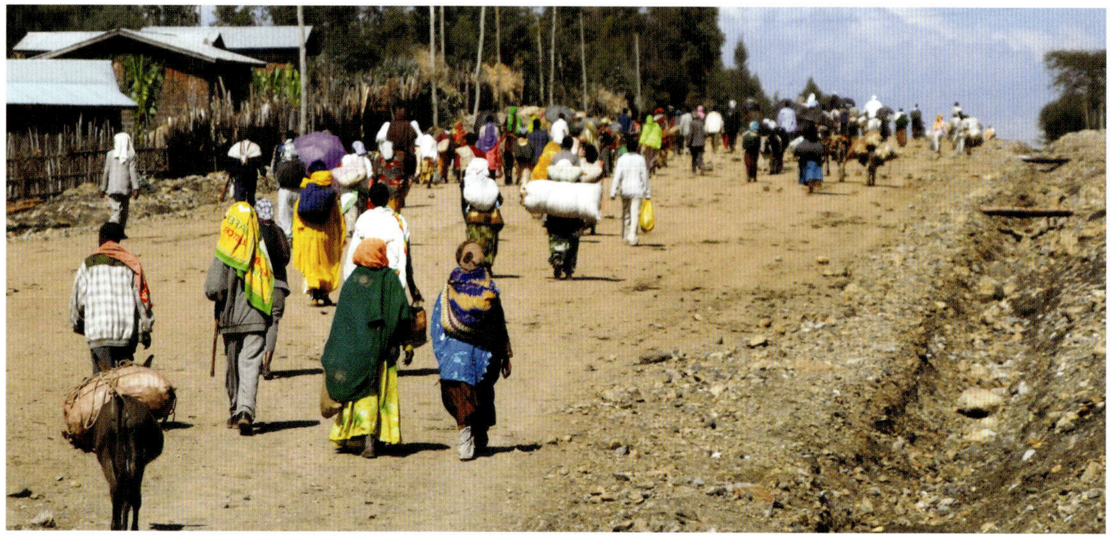

A large mass of people walk towards the local market

Day Twenty-two

By far our longest day's drive, so we are up bright and early for breakfast – Sami tempts us with porridge and omelette made with the Ethiopian equivalent of Dairylea triangles! We stroll down to the gate at the end of the track which leads to the EWCP lodge while our crew are busy loading the last of the cooking equipment and leave shortly after 7.00 am, as this time we are driving back to Addis without breaking the journey.

After driving for about two hours we start to see hoards of people walking along the road towards a local market. When we eventually reach it we are keen to stop to have a look round. Tariku wants to press on and get some mileage behind us, but we want to shop. The immense market is loosely divided into areas for different commodities – sugar cane, shoeshine, pottery, eggs, veggies, fruit, spices, meat (this area has its own very distinct smell) spices, *injera* pots, baskets, skins, etc. There is parking for your donkey – and tens of thousands of people. More people than I have ever seen gathered in one place before – the market makes a full crowd at Wembley stadium look like a gathering in a rural village hall.

Among all of these people we are the only ones with white skin, so it is fair to say that we attract a little attention. We quickly gather a retinue of children (and a few adults) who trail behind us as we weave through the crowds to look at the amazing range of goods on display. Every time I stop to examine the lid of an *injera* pot made from baked dung, or to make a purchase of an 'Obama – Yes We Can' T-shirt I am aware of women timidly putting out a hand to gently stroke my hair. Pollyanna stops to have her Converse boots washed by a shoeshine boy, and leaves with them slightly grimier than before. We could have wandered round the market all day, but as we have a long journey we force ourselves to drive on. And on... and on...

We have a very good photo stop when we spot a troupe of olive baboons – the first we have been able to get close to. Their coat is greenish-grey, giving rise to their name, and the males have a dark ruff of fur. We count about twenty-five baboons in the group we have found – but some troupes can number as many as one hundred individuals.

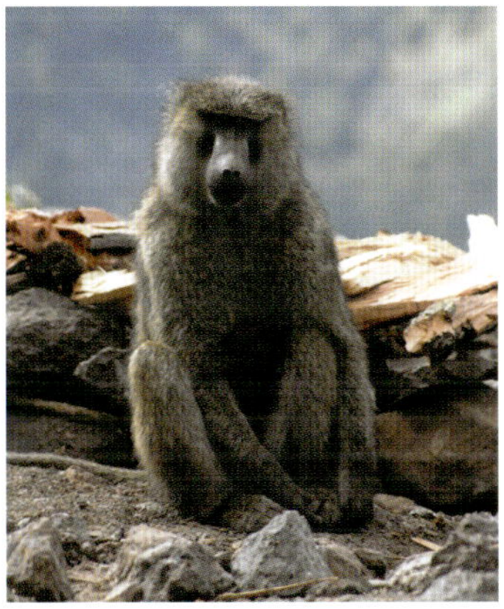
Olive baboon

Eventually we come to Shashamane, where we stopped for a welcome fruit juice on our drive to Goba. This time we visit the South Rift Valley Hotel for lunch, a stopover popular with backpackers. Over the years we have found that 'travellers' have a tendency to sneer at 'tourists' who go on holiday and flatly refuse to eat the local cuisine, preferring to stick to a diet of Full English Breakfasts and MacDonalds. All I can say is that if the travellers have been travelling through Ethiopia and been faced with mountains of *injera*, they might lose a little of their food snobbery – we are more than happy to have the opportunity to order toasted cheese sandwiches and chips. In fact we accidentally order an extra mountainous bowl of chips in our excitement – so

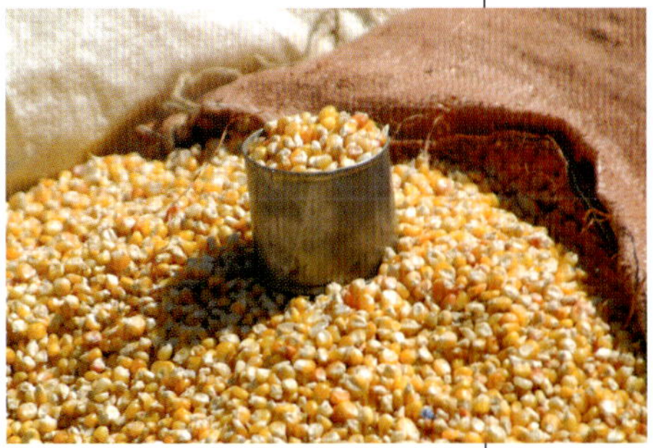

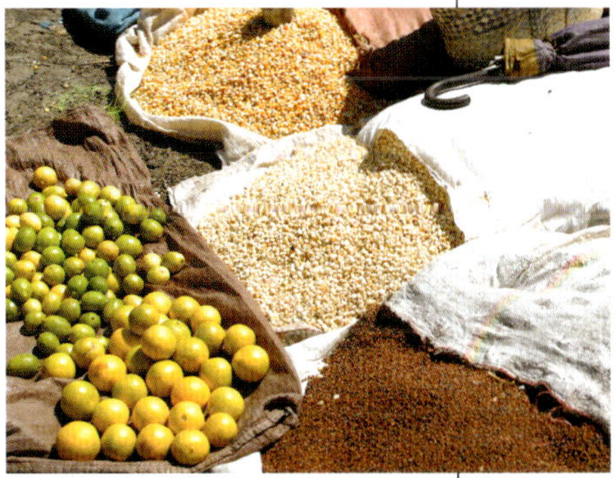
Colourful spices

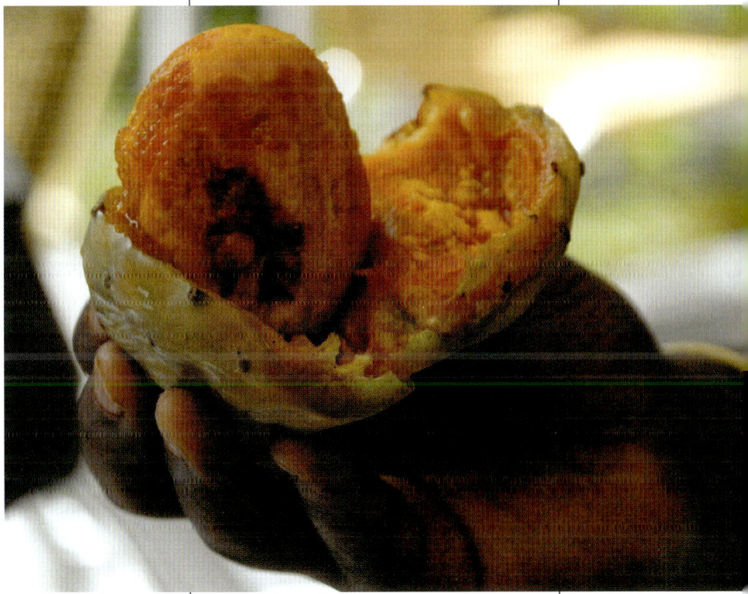
Cactus fruit – freshly picked at the roadside

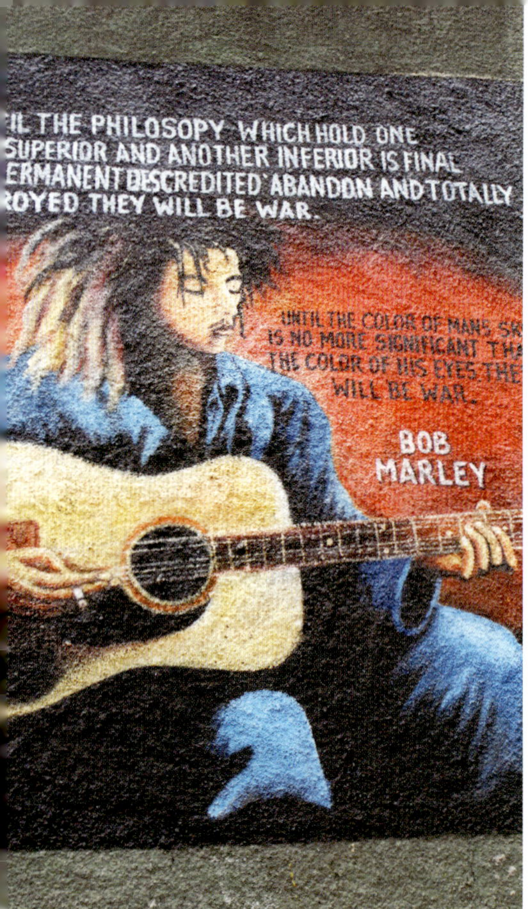

The hotel wall is decorated with murals of Bob Marley...

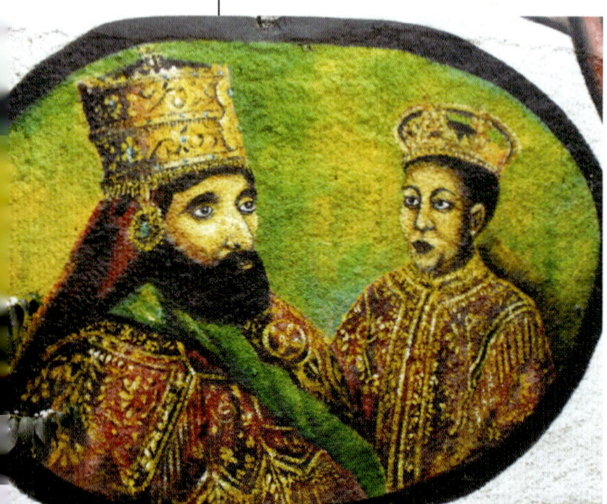

... and Haile Selassie

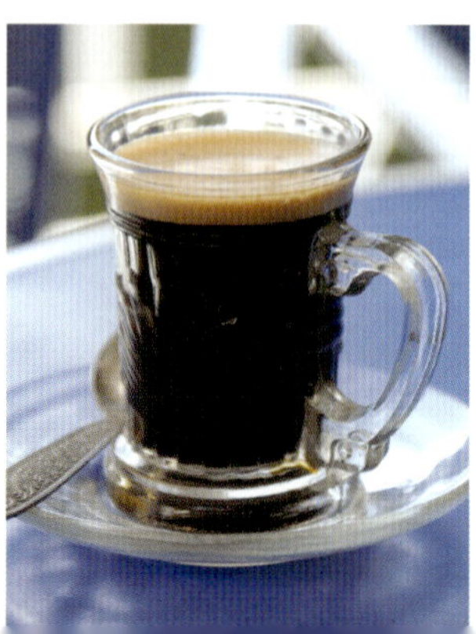

once the crew have finished their *injera* they help us polish them off. However, for pudding we enjoy some fresh cactus fruit which Tariku has picked at the roadside and split open to reveal the orange pulpy flesh under the tough green skin – so I do at least try one food I have never tasted before. The hotel is very cool – the exterior walls are decorated with huge murals of Bob Marley, Martin Luther King and Haile Selassie – and a recently added Barack Obama.

On and on and on – but after leaving Shashamane we quickly rejoin a tarmaced road. It actually feels strange to have stopped jolting, bouncing and vibrating. This is good in as much as the ride is smooth and much quicker – but also bad as we have discovered that the unpaved roads in Ethiopia are actually much safer, probably due to the comparative lack of traffic and because it is impossible to drive too fast on them. Abbi is an excellent driver and we feel very safe in his hands, but his skills do not seem to be shared by many Ethiopian drivers. Ethiopia has the second highest per capita rate of roadside fatalities in the world. Experts from the United Nations and World Health Organization estimate road crash deaths in Africa will surpass AIDS deaths within nine years – a statistic which sounds shocking until you have spent any time driving in Ethiopia, when it quickly starts to seem very plausible. We pass several overturned trucks as we near the capital. In Ethiopia it is not illegal to drink and drive and many of the drivers still think it is taboo to wear seat belts – although the custom is slowly gaining favour. Drivers actually used to be ashamed of wearing seat belts, thinking that they would be seen as a less capable driver by others.

All the roads in Ethiopia, whether dirt tracks or tarmac, are crowded with vehicles (donkey carts in the rural areas, trucks and cars near the towns), but also with animals and people. We notice that the animals have learnt to get out of the way when they are hooted at – although sometimes the cows lying in the middle of the road will look at us balefully before getting to their feet as though they are highly offended at having to move. The pedestrians, however, are not so keen to give way, and we see many cars physically nudging or bumping people off the road.

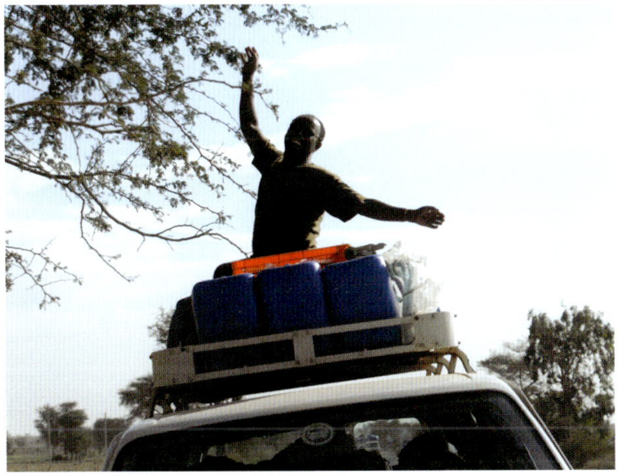

Sami on the roof of our car

We are therefore grateful to survive to reach our overnight accommodation. As we will now not be staying in Addis, we have booked ourselves into a hotel a little way out of the capital city – and after the joys of sleeping in our EWCP cell have treated ourselves to a night in Debra Zeit, in a hotel reputed to be the best in the country. We feel a little sorry for the staff as we arrive in reception and immediately lower the tone of the establishment – we are dusty, pouring sweat and haven't been able to shower properly in days. The receptionists sweetly ignore this and present us with welcome glasses of juice while we check in. However, to add to the sophisticated image we are presenting, Pollyanna fails to find her passport which is required at check in – and starts whirling around like a demented banshee opening bags and scattering clothes, water bottles, paints, brushes and sketches asunder. Still no passport, so we are escorted down to our room with instructions to return once we have it. Into room. Find passport immediately. Sheepishly return.

We are then free to explore our room. It is unbelievable – a quiet oasis in the chaos of Ethiopia. We are overlooking a lake and our room has a beautiful patio with seats and a fireplace. A pair of swallows is building a nest in the corner, busily flying back and forth with beaks full of sticky black mud. I spend much of the rest of the afternoon photographing them as they fly in and out – which means that when a member of the hotel staff returns with Pollyanna's passport he steps out on to the balcony to find me standing on a chair with my long lens firmly trained on what appears to be a small lump of mud on the wall. He seems unphased.

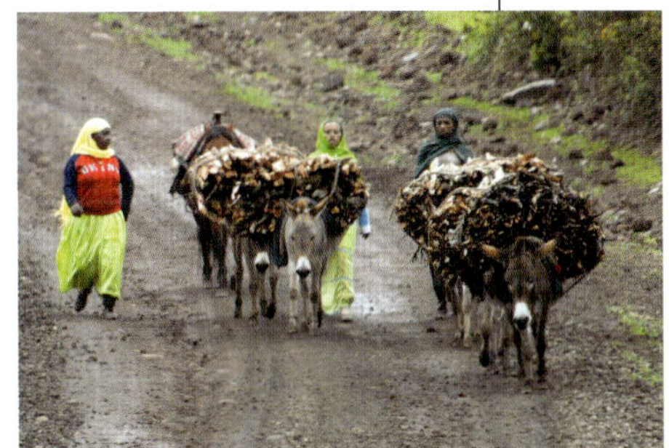

Rural roads are crowded with pedestrians and animals

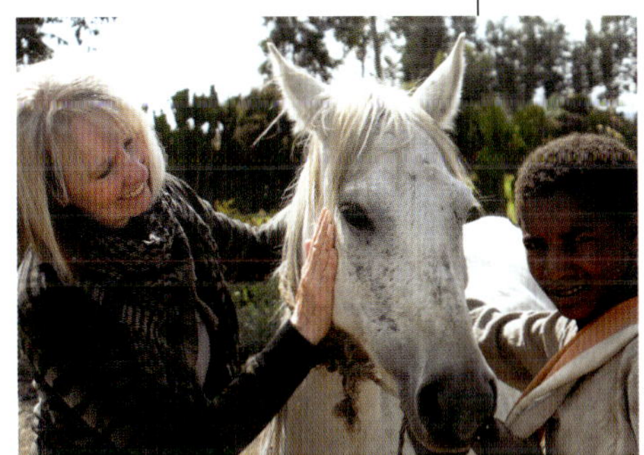

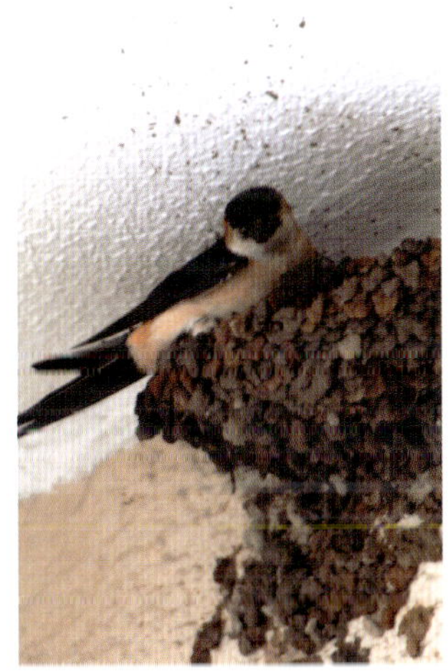

A swallow builds a nest of mud

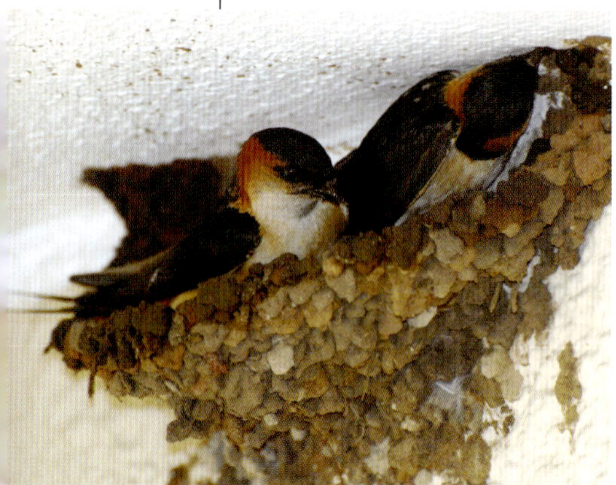

A pair of swallows is nesting above our patio – Pollyanna takes the opportunity to complete some sketches

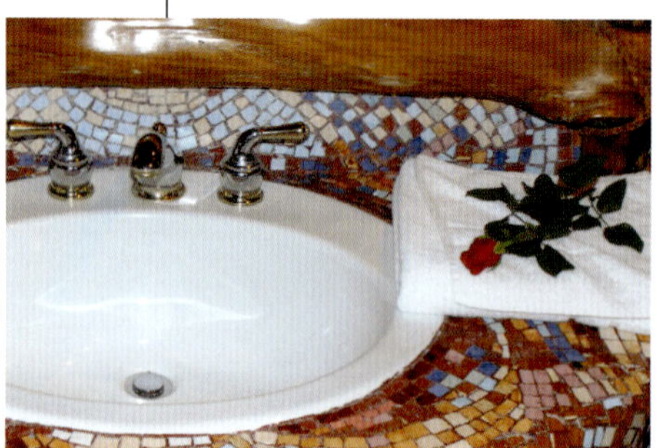

Our beautiful clean basin

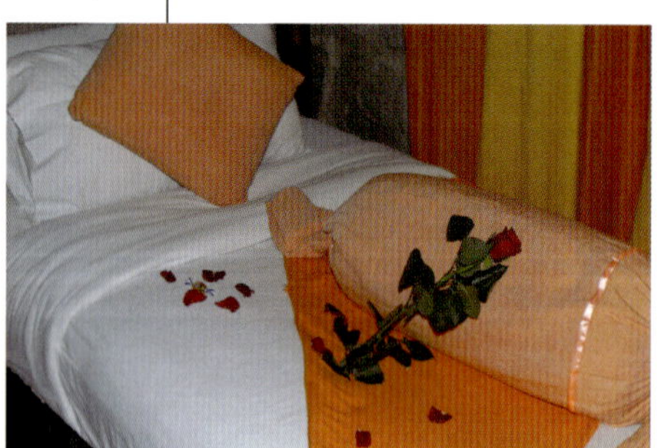

A beautiful clean bed

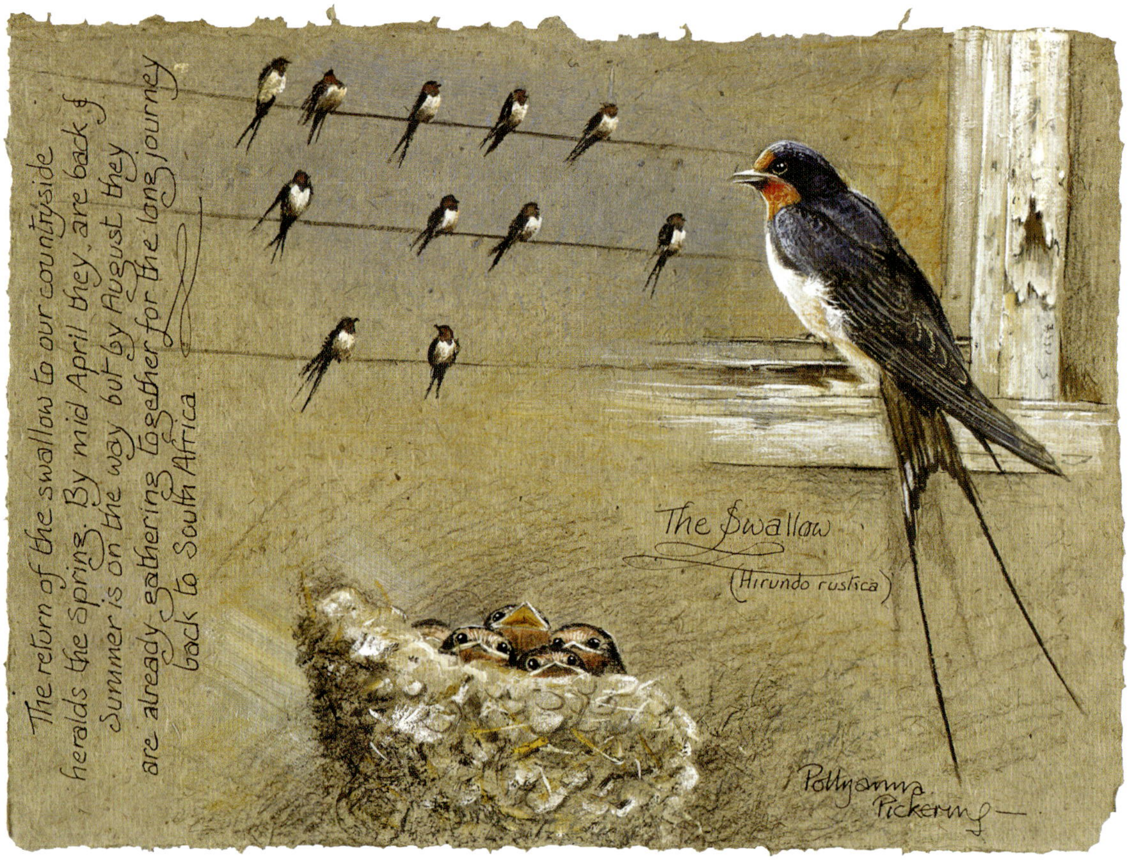

The beautiful bedroom has two beautiful beds with beautiful clean sheets. In the beautiful bathroom there is a huge 'rain style' shower, and the immaculate sink has fresh roses placed on it, instead of a soggy dead spider. When I see the gleaming loo I almost burst into tears. I think I am slightly overwrought and very dehydrated after the very long drive. To add to the effects of altitude, the fact that I have barely drunk anything during the day due to the lack of suitable places for bush stops, and the heat of the lowlands is causing me to sweat out my own body weight in fluid and I am feeling a little fey. I drink two cups of green tea and two bottles of water and a glass of rehydration salts in quick succession and feel much better, then I have the best shower of my entire life.

Day Twenty-three

No chance to luxuriate in our beautiful hotel beds, we have to be up at 4.45 am to meet Tariku and Abbi in the lobby at 5.30 am to transfer to the airport to catch the 7.00 am flight to Harar. We check in, head through security and have a welcome cup of green tea in the cafe. It takes us a very short time to stroll around the few shops – where they refuse to sell Pollyanna a necklace (because she is 'internal') and refuse to sell me a lovely mug (because it comes in a set with five others – which are awful). We learn the flight has been delayed by two hours, so we return to the café for more tea and a croissant.

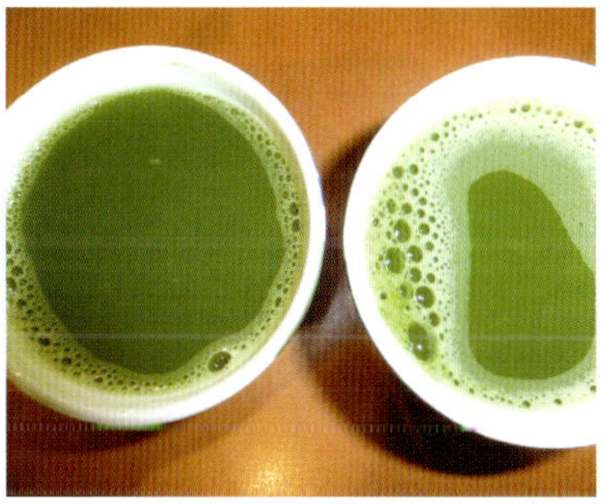

WELCOME CUPS OF GREEN TEA

Once the allotted two hours have passed we go to a departure gate and through another security check (shoes off, watches off, bags x-rayed, bags emptied, patted down) and then perch on uncomfortable plastic chairs waiting for the flight to be called, while attempting to eke out the reading matter in an ancient reader's digest. Time passes, and passes. A further two hours later they announce that the first leg of the three-leg flight has been cancelled, and a large group of the passengers get up and leave. Those of us who remain are presented with a plastic cup of tepid water and a slice of Ethiopian dry cake (the name says it all). A couple more hours pass and we are all herded away from the flight gate into the restaurant where we are to be given lunch – a pretty decent hot buffet of *injera,* spaghetti and veggies. More time passes. I read a three-week-old Ugandan newspaper. More time passes. The second leg of the flight is cancelled, more people leave. Finally, at 3.00 pm, an official appears and announces that the flight has now been cancelled, but we can join a 4.00 pm flight to the same destination. From here on chaos ensues. The 4.00 pm flight is called, so we head to the new departure gate with our boarding passes, only to be told we need new passes. Head to nearest official, who says we do not need new passes. Return to gate to be told yes the passes are OK, but we can't go through yet. Wait. Wait some more. A new official appears to announce that

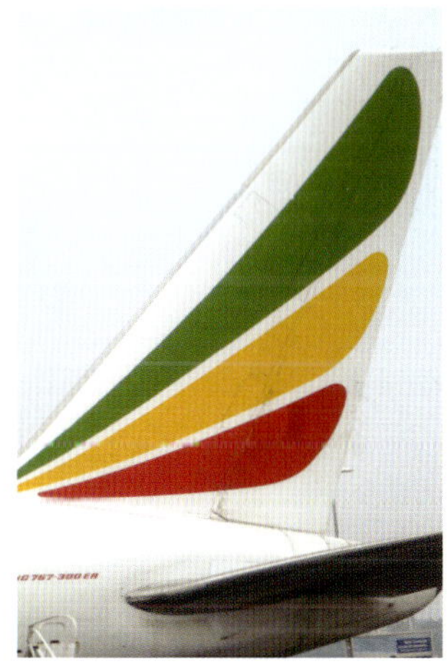

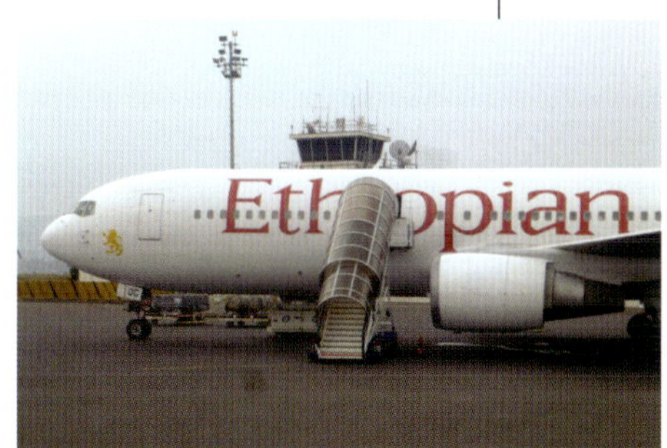

ETHIOPIAN AIRWAYS

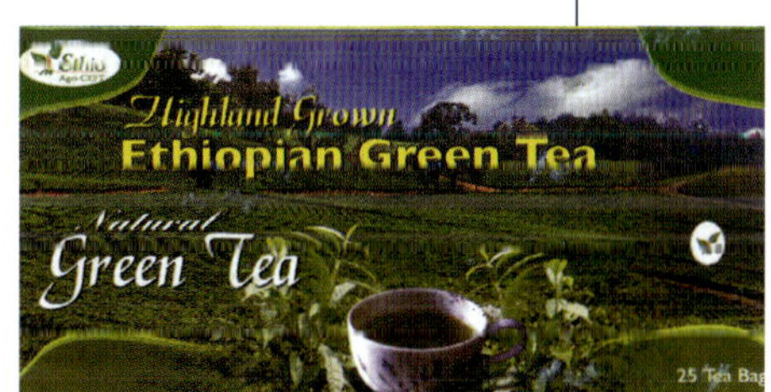

A BOX OF ETHIOPIAN TEA

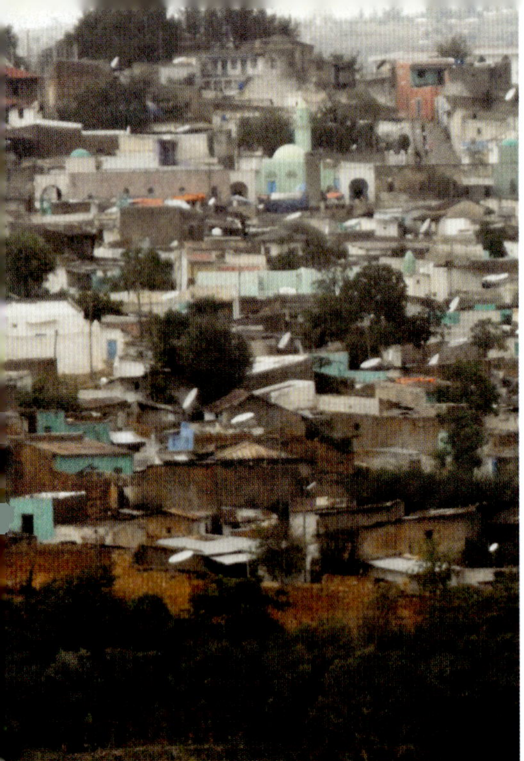

Greater Harar

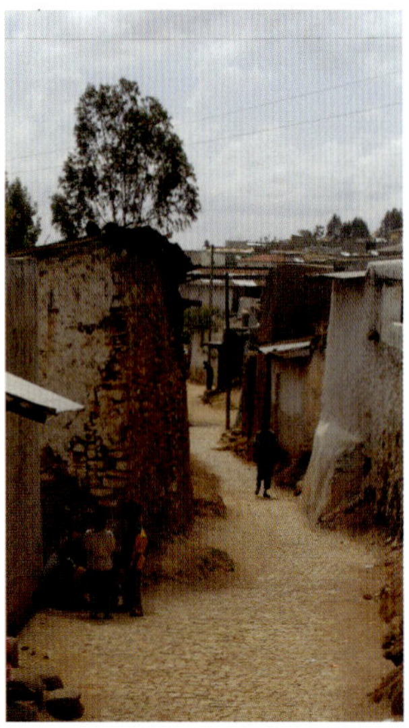

Alleyways wind through Harar

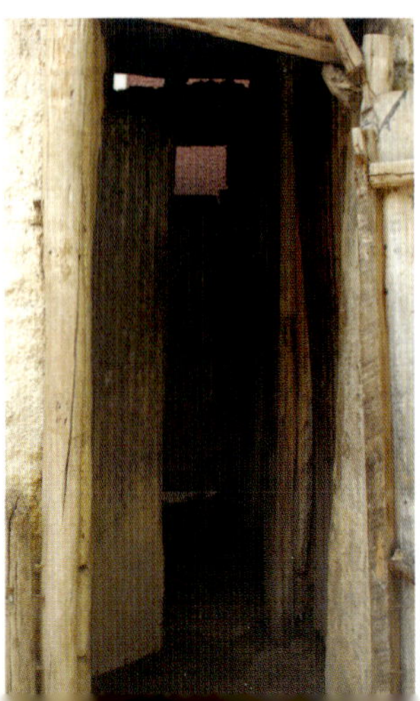

in fact this flight is already fully booked, so the passengers from our flight can't join it after all, BUT they have now fixed the mechanical fault which lead to our original flight being cancelled, so that plane will now be taking off after all... So, ten hours after we should have left, we finally go through security again (shoes off, watches off, bags x-rayed, bags emptied, patted down), perch on uncomfortable chairs, more water and dry cake, have slight altercation with guard who doesn't want to let me visit the 'Ladies' directly opposite the departure gate, and then finally on to the plane where we are presented with pineapple juice and yet more dry cake.

I am not usually a nervous flier, BUT I am tired after an early start and eleven hours in an Ethiopian airport terminal, and we have been put back on to the original plane which was grounded this morning due to a mechanical fault, so it is not too reassuring as we walk across the tarmac to see a bloke in overalls up a stepladder half-heartedly poking at one of the engines with a screwdriver and wiping it on an oily rag. The plane is an ancient Fokker 50. As we take our seats we discover the only other two English-speaking passengers on the flight are sitting directly behind us – an American man and his companion who are loudly discussing the fact that they are about to retire all of these aircraft from the fleet. As we taxi down the runway they go on to discuss why planes that crash-land on water always fall apart, and as we take off are cheerily describing footage of plane crashes they have been watching on Youtube. I grimly grip on to the headrest of the seat in front of me.

One way or another I am relieved to land in Dire Dawa. Tariku is leaving us here – but he has arranged for us to meet with a friend and colleague of his, Abdul. We say our farewells and thanks and head off with Abdul towards Harar. We pass through a local town, which is famous for selling khat – the drug of choice for Ethiopian men in this region. Even at this late hour it is thronged with people buying and selling the wilting leaves.

Harar is a walled city and a second town has been built on the outskirts. We drive through the new town on increasingly narrower roads, until we finally turn into a small courtyard and park. Our bags are unloaded – the roads within the walled city are too narrow to allow cars to drive through, so we walk down a series of narrow dark steep alleyways until we finally reach a grey metal door set into a high wall. Abdul has arranged for us to stay in a family home within the walled city. We meet the lady who will be our hostess, and she opens the door to her home and shows us in. Stepping out of the darkness of the alleyways we are confronted with an absolutely eye-watering blaze of colour. Every inch of the whitewashed walls is covered with a multi-coloured array of woven baskets, bowls, brightly coloured trays, pictures and photographs. Sunken alcoves are laden with glassware and china. Cooking pots are proudly arranged on high shelves with barely an inch between them. Every single bowl, tray, glass and ornament is immaculately clean and dust free. The floor is on three levels, covered in bright scarlet rugs and surrounded by bright embroidered satin cushions. After the dark street the effect is dazzling and almost overwhelming.

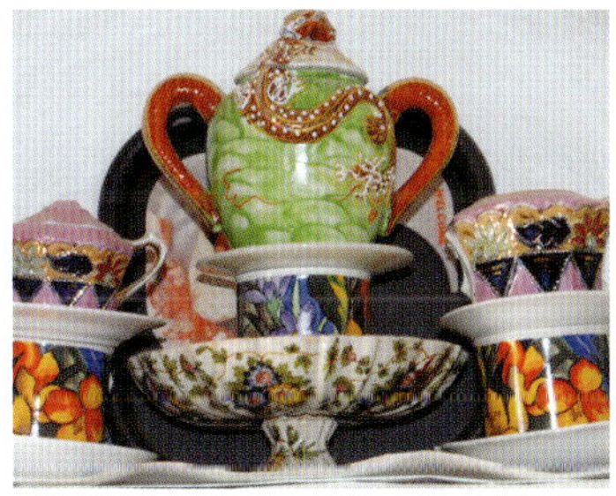

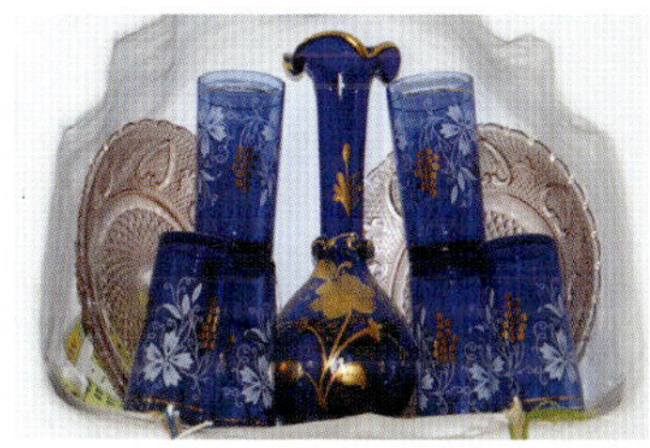

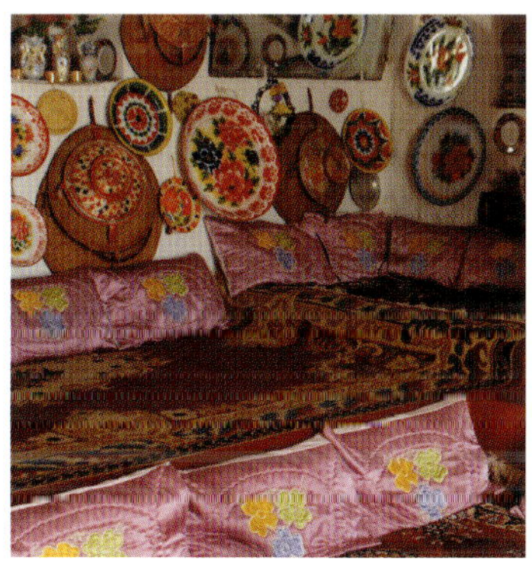

Every shelf is full of brightly coloured cups and ornaments and the walls are a riot of richly painted objects

Pollyanna has a bedroom at the back of the house, I have a small bed in a raised alcove in a broad corridor. Some time later I am stirred from my pre-sleep doze by something buzzing over my head more loudly than a Fokker 50, and landing somewhere close behind me with a resounding THWACK. While I am clattering around trying to find the lights, I rouse Pollyanna and between us we locate a fairly large scarab beetle which has alighted (much too delicate a word for the sound it made on landing) on the top of my furled mosquito net. I remove it with a glass (not a difficult thing to find in this house) and a piece of card. I get back into bed with the newly unravelled mosquito net firmly tucked underneath my mattress.

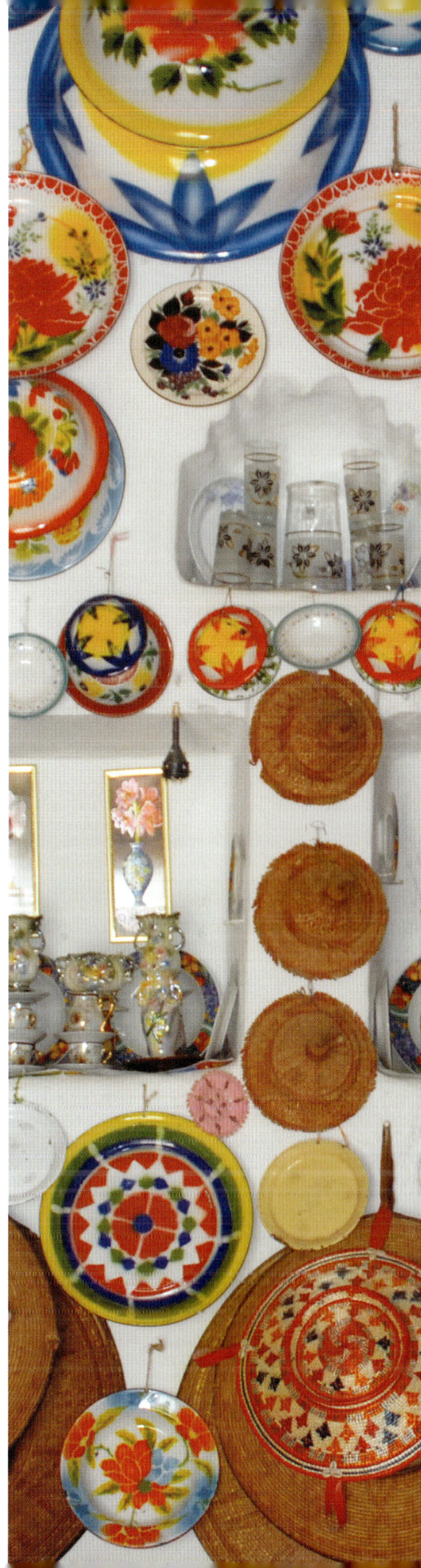

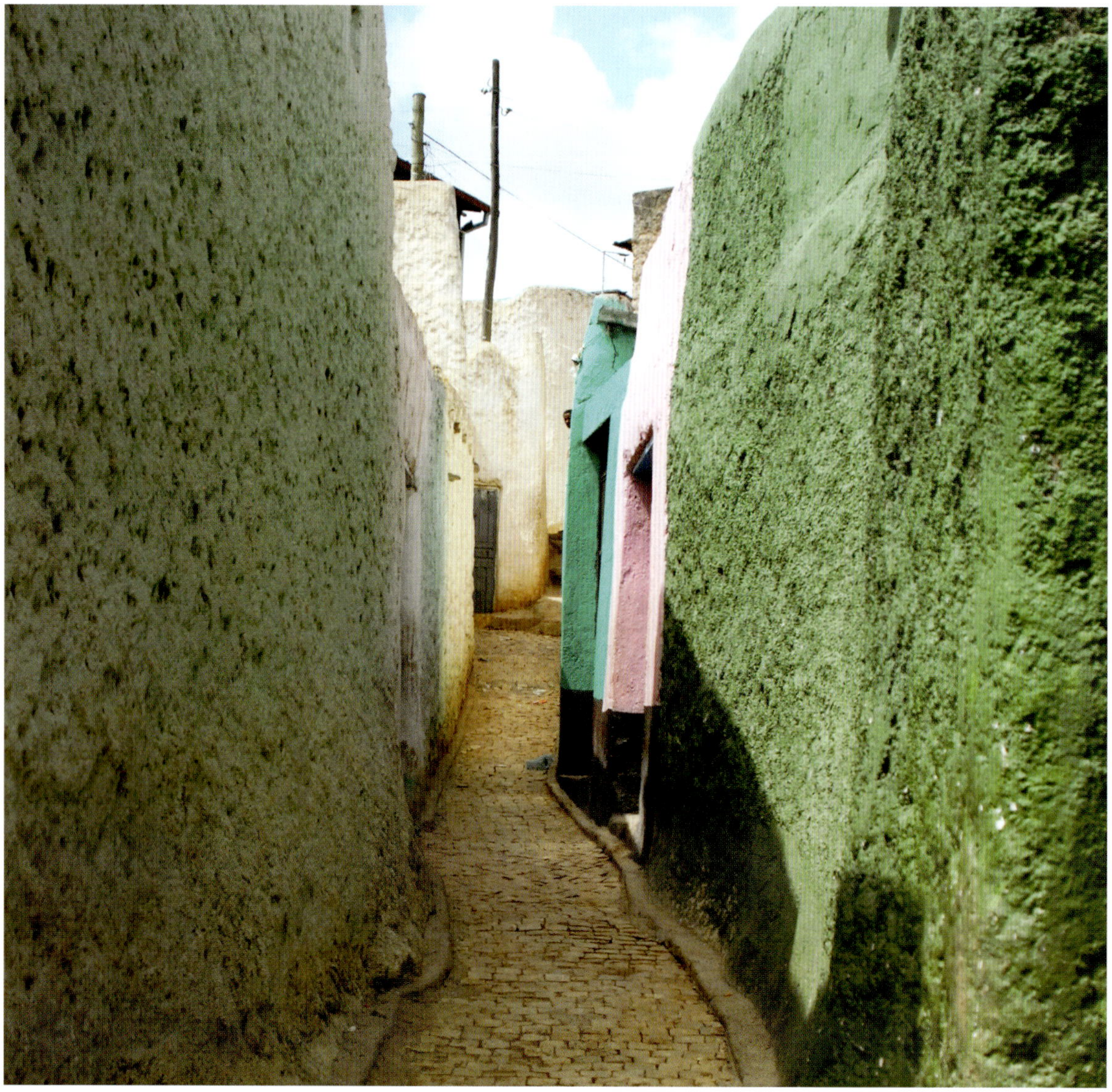

Chapter Seven

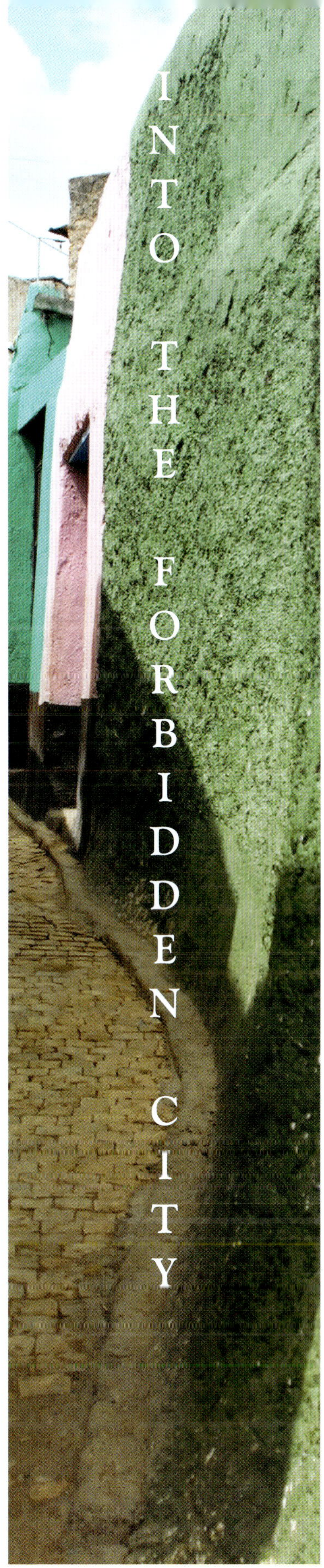

'True alchemy lies in this formula: Your memory and your senses are but the nourishment of your creative impulse.'

Arthur Rimbaud

Harar was completely closed to visitors until 1887. It still has its own unique language, A'dare, which is known only to the people of the city. Founded between the seventh and the eleventh century, in the sixteenth century Harar became the capital of an independent Muslim kingdom. Great walls with five gates were built to keep out powerful neighbouring hostile groups. This was the Golden Age of Harar. The local culture flourished and many poets lived and wrote there. It also became known for coffee, weaving, basketry and bookbinding. The city even issued its own currency.

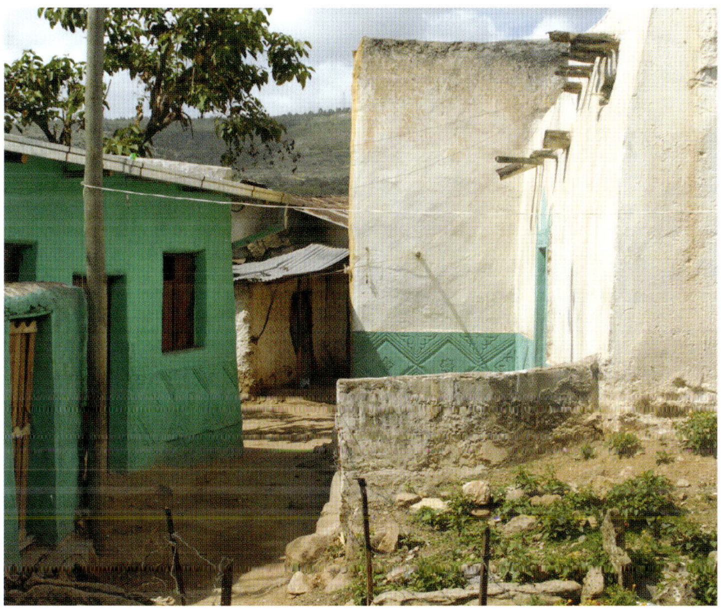

One of Harar's twisting alleyways

Harar remained independent until 1875, when it was conquered by Egypt. Ten years later it regained its independence, but this lasted only two years when it was incorporated into Ethiopia's growing empire.

Harar has for centuries been the main centre for Islamic learning and culture in Ethiopia, and a prosperous trade centre.

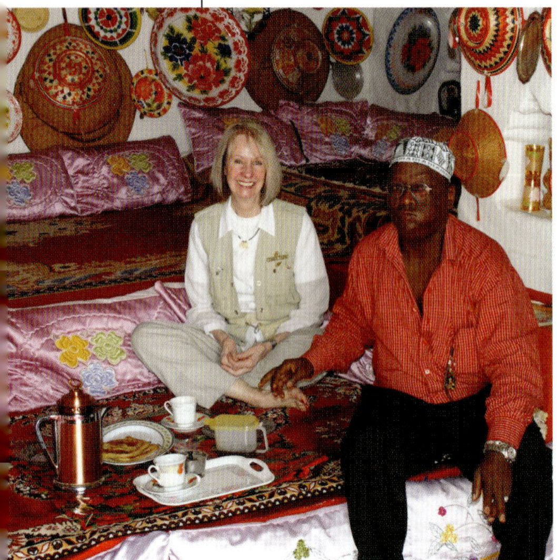

Abdul finishes our breakfast

Day Twenty-four

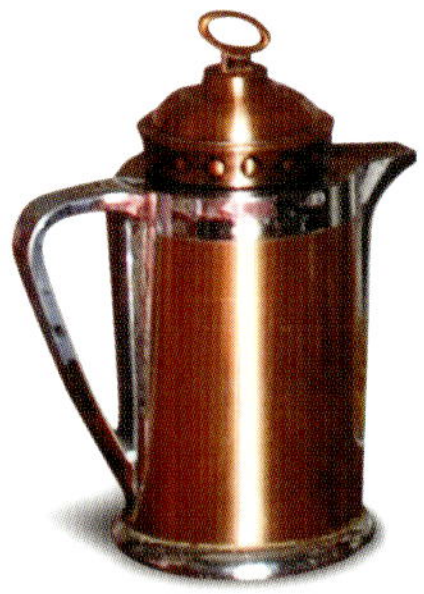

I rouse myself to find a rattled Pollyanna has barely slept – freaking herself out by overthinking the fact that she has been spirited down a series of dark alleyways by a man we have never met before into a small room behind a high wall in the walled area of a Muslim town a mere hour and a half's drive from the Somali border where conflict continues and anarchy reigns. She has realised we have not told anyone back home of our change of plans – and no-one knows we are here. I, meanwhile, have slept well in my corridor, woken only by the occasional thud of a scarab.

In the light of day she soon starts to feel much perkier – and by the time our hostess brings breakfast she is back to her usual self. Breakfast is traditional deep-fried Arabic flat bread with clear honey and strong coffee – and absolutely delicious. We sit cross-legged to eat among the brightly coloured hand-woven rugs and satin cushions. Abdul arrives to polish off the remains of our breakfast. As we missed our planned visit to the child care centre the previous day due to all the flight delays, he has contacted them and rearranged so that we can visit this afternoon instead. He suggests we might enjoy a walk around Harar this morning –and we leap at the opportunity. When Tariku introduced us to Abdul he described him as 'Someone who knows everyone – the unofficial mayor of Harar' – and this turns out to be very apt. As we head out through the narrow twisting alleyways practically everyone we meet greets Abdul smiling, coming up to shake his hand. We follow in his wake smiling and waving genially, shaking the occasional proffered hands, patting small children on the head and generally trying to look benign.

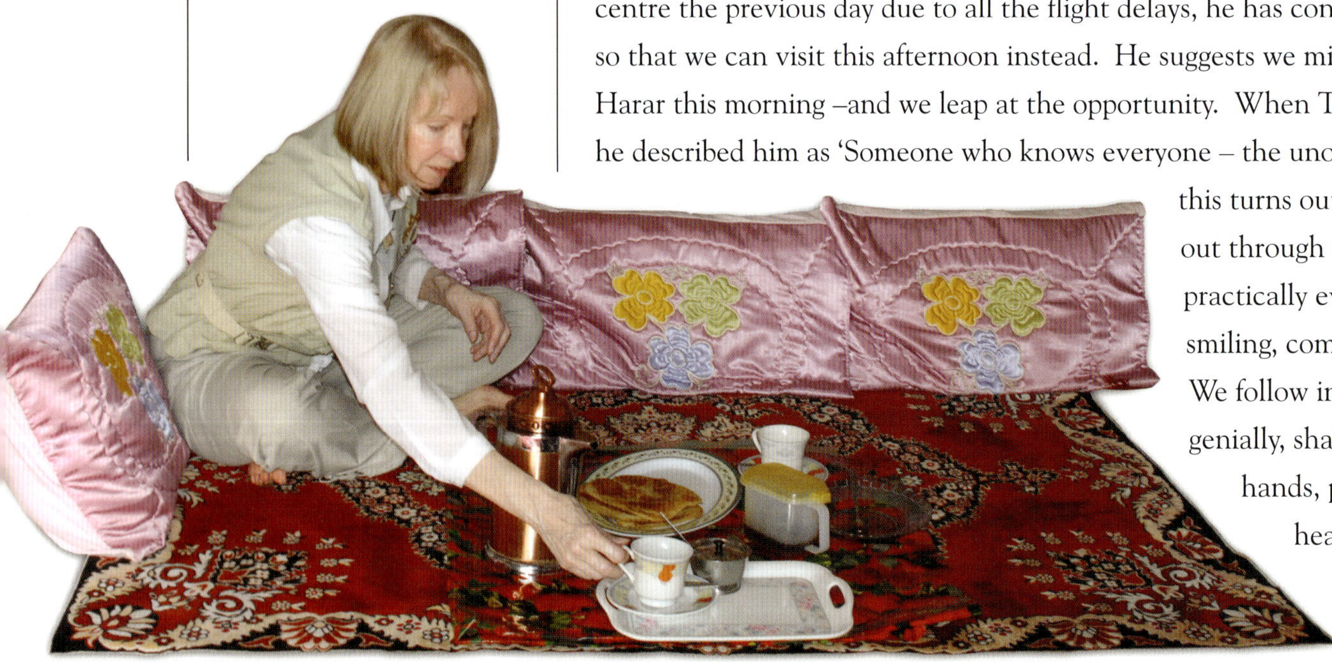

We sit cross-legged to eat

A town has stood here for more than a thousand years and for centuries Harar was the centre for trade in the region, with bustling markets at each of the five gates. Four of the markets still thrive and we follow Abdul through their maze of alleyways – huge arrays of fruit, vegetables, chickens, incense, livestock, coffee, baskets, spices, firewood, sugar cane, potatoes, frankincense, myrrh and people, people, people everywhere. Here in Harar they seem to greet us with a shout of *faranjo* rather than *farangi* – and as soon as the cry goes up more and more people arrive out of nowhere to inspect us.

We meander through the market and are tempted to try a taste of *berebere* – this red pepper spice is used in just about every dish here, providing the rich colour and traditional spiciness of the food. We have not tried it 'neat' however – and I spend much of the rest of

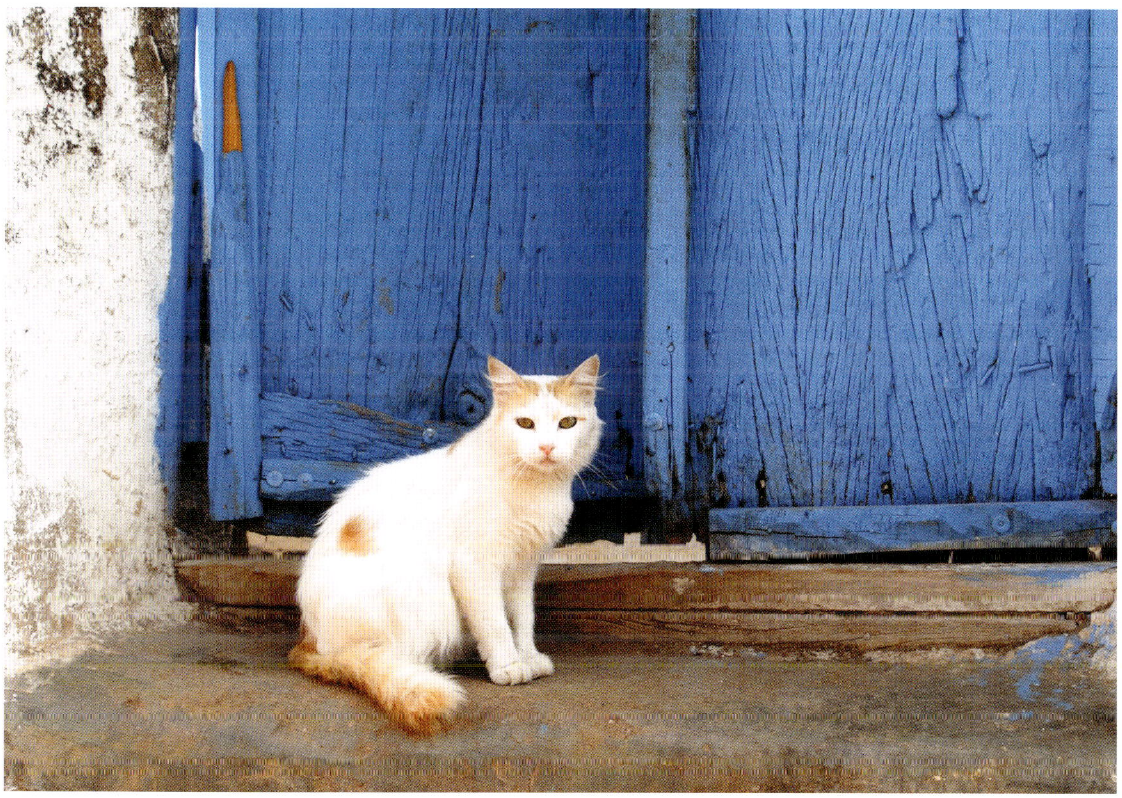

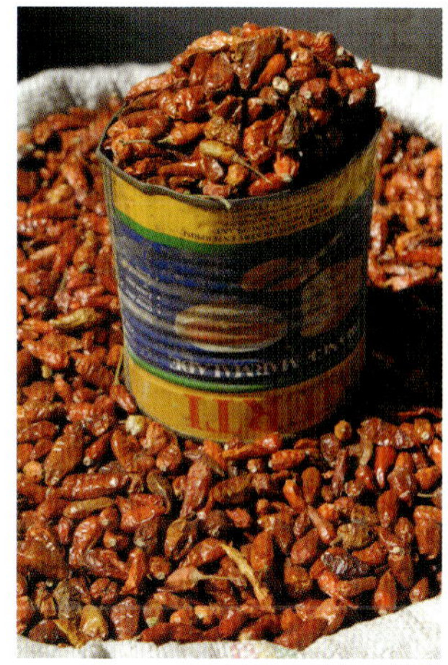

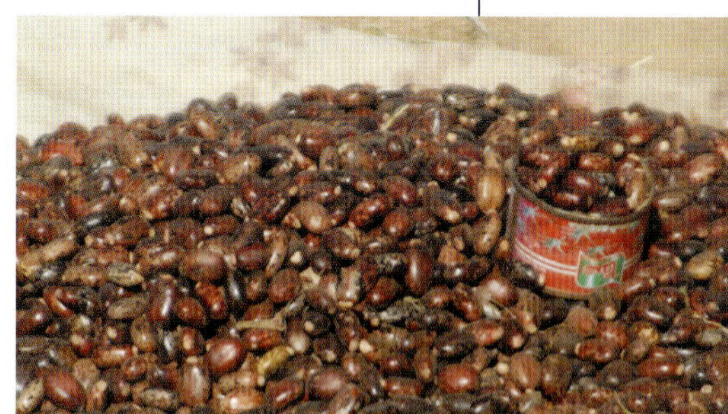

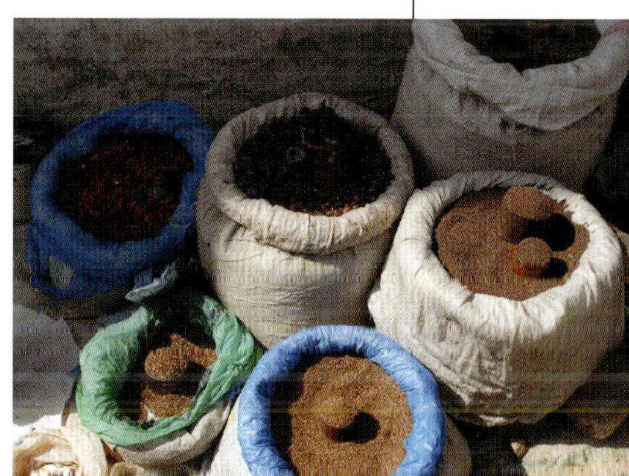

An array of spices and foods for sale in the market

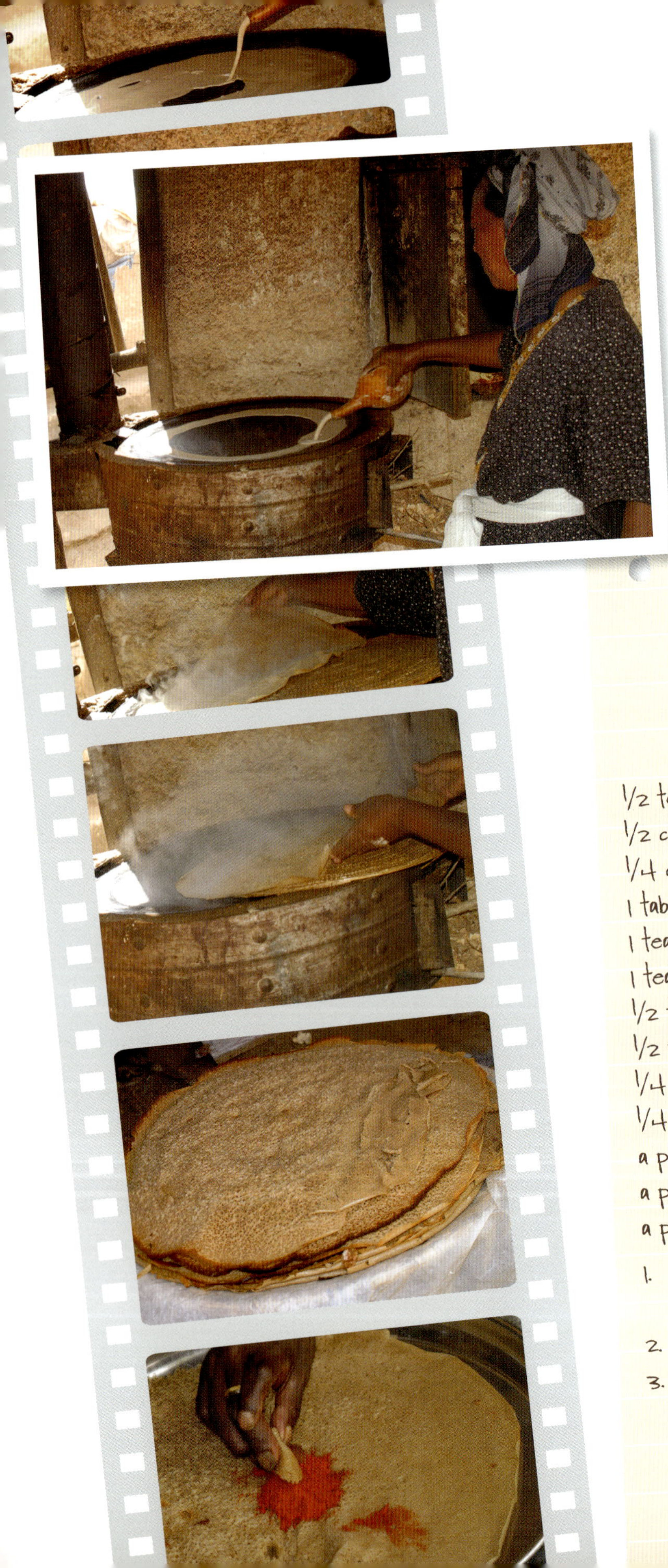

the morning feeling as though my mouth is on fire. A little further on we come across a lady cooking fresh *injera* in a huge black pan. This is the best *injera* we have tried yet – it is thin and pale golden, with a milder taste. Abdul tells us the cook has added sorghum into the mix to reduce the acidity. Perhaps I am finally getting a taste for the national dish.

Recipe Berebere

Berebere is a chilli—and—spice blend used to season many Ethiopian dishes

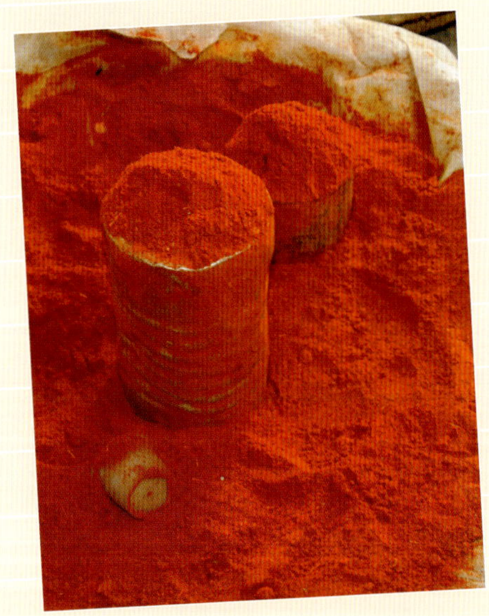

1/2 teaspoon fenugreek
1/2 cup ground dried chillis
1/4 cup paprika
1 tablespoon salt
1 teaspoon ground ginger
1 teaspoon onion powder
1/2 teaspoon ground cardamom
1/2 teaspoon ground coriander
1/4 teaspoon ground nutmeg
1/4 teaspoon garlic powder
a pinch of ground cloves
a pinch of ground cinnamon
a pinch of ground allspice

1. Finely grind fenugreek seeds in an electric coffee or spice grinder
2. Stir together with remaining ingredients until combined well
3. Berebere keeps in an airtight container, chilled, for 3 months

Through the 'recycling' market where you can buy any imaginable part of an engine, into the 'wholesale' market of clothes and fabrics, past 'tailor's row' with its constant background hum of sewing machines and into the 'smuggler's' market – goods brought in illegally from the Middle East and China – Abdul tells us that if the smugglers manage to get these items past the customs guards at the border they can legally sell them quite openly in the markets. Among the goods to be found in this quarter are belts with D&G and Armani buckles, Barack Obama and football logos – and bizarrely one buckle which combines these two national passions, showing an image of Obama playing soccer.

Abdul walks us past streets of 'mini-mosques'. Many Muslims consider Harar the fourth most sacred Islamic city and it is reputed that two hundred holy graves of saints and ninety-nine mosques can be found within its walls – an area of just one square mile. Peering inside the open doors of houses we see the modern 'pared down' version of the traditional decorations. We stop for coffee in the only café in town which boasts a gas ring to heat water – needless to say there is another electricity blackout, we are following the power cuts around the country. Throughout the time we have been in Ethiopia we have been without electricity for the majority of the time.

EMBROIDERED SILKS IN THE MARKET

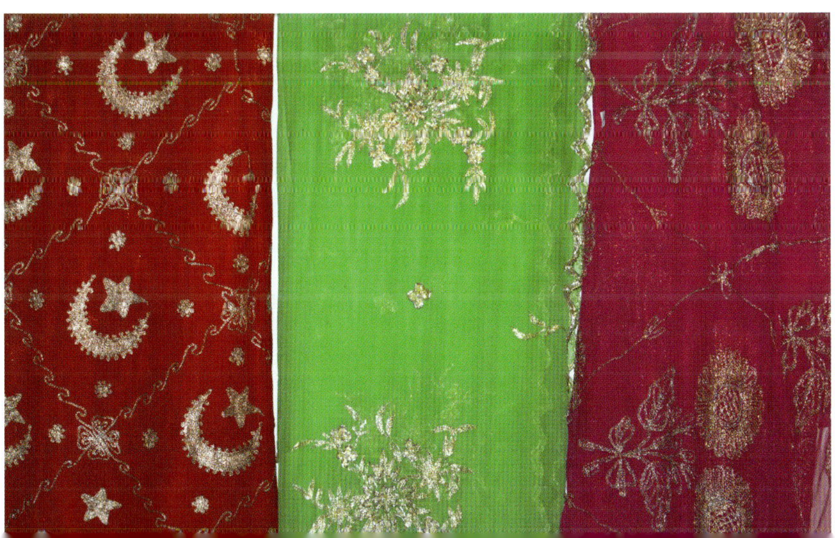

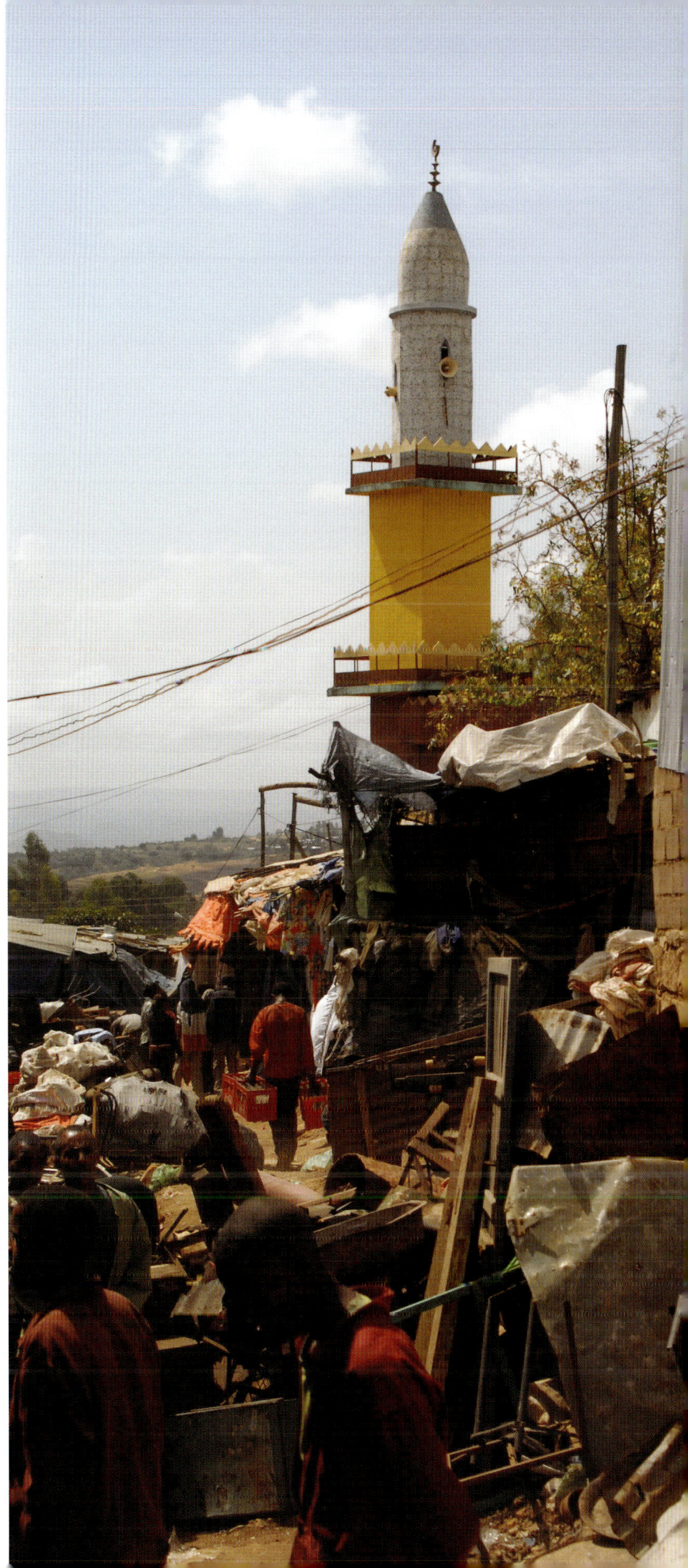

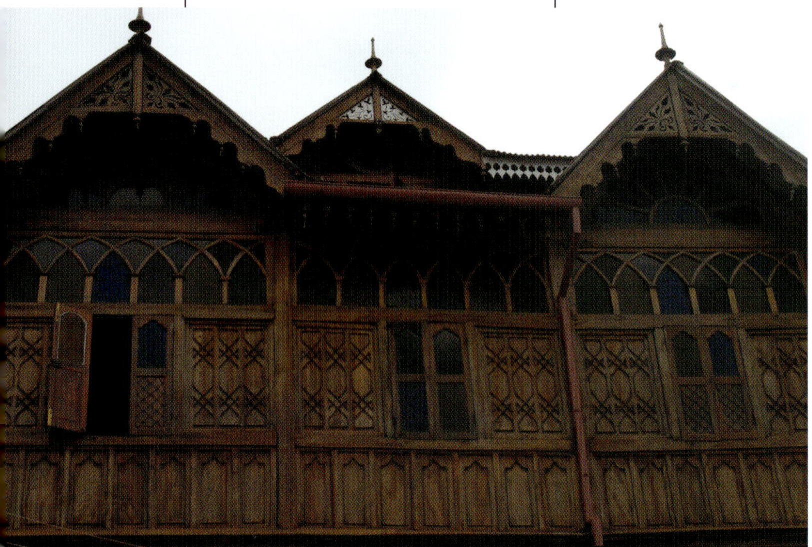

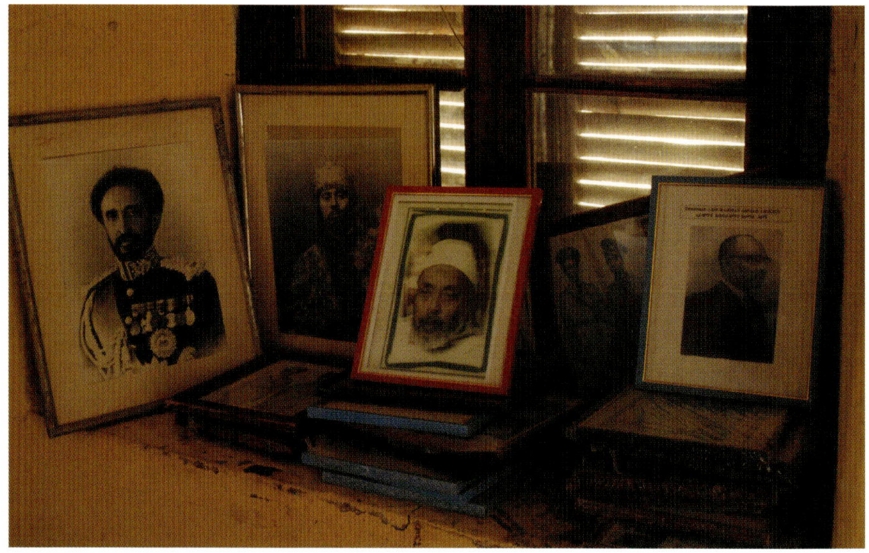

RIMBAUD'S HOUSE
ABOVE RIGHT: THE INTERIOR

QUOTATIONS LINE THE WALLS OF RIMBAUD'S HOUSE

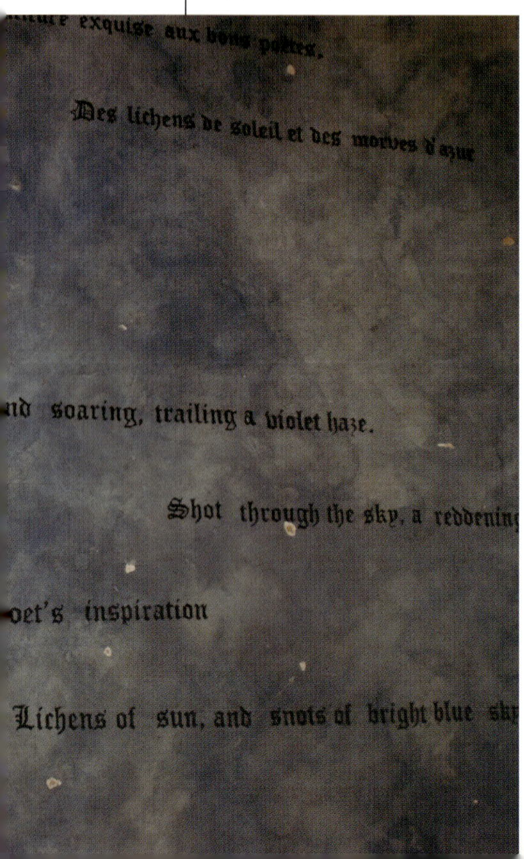

For the last ten years of his highly eventful life the French poet Arthur Rimbaud had a home in Harar which has now been beautifully converted into a museum dedicated to his memory. A child prodigy, by the age of fifteen he was composing original verses in Latin and went on to write hundreds of mystical and surreal poems, enthralling Paris with both his talent and debauched lifestyle. In 1874, at the age of twenty, he stopped and never wrote another poem again. Seized by an insatiable wander lust he began to travel throughout Europe, often on foot. He enlisted in the Dutch army in 1876 in order to gain free passage to Java, but deserted upon his arrival. Rimbaud lived for a time in Cyprus and Yemen before arriving in East Africa, where he became a coffee dealer, an ivory trader, a gun runner, a photographer and an explorer before settling in Harar.

Downstairs the building houses a small museum of books and letters; upstairs we find stunning collections of Rimbaud's photographs of Harar, his poetry and a collection of modern sketches inspired by his work.

Close by is a museum located in the former home of Ras Makonnen, father of Ras Tafari, who would become Emperor Haile Selassie. The displays here include a remarkable collection of centuries old hand-transcribed copies of the Koran, beautifully illustrated and scripted.

After lunch and coffee (accompanied as usual by smouldering frankincense and crunchy popcorn) we make our way to the child care centre. We are introduced to Lily Mersha, the director of the care centre and she tells us more about their work. The centre was set up to care for one hundred children aged between five and fifteen. Most have lost both parents, others have one or both parents who are totally unable to support them. In this region life expectancy does not exceed forty-two. It is estimated that a million Ethiopian children have lost one or both parents to the AIDS epidemic of the last two decades, while others have been killed in the fighting on the Somalian borders, or by the spread of disease caused by poor water and sanitation – which remains Ethiopia's largest health problem. Walking through Harar we have had to navigate our way around the open sewers which still run down the centre of many of the streets. The Red Cross estimate that seventy-six per cent of the general population do not have access to safe water and that rises to eighty-seven per cent in rural areas. Ethiopia remains one of the poorest sub-Saharan countries and the children

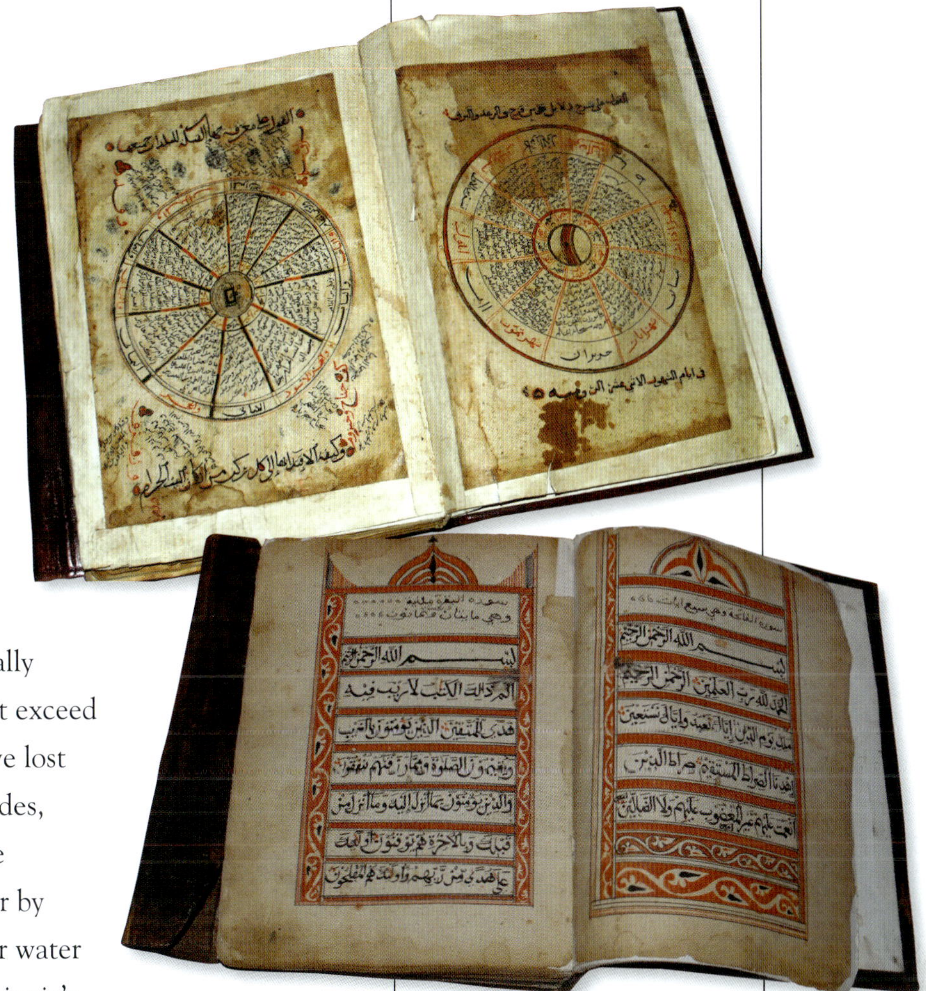

Ancient hand-transcribed books

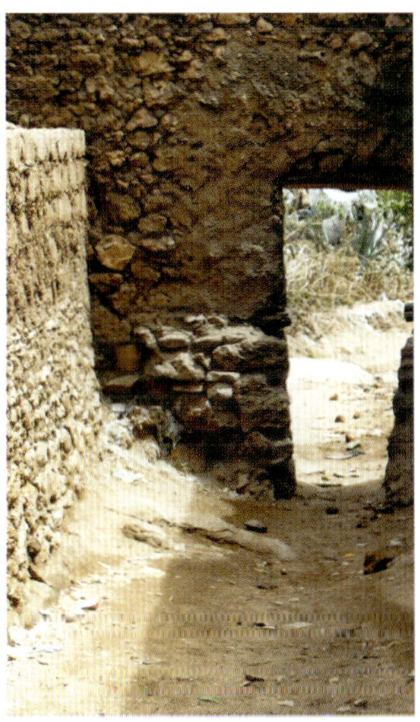

The streets of Harar

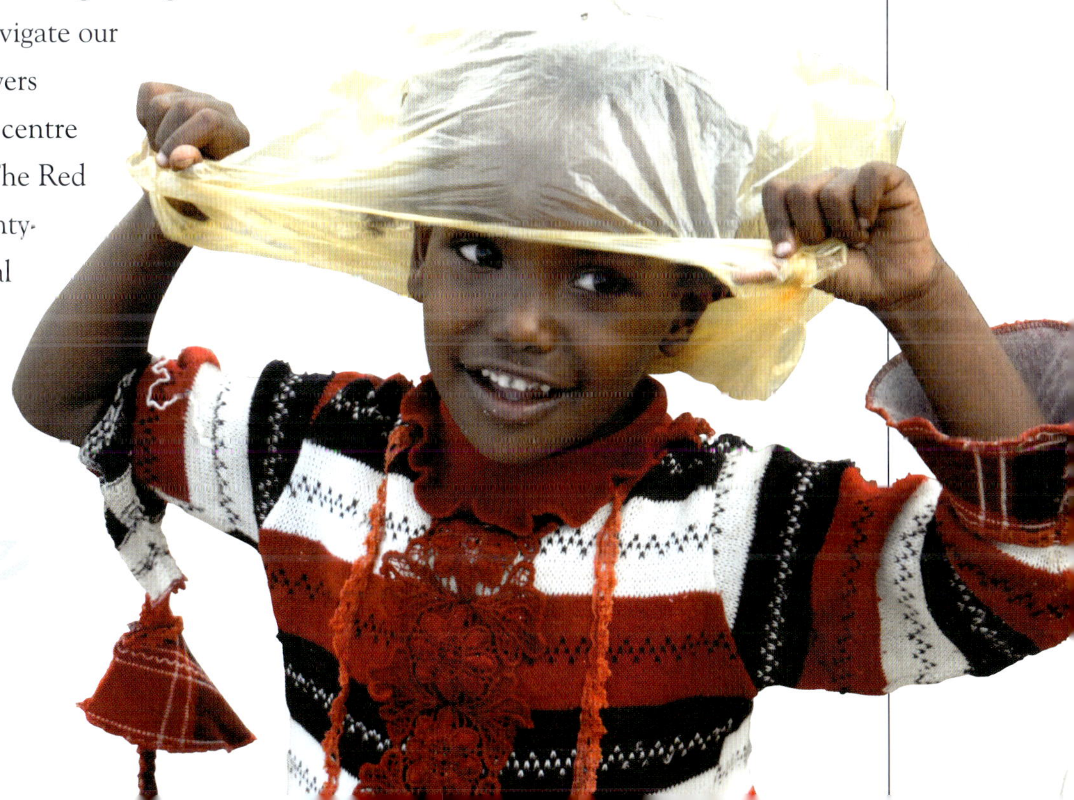

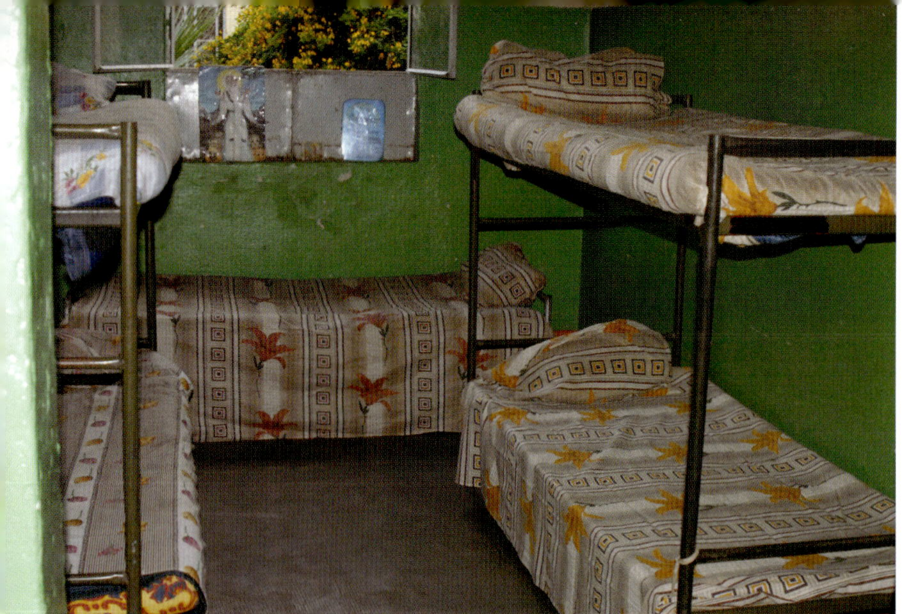

remain the most vulnerable members of society. Large numbers live in poverty, with no access to schooling and are victimized by harmful so-called 'traditional' practices including marriage before adulthood, female genital mutilation and exploitation.

This centre offers the children it cares for an escape from the serious physical threats to their wellbeing they would otherwise face. They live in a safe and secure environment, beside the local school. Many go on to higher education, giving them the opportunity to secure better living conditions for themselves and their families, and helping them to break out of a cycle of illiteracy and poverty.

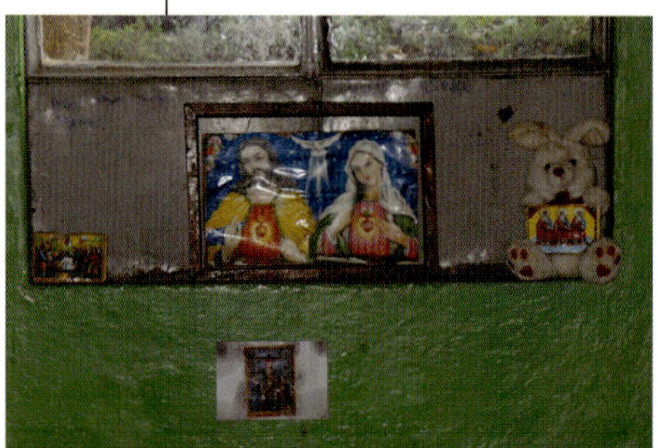

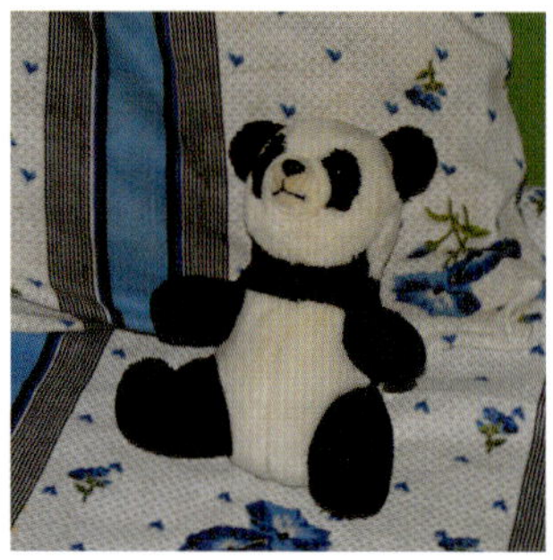

The children possess very few toys
Above top: a dormitory in Harar's child care centre

Lily takes us on a tour of the centre. We are very impressed. The rooms are basic but very clean and well cared for – simple little dormitories, each with five beds, larger communal rooms and, outside, a playing field. The washing facilities consist of two large stone sinks outside – all the laundry is washed here as well. With only three adult members of staff the children have to learn to be very self sufficient – they wash, cook and clean for themselves. Although this seems a harsh routine it is also very practical – they learn valuable life skills at a very early age. Although we are impressed by the facilities, we are also struck by how extremely basic they are. The children have the bare minimum and nothing else. On our tour we hardly see any toys – certainly not enough to allow for one per child. The centre received a grant to build a library, but the shelves remain bare – they can't

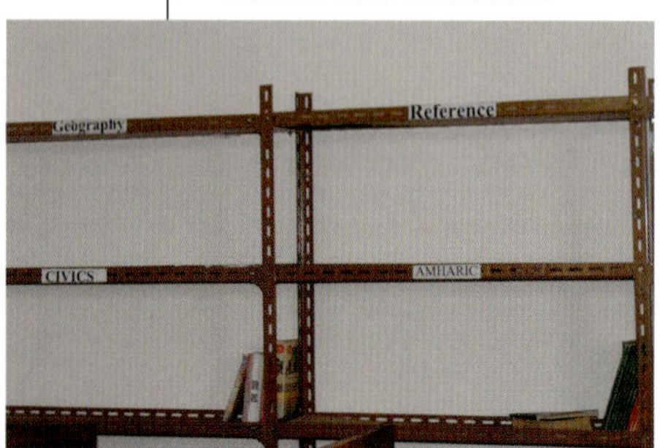

The library shelves are empty
Below: the washing facilities

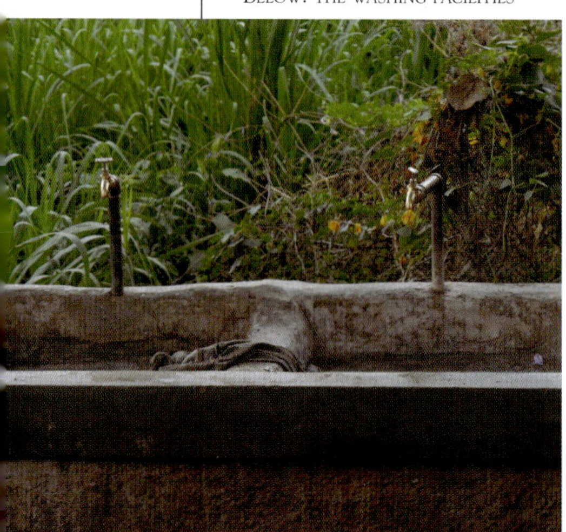

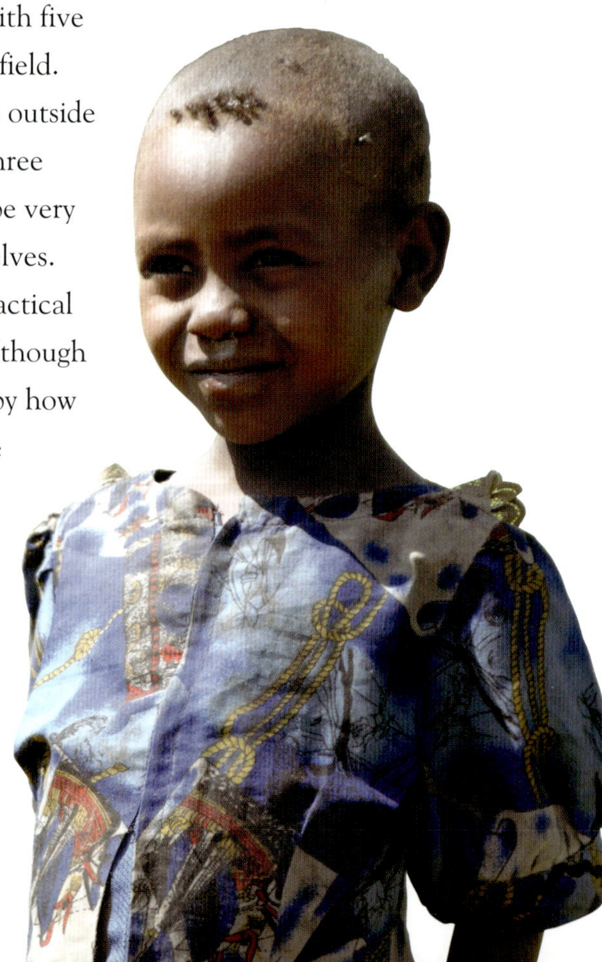

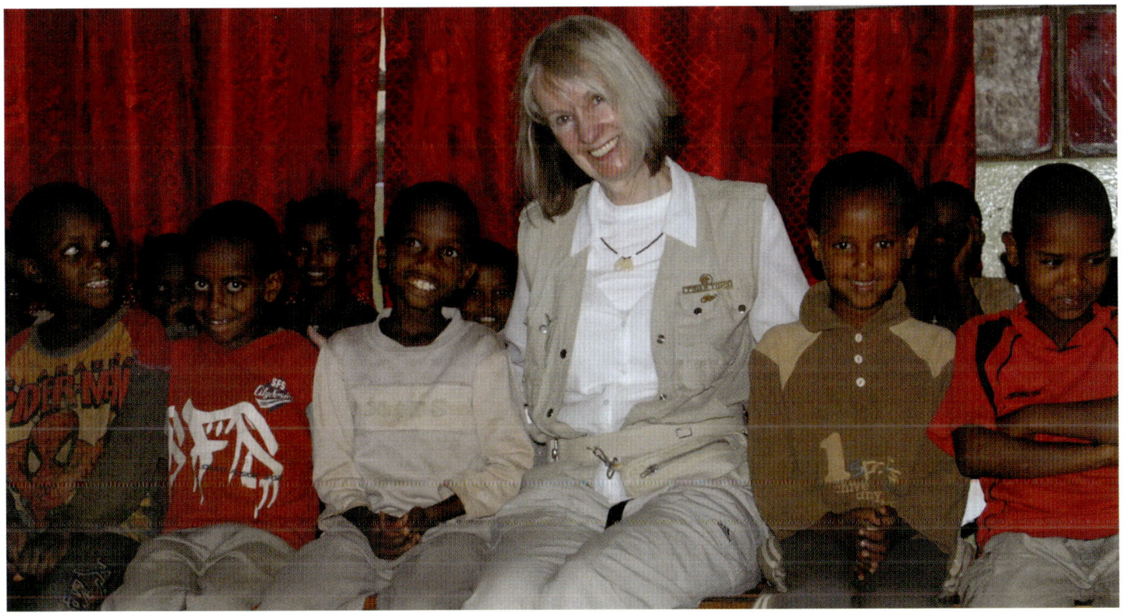

Pollyanna met many smiling faces

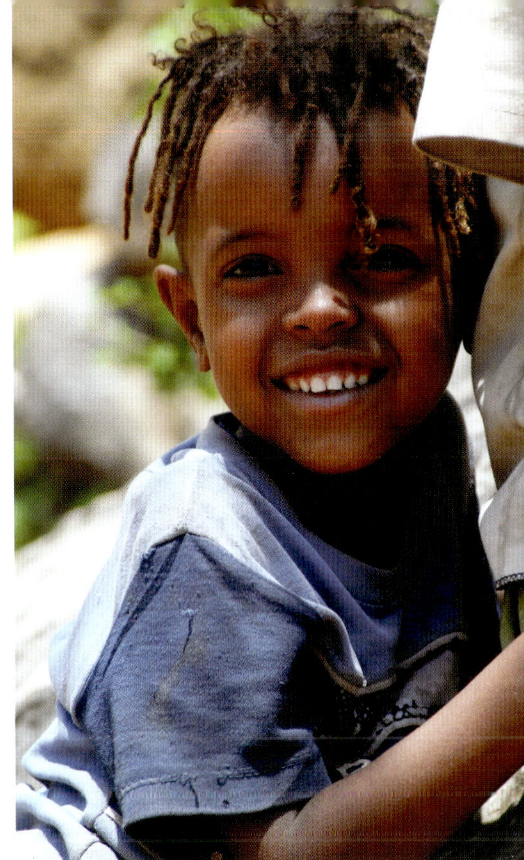

Children are healthy and well cared for at the centre

afford books. And Lily tells us that many of the children arrive with only the clothes they are standing up in; when these need washing they don't even have a spare set to change into. Aside from the money raised by the charity Tariku works with, Citizens for Humanity, which has a large supporter base in Greece, the centre receives almost no outside help. Tucked in a remote corner, hardly anyone in the outside world is aware of its existence. We promise to help if we can, at the very least by sending regular packages of clothes and toys from the UK.

Finally Lily takes us into the dining-hall to meet the children – all one hundred and thirty of them! They are singing – but as soon as we enter leap to their feet to meet their guests. Once the song has ended Lily introduces us to some of the children and we chat to them before requesting a couple more songs. We wave goodbye knowing that our afternoon at the care centre will be one of the experiences which will remain in our thoughts and hearts long after we have flown home.

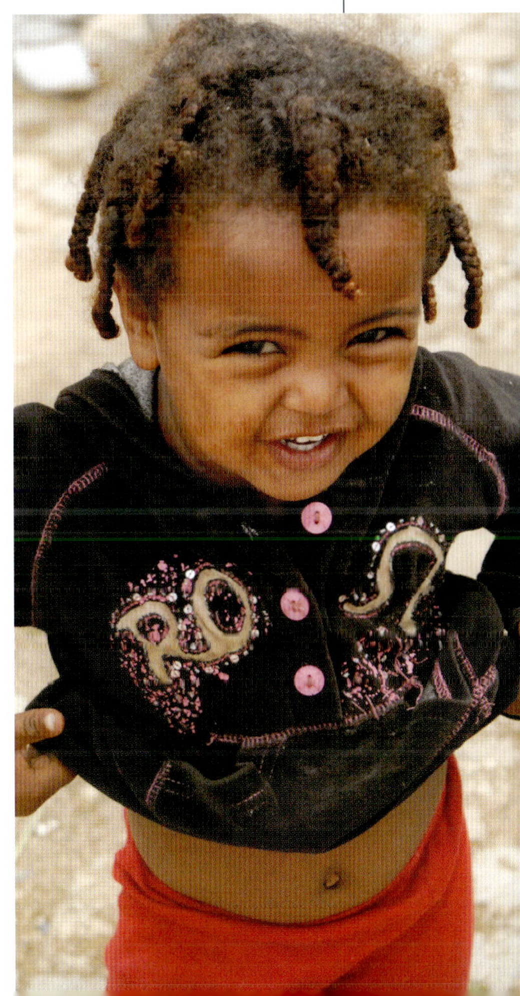

These happy resilient children have made a lasting impression on us

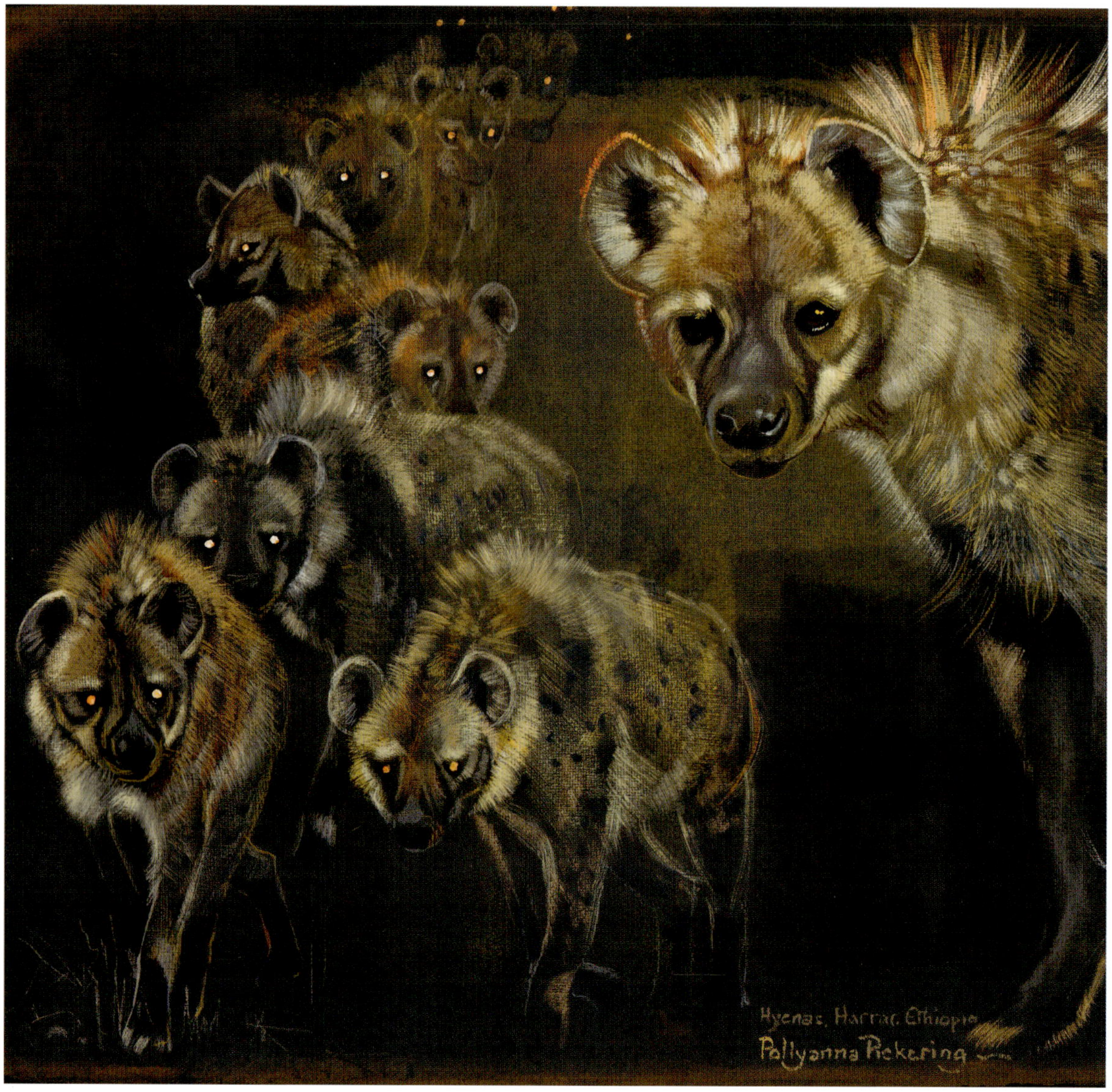
Hyenas, Harrar, Ethiopia
Pollyanna Pickering

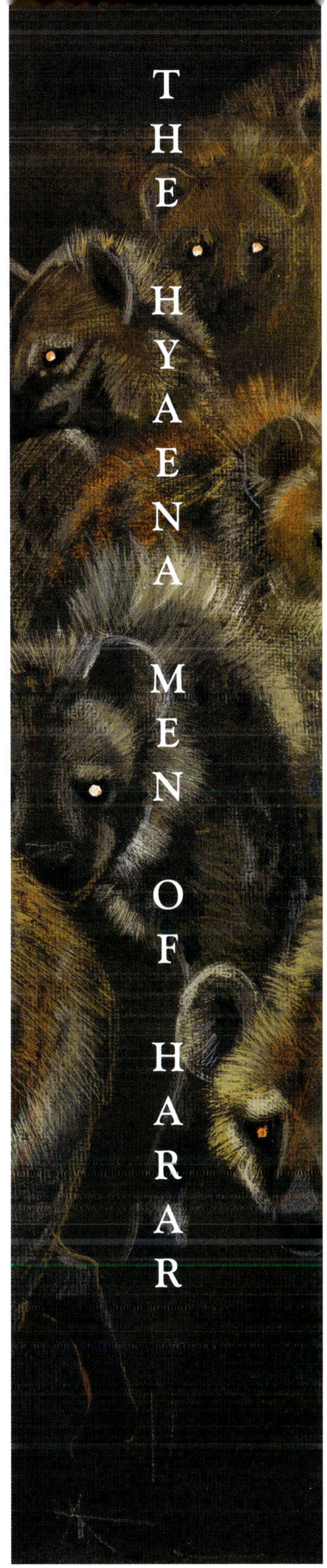

Chapter Eight

'I turned silences and nights into words.
What was unutterable, I wrote down.
I made the whirling world stand still.'

ARTHUR RIMBAUD

Like Transylvania, Ethiopia has a rich history of myths and folklore, and it is in the hyaenas the two mythologies collide. In Europe we have vampires and werewolves. In Ethiopia the two combine into a shape-shifting blood-drinking werehyaena. Most countries in the world have a 'were' myth, humans who take the form of an animal, and this is usually the largest fiercest predator known to the region, hence wolves in Romania, jaguars in Argentina and bears in Canada. In Ethiopia the 'Bouda' is a blacksmith by trade who can transform himself into a hyaena; he is still recognisable as he wears a piece of jewellery or a talisman which he made when in human form. If he should lose this amulet the Bouda is unable to regain human form, and eventually becomes a true hyaena, losing all of the human cunning he possessed. While in the form of a hyaena the Bouda is said to drink the blood of his human victims. Though fierce a werehyaena does not possess supernatural strength and is not immortal, so he can be killed by any ordinary means either in human or animal shape.

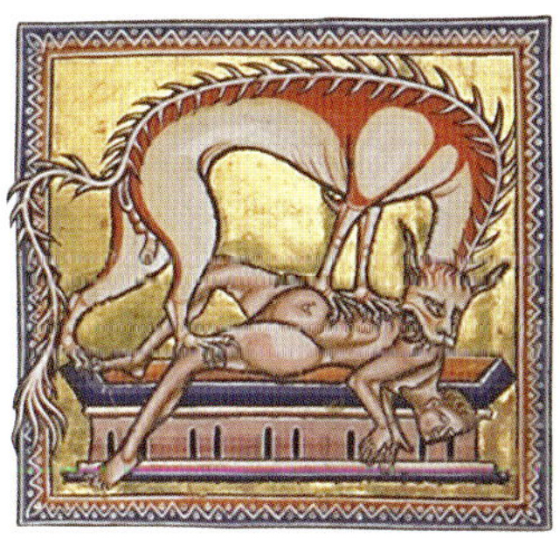

THE MYTHICAL BLOOD-DRINKING WEREHYAENAS

In the former Kingdom of Kaffa (now part of south-western Ethiopia), special laws were passed to outlaw werehyaenas. Those accused of turning into hyaenas at night were arrested and presented to a priest, who would determine if the accused was a werehyaena. If found guilty, the individual would be sentenced to slavery, death, or exile. Although only a priest of Docco could pass sentence,

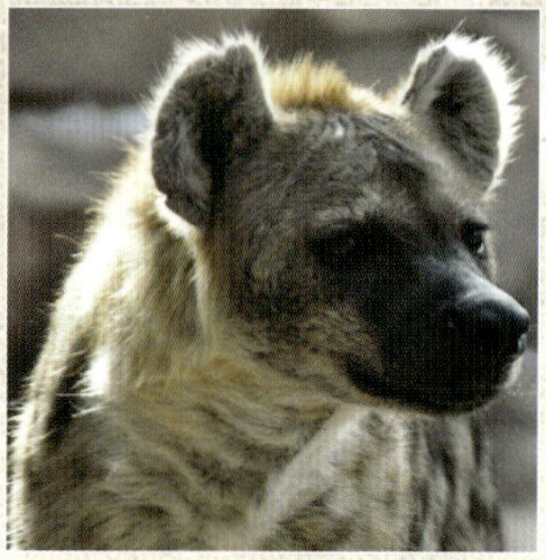

THE HYAENA

Although hyaenas look like large wild dogs, they are actually from a separate biological family which is most closely related to mongooses and meerkats. The spotted hyaena or 'laughing hyaena' is the largest of the species. Its coat is sandy, yellowish or grey with dark brown spots on most of its body. A group of spotted hyaenas is called a clan or a 'cackle' and can number between 5 and 90 individuals led by a single alpha female. A complicated social hierarchy governs the clan, which is lead by alpha females, followed in rank by cubs, while adult males rank lowest. The males are smaller than the females, weighing between 45 and 60 kilos, while females weigh 55 to 75 kilos on average but have been recorded weighing up to 90 kilos.

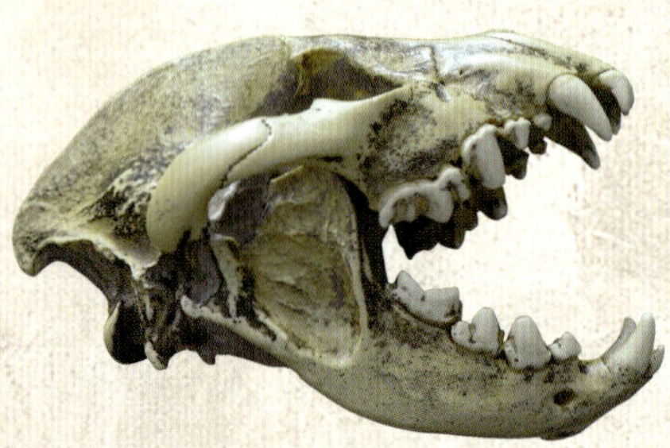

THE IMMENSELY POWERFUL JAW OF THE SPOTTED HYAENA

anyone could accuse another of this crime. Following his interviews with elderly Kaffa men, the anthropologist GWB Huntingford commented, 'This led to much injustice, and... the law was often set into motion'.

Intermingled with the legends of the Boudas are stories of Ethiopian 'Death Birds' said to 'feed on the blood of animals and humans, causing puncture wounds and debilitating sickness'. In the 1930s this folklore intrigued the adventurer and archaeologist Count Byron de Prorok who 'explored a cave said by the locals to be haunted by hyaena-men and a death bird. The hyaenas proved real enough, and so did the death birds, in the form of a huge swarm of bats'. However, there are no vampire bats to be found outside the Americas and the Ethiopian 'death birds' are described as having wingspans of about thirty centimetres, twice as big as the common vampire bat. If their existence were to be confirmed by zoologists they would be a new species. However, it has been assumed that their blood-drinking habits are folklore; the bats are actually contaminating the water supply with their excrement, spreading Weil's disease. This illness causes haemorrhages to break out on the skin, which could look like the bites of vampires.

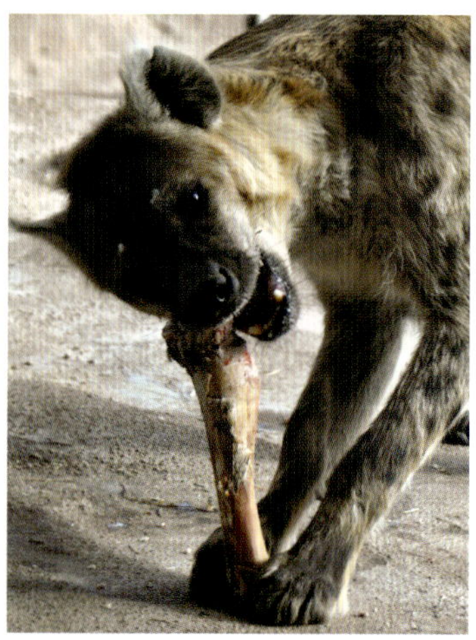

HYAENAS ARE SUPREME PREDATORS

It is easy to understand why hyaenas would have become fearsome monsters in local mythology. Despite common belief, only some species belonging to this family are scavengers – the spotted hyaena found in this region is not only a real predator, but also the most effective predator on the African savannah, with one of the strongest jaws in the animal kingdom. Hyaenas are also highly intelligent – some scientists who have observed them claim they are of equal intelligence to certain apes.

Day Twenty-five

Our final day in Harar – tomorrow we will fly back to Addis and begin making our way back to the UK after meeting with James Young, Born Free's representative in Ethiopia. The Born Free Foundation is one of the major funders of the EWCP, and it was through the Foundation that we were originally introduced to Claudio. Pollyanna has been a supporter of the work of Born Free for many years, and proud to have been able to support its work through her exhibitions and artwork. At the time of our arrival, Born Free is in the middle of raising funds to open a new wildlife sanctuary on the outskirts of Addis Ababa – the first of its kind in the country. The sanctuary will provide a home for rescued big cats and other wildlife, as well as housing an education centre. We have pledged to raise funds for this project and are looking forward to learning more about the progress.

Meanwhile we are keen to make the most of our last day in this region. We have learnt that Harar is notorious for a very unusual reason – the so-called hyaena men. We have heard that these daring individuals feed packs of wild hyaenas just outside the city walls every night. We have told Abdul we would love to witness this and he has agreed to take us when darkness falls.

Before then, however, he is keen to show us more of the sights of his home city. Today he starts by driving us outside the city walls, to a high point where we can get an overview of the tangled warren of roofs and alleyways. A tall impressive statue of Ras Makonnen stands here, outside the supreme court. Abdul's obvious influence sends a minion scurrying to unlock the court building so that we can see the impressive stained-glass windows. The courtroom is currently being used as a grain and spaghetti store by the staff who are buying these staple commodities wholesale and are unloading sacks from a lorry.

We divert to the nearby khat market. Khat is a flowering evergreen shrub native to this region, and the fresh leaves have been chewed as a mild narcotic since at least the eleventh century. Many Ethiopian men will spend several hours each day sitting and methodically chewing the leaves. They seem to chew to stave off hunger and pass the time

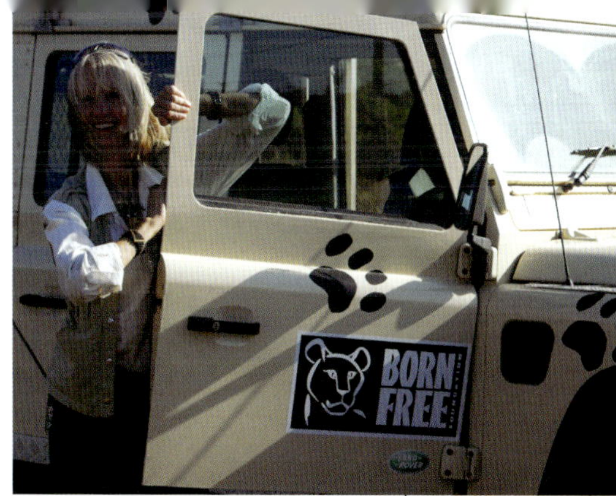

Visiting Born Free projects in Africa

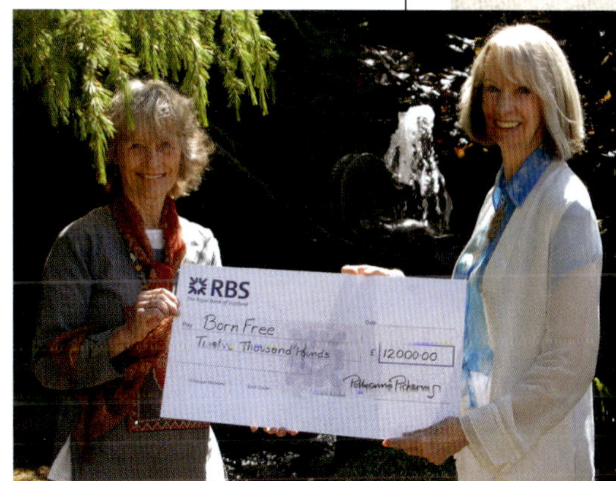

Presenting Virginia McKenna with a cheque from The Pollyanna Pickering Foundation to fund the building of a sanctuary near Addis

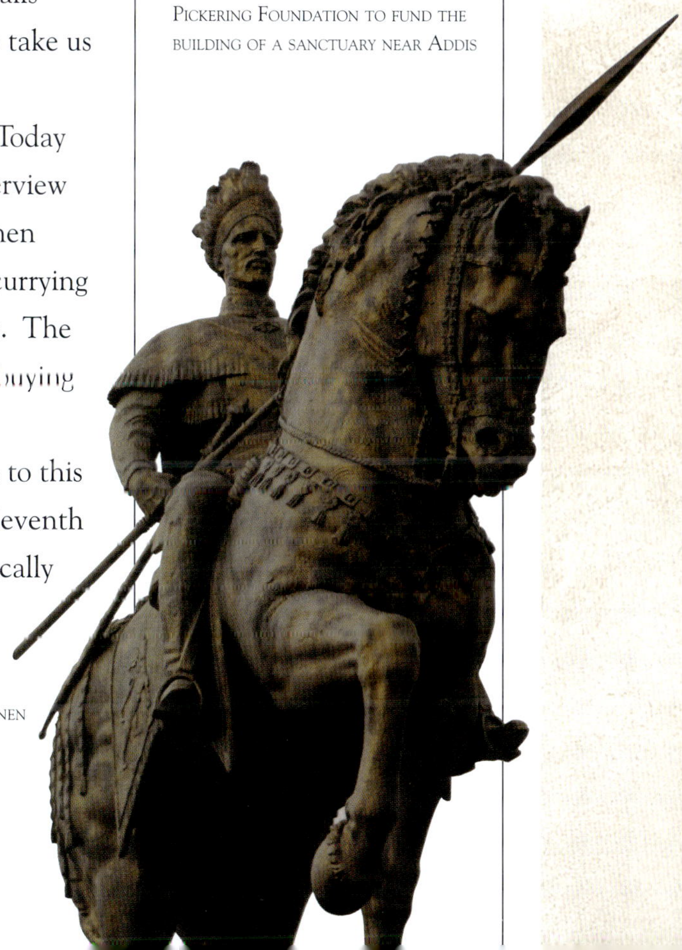

Ras Makonnen

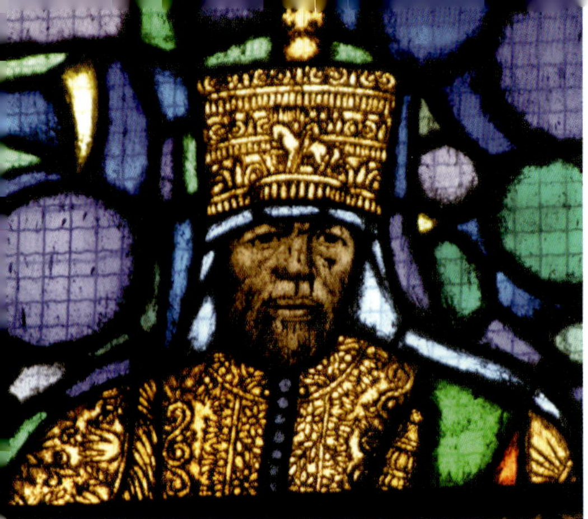

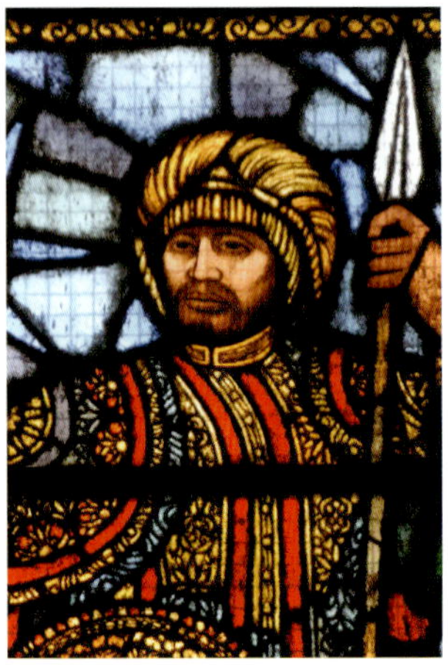

IMPRESSIVE STAINED-GLASS WINDOWS IN THE COURT BUILDING

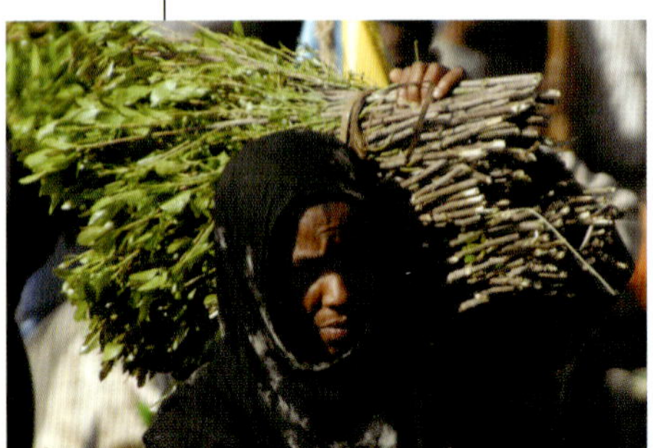

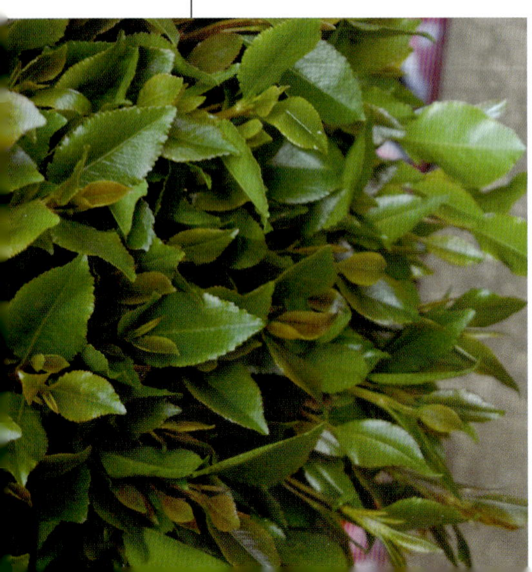

rather than for the stimulant effect – which was described to us as bringing a similar buzz to drinking a pint of beer after an afternoon's worth of chewing. The fact that many Ethiopian drivers are high on khat undoubtedly contributes to their appalling road safety statistics.

In this Muslim city some of the devout chew to focus their prayers. Khat is even more popular sixty miles away in Somalia, where it is estimated that seventy-five per cent of men use the drug. Convoys of Toyota flatbed trucks packed with bundles of fresh khat drive from Harar towards the border, while smuggled cars, televisions, radios and even weapons head the other way. Khat is experiencing a boom in trade and, with the falling price of coffee in recent years, has become the country's second biggest export, with many farmers in these mountainous, rainy areas

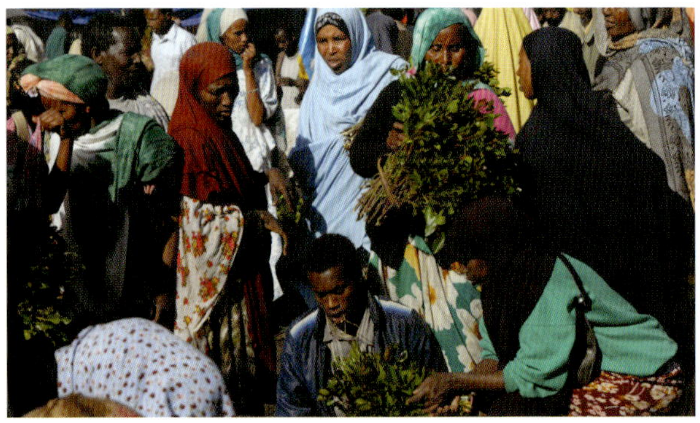

THE KHAT MARKET

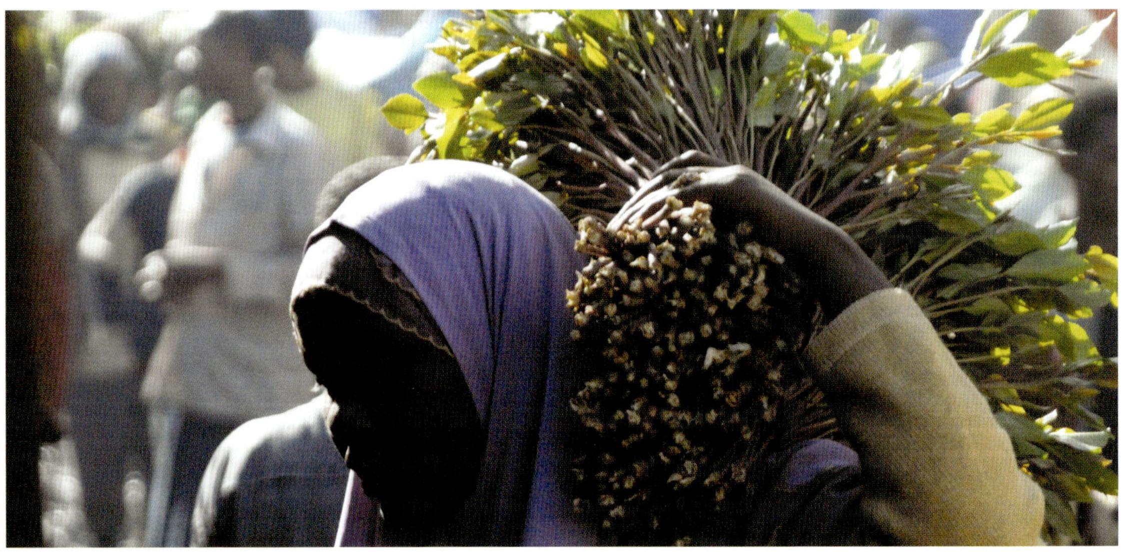

SCENES FROM THE KHAT MARKET. KHAT IS NOW ETHIOPIA'S SECOND BIGGEST EXPORT. LEFT: KHAT LEAVES

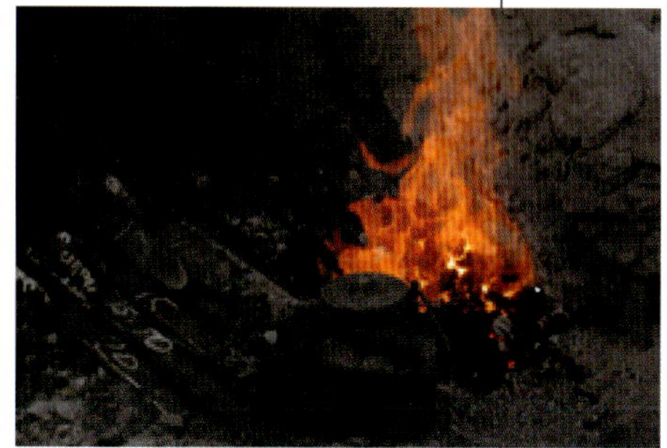

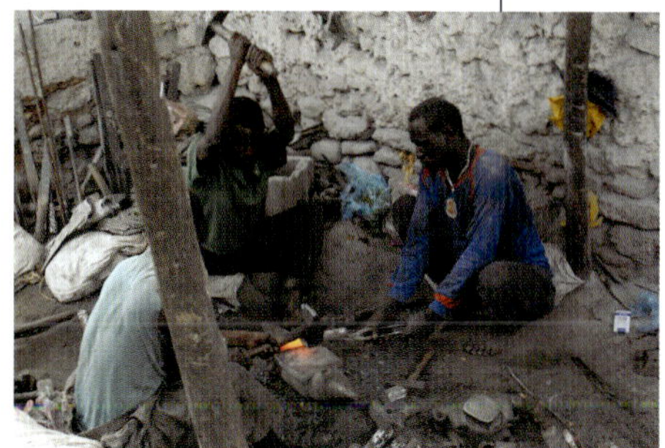

The smithy

The owner of a chaotic gift shop cuts an imposing figure

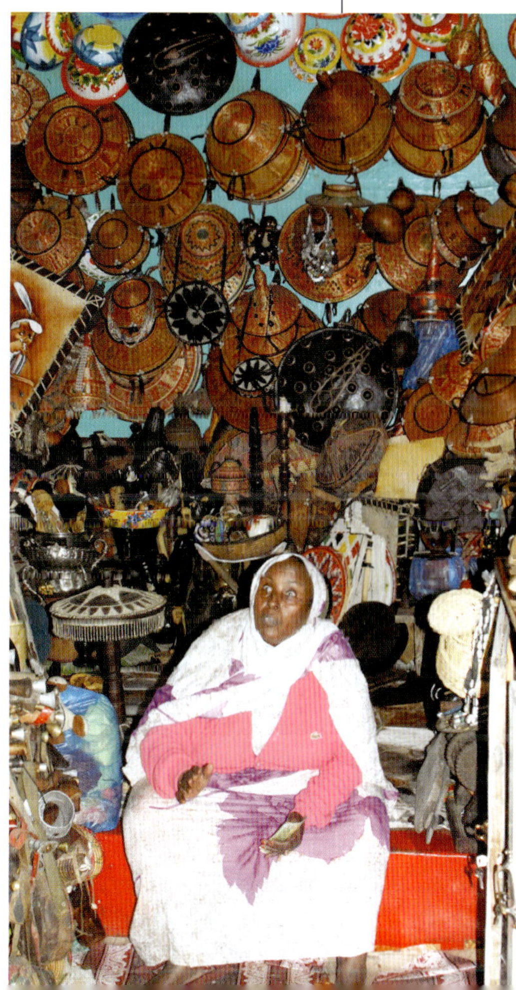

converting their land from coffee plantations. It is such a part of the country's culture that the Ethiopian Postal Service even issued three stamps featuring khat leaves!

The market is chaotic as people barter over the piles of fresh leaves. Around the edges men lean against the walls of the buildings, already pulling khat leaves from the bags they have purchased and chewing rhythmically. Oddly we attract less attention here than we have anywhere we have walked in Ethiopia – people are so focused on buying and selling the dark oval leaves that many of them give us little more than a cursory glance as we pick our way through the sellers. Underfoot the ground is a sticky, viscous, smelly mud, presumably mixed with excrement from the goats which are running around the market place.

Back in to the city of Harar we walk through the streets, slowly making our way back towards our accommodation. Every open door offers a glimpse of Ethiopian life – we step into the burning hot interior of a smithy, watching as one man operates an ancient pair of bellows, while his three workmates hammer molten metal over the fire. A dark Aladdin's cave of a gift shop is presided over by a huge lady swathed in shawls who sits in the centre of her wares like an enormous spider in the middle of her web. We select a bracelet each from the bewildering array of goods. In a jeweller's shop they are welding beads into delicate necklaces; we squeeze into the crowded back room where the owner offers us a smoke of his hookah pipe – and then find ourselves in the Muslim market square. This open space is surrounded by rows and rows of black kites lining the roofs overlooking the butchers' shops. Abdul throws scraps of meat for them and suddenly the air is filled with flocks of kites wheeling around us – like pigeons in Trafalgar Square, but MUCH MUCH bigger. Kites are scavengers and their black silhouettes can be seen in the skies above every town in Ethiopia. They perform a vital role in keeping the city streets clean. It is hard to imagine that centuries ago we would have seen a similar sight back home in England – red kites were widespread in the Middle

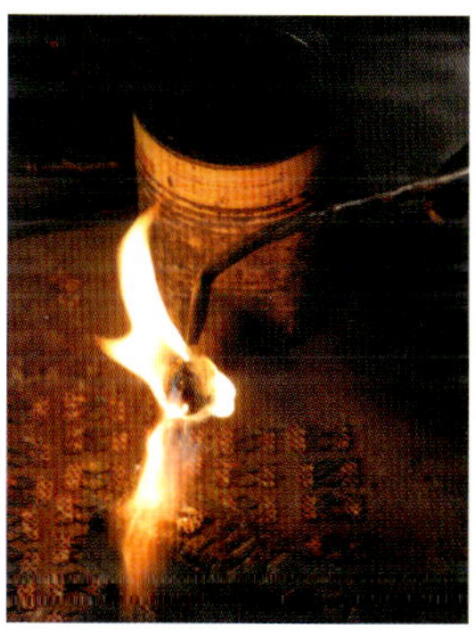

Molten metal is hammered into shape

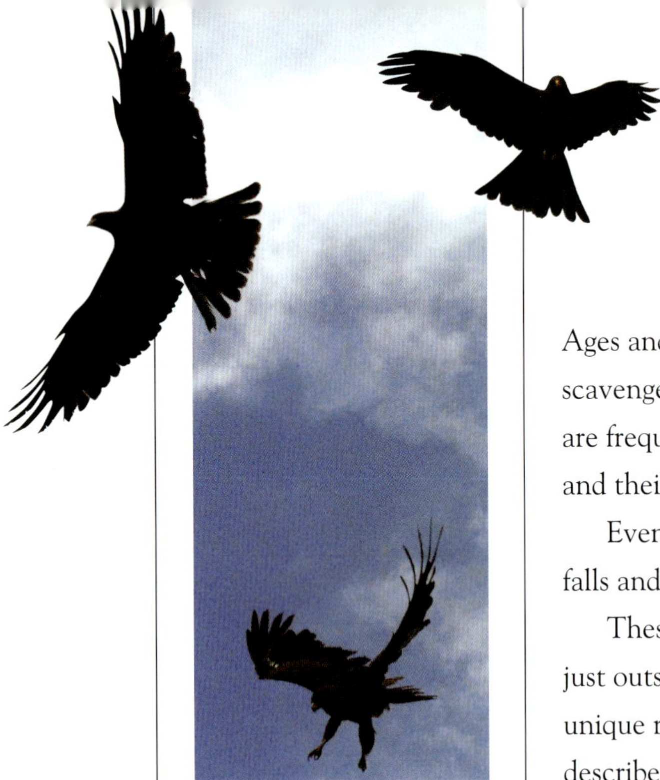

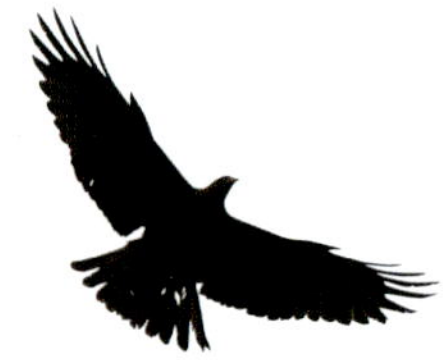

Ages and they were protected by royal decree in order to carry out their vital role as scavengers. Evidence of their ubiquity can be found in Shakespeare's plays where they are frequently mentioned. However, by the sixteenth century they were classed as vermin and their numbers began to decline.

Eventually we return to our lodgings, where we have time for a coffee before night falls and Abdul comes to take us to see the legendary hyaena men.

These daring individuals collect offal and bones to feed to the wild spotted hyaenas just outside the old city walls of the town. No-one is certain when this strange and unique ritual began. The Hararis have had a long relationship with hyaenas, which they describe as their neighbours and spiritual guides. The animals rid the streets of decaying meat and bad spirits and some say they can tell the future – and if they are not treated well the hyaenas will exact their revenge. It is believed that the feeding ritual may have existed in some form for nearly eight hundred years, and oddly may be related to a separate practise of feeding hyaenas with bowls of porridge.

Until 1887 Harar was a forbidden city, closed to visitors for centuries and many unique beliefs and rituals were able to grow and flourish in isolation. To many of the Muslims in this historically Islamic city, hyaenas are seen as ancient spiritual guides that predict hardship or prosperity. During the annual festival of Ashurer hyaenas are summoned with drums. Lead by the 'hyaena king', the animals join the party and are offered a bowl of porridge smothered in rich butter. If the hyaenas refuse the porridge, or conversely eat the whole bowlful, the coming year will bring famine or pestilence. If they eat over half the porridge the year will bring peace and prosperity. It is difficult to imagine the reaction of the population to the first person who suggested it might be possible to foretell the future success of crops by measuring the specific amount of porridge eaten by a wild hyaena. However, the suggestion was clearly accepted by the people and became part of the annual festival. It is also quite difficult to imagine how this could later transform into the practise of hand feeding them every night.

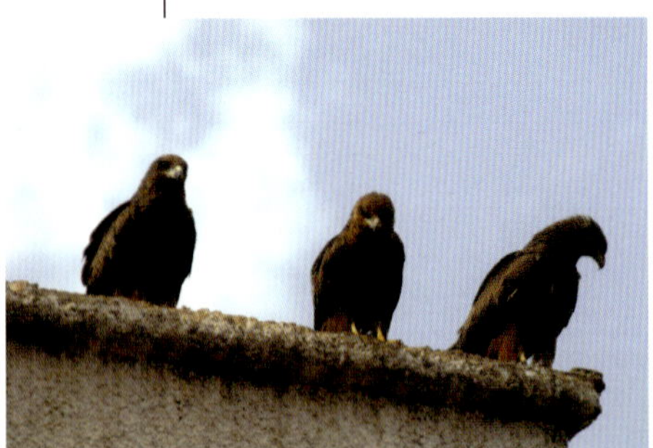

Black kites

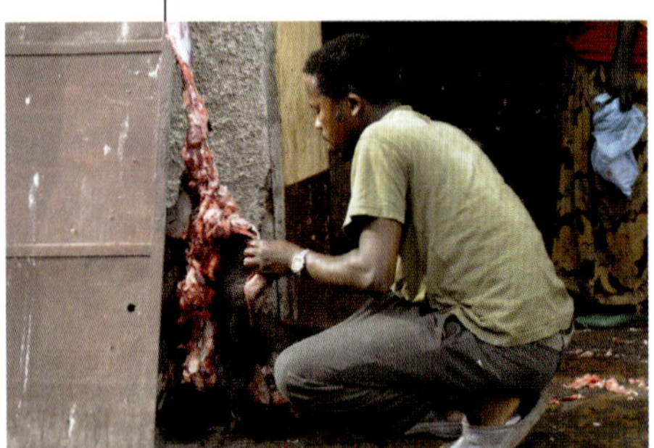

A butcher's shop in the market square

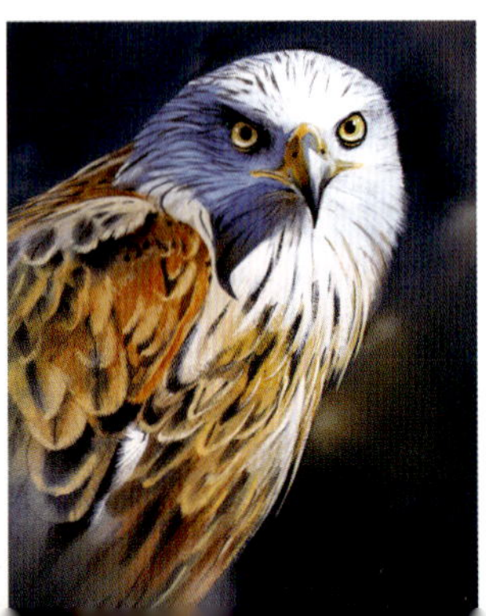

Painting of a red kite

None the less this is precisely the ritual we have come to witness.

Abdul draws our car to a halt in a dusty patch of scrub outside the city walls. There are a few small groups of locals standing nearby, also here to see the hyaena man. The only light comes from the headlights of our car. As we stand there scarab beetles smack into us, landing on our clothing and in our hair. They are actually quite beautiful beetles, but they seem to fly almost blindly through the air until they hit an object with a resounding thud, at which point they alight on whatever it may be. After darkness falls we are constantly removing them.

A man appears, bringing with him a huge bucket full of scraps of meat. He starts calling in a high-pitched chant. We see a spotted hyaena skulking in the shadows. The hyaena man throws it some scraps of meat, which it nervously snaps up from the dusty ground, before sloping away into the darkness. It is fantastic to see a wild hyaena, but after the build up I had been expecting something more dramatic from the legendary 'hyaena man', and I have to confess to feeling a little underwhelmed.

But then another hyaena appears, red eyes glowing in the headlights. Another arrives, and another. They circle at a distance, and two more come, then three. Suddenly there are at least twelve of the animals around him – now that there are more of them they grow in courage and venture closer, until he begins to hand feed them strips of meat from a short stick – maybe just six inches long. Now I am impressed.

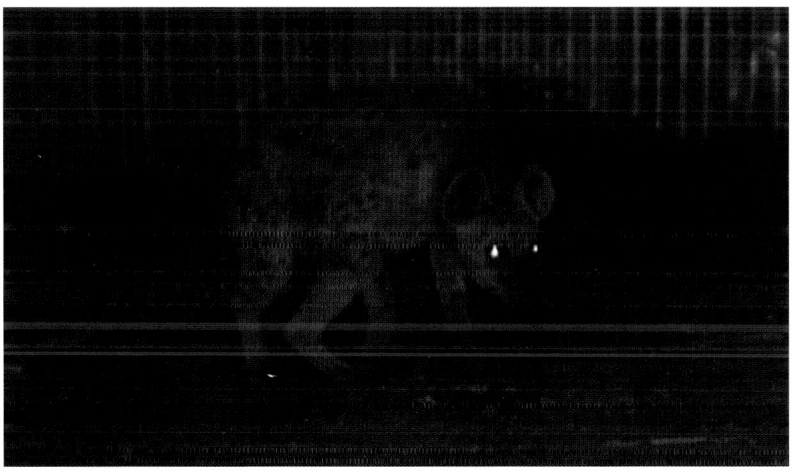

A DOZEN HYAENAS COME INTO VIEW, THEIR EYES GLOWING RED IN THE HEADLIGHTS

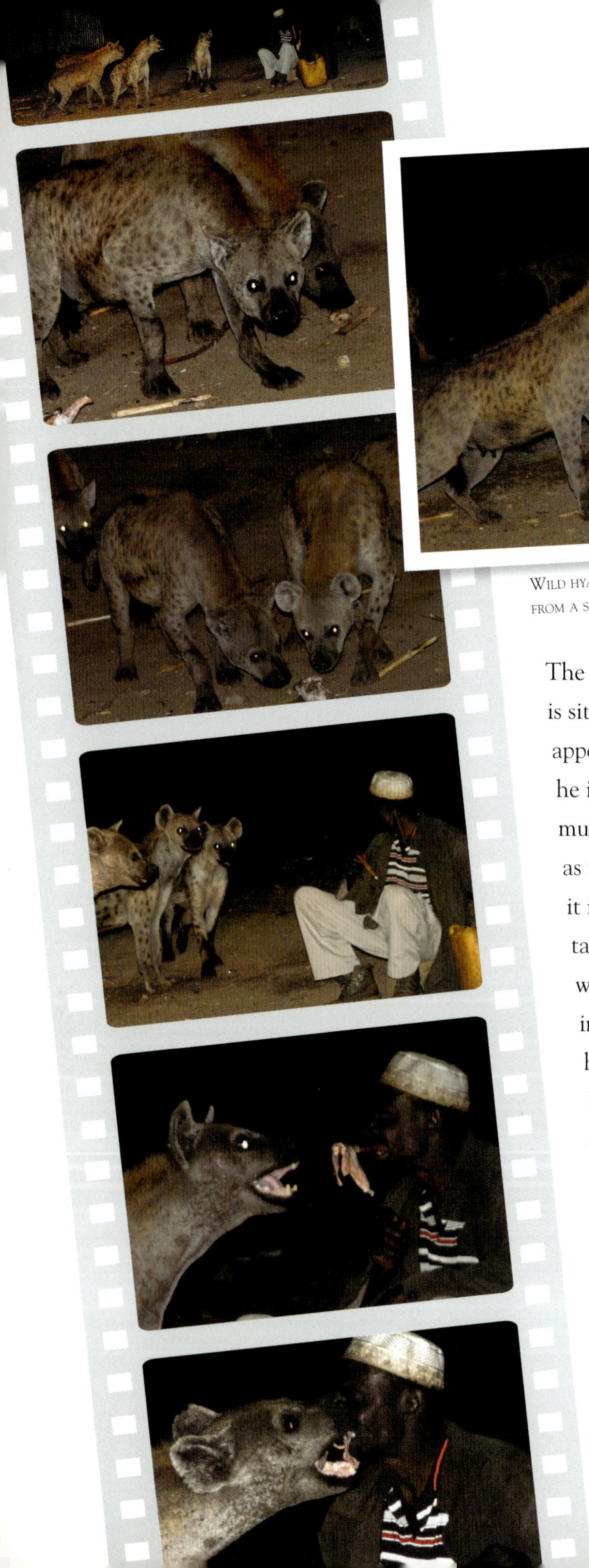

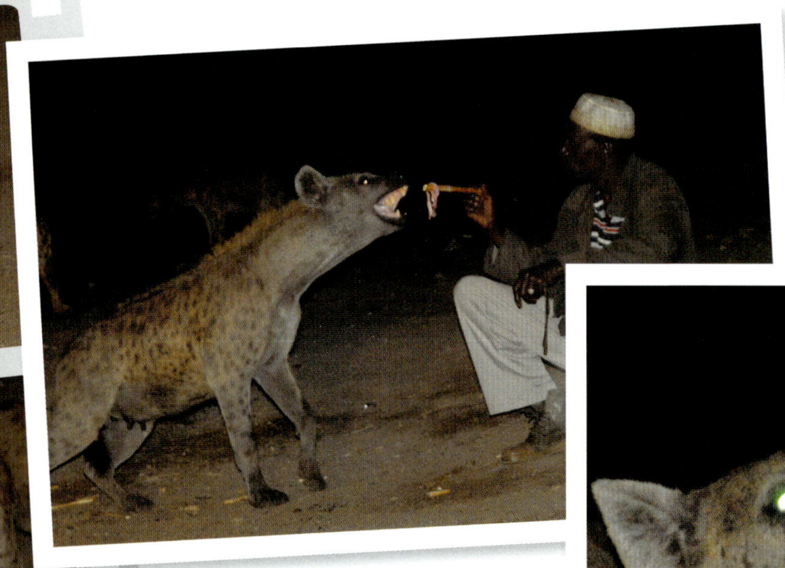

Wild hyaenas take meat from a short stick

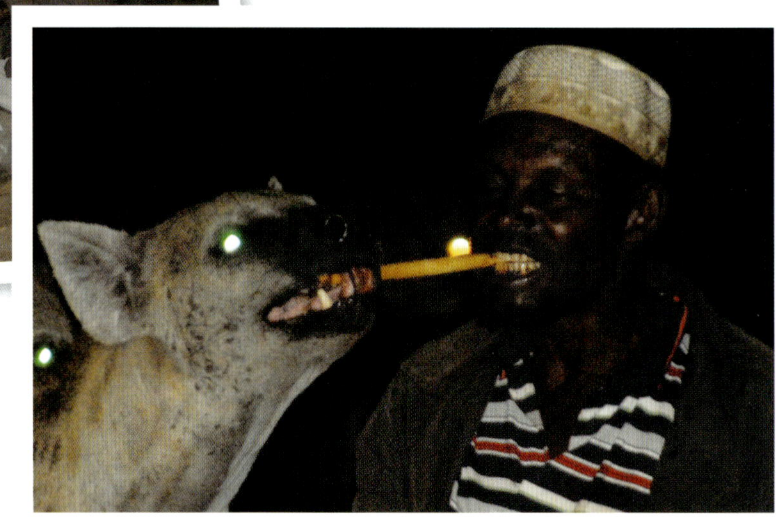

The legendary hyaena man of Harar

The hyaenas are huge – he is sitting down and they appear much bigger than he is. With their sloping muscular shoulders and low-set heads they look every bit as powerful and dangerous as their reputation would suggest. Then the hyaena man takes the stick and places it in his mouth, with a strip of meat hanging from it. A hyaena comes right up and takes the meat from his mouth. It is one of the most remarkable sights I have ever witnessed. He repeats this several times with different hyaenas and the number in the clan continues to grow until there are as many as twenty animals circling him. Some make an excited yipping noise while others make their characteristic laughing sound which echoes eerily in the darkness. We can see clearly that they are pack animals – as individuals they were wary hanging back in the darkness, but as a group they are far more fearless and aggressive.

Sketching on location is often challenging – circumstances are rarely ideal. Sometimes strong wind blows Pollyanna's sketch paper; rain and drizzle dampen her page as she works; cold conditions stiffen her fingers making it hard to move her pencil; while in the heat she is sometimes plagued by flies and insects. However, these are some of the most challenging conditions she has ever worked in. Leaning on the bonnet of Abdul's car, the scene in front of her is lit only by the glow of the headlights – but her sketchpad is in almost complete darkness. The hyaenas are in almost constant movement, dark shapes looming in and out of the shadows. Grey dust from the unpaved area blows across her paper as she works, and huge scarab beetles blunder across her page. Despite this, Pollyanna manages to outline the muscular forms of the hyaenas in a sheet of images which capture perfectly the tension in the atmosphere. On her return to our accommodation she works further on the simple outlines while the image is fresh in her mind, using pastels and watercolours to add colour and form.

Meanwhile I am taking grainy photographs from a safe distance, when Abdul comes up behind me: 'Would you like a go?' he asks. I laugh merrily, assuming that he is joking. He is not.

A moment later I find myself crouching beside the hyaena man with a short stick in my outstretched hand and a wild hyaena eyeing me warily before darting in and snatching the meat speared on the end. I repeat the experience and feed another, then more. It is one of the most remarkable moments of my life – for one brief moment the whirling world stands still, as I focus solely on this wild creature, his burning eyes locked on mine.

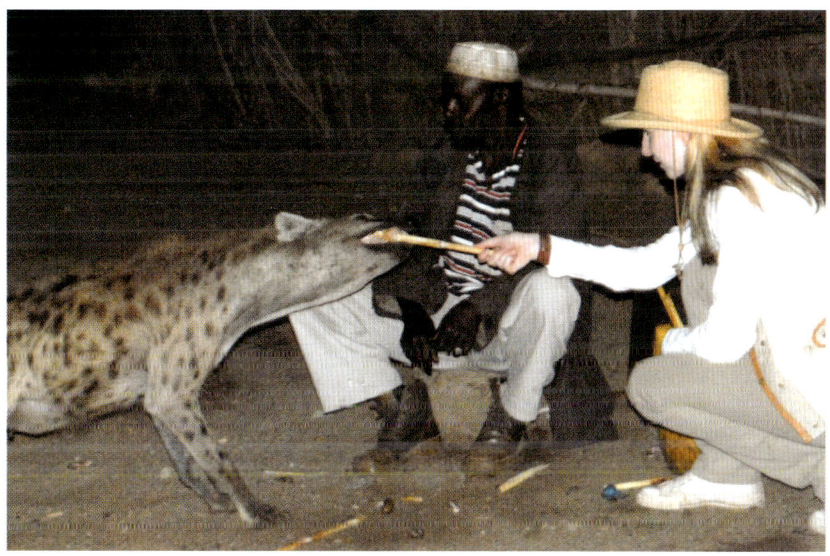

An unforgettable experience

Sometimes we imagine that globalisation, tourism and modern

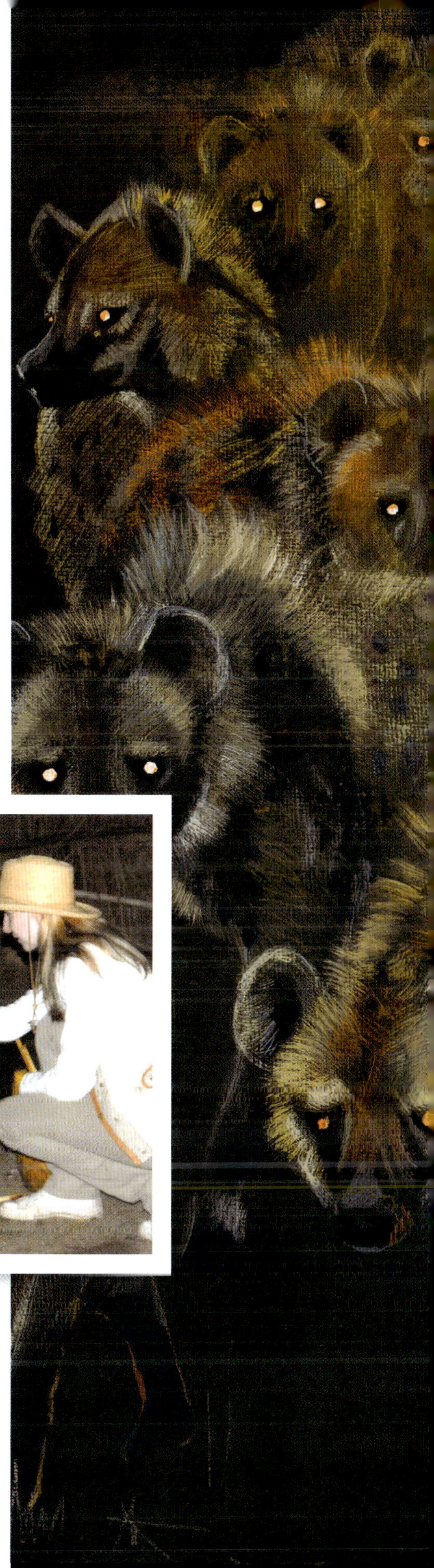

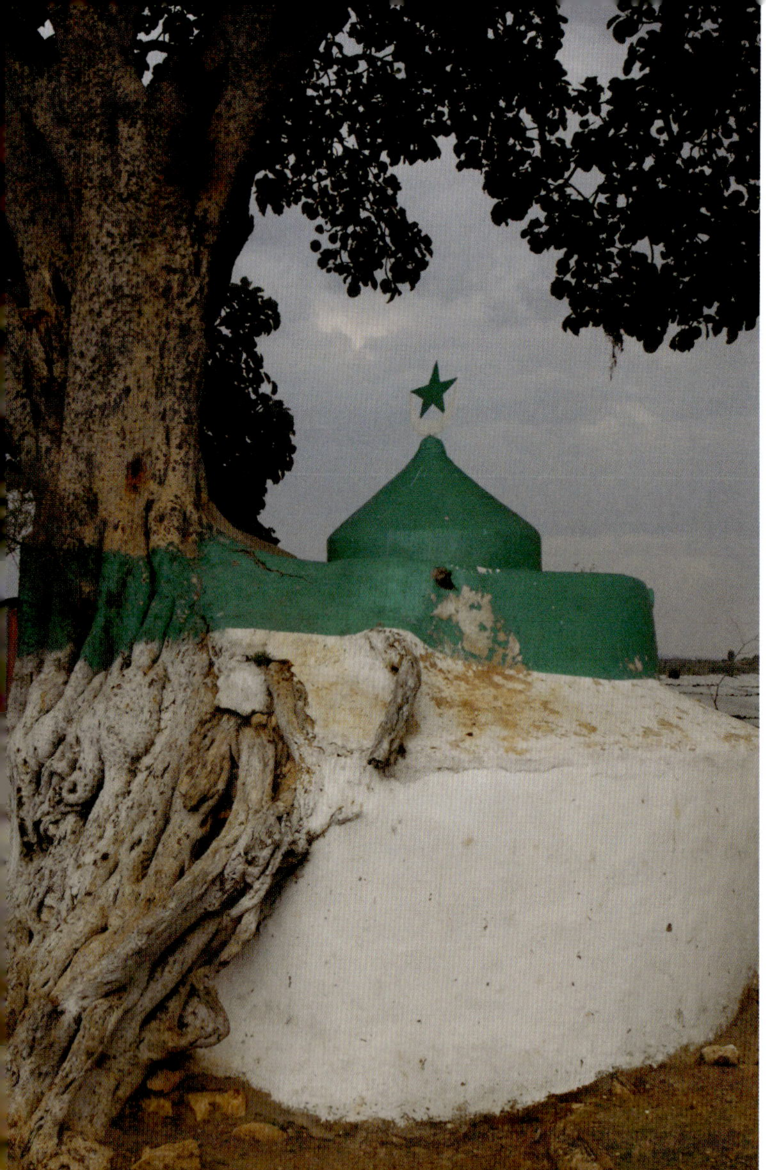

technology has shrunk the world. We think that no matter where we go it is hard to find anywhere truly different, truly 'other', in this modern age of the internet and MacDonald's, satellite television and a Starbucks on every corner. But a place exists where once a year bowls are placed on an ancient shrine – a whitewashed edifice completely intertwined with the roots and trunk of a giant fig tree and people wait to divine the prosperity of the coming year from offerings of porridge made to packs of hyaenas. Here coffee is served with fragrant smouldering frankincense and a sour staple food is created from the smallest grain in the world. And every night outside an ancient walled city men feed wild hyaenas with food from their mouths.

Ancient shrine outside the city walls

We planned a visit to Ethiopia to search for beautiful red wolves which live in high mountains almost unknown to the outside world. Our quest has allowed us to observe the most endangered wolf in the world in its natural habitat, a surreal mist-shrouded plateau of giant plants and red rocks. But during our journey we have discovered and experienced so much more. A land which has defied every expectation we held before our visit. The kingdom of Prester John, the homeland of Rastafarians, protector of the legendary Ark of the Covenant, a nation of exceptionally beautiful people. We have been offered generous hospitality in the midst of poverty, but travelled through a lush fertile landscape. We have seen camels walking through a biblical landscape and eagles soaring above forested cliffs. And right at the end of our journey I have sat in the dust and darkness and hand fed a dangerous wild predator, eyes fixed on mine as he took the meat, both of us acting out a modern ritual with its roots lost in the mists of time.

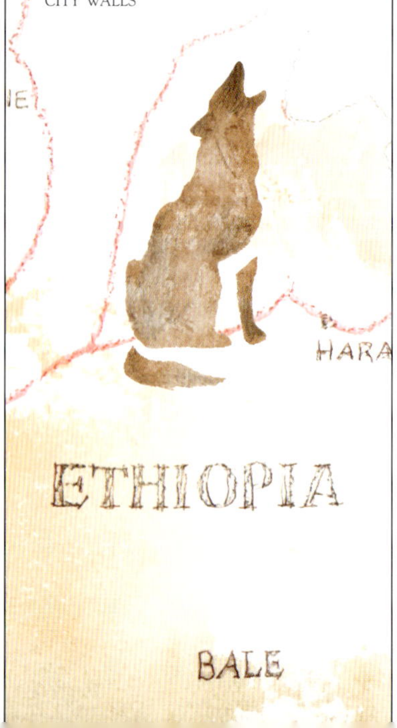

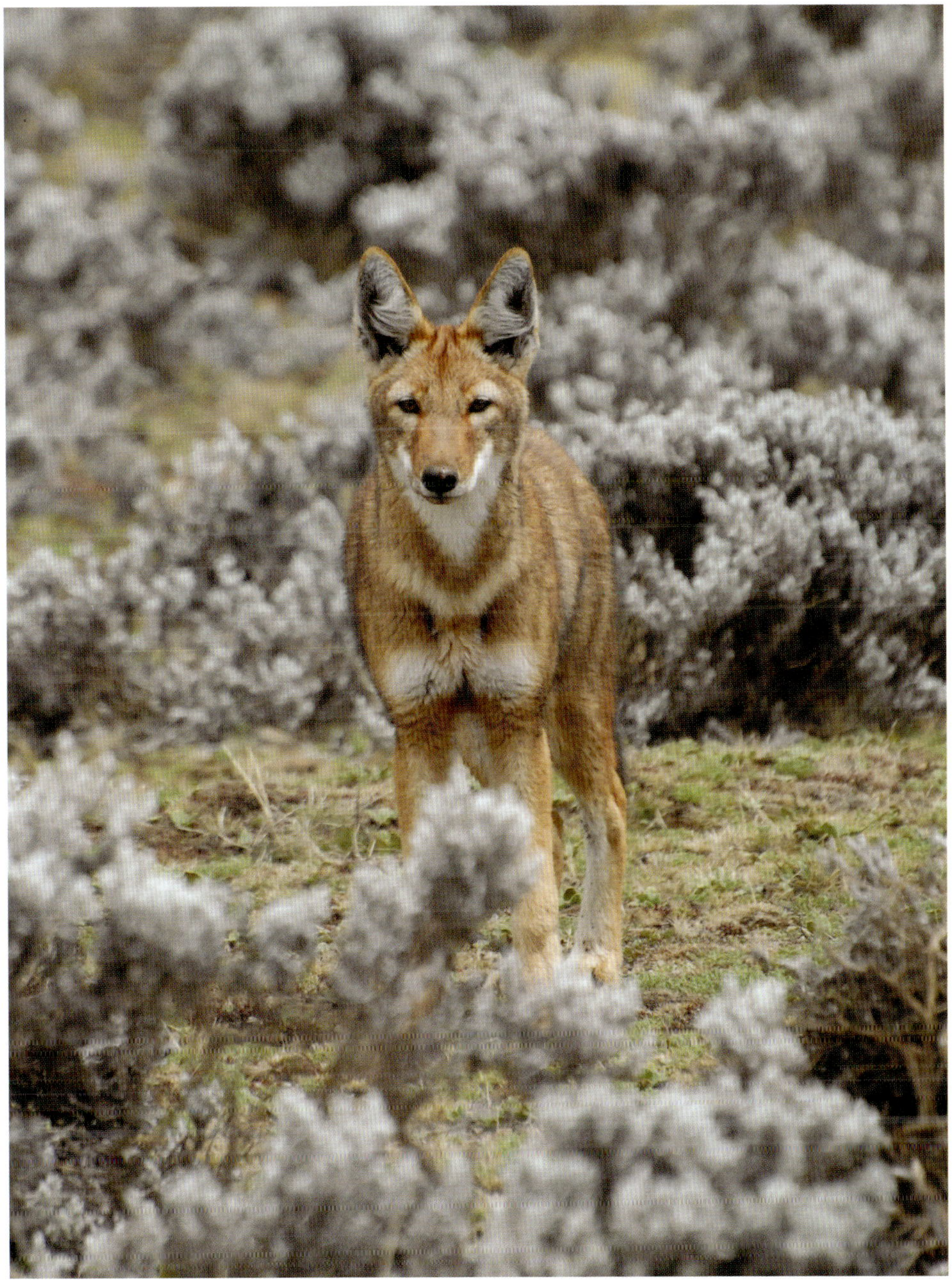

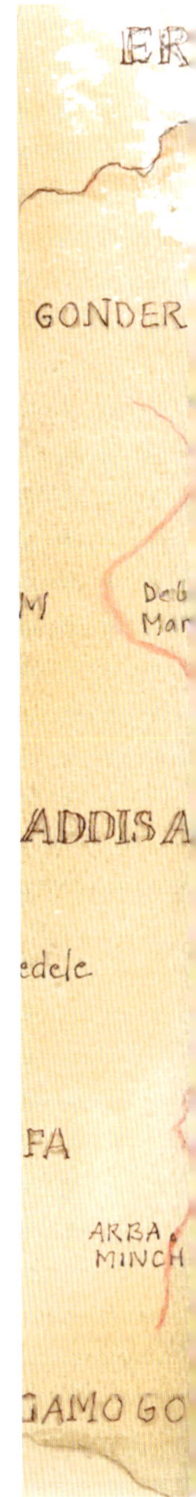
ER
GONDER
M
Deb
Mar
ADDIS A
edele
FA
ARBA
MINCH
GAMO GO

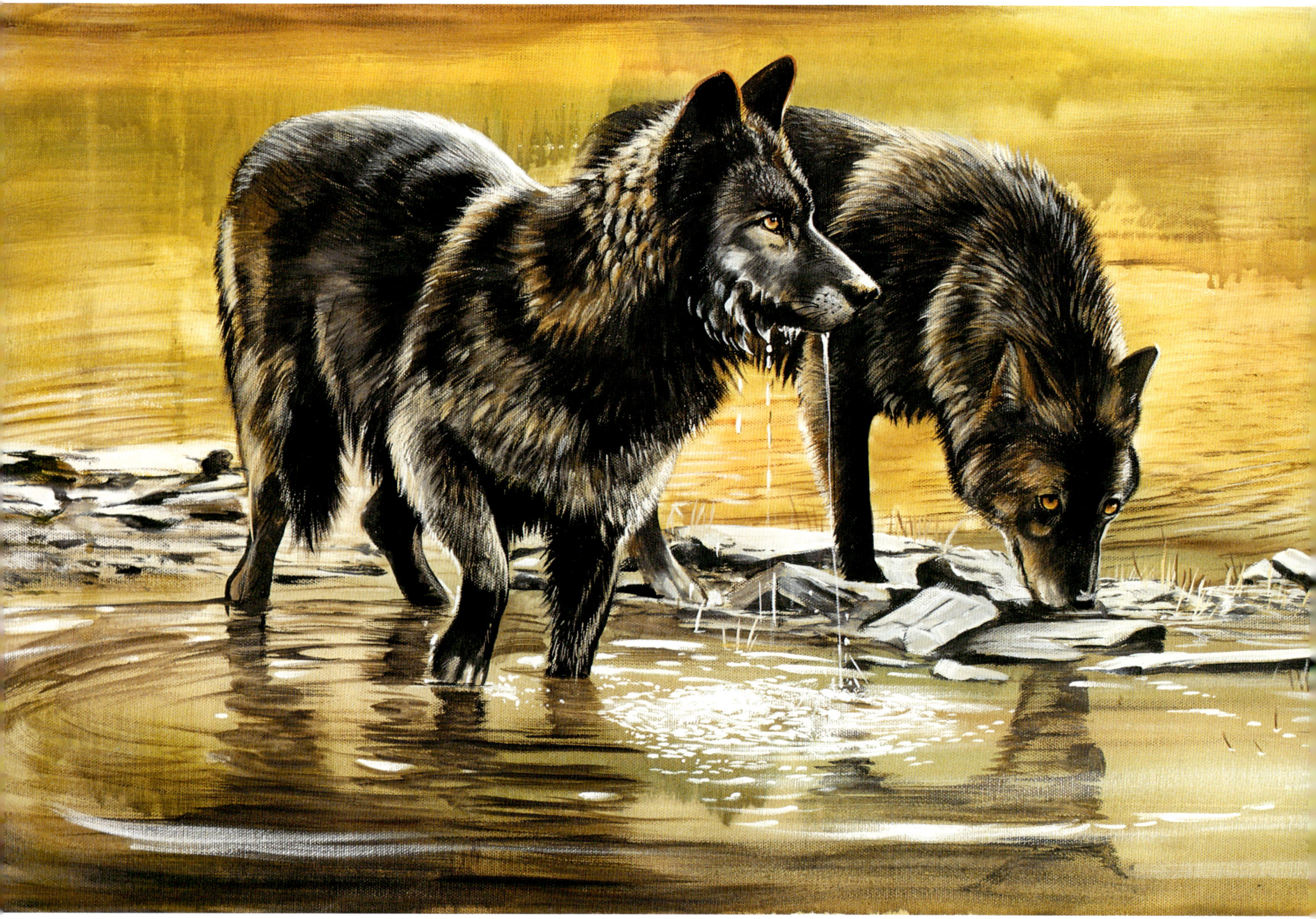

Unchartered Territory

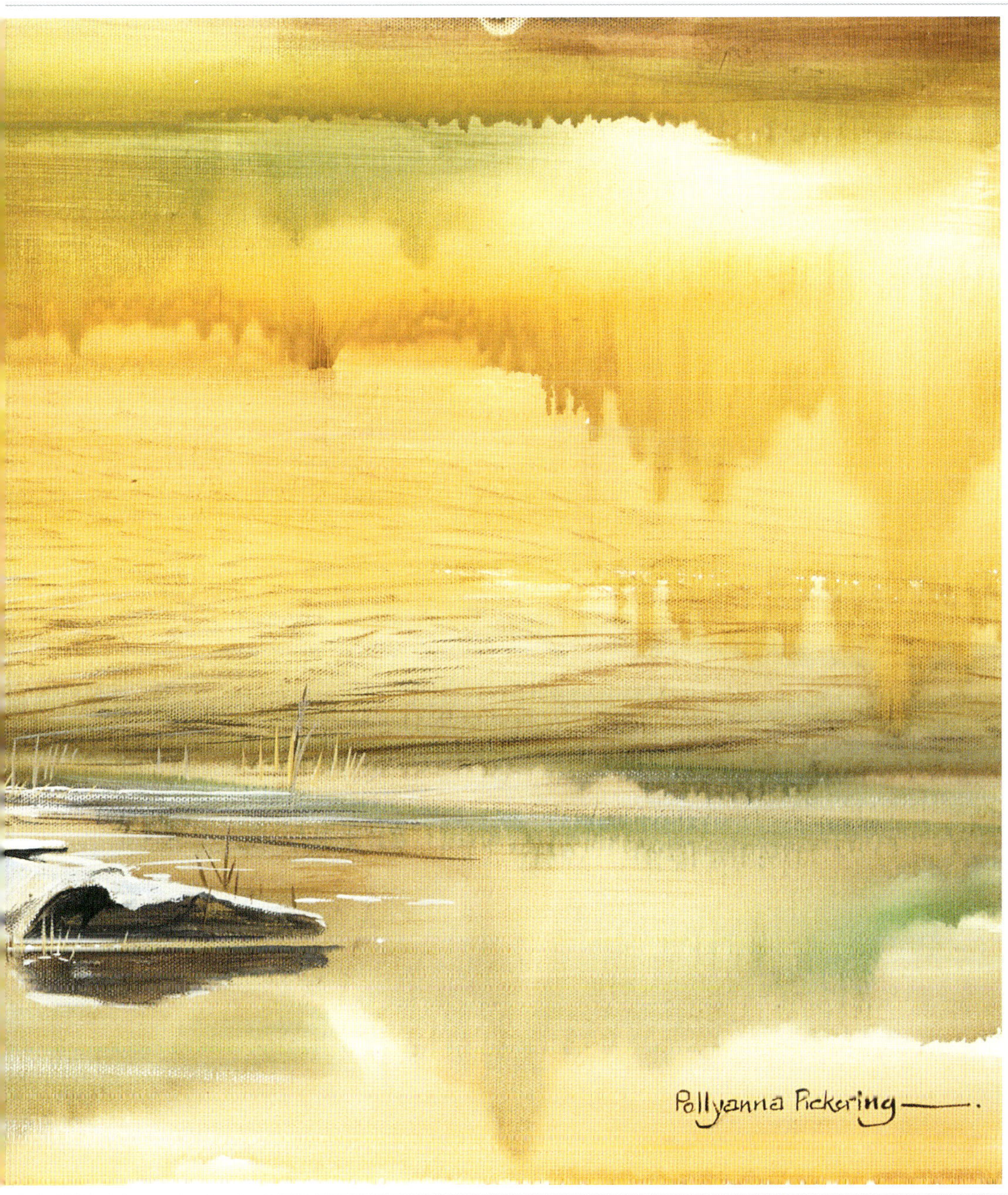

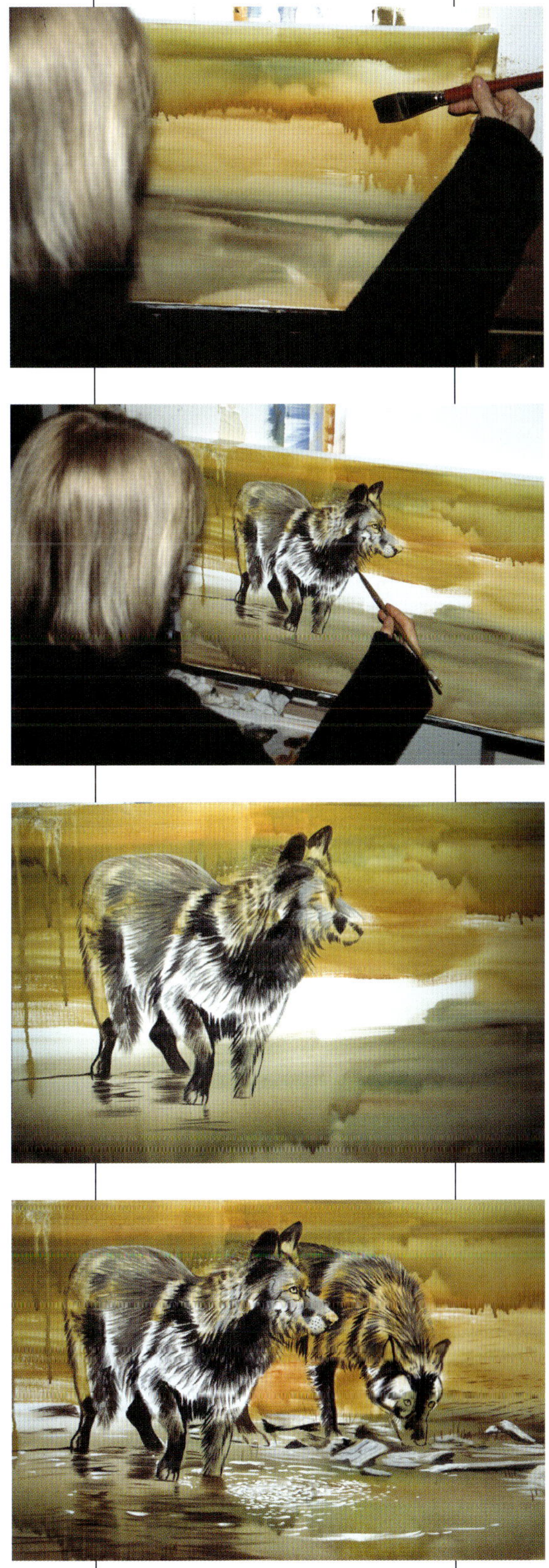

Pollyanna works on a painting of wolves in her studio

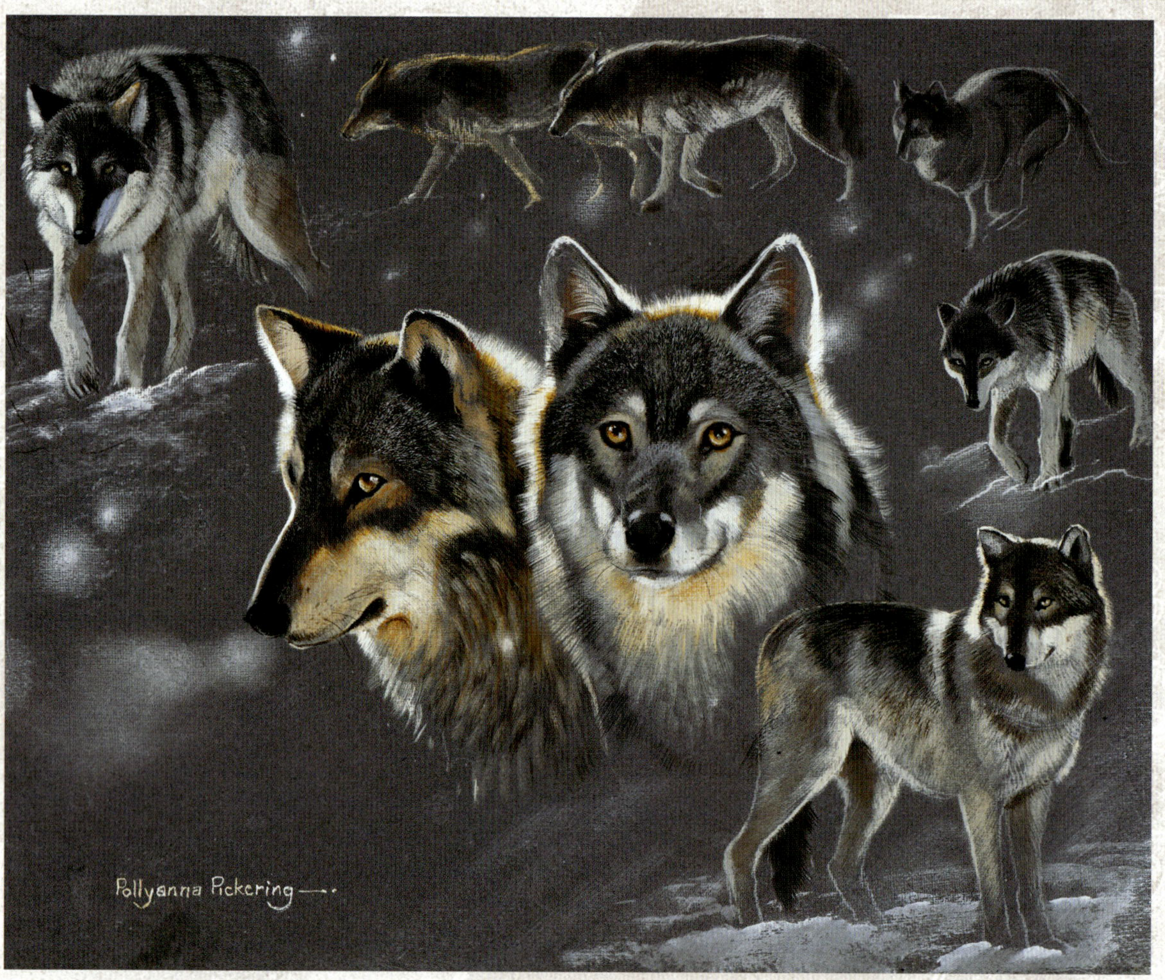
Pollyanna Pickering

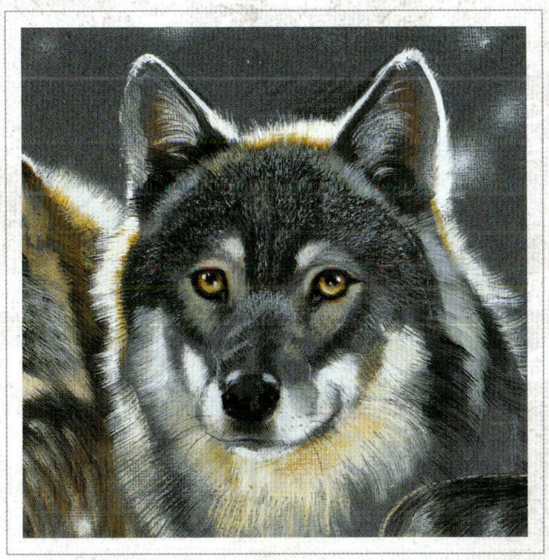

THE PAINTINGS

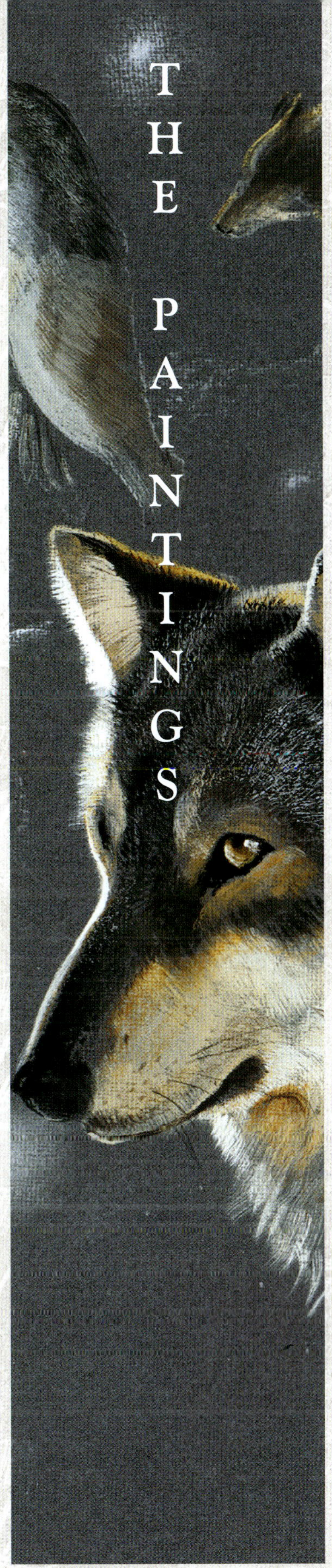

'For a long time, wolves were seen but not looked at. Grown scarce they made themselves elusive, and their actions were interpreted more than observed. Recently though they have begun to emerge from their night.'

PAUL-EMILE VICTOIRE: *L'EMPIRE DES LOUPS*

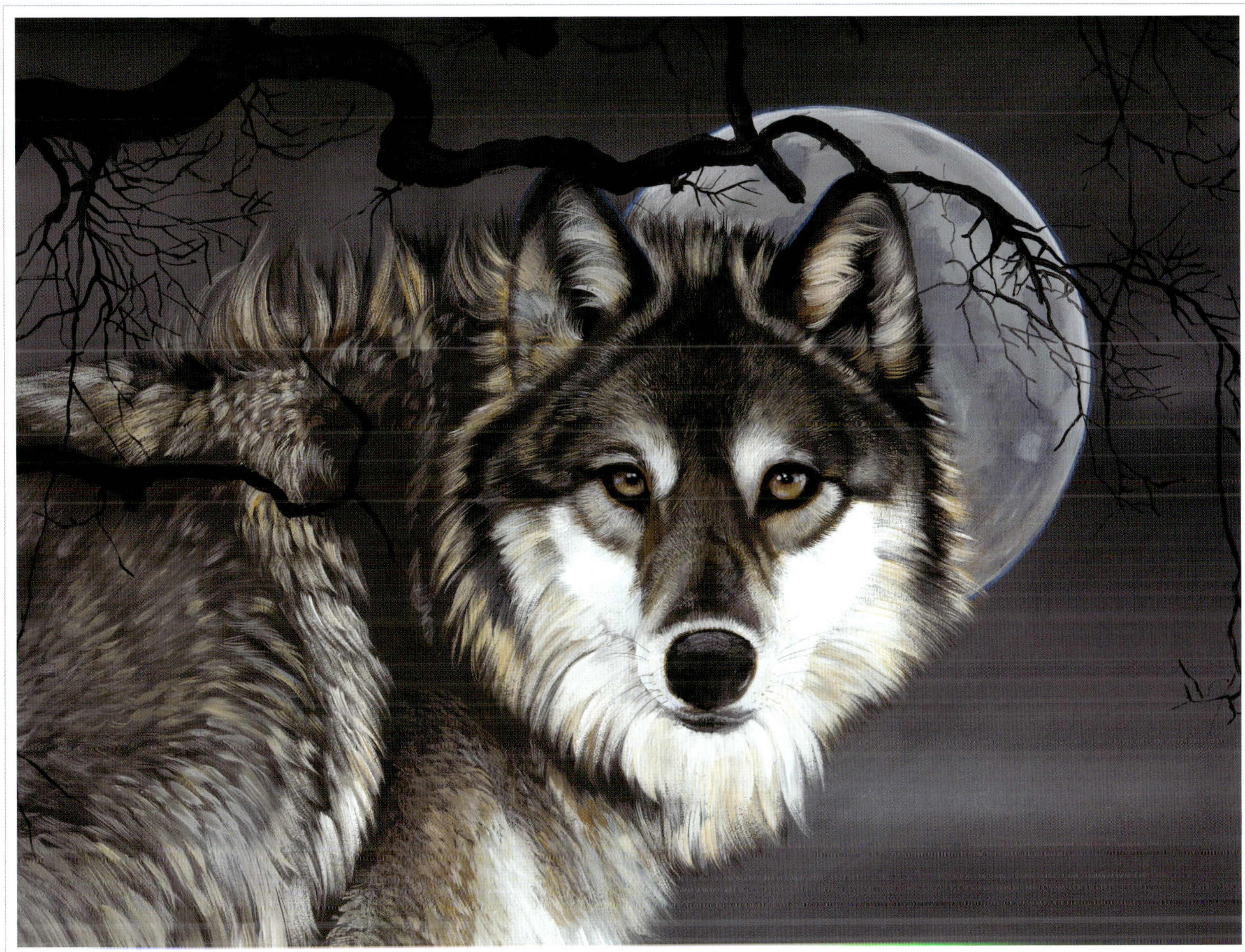

The Dark Side of the Moon

'As I listened, I heard as if from down below
in the valley the howling of many wolves.
The Count's eyes gleamed, and he said
"Listen to them, the children of the nyght.
What music they make!" '

JONATHAN HARKER'S JOURNAL: *DRACULA*

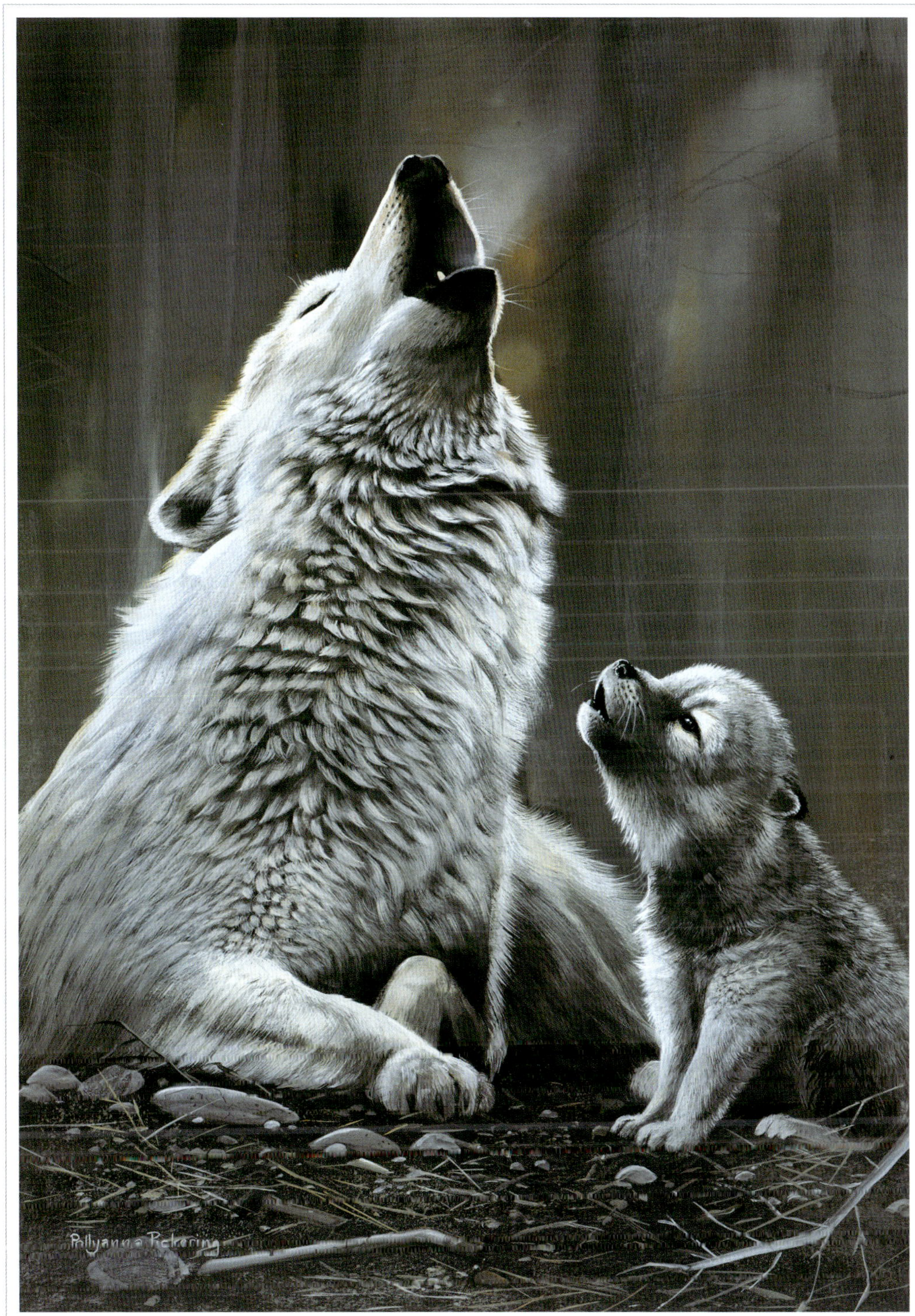

The Singing Lesson
Private collection (UK)

'Wolf I am
In Darkness
In Light
Whenever I search
Wherever I run
Whenever I stand.'

KARL U. SCHLEISER: *THE WOLVES OF HEAVEN*

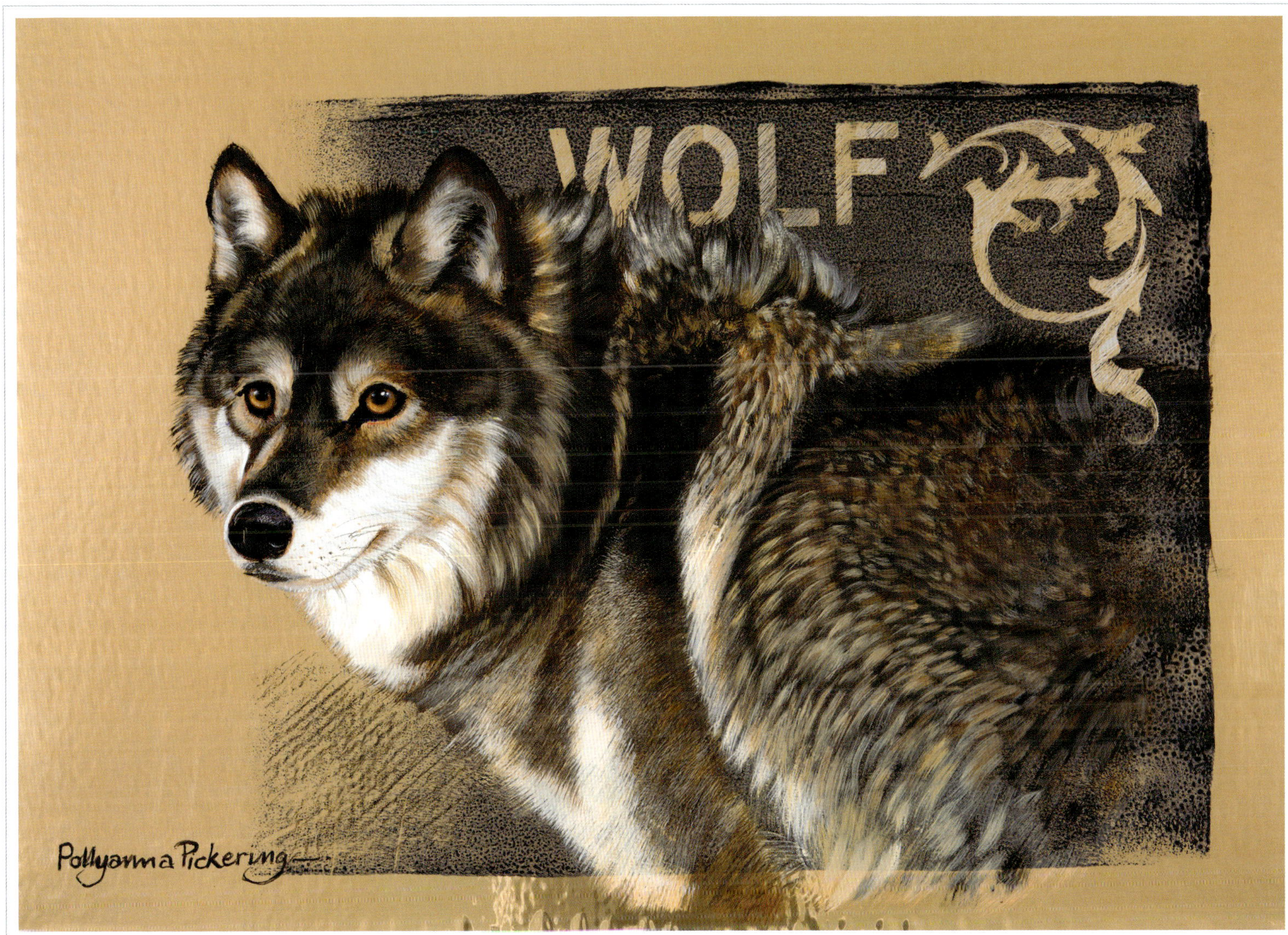

The Watcher
Private collection (UK)

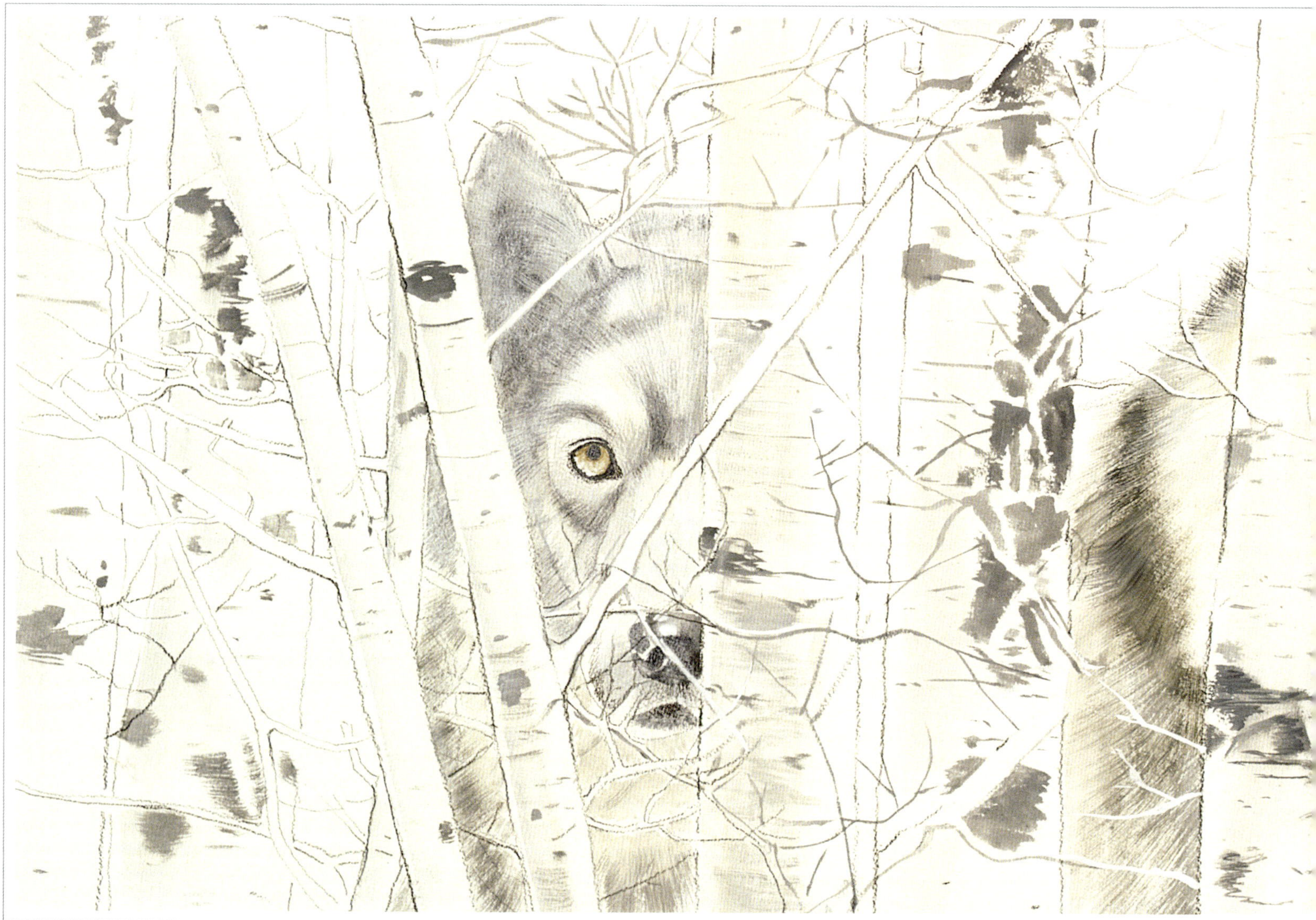

'It is not so much for its beauty that the forest
makes a claim upon men's hearts,
as for that subtle something, that quality of air
that emanation from old trees, that so wonderfully
changes and renews a weary spirit.'

Robert Louis Stevenson

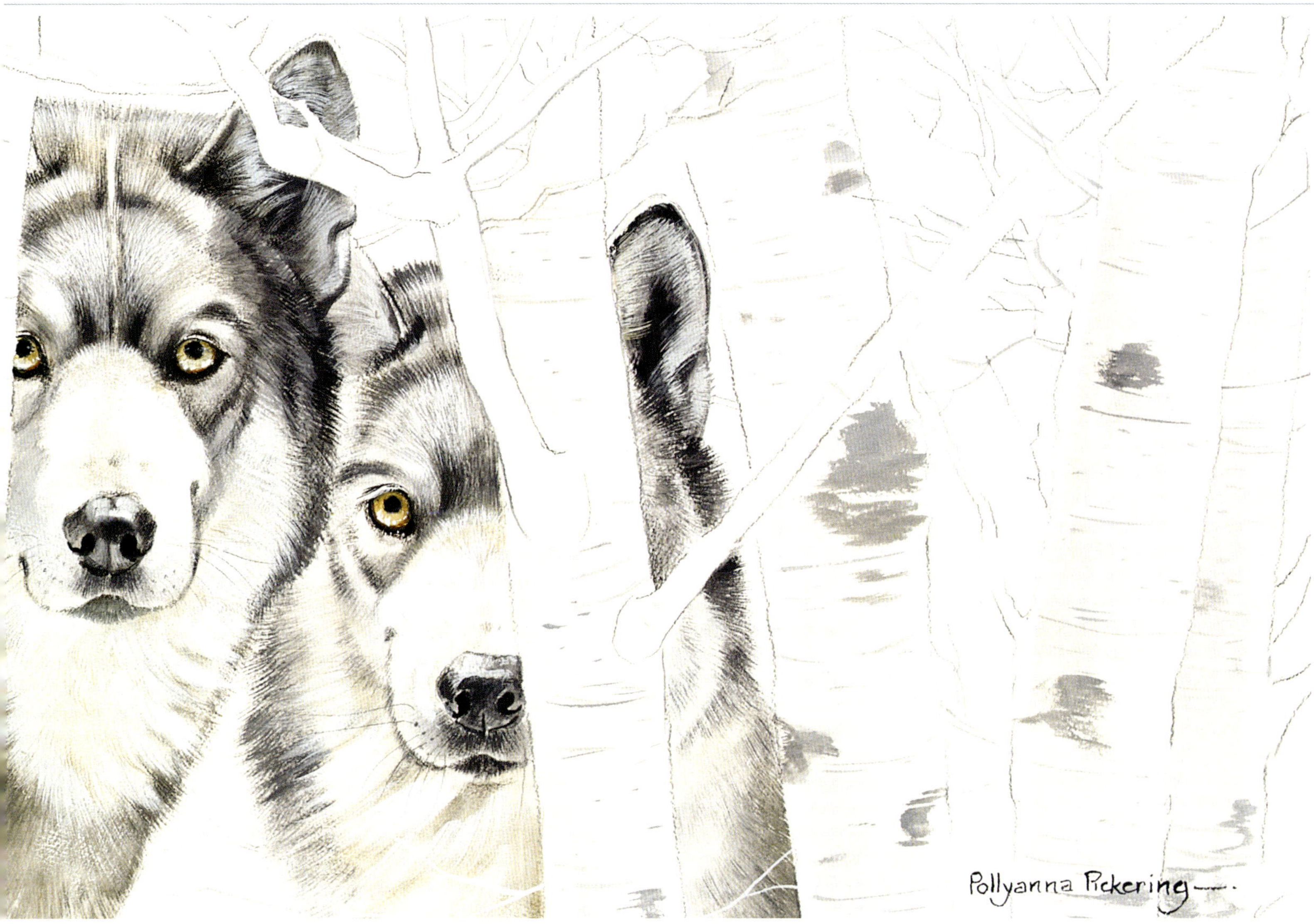

Into the Woods
Private collection (USA)

'It is a green hollow where a stream gurgles,
Crazily catching silver rags of itself on the grasses;
Where the sun shines from the proud mountain:
It is a little valley bubbling over with light.'

ARTHUR RIMBAUD

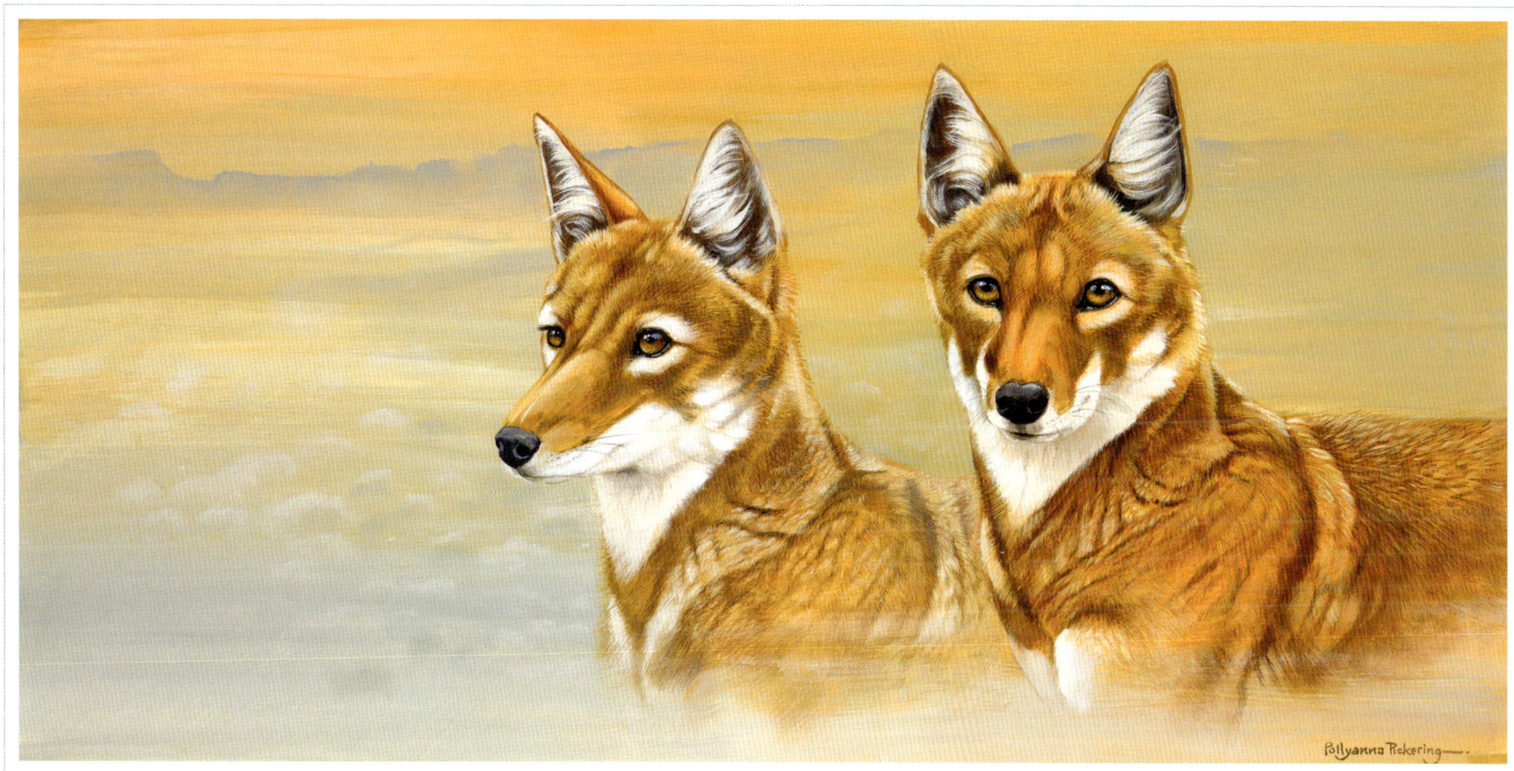

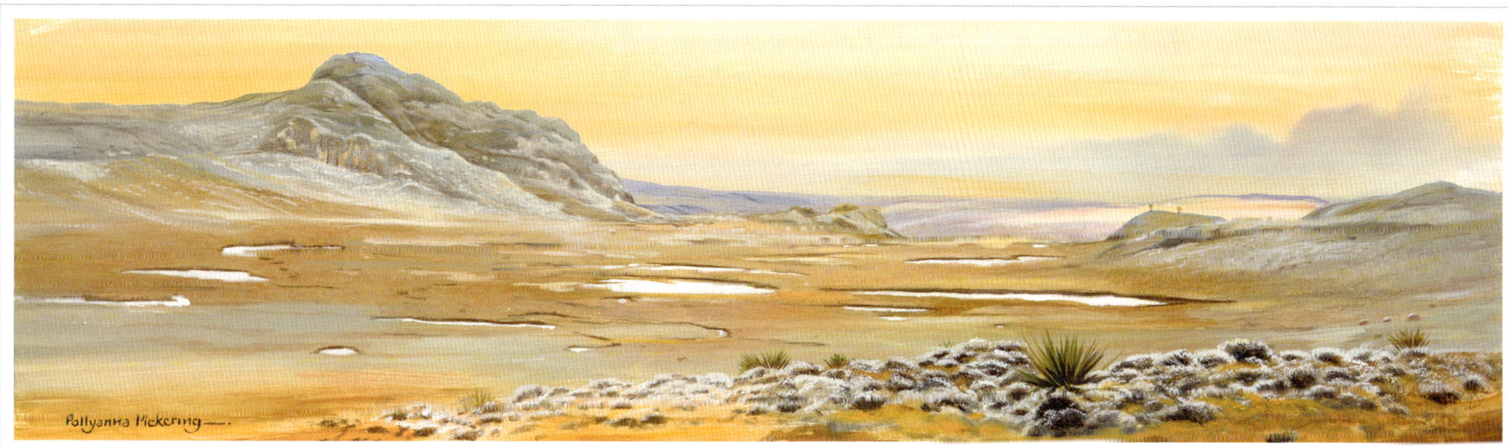

In the Bale Mountains

'We rustled through the leaves like wind,
Left shrubs, and trees, and wolves behind;
By night I heard them on the track,
Their troop came hard upon our back,
With their long gallop, which can tire
The hound's deep hate, and hunter's fire:
Where'er we flew they followed on,
Nor left us with the morning sun;
Behind I saw them, scarce a rood,
At day-break winding through the wood,
And through the night had heard their feet
Their stealing, rustling step repeat.'

LORD BYRON: *MAZEPPA*

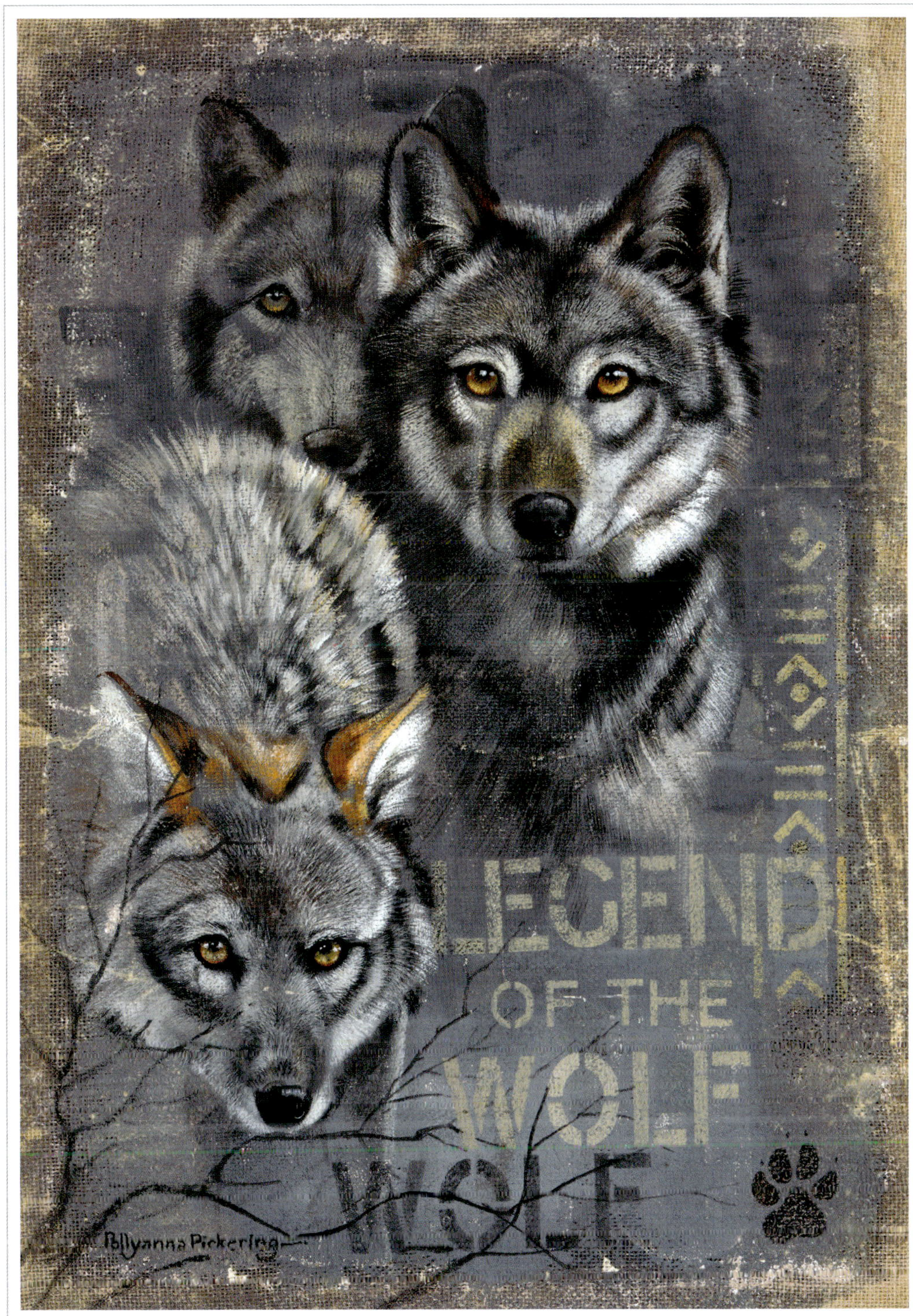

Legend of the Wolf

'Quoth the raven, "Nevermore".'

EDGAR ALLEN POE

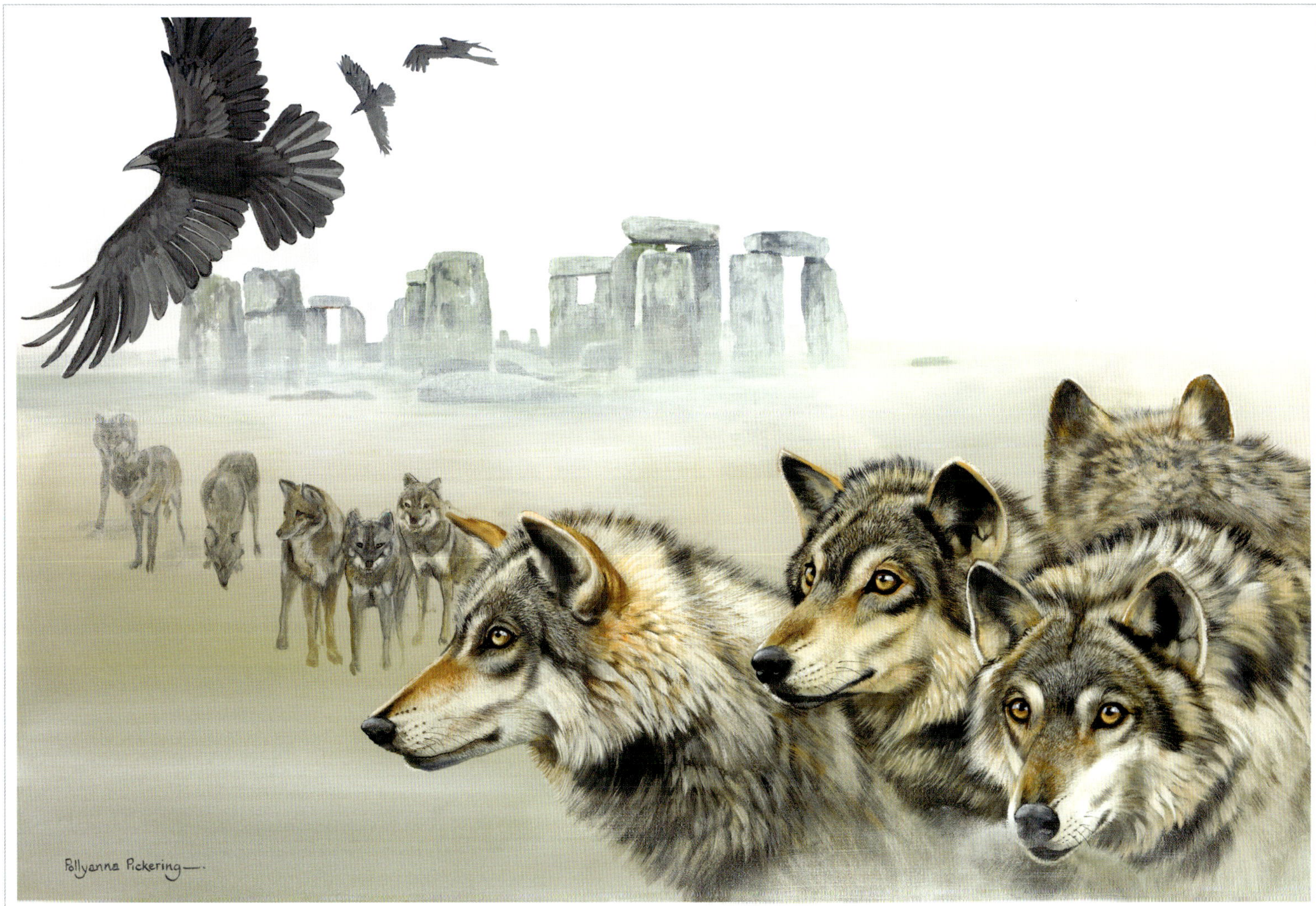

NEVERMORE

'For the strength of the Pack is the Wolf,
and the strength of the Wolf is the Pack.'

Rudyard Kipling

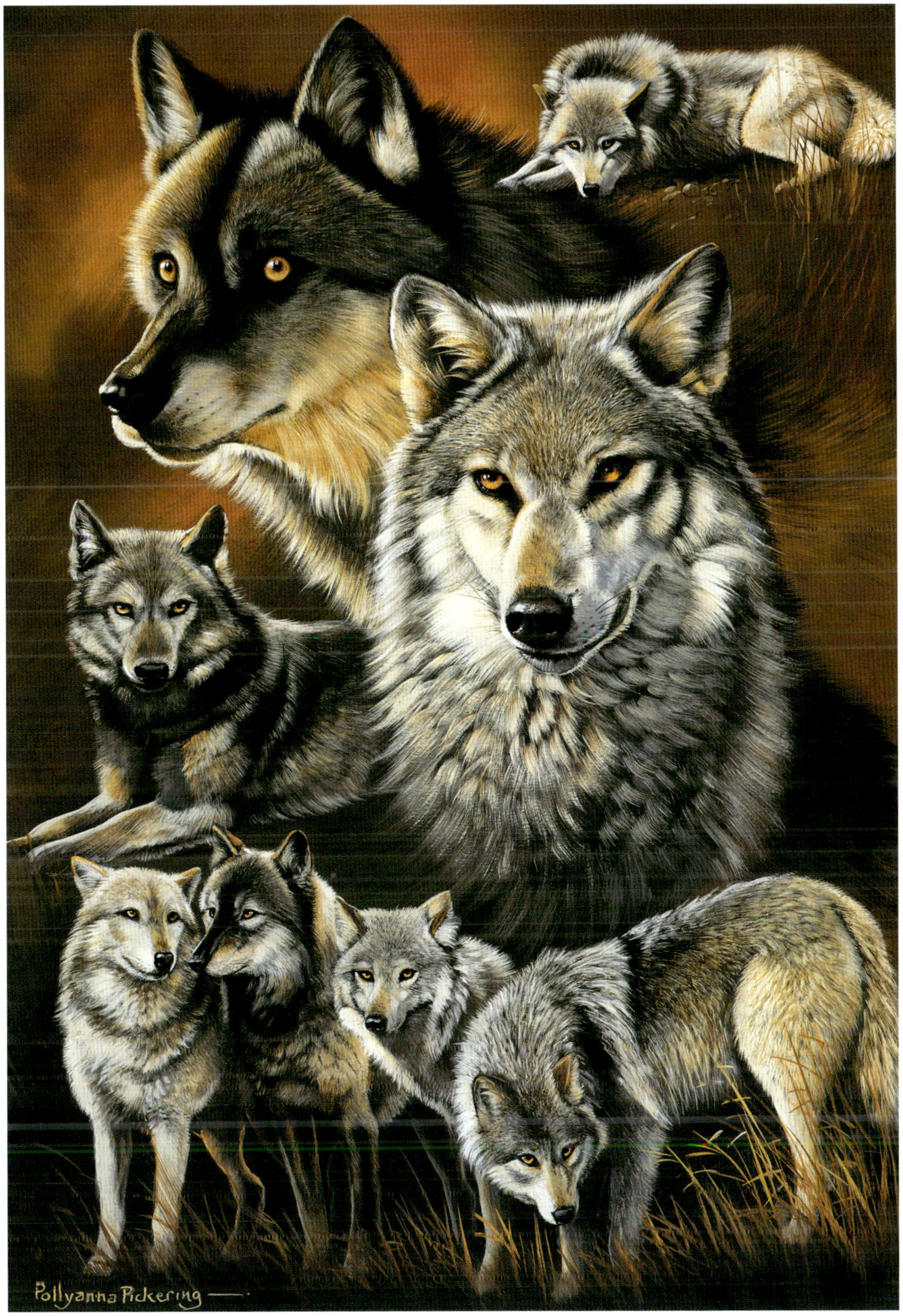

Wolf Watch
Private collection (UK)

'Transylvania had been a familiar name as long as I could remember. It was the very essence and symbol of remote. Leafy, half-mythical strangeness; and, on the spot, it seemed remoter still, and more fraught with charms.'

PATRICK LEIGH FERMOR: *BETWEEN THE WOODS AND THE WATER*

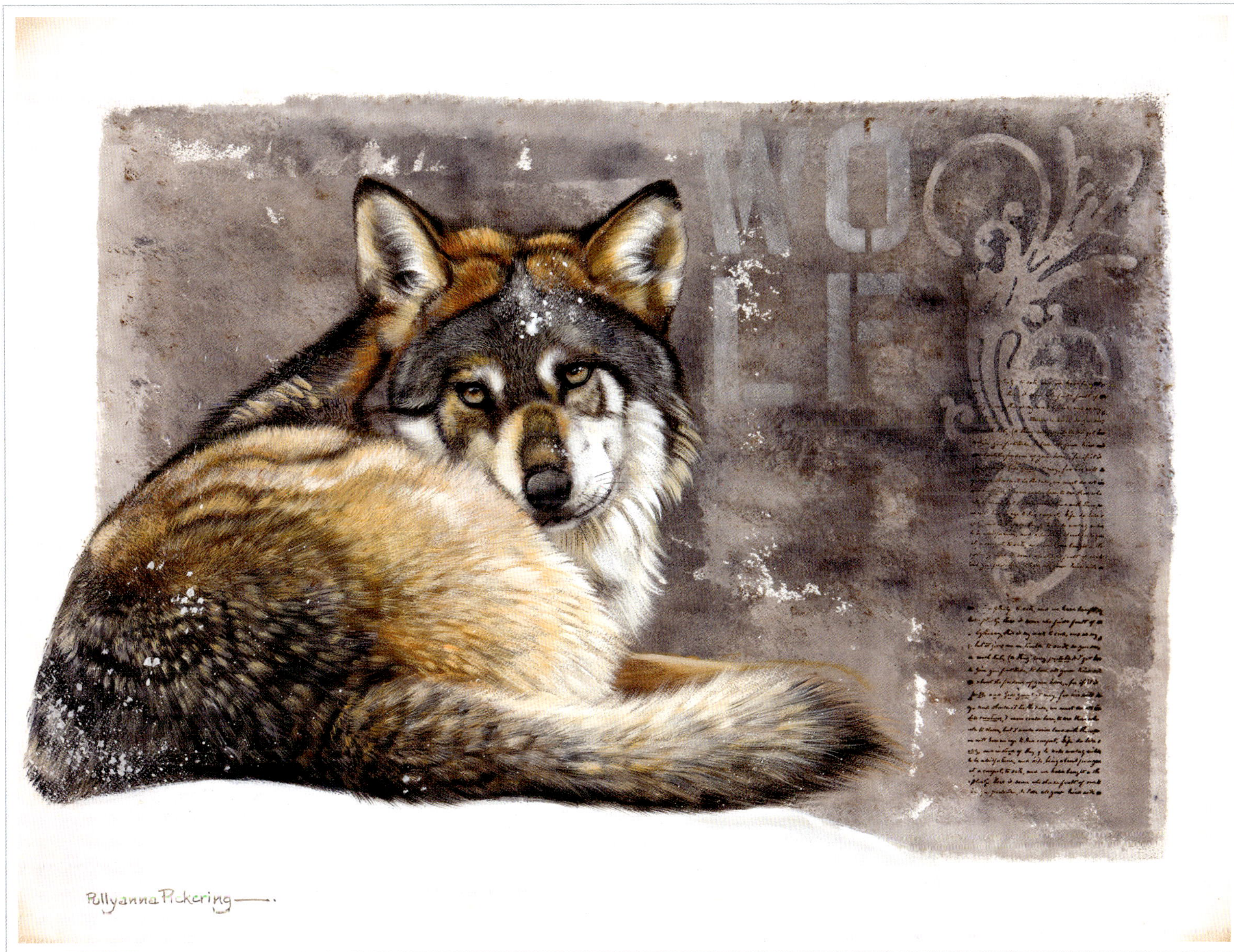

Myths and Legends
Private collection (UK)

'With the still of the frost comes the wolf, wolf, wolf,
The gaunt red wolf at my door.
He's as tall as a Great Dane, with his grizzly russet mane;
And he haunts the silent woods at my door.'

BLISS CARMAN: *THE RED WOLF*

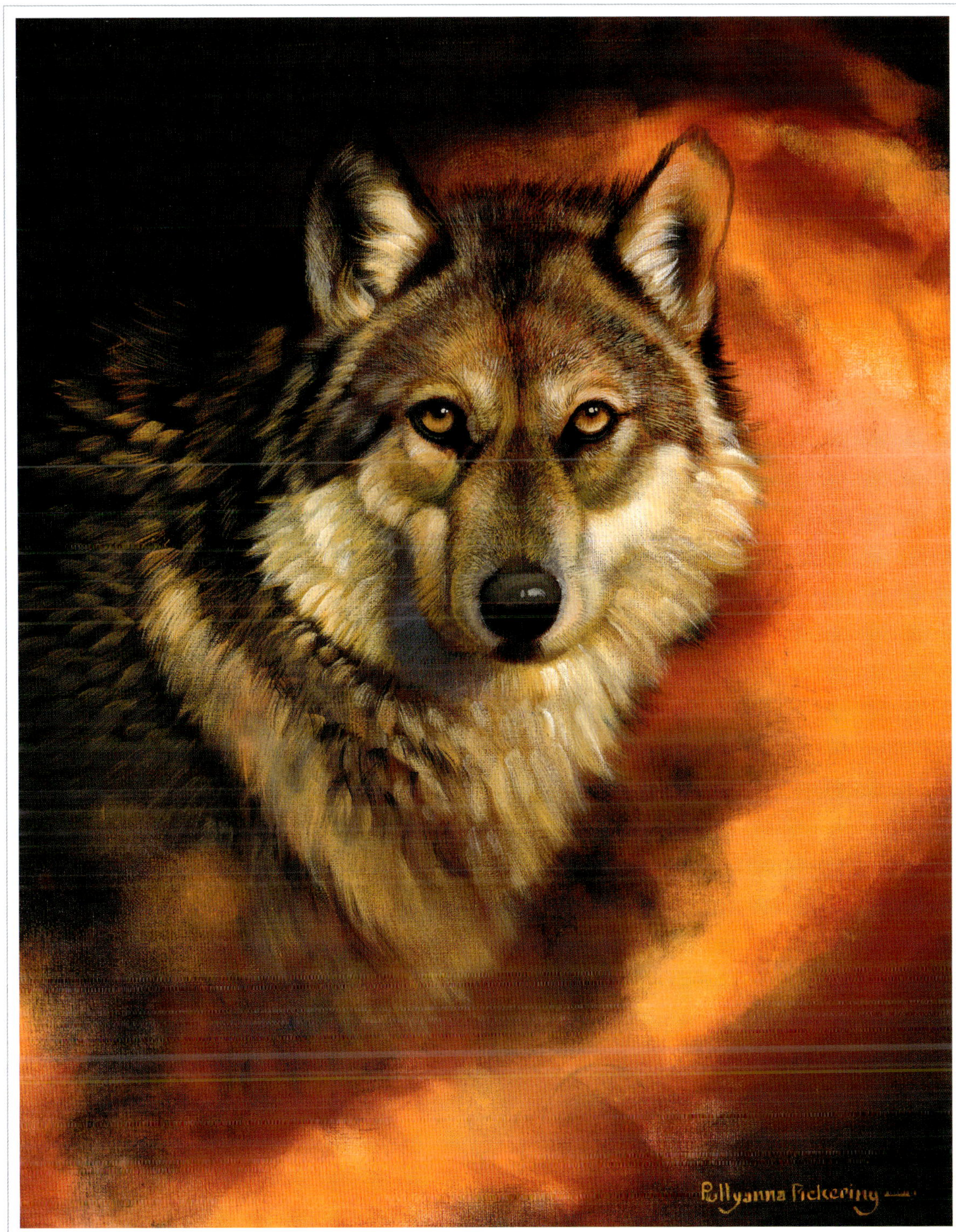

Spirit of the Wolf

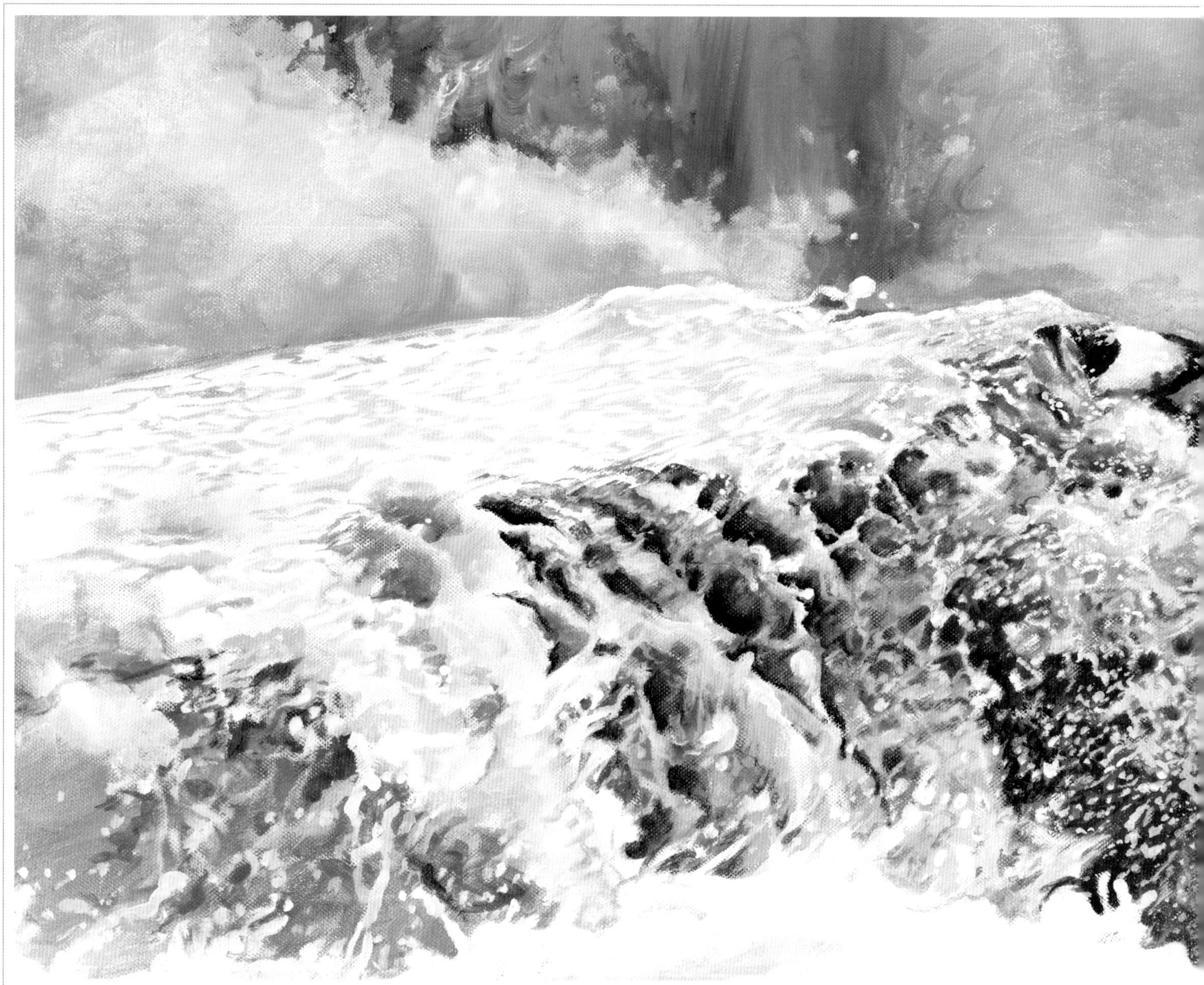

'A mountain with a wolf on it stands a little taller.'

EDWARD HOAGLAND: *RED WOLVES AND BLACK BEARS*

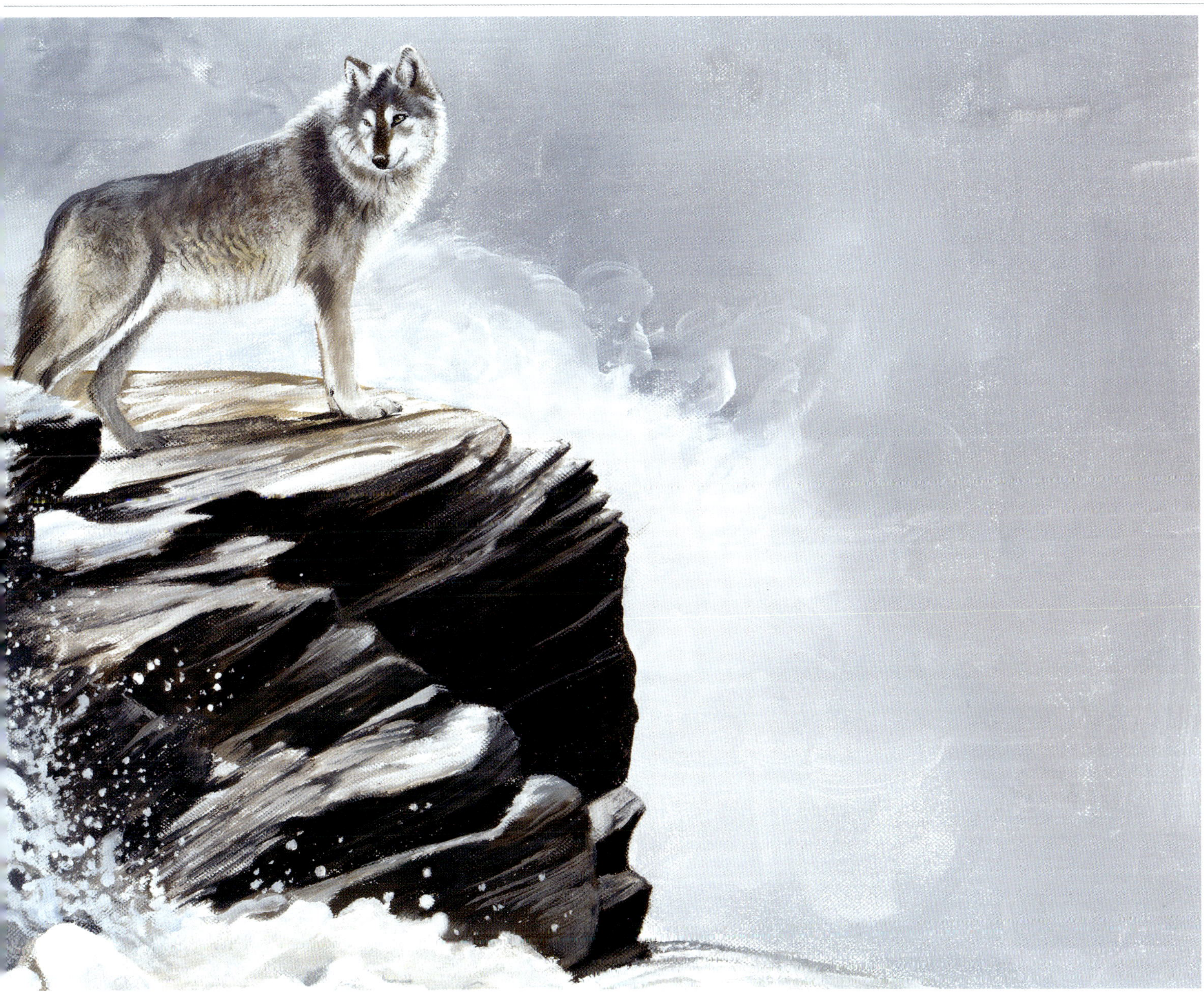

Wild Water

'Our great woods would smell of sap,
And the sunlight
Would dust with fine gold their great
Green and bronze dream.'

ARTHUR RIMBAUD: *LES REPRATIES DE NINA*

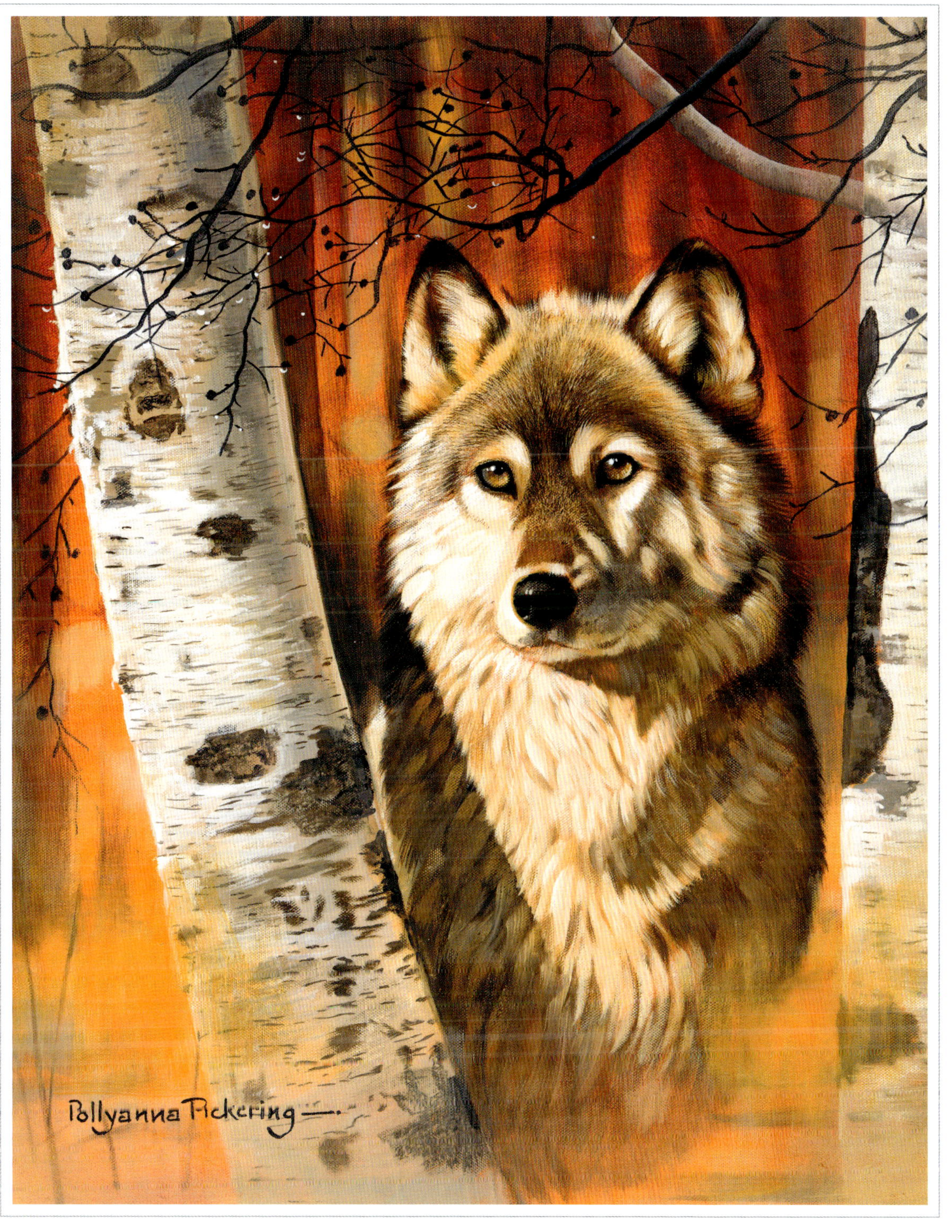

The Birchwoods
Private collection (UK)

*'...each day mankind and the claims
of mankind slipped farther from him.
Deep in the forest a call was sounding,
and as often as he heard this call,
mysteriously thrilling and luring,
he felt compelled to turn his back
upon the fire and beaten earth around it,
and to plunge into the forest, and on
and on, he knew not where or why;
nor did he wonder where or why, the call
sounding imperiously, deep in the forest.'*

JACK LONDON: *WHITE FANG*

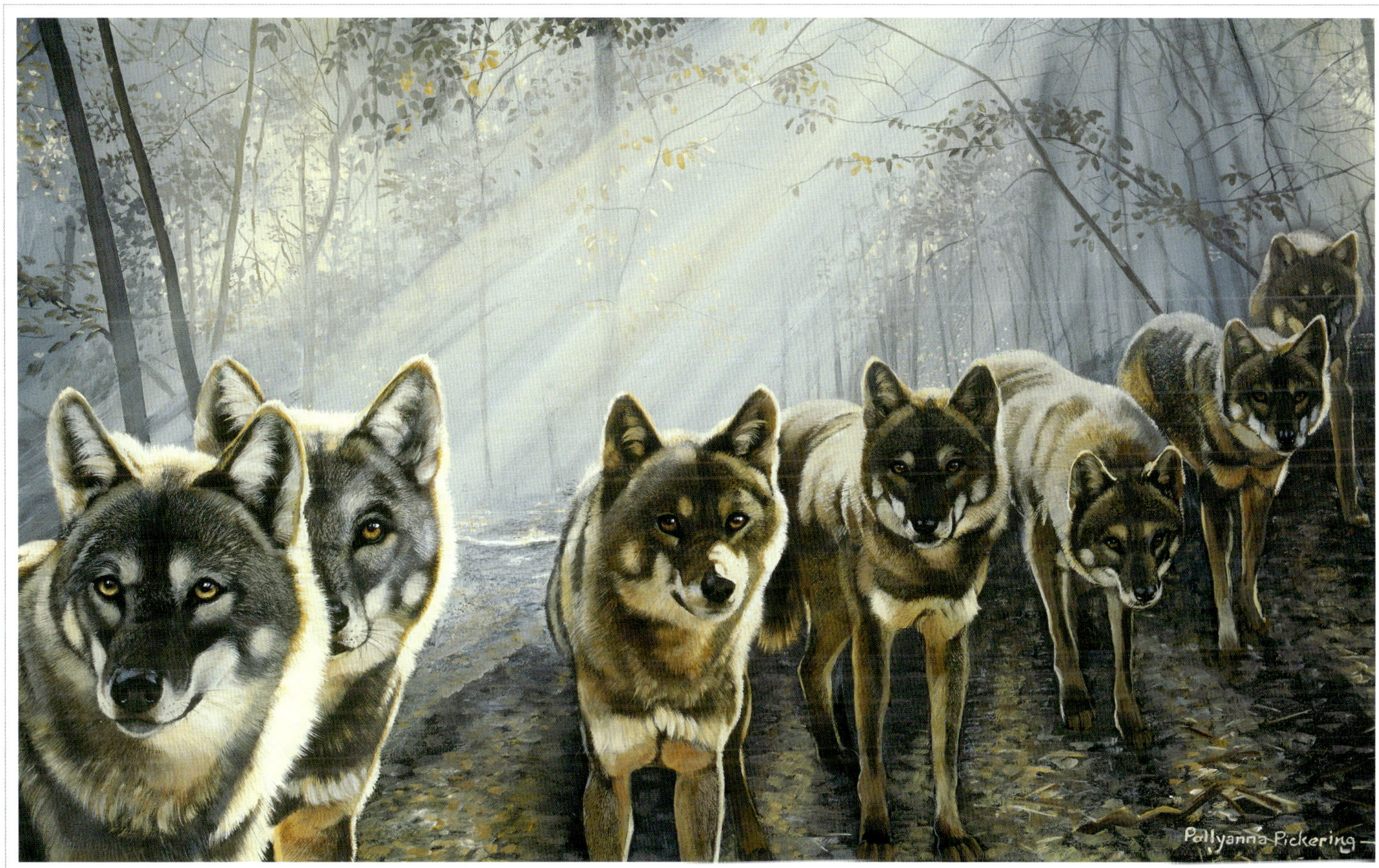

WHERE THE WILD THINGS ARE

'We call the wolves
The powerful spirits
Who look like dogs
When they climb the hills
To lift their heads and sing
Toward that road of stars
Their songs grow stronger
As they join their voices.'

JOSEPH BRUCHAC AND JACK LONDON: *THIRTEEN MOONS ON TURTLE'S BACK*

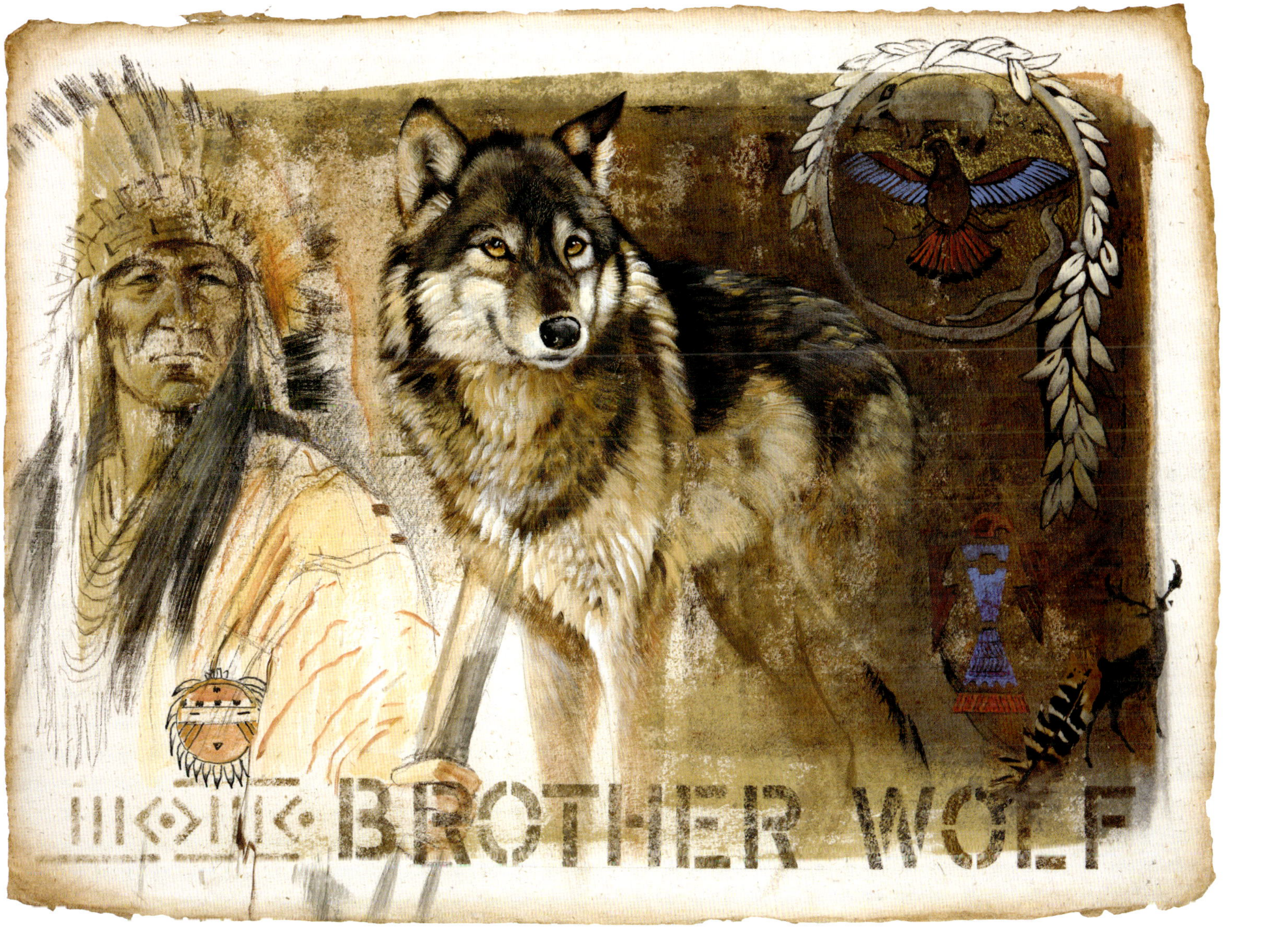

Brother Wolf

'They live within the wildness still
Beyond the boundaries of our will
Not like our faithful stay at homes
But wraithful – hunting on their own
They are not tied by lead or lash
The wildness always calls them back
Across the snows they scent the scat
Of caribou far down the track
And lope across the endless miles
A bounding ring of snarling smiles
All through the starlit Arctic night
Their borealis eyes alight
They drive on through the drifting flak
A hydra headed hunting pack
All harnessed by a single will
To overtake and close and kill
To howl across the starry dome
Of homelands written in their bones
They will not curl up by the fire
For wildness is their true desire
They live their lives beyond the pale
Their liberty is not for sale
They thrive way off the beaten track
Where White Fang climbs the mountain's back
To howl across her vast domain
A landscape hung with curtained flames.'

RICHARD BONFIELD: *WOLVES* (*FOR* JACK LONDON)

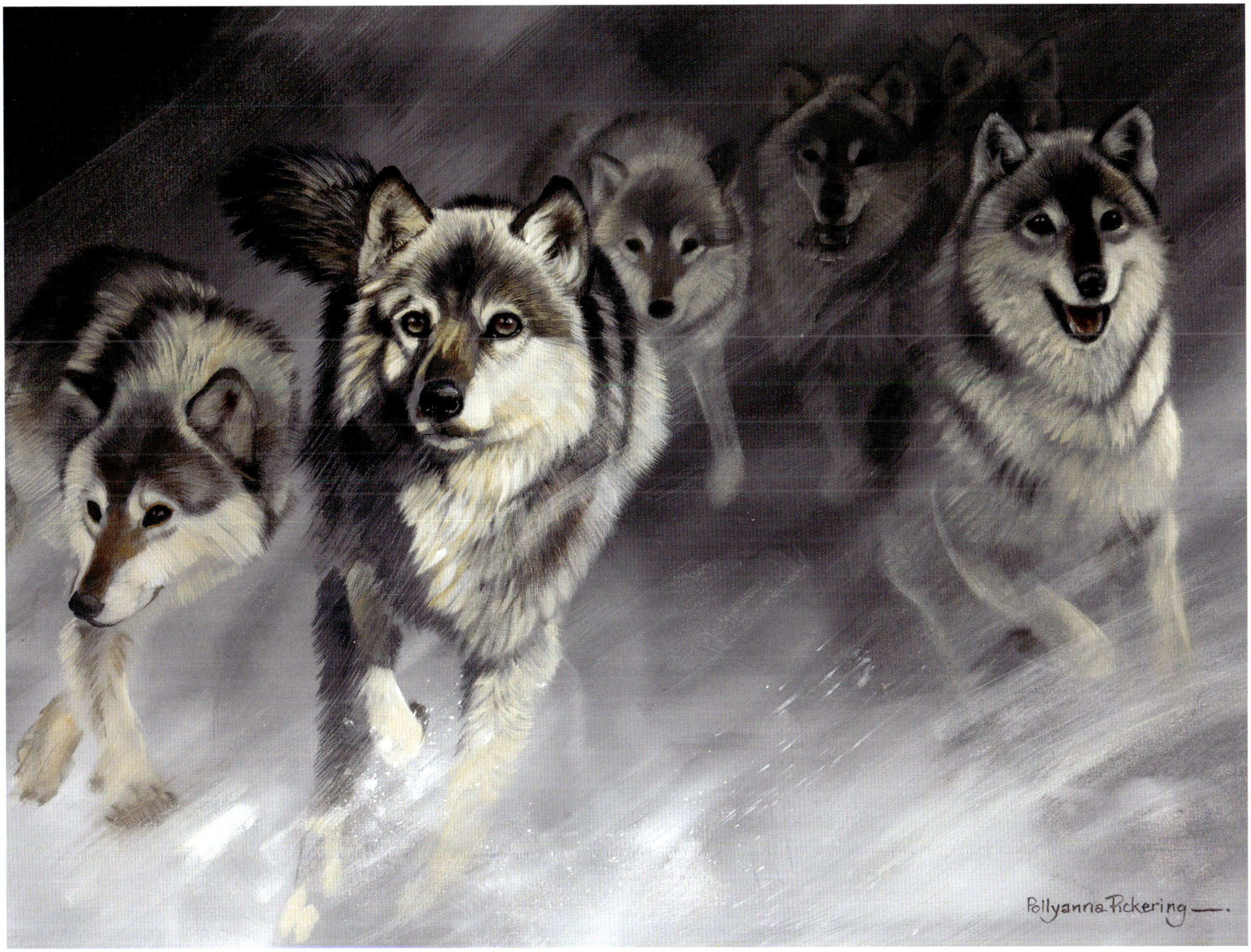

THE COMPANY OF WOLVES

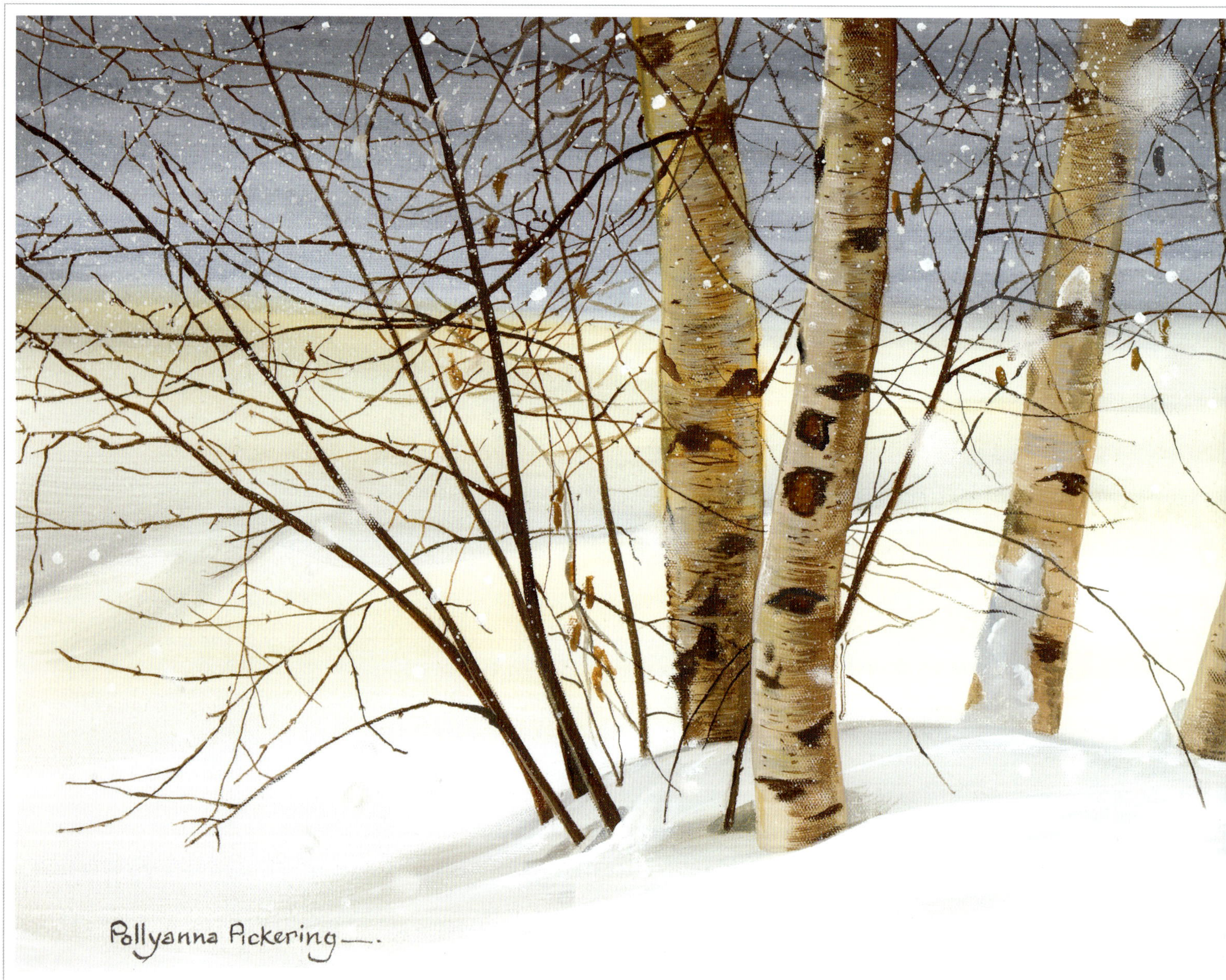

'The wolf trots to and fro
The world lies deep in snow
The raven from the birch tree flies,
But nowhere a hare, nowhere a roe...'

HERMANN HESSE: *STEPPENWOLF*

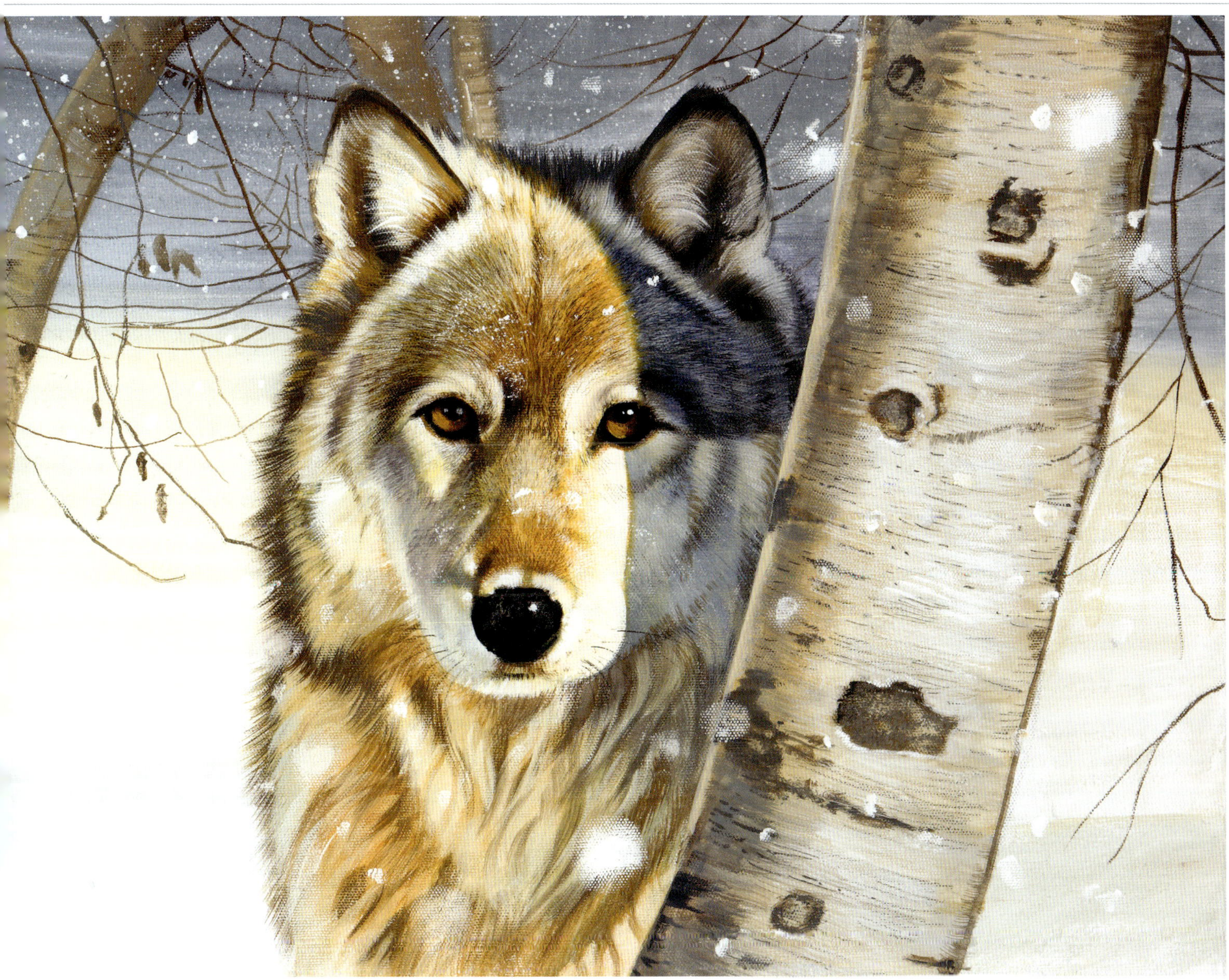

The Enchanted Forest
Private collection (UK)

'Wolf is the Grand Teacher.
Wolf is the sage, who after many winters
upon the sacred path and seeking the ways
of wisdom, returns to share new knowledge
with the tribe.
Wolf is both the radical and the
traditional in the same breath.
When the Wolf walks by – you will remember.'

ROBERT GHOST

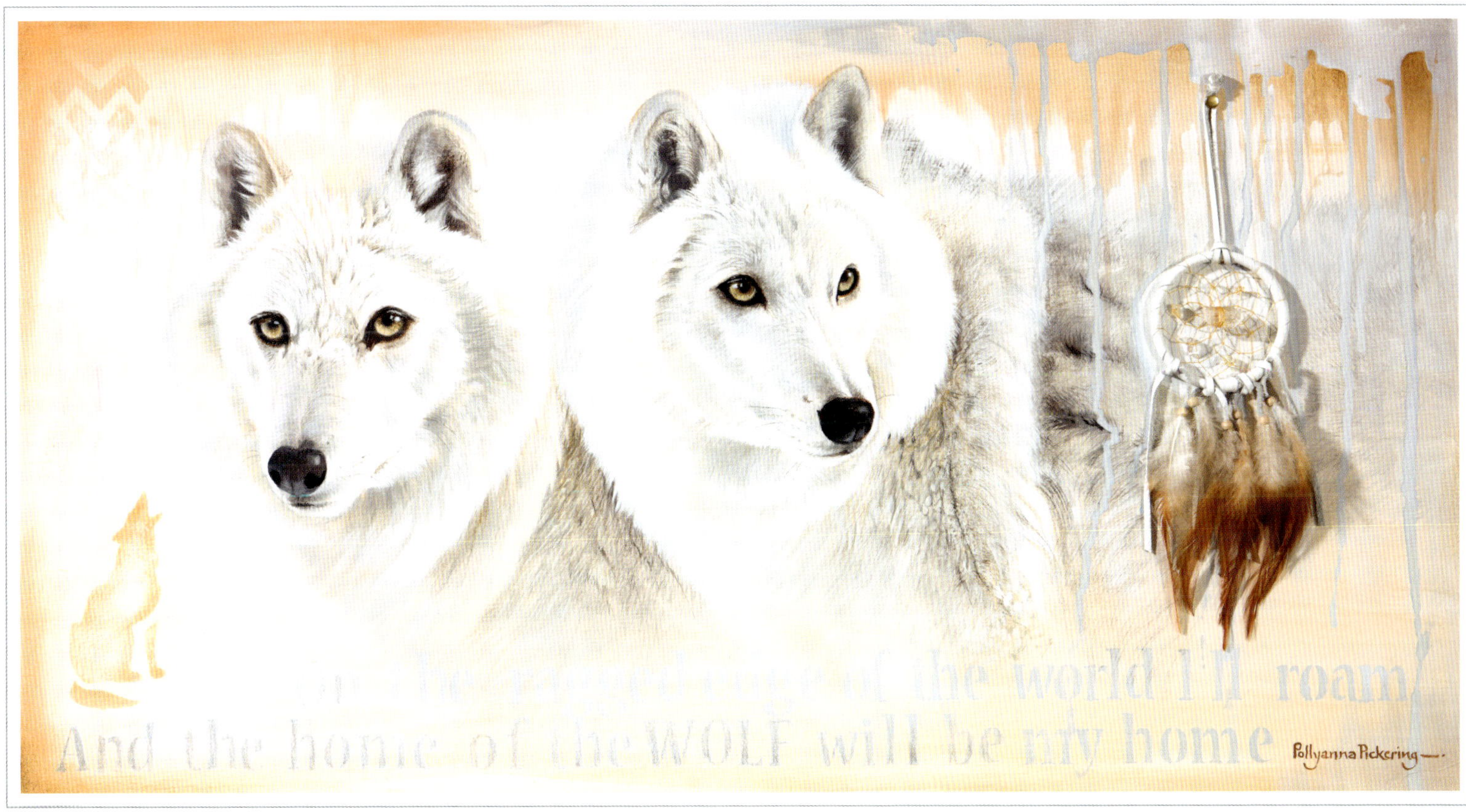

THE HOME OF THE WOLF

'The origin of Ethiopia is largely the origin of us all.'

Javier Gozalbez

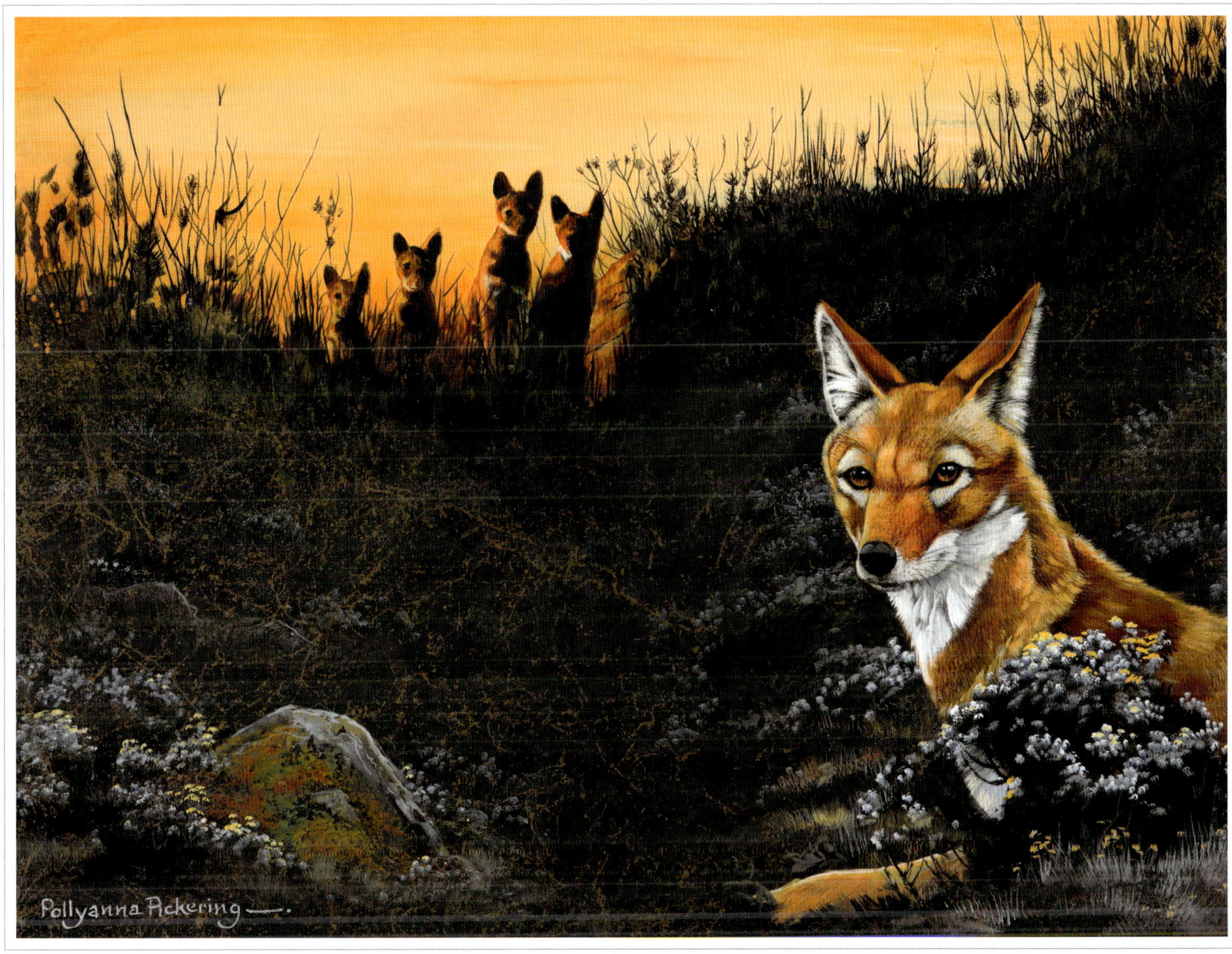

The Future of Ethiopia

'The world needs an embodiment of the frontier mythology, the sense of horizons unexplored, the mystery of uninhabited miles. It needs a place where wolves stalk the strand lines in the dark, because a land that can produce a wolf is a healthy, robust and perfect land.'

Dr Robert Weeden: *Man in Nature: A Strategy for Alaskan Living*

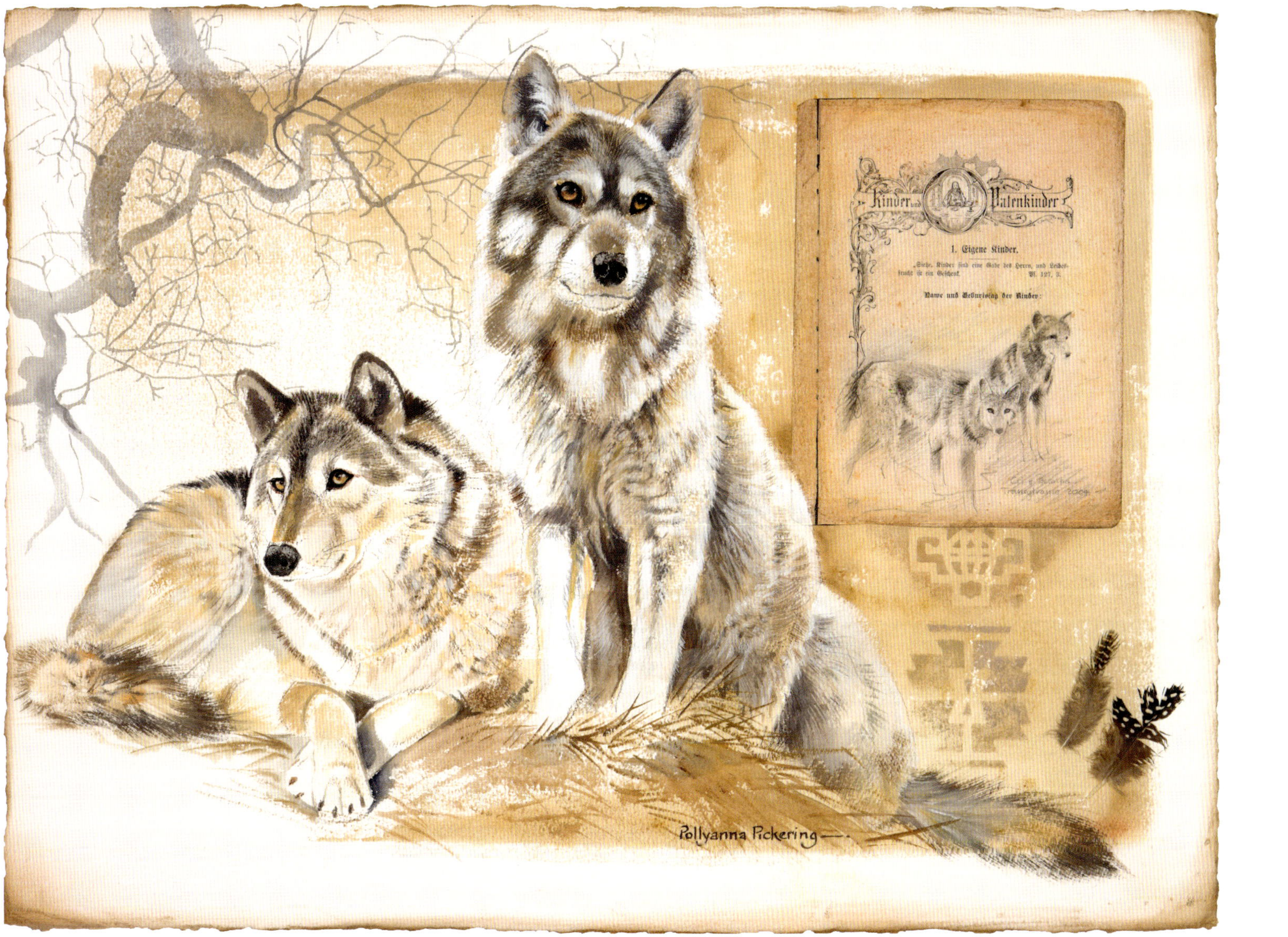

Once Upon a Time...

'When, standing on the plain,
he heard round about him
Living Nature answer his call;
When the silent trees
cradling the singing bird,
Earth cradling mankind,
and the whole blue Ocean,
And all living creatures loved.'

ARTHUR RIMBAUD

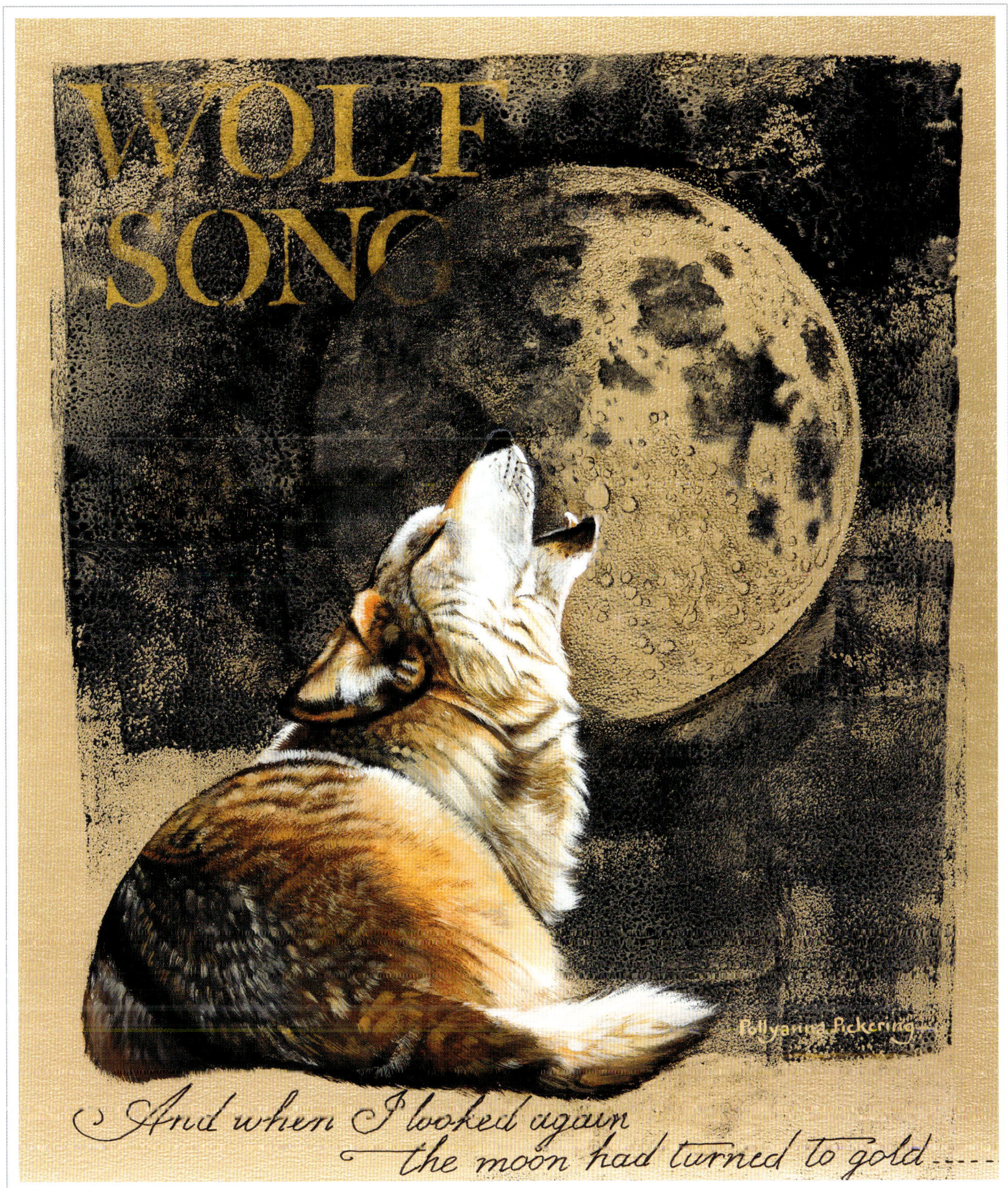

Wolf Song
Private collection (UK)

'Once in the dream of a night I stood
Lone in the light of a magical wood,
Soul-deep in visions that poppy-like sprang;
And spirits of Truth were the birds that sang,
And spirits of Love were the stars that glowed,
And spirits of Peace were the streams that flowed
In that magical wood in the land of sleep.'

SAROJINI NAIDU

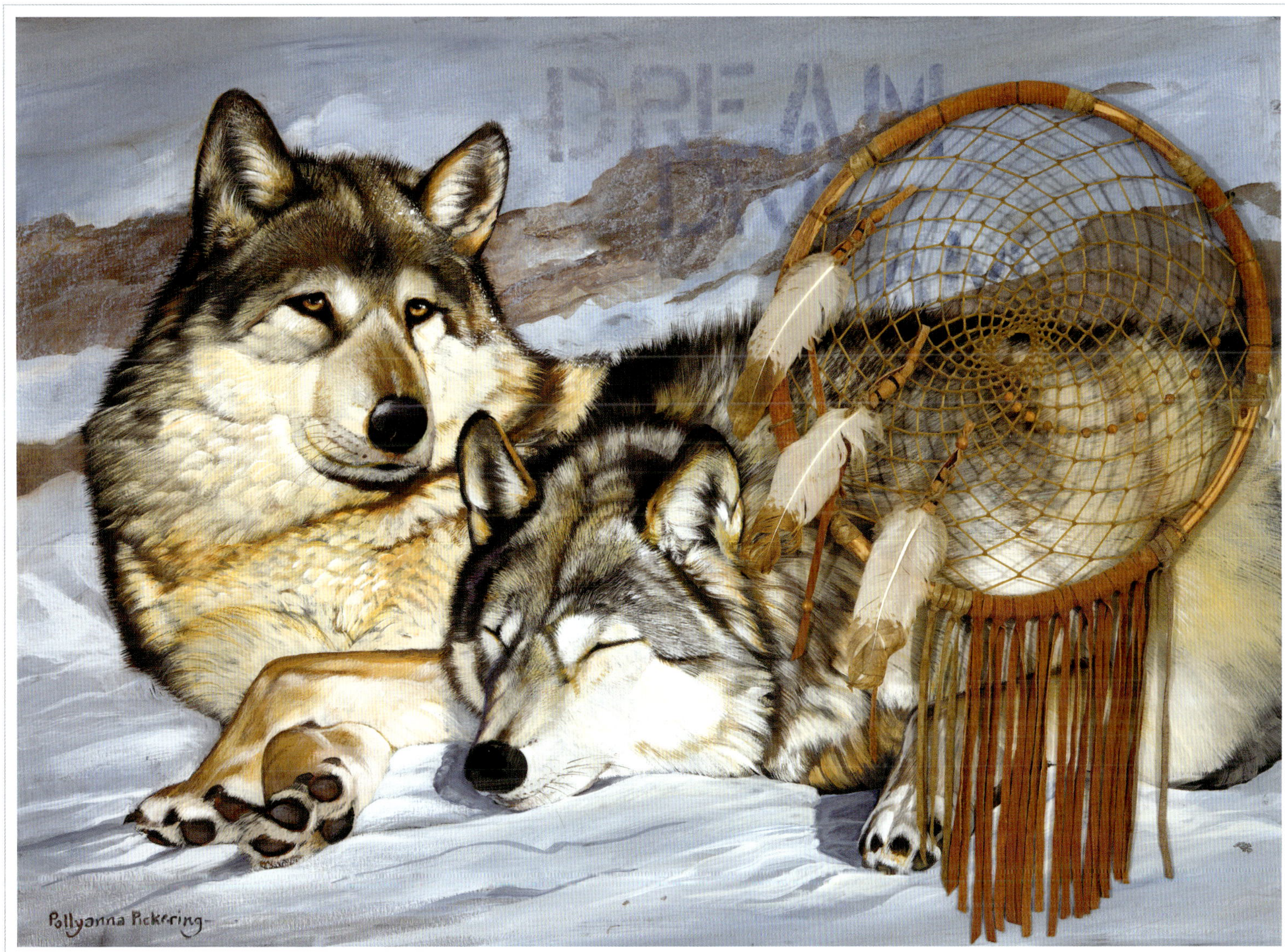

DREAMCATCHER

'Wolves may feature in our myths, our history, our dreams. But they have their own future, their own loves, their own dreams to fulfill.'

ANTHONY MILES

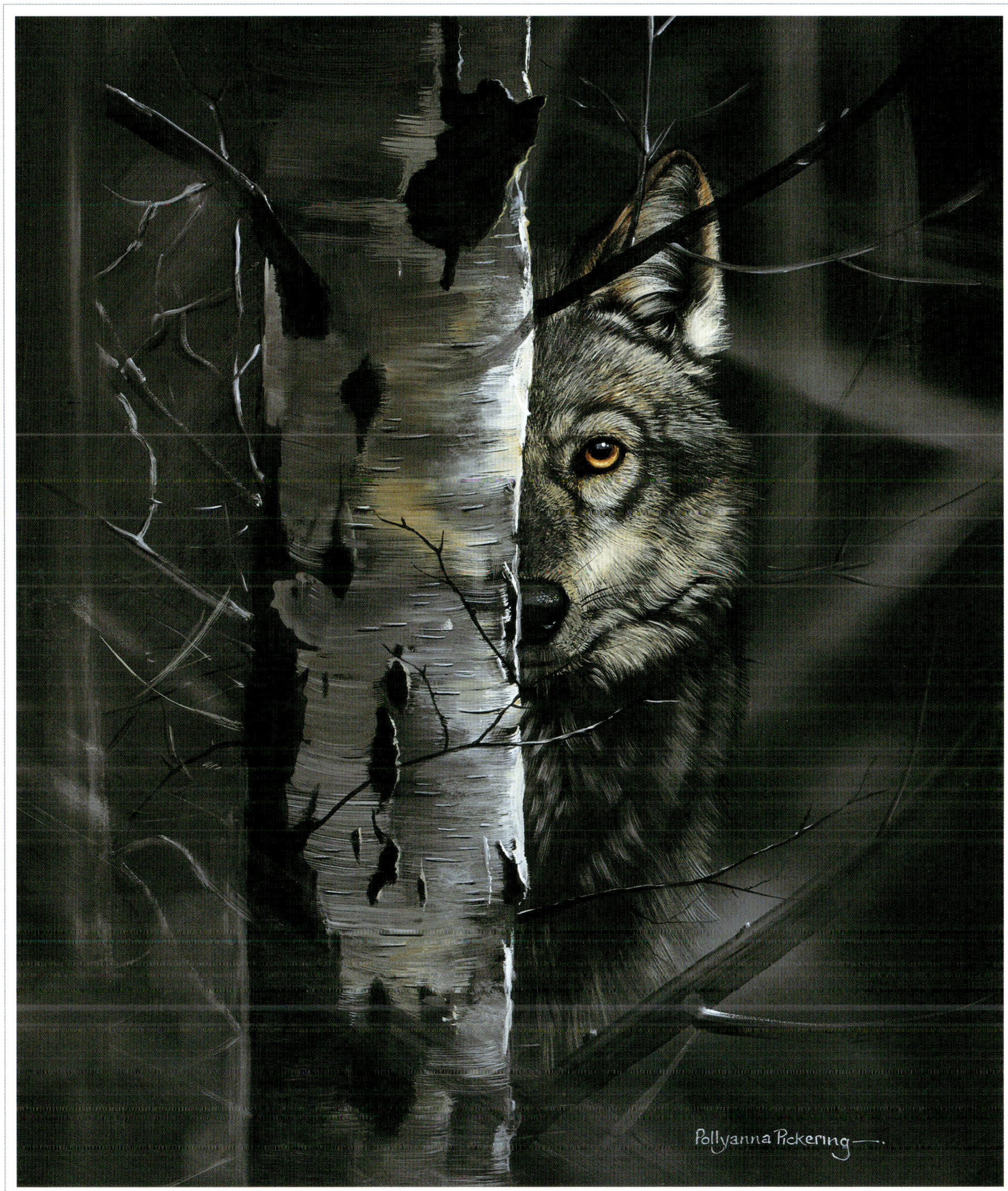

Midnight Hour
Private collection (UK)

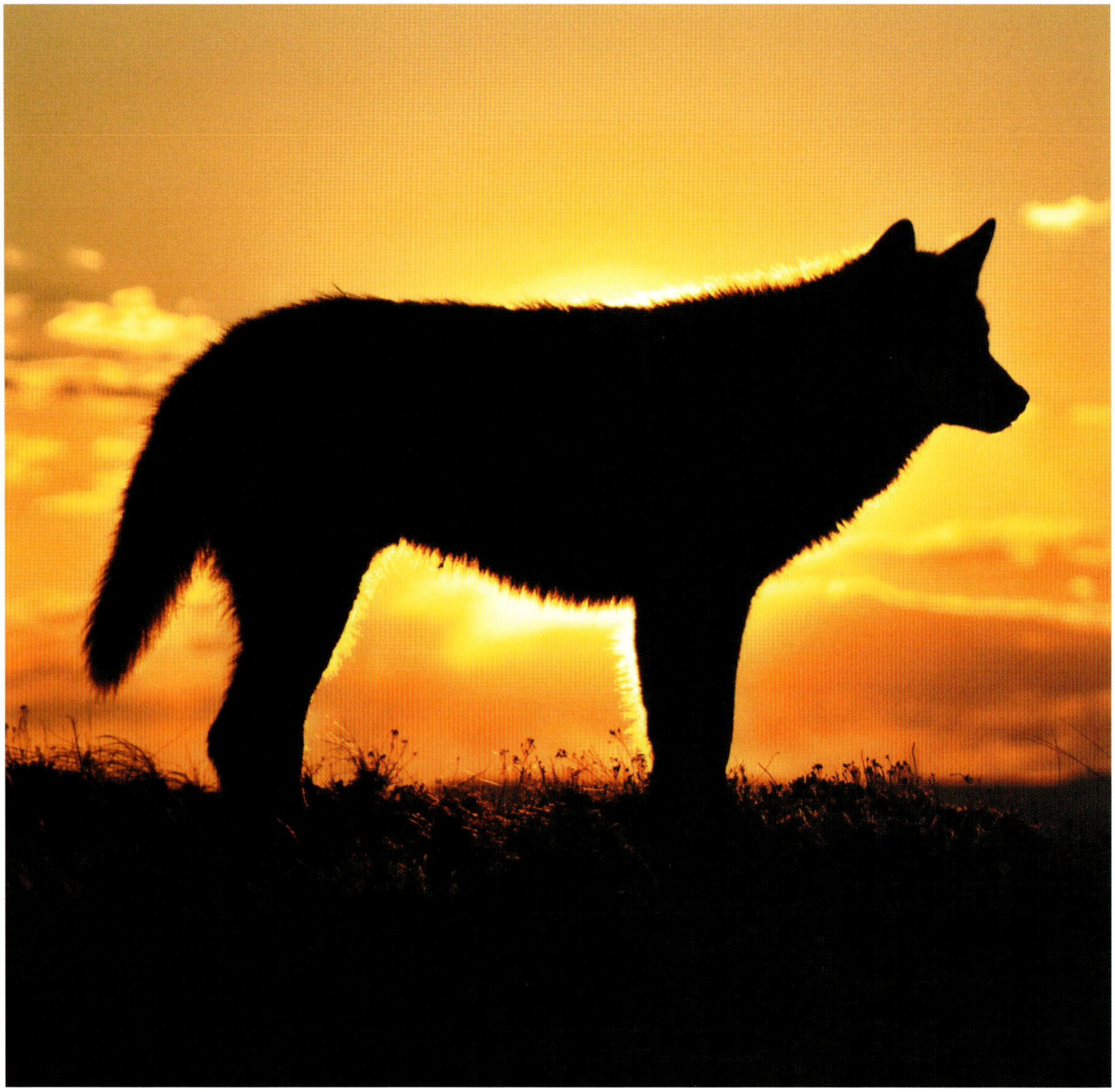

Afterword

'Foremost, in my view, is nurturing a closeness with nature, and resisting the notion that it is too late to stop the humankind's juggernaut destroying our wonderful biosphere. Against the odds, we should continue to work and campaign for a change in the current trends of biodiversity loss, and secure a future, at least for some of this grandeur of life, not just for us, but for generations to come.'

Professor Claudio Sillero

With less than five hundred individuals remaining, the Ethiopian wolf could easily be the next of earth's large mammals to be become extinct. They are threatened by persecution, loss of habitat and the spread of diseases carried by domestic dogs. Sometimes people ask why we invest so much time, emotion and money in creating a safe haven for a species like the Ethiopian wolves, which are teetering on the brink of oblivion. If they vanish the vast majority of people in the world will never even know that they once existed. Does it really matter if they slip into extinction, will our world be any different without them? The stark truth is, as with many wild creatures, their survival is inextricably linked to our own. In our modern society, where we are both protected from and deprived of wilderness by a ring fence of computers and televisions, indoor plumbing and convenience foods, it is easy to forget that we are still part of one gigantic ecosystem. If, as the chaos theorists hold, the flap of one butterfly's wings in the rainforest can cause a tornado half a world away, then the extermination of an entire species can certainly bring about a cascade of events which could blight an enormous part of a continent.

Mountains influence both local and global climates. Their dense forests absorb atmospheric carbon and replenish the oxygen in the air we breathe and serve as vital water catchments. In areas of pristine habitat the vegetation absorbs excessive rainfall which is then gradually released. Stream flows are regulated by the absorption of the water into the underlying soil, followed by its gradual release into the streams and rivers of the watershed. This minimizes both downstream flooding and drought conditions.

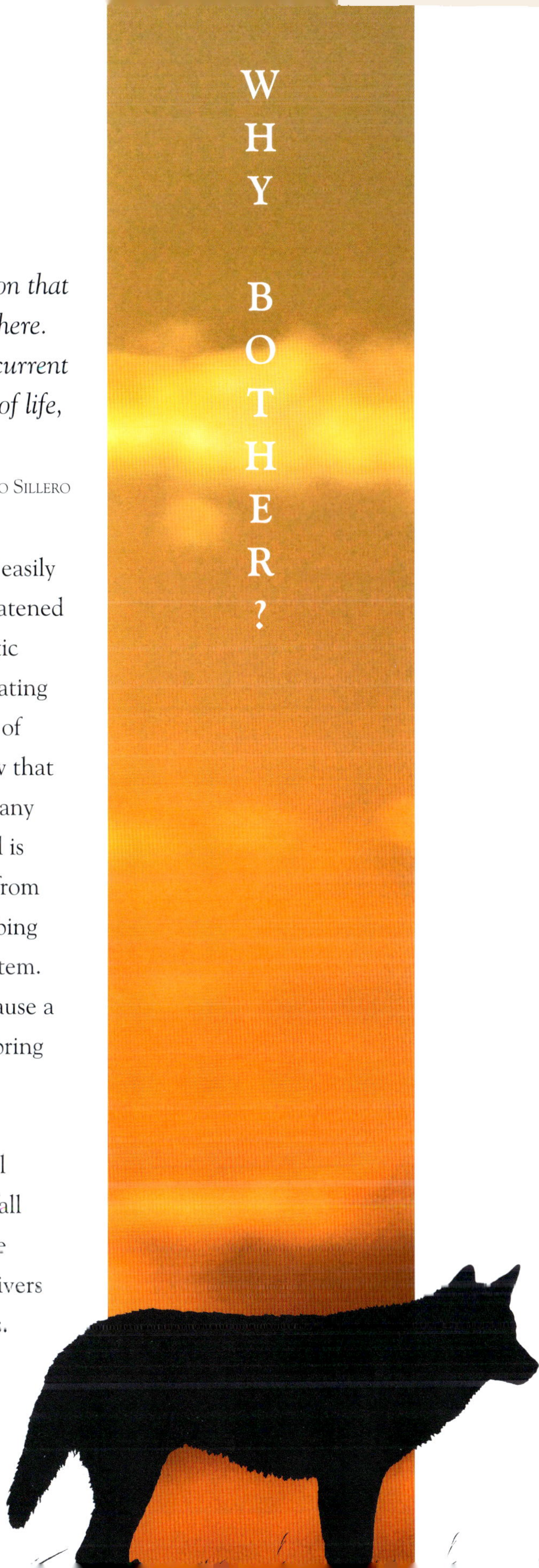

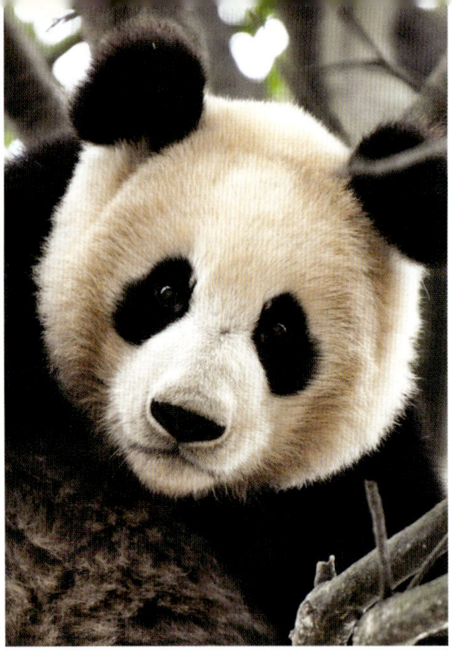

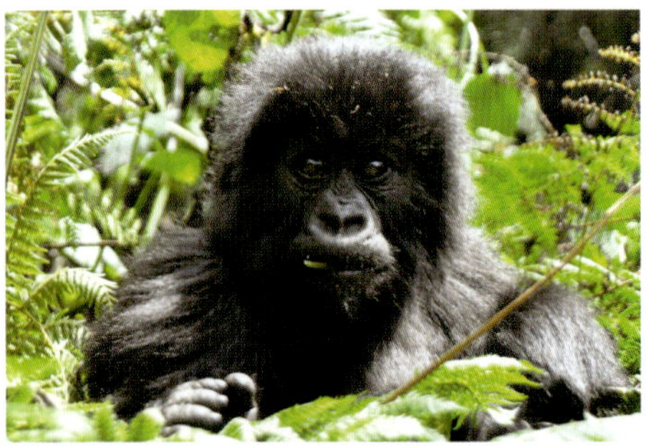

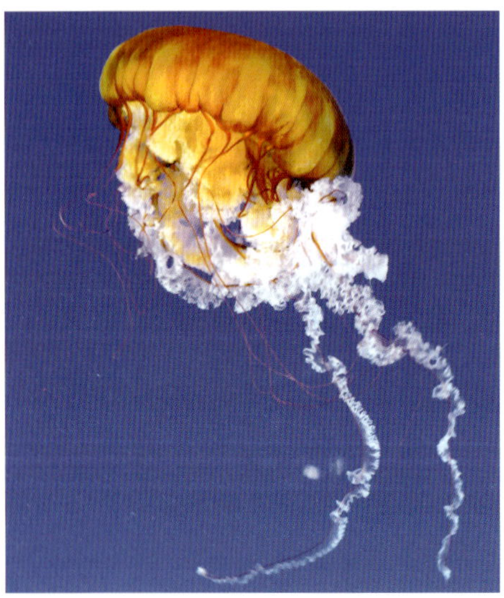

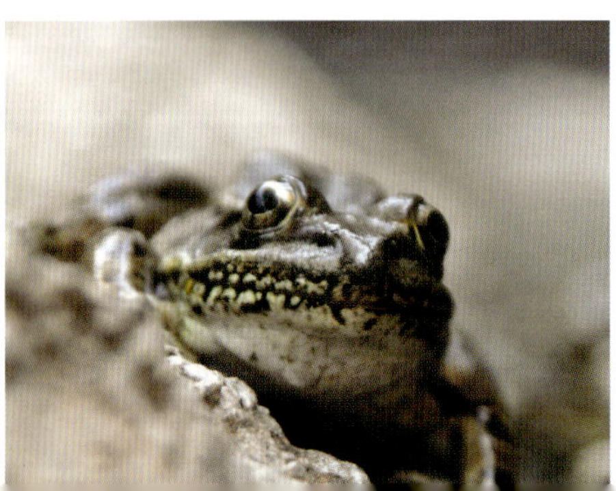

Deforestation causes changes to climatic conditions, potentially resulting in eventual desertification.

Ethiopia is sometimes referred to as East Africa's 'water tower', because of the rivers which flow down from its high tableland. The mountains and forests of the Bale Mountains ecosystem create a critically important water catchment area for southern Ethiopia, western Somalia and Kenya, supplying perennial water to twelve million people in the lowlands. If this unique habitat, which supports the wolves, were to be lost it could lead to a drought and subsequent famine more far reaching and potentially disastrous than any we have previously witnessed. This region has frequent wind-driven clouds from which the forest can 'harvest' atmospheric moisture. If the scrub and vegetation were to be cleared from the plateau and mountain sides for cultivation or housing, the environmental change would result in the vital rain-giving clouds passing over without releasing their precious water.

Claudio Sillero sees the Ethiopian wolf's survival as key to the continuation of the whole highland ecosystem. As the top carnivore, it is responsible for controlling the population of smaller grazing herbivores, especially rodents, and preserving the delicate balance of the region: 'The wolves reign there; I like to think of them as the guardians of the high mountains of Africa'.

In most wild habitats the continued survival of the 'flagship' species, the large predators at the top of the food chain, is the best indicator that biodiversity is working as it should. That means healthy rivers are supporting thriving indigenous plant life, which supports insects and small mammals, which in turn feed birds and larger predators. All the parts of the ecosystem are inextricably linked – if any one species is lost, a chain reaction occurs which leads to the eventual collapse of the whole.

Part of our problem is that we still don't understand entirely how the system works. Asian forests regulate rainfall and combat the greenhouse effect. After the monsoons they regulate water flow, their tree roots preventing the erosion of river banks. Without the forest the waters would flood unchecked, crops would disappear and millions of people would die of famine or be swept away in raging waters. The key to the survival of an entire forest in India might be the tadpoles of a tiny uninteresting brown frog that, by eating the swimming larvae of a specific kind of insect, prevents the explosion of a plague which would strip the foliage from the forest, which in turn would lead to a famine for the deer which are the prey of the tiger. This is a highly over simplified model but it illustrates the point that we don't necessarily know which are the most vital links in the chain – we need to preserve entire habitats to protect eco-diversity.

Conservation purists sometimes complain that too much money and attention is focused on the flagship species at the top of the chain – the wolves, pandas, tigers and lions. In their world lichen and beetles are of equal importance – and I wouldn't argue with the accuracy of that opinion for a moment. But the 'glamorous' predators can only exist in the wild in a healthy ecosystem which supports a wide variety of species, from microscopic insects to thriving herds of deer. And if we think it is difficult to raise funds, change opinions and lobby governments to try to ensure the protection of a jungle which is home to tigers, imagine trying to do the same thing when the tigers have gone, but we still need to protect the uninteresting brown frog and its tadpoles. We can imagine an ecosystem like a tall tower of bricks: we can remove several bricks and the tower will remain standing, but ultimately taking one brick too many will cause the entire structure to come crashing down. And crucially we have no way of knowing for sure which is the critical tipping point, which individual brick will lead to the final collapse.

All of these arguments ultimately reflect a human self-interest – we need to preserve the wolves so humans don't starve, protect tigers because we think they are beautiful and save lions because they create tourism and wealth for poverty stricken countries. Personally I struggle with the idea that wildlife has to justify its existence in these terms – that it in some way needs to be of benefit to us to be worth saving. Surely all these creatures have the right to exist simply for their own sakes, each species, each individual is a miracle of evolution. With our opposable thumbs and creative intelligence, as we have outgrown our

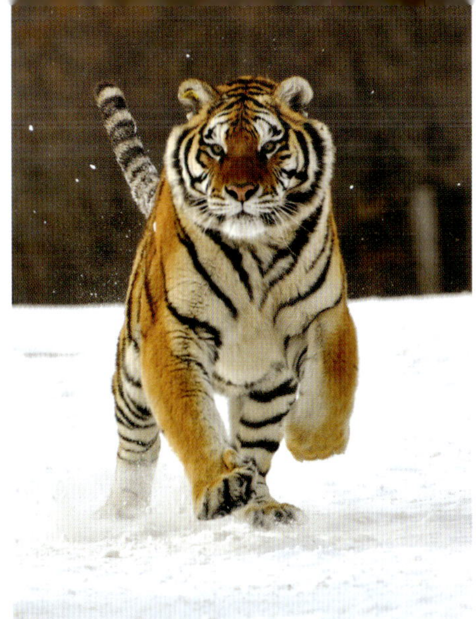

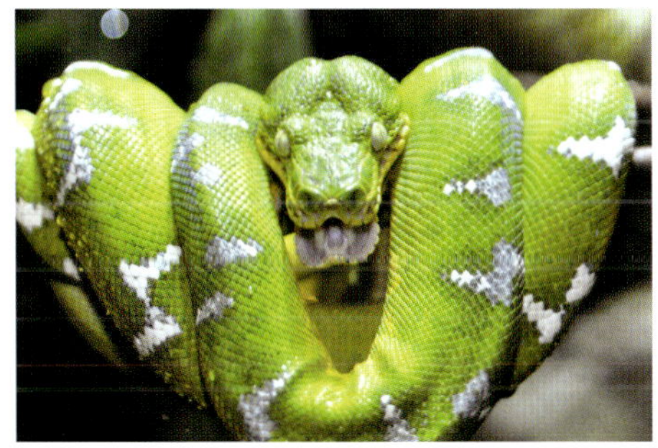

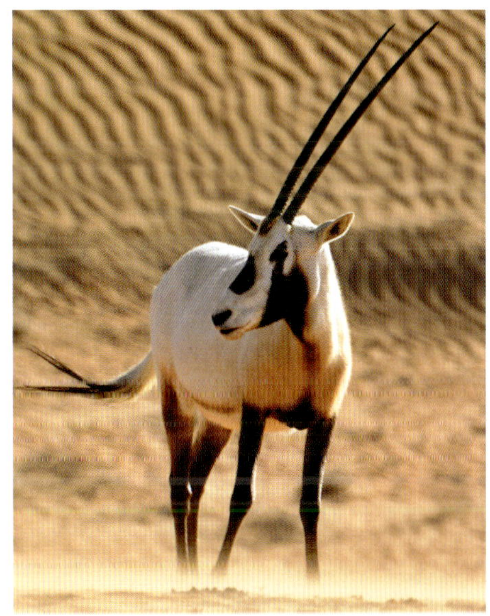

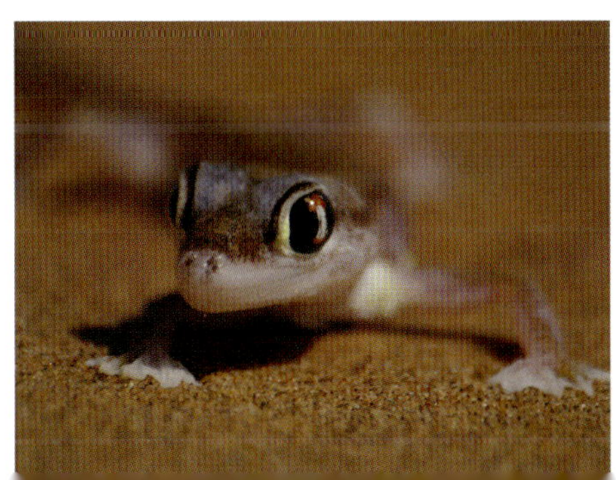

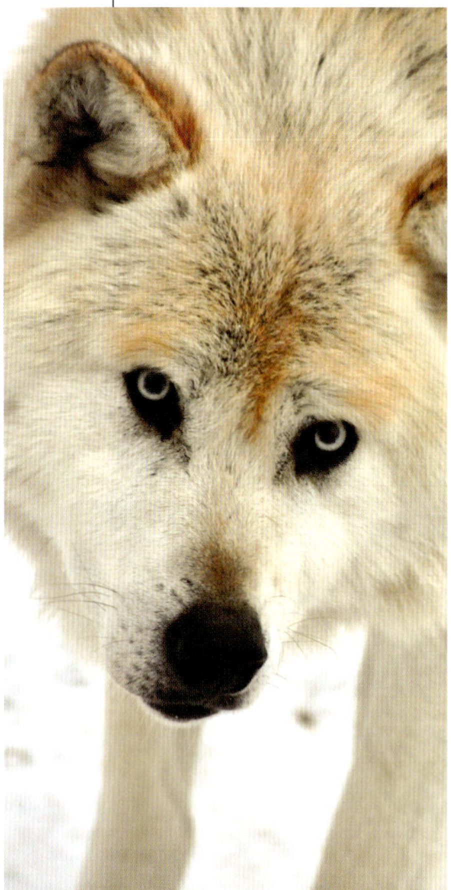

original niche in the ecosystem, we have ravaged the rest of creation through fear, ignorance, greed and need. After centuries of negative interference, when we have hunted species to extinction, slashed and burned forests, poached endangered species for their skins or to stick trophies on the wall, surely we can support people and organisations who are trying to redress the balance in even a small way through some positive interference. With our weapons, our technology and our civilisation we are the undisputed new kings of the jungle – don't we have a duty of care to our kingdom? After all, 'With great power comes great responsibility'.*

In the end, however, it probably doesn't matter why we conserve wildlife – whether for noble ideals or purely out of self-interest – it doesn't make any difference to the wolves what our motives might be. In his book *Last Chance to See* the writer Douglas Adams expressed this thought far more eloquently than I could ever hope to: *'There is one last reason for caring and I believe no other is necessary. It is the reason why so many people have devoted their lives to protecting the likes of rhinos, parakeets, kakapos and dolphins. And it is simply this: the world would be a poorer, darker, lonelier place without them.'*

Organisations like the EWCP offer a glimmer of hope amidst all the negative stories about disappearing wildlife. They have shown that with positive intervention we have the opportunity to save an entire species and that conservation projects can work with local communities for their benefit as well as that of the wider world. They are proof that conservation initiatives can work.

**This quote doesn't come from a respected conservationist, but from the film* Spiderman. *I included it here because it seems so apt, and if nothing else it illustrates that inspiration can be found in unexpected places. I didn't attribute the quote immediately in the text as that might have meant stepping off my conservation soapbox for a moment to consider radioactive spider bites. Incidentally, when I say all species have the right to exist for their own sake, obviously I am not including Daddy Long Legs, which in my opinion are entirely unnecessary.*

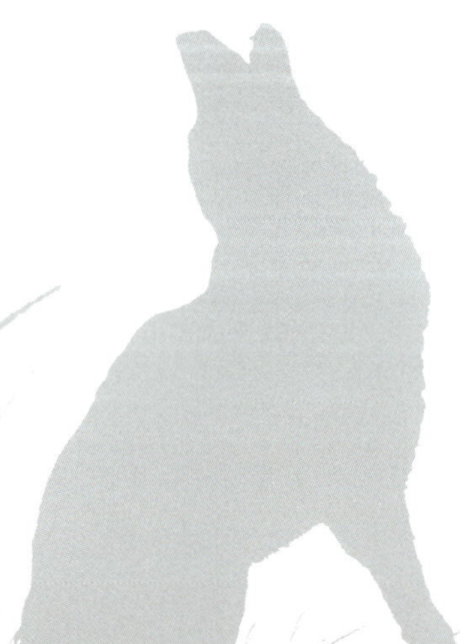

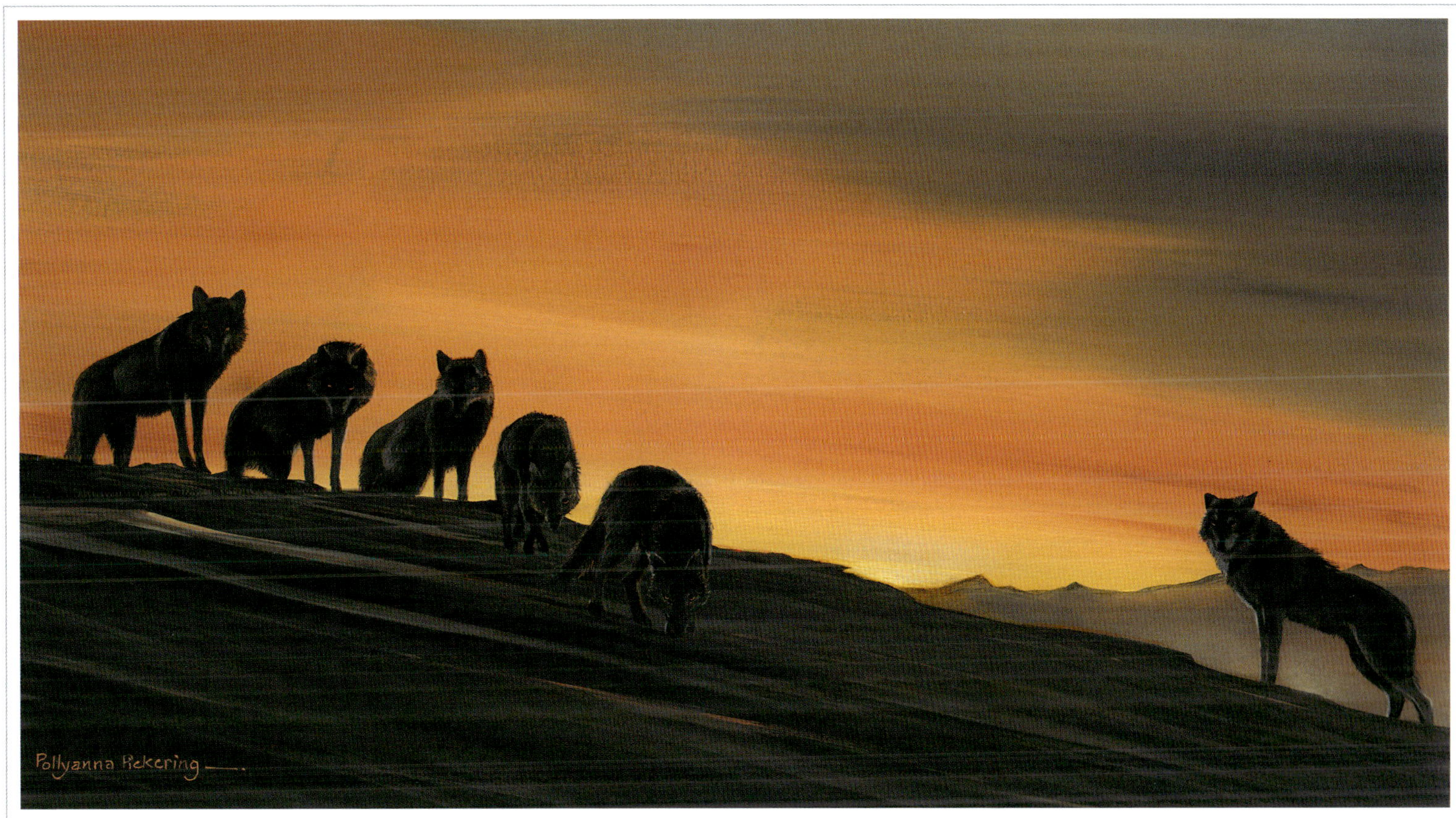

'With great power comes great responsibility'

SPIDERMAN

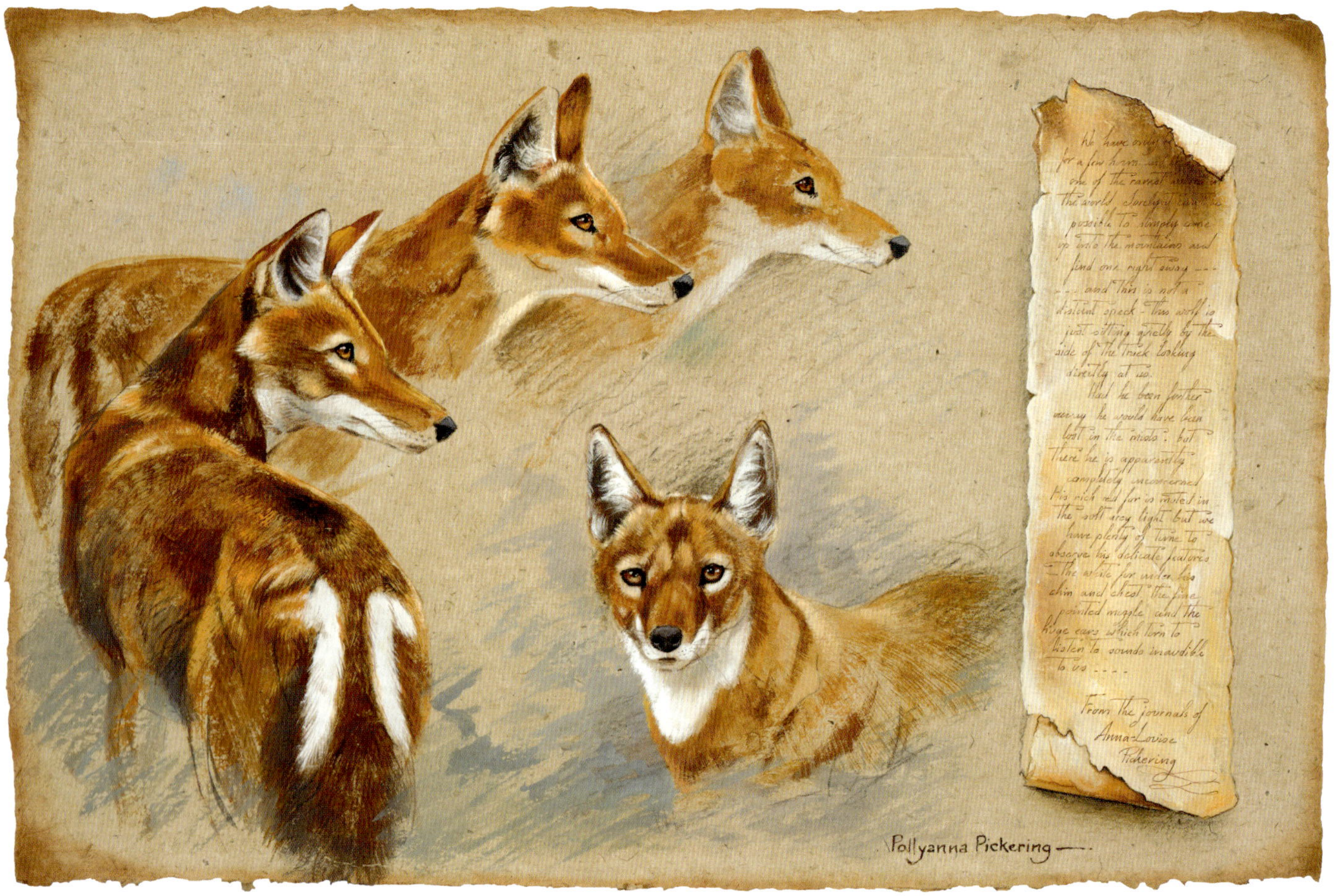

one of the rarest
possible to simply come
up into the mountains and
find one right away - - -
- - - and this is not a
distant speck - this wolf is
just sitting quietly by the
side of the track looking
directly at us.
Had he been further
away he would have been
lost in the mists - but
there he is apparently
completely unconcerned
His rich red fur is muted in
the soft grey light but we
have plenty of time to
observe his delicate features
- the white fur under his
chin and chest, the fine
pointed muzzle and the
huge ears which turn to
listen to sounds inaudible
to us - - - -
From the journals of
Anna-Louise
Pickering
Pollyanna Pickering

Acknowledgements

Although we received a great deal of help from conservation groups in arranging both journeys, we travelled independently, and have no formal links with any scientific or zoological organisation. Where appropriate I have included facts and figures about wolves and other wildlife, to provide background information about the animals Pollyanna set out to paint. Doubtlessly any mistakes or factual errors in the text are mine, not those of the wildlife experts with whom we stayed or accompanied us on our travels. I have cross checked as much as possible to ensure the accuracy of the information included – but as with all wildlife our assumed knowledge about wolves is constantly being challenged as more research and observation is carried out. Ever adaptable, wolves will also change their behaviour patterns in the quest for survival. For those who wish to read more detailed biological information, there are many excellent books which give extremely detailed accounts of the natural history of canids including, *Wolves* by Todd Fuller (World Life Library 2004).

Ethiopia and Transylvania are both regions which have recently undergone significant political changes and upheavals. In many places the traditional way of life is changing rapidly, and this book captures a portrait of a specific place at a specific time. Visitors in future years will doubtless find much has altered.

There are so many people we need to thank for enabling us to make two such diverse journeys through areas which both present their own unique challenges to travellers. Mike and Angela Morton of Transylvania Uncovered provided invaluable help, advice and contacts for our journey into Transylvania. Thank you to Colin Shaw for his unflappable organisation in transporting us through the Danube Delta and into the Carpathians. Count Kalnoky was a gracious host, committed to preserving the wildlife and culture of the region, and all of his staff made our stay in Miklósvár so comfortable. We were very lucky to enjoy the expertise of both Dan Marin and Robert Zeitz while trekking in the mountains. Thank you to Roger Diski and all at Rainbow Tours for arranging ground transportation and assisting with our itinerary in Ethiopia. Our Ethiopian 'crew' made our journey huge

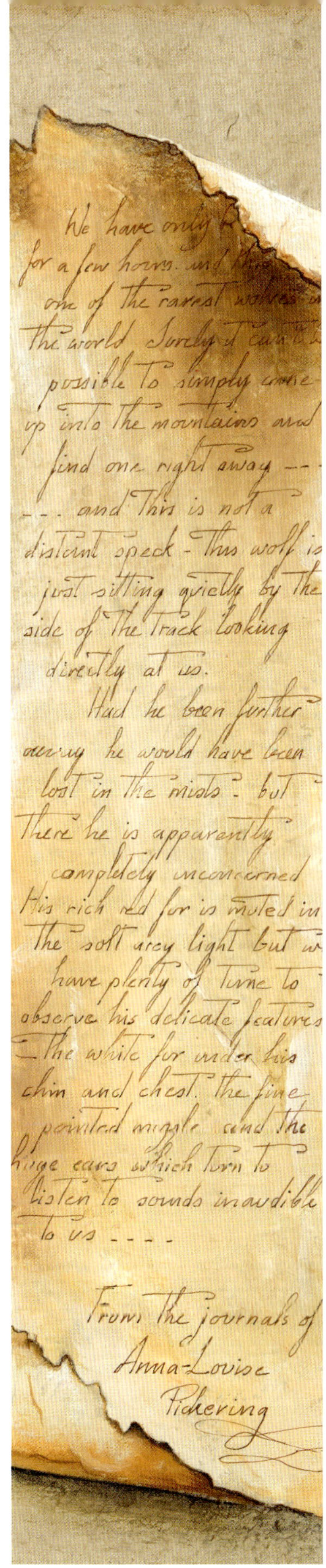

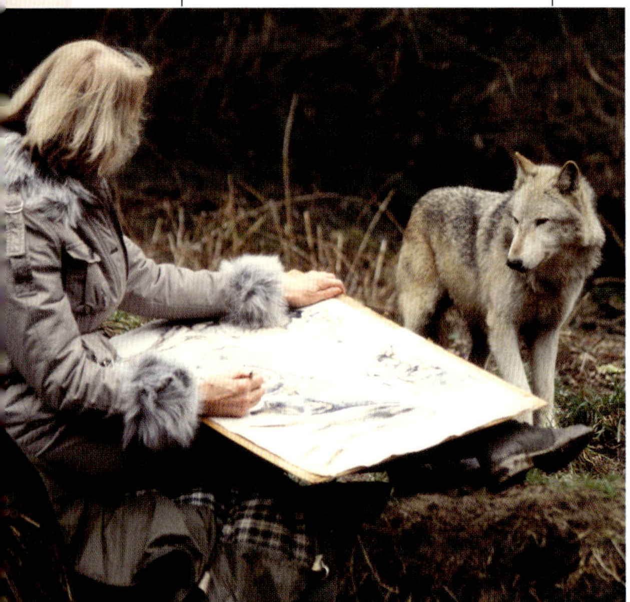

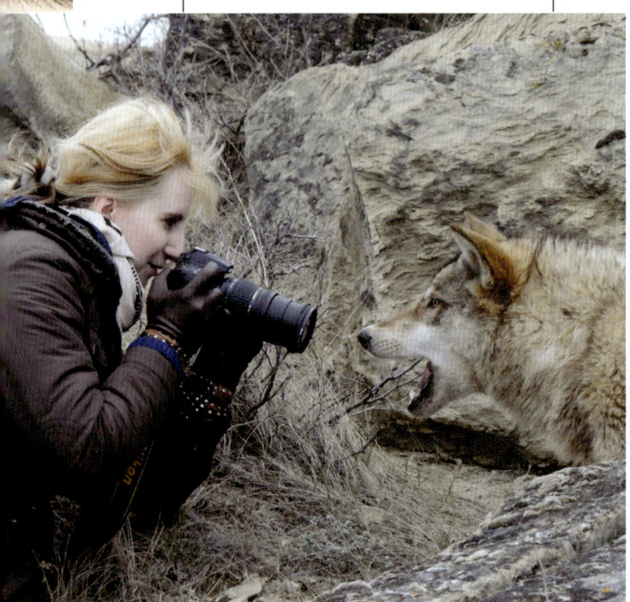

fun – I can't imagine we could have had a better guide and interpreter than Tariku Gudissa and we greatly admire his work with Citizens for Africa. Abbi drove us safely through the many challenges presented by the Ethiopian road system, and Sami was the best cook in the most difficult of circumstances. We shall never forget the hyaena men of Harar – thanks to Abdul for showing us the many unforgettable sights of his amazing city. Thank you to Claudio Sillero – it was a privilege to be able to visit and stay at the EWCP and see the wolves and learn about the conservation work first hand. Thank you to Chris and Anne-Marie for their hospitality, and most especially the sleeping bags, and to Graham for the wonderful and informative drive into the Web Valley to see the wolf pups. As ever many more people helped us with our travels and I apologise to all those who haven't been mentioned here.

Special thanks to Virginia McKenna, Will Travers, Alison Hood and everyone at Born Free – both for the initial introduction to Claudio and for all their work on behalf of endangered species. A percentage of sales from Pollyanna's summer exhibition immediately following her journey to Ethiopia raised £12,000 which was donated towards the construction of one of the enclosures at the proposed wildlife rescue centre 'Ensessakotteh' near Addis Ababa. We are delighted that the centre is now operational, and look forward to visiting in the future!

Once again Otter House have shown great support during the lengthy writing and production of the book. Thank you Alyson Gregory for your thoughtful and patient editing and Sue Stainton and Victoria Selby for design. Carl Whitham has once again patiently scanned several hundred slides, and provided the technical photography of the paintings and sketches. Thank you to Richard Bonfield for allowing us to use his poem *Wolves*. And as ever a huge thank you to our friends and family for all their support and advice, especially Andrew Marshall, Steve and Elaine Naylor, Pat and Stuart Wells, Lisa and Alan Jackson, Scott and Christelle Wells and Linda and John Soto.

Charitable Contacts

The Pollyanna Pickering Foundation raises funds and campaigns for conservation and animal welfare in the UK and worldwide.

The Pollyanna Pickering Foundation, Brookvale House, Oaker, Matlock, Derbyshire DE4 2JJ
www.pollyannapickering.co.uk/foundation.htm

Count Kalnoky's Ecotourism in Transylvania
For more information and contact details:

Tel: +40-742-202586
www.transylvaniancastle.com

The Milvus Group Bird and Nature Protection Association is a non-profit, non-governmental organization, acting in the fields of education, research and advisory, in order to make Romania a better place for birds, wildlife and people. Count Kalnoky is a patron of the group.

Milvus 540620 Tirgu Mures OP3 CP39
Tel/Fax: +40-265-264726
www.milvus.ro

The Born Free Foundation is an international wildlife charity working throughout the world to stop individual wild animal suffering and protect threatened species in the wild.

Born Free Foundation, 3 Grove House, Foundry Lane, Horsham, West Sussex RH13 5PL
www.bornfree.org.uk

The Ethiopian Wolf Conservation Programme targets the greatest threats to the survival of Ethiopian wolves and their Afroalpine habitat. It promotes this charismatic species as a flagship, so protecting many of Ethiopia's endemic species and natural resources.

www.ethiopianwolf.org

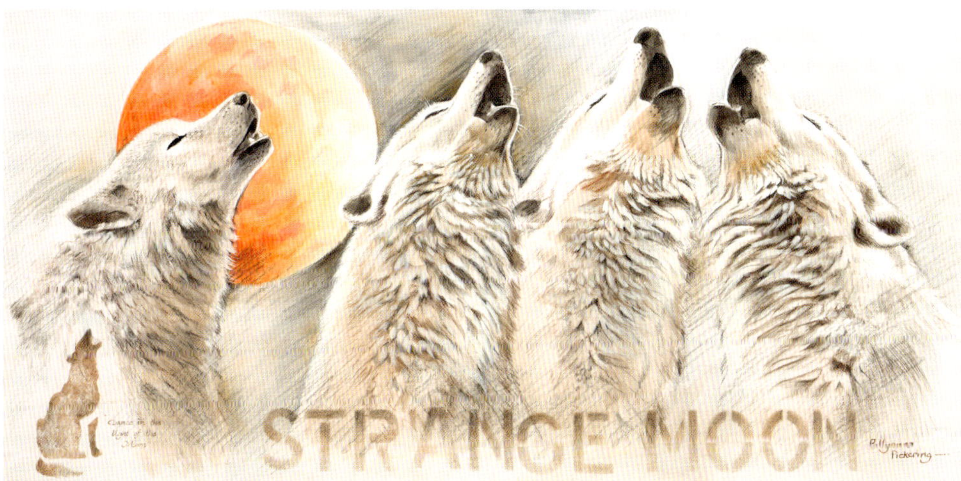

Strange Moon

Recommended Reading

Wolves : Todd Fuller (World Life Library 2004). A very comprehensive book covering wolf ecology.

Bibliography

Ethiopian Wolf : Sillero-Zubiri, C and J Marino (IUCN 2004)

Afroalpine ecology, solitary foraging and intense sociality amongst Ethiopian wolves : Sillero-Zubiri, C, J Marino, D Gottelli and D W Macdonald (Oxford University Press)

The Ethiopian Wolf: status survey and conservation action plan : Sillero-Zubiri, C and D W MacDonald (Eds) (IUCN 1997)

Wolf Within : Shaun Ellis with Penny Junor (Harper Collins 2010)

Brother Wolf – A Forgotten Promise : Jim Brandeburg (Northwood Press 1993)

The Wolf Watchers : Alison Hood and Claudio Sillero-Zubiri (Templar Publishing 1997)

Wolf – Spirit of the Wild : Edited by Diana Landau (The Nature Company 1993)

Wolves of the High Arctic : L David Mech (Voyager Press 1992)

Wolves: A Portrait of the Animal World : Leonard Lee Rue III (Smithmark 1994)

Wolves : Kit Coppard (Greenwich Editions : 1999)

Wild Dogs : Erwin A Bauer (Chronicle Books 1994)

The Nature of Wolves : Candace Savage (Greystone books 1996)

Little Red Cap – Selected Tales : Jacob and Wilhem Grimm (Penguin Classics 1982)

Wolves of Ethiopia : Morell, V

National Geographic Magazine, March 2006

Ethiopian Wolves on High : Williams, S

BBC Wildlife Magazine, July 2004

On the Trail of Dracula : Andrew Eames

The Times May 2006

Ethiopia's Artful Weavers : Judith Olmstead
National Geographic Magazine, January 1973

Bale Mountains: A Guidebook : Stuart Williams (EWCP 2002)

Ethiopia : Philip Briggs (Bradt 2005)

Ethiopia and Eritrea : Matt Philips and Jean-Bernard Carillet (*Lonely Planet*)

Romania and Moldova : Nicola Williams and Kim Wildman (*Lonely Planet*)

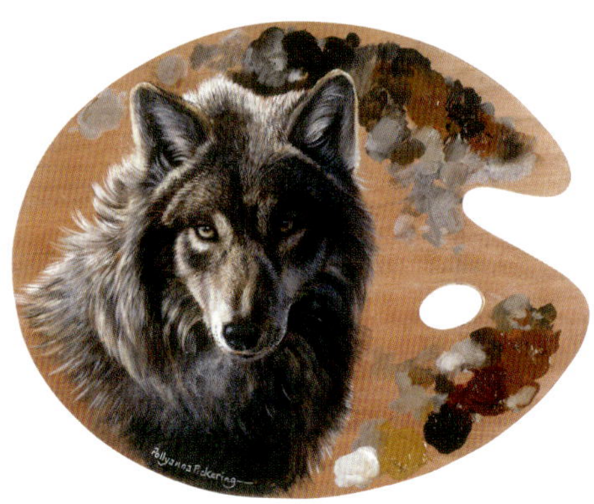

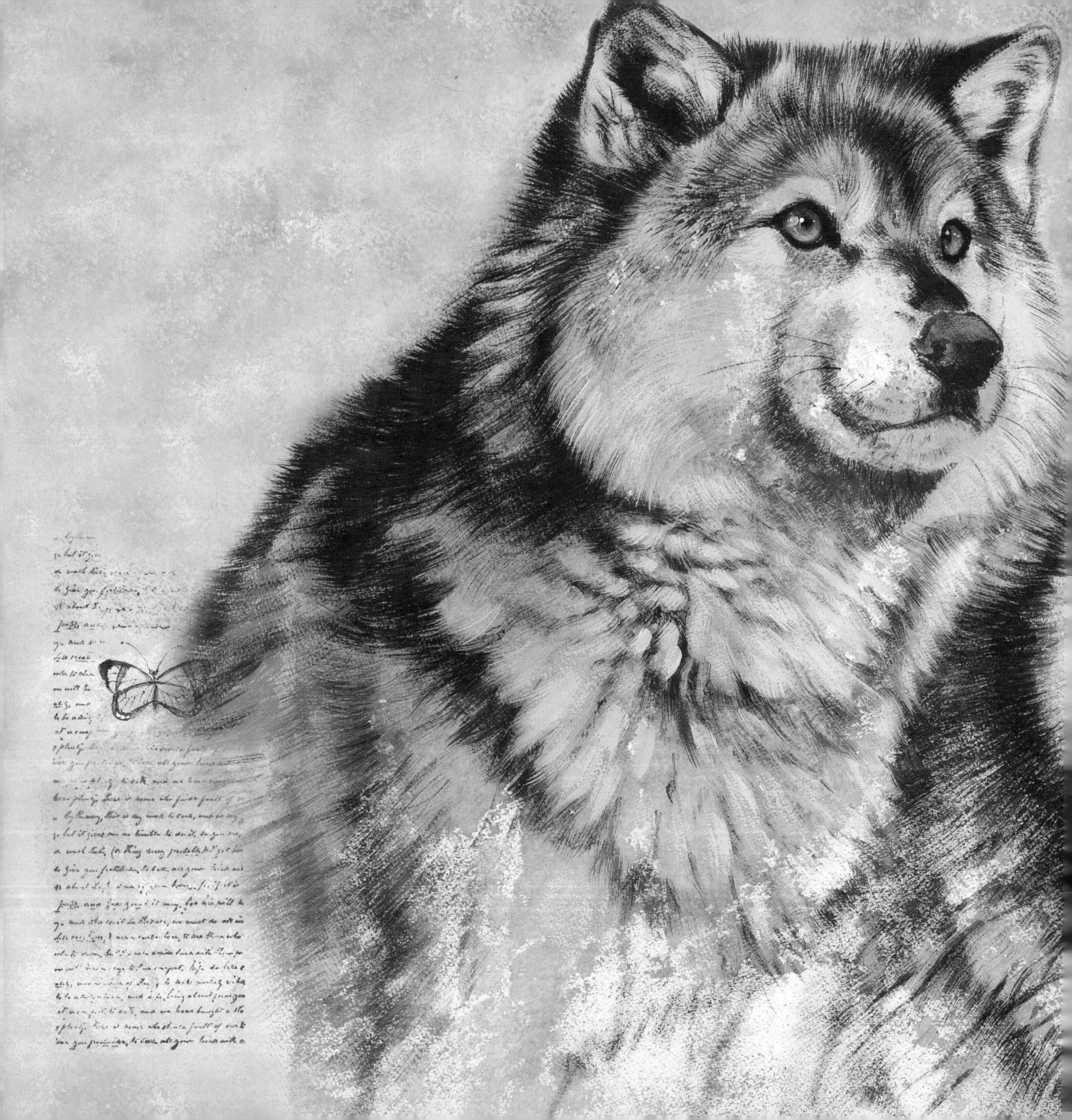